T0142280

Advances in Intelligent Systems and Computing

Volume 698

Series editor

Janusz Kacprzyk, Polish Academy of Sciences, Warsaw, Poland
e-mail: kacprzyk@ibspan.waw.pl

The series "Advances in Intelligent Systems and Computing" contains publications on theory, applications, and design methods of Intelligent Systems and Intelligent Computing. Virtually all disciplines such as engineering, natural sciences, computer and information science, ICT, economics, business, e-commerce, environment, healthcare, life science are covered. The list of topics spans all the areas of modern intelligent systems and computing such as: computational intelligence, soft computing including neural networks, fuzzy systems, evolutionary computing and the fusion of these paradigms, social intelligence, ambient intelligence, computational neuroscience, artificial life, virtual worlds and society, cognitive science and systems, Perception and Vision, DNA and immune based systems, self-organizing and adaptive systems, e-Learning and teaching, human-centered and human-centric computing, recommender systems, intelligent control, robotics and mechatronics including human-machine teaming, knowledge-based paradigms, learning paradigms, machine ethics, intelligent data analysis, knowledge management, intelligent agents, intelligent decision making and support, intelligent network security, trust management, interactive entertainment, Web intelligence and multimedia.

The publications within "Advances in Intelligent Systems and Computing" are primarily proceedings of important conferences, symposia and congresses. They cover significant recent developments in the field, both of a foundational and applicable character. An important characteristic feature of the series is the short publication time and world-wide distribution. This permits a rapid and broad dissemination of research results.

More information about this series at http://www.springer.com/series/11156

Hasmat Malik · Smriti Srivastava
Yog Raj Sood · Aamir Ahmad
Editors

Applications of Artificial Intelligence Techniques in Engineering

SIGMA 2018, Volume 1

 Springer

Editors
Hasmat Malik
Department of Instrumentation and Control
  Engineering
Netaji Subhas Institute of Technology
New Delhi, Delhi, India

Smriti Srivastava
Department of Instrumentation and Control
  Engineering
Netaji Subhas Institute of Technology
New Delhi, Delhi, India

Yog Raj Sood
National Institute of Technology
Puducherry, India

Aamir Ahmad
Perceiving Systems Department
Max Planck Institute for Intelligent Systems
Tübingen, Germany

ISSN 2194-5357   ISSN 2194-5365 (electronic)
Advances in Intelligent Systems and Computing
ISBN 978-981-13-1818-4   ISBN 978-981-13-1819-1 (eBook)
https://doi.org/10.1007/978-981-13-1819-1

Library of Congress Control Number: 2018949630

This Springer imprint is published by the registered company Springer Nature Singapore Pte Ltd.
The registered company address is: 152 Beach Road, #21-01/04 Gateway East, Singapore 189721,
Singapore

Preface

This Conference Proceedings Volume 1 contains the written versions of most of the contributions presented at the International Conference SIGMA 2018. The conference was held at Netaji Subhas Institute of Technology (NSIT), New Delhi, India, during February 23–25, 2018. NSIT is an autonomous institute under the Government of NCT of Delhi and affiliated to University of Delhi, India. The International Conference SIGMA 2018 aimed to provide a common platform to the researchers in related fields to explore and discuss various aspects of artificial intelligence applications and advances in soft computing techniques. The conference provided excellent opportunities for the presentation of interesting new research results and discussion about them, leading to knowledge transfer and the generation of new ideas.

The conference provided a setting for discussing recent developments in a wide variety of topics including power system, electronics and communication, renewable energy, tools and techniques, management and e-commerce, motor drives, manufacturing process, control engineering, health care and biomedical, cloud computing, image processing, environment and robotics. This book contains broadly 13 areas with 59 chapters that will definitely help researchers to work in different areas.

The conference has been a good opportunity for participants coming from all over the globe (mostly from India, Qatar, South Korea, USA, Singapore, and so many other countries) to present and discuss topics in their respective research areas.

We would like to thank all the participants for their contributions to the conference and for their contributions to the proceedings. Many thanks go as well to the NSIT participants for their support and hospitality, which allowed all foreign participants to feel more at home. Our special thanks go to our colleagues for their devoted assistance in the overall organization of the conference.

It is our pleasant duty to acknowledge the financial support from Defence Research and Development Organisation (DRDO), Gas Authority of India Limited (GAIL), MARUTI SUZUKI, CISCO, Power Finance Corporation (PFC) Limited, CommScope, Technics Infosolutions Pvt. Ltd., Bank of Baroda, and Jio India.

We hope that it will be interesting and enjoying at least as all of its predecessors.

New Delhi, India	Hasmat Malik
New Delhi, India	Smriti Srivastava
Puducherry, India	Yog Raj Sood
Tübingen, Germany	Aamir Ahmad

Contents

About the Editors

Hasmat Malik (M'16) received his B.Tech. degree in electrical and electronics engineering from GGSIP University, New Delhi, India, and M.Tech. degree in electrical engineering from NIT Hamirpur, Himachal Pradesh, India, and he is currently doing his Ph.D. degree in Electrical Engineering Department, Indian Institute of Technology Delhi, New Delhi, India.

He is currently Assistant Professor in the Department of Instrumentation and Control Engineering, Netaji Subhas Institute of Technology, New Delhi, India. His research interests include power systems, power quality studies, and renewable energy. He has published more than 100 research articles, including papers in international journals, conferences, and chapters. He was a Guest Editor of Special Issue of Journal of Intelligent & Fuzzy Systems, 2018 (SCI Impact Factor 2018:1.426) (IOS Press), and Special Issue of International Journal of Intelligent Systems Design and Computing (IJISDC).

He received the POSOCO Power System Award (PPSA-2017) for his Ph.D. work on research and innovation in the area of power system in 2017. His interests are in artificial intelligence/soft computing applications to fault diagnosis, signal processing, power quality, renewable energy, and microgrids.

He is a life member of the Indian Society for Technical Education (ISTE); International Association of Engineers (IAENG), Hong Kong; International Society for Research and Development (ISRD), London; and he is a member of the Institute of Electrical and Electronics Engineers (IEEE), USA, and MIR Labs, Asia.

Smriti Srivastava received her B.E. degree in electrical engineering and her M.Tech. degree in heavy electrical equipment from Maulana Azad College of Technology [now Maulana Azad National Institute of Technology (MANIT)], Bhopal, India, in 1986 and 1991, respectively, and Ph.D. degree in intelligent control from the Indian Institute of Technology Delhi, New Delhi, India, in 2005. From 1988 to 1991, she was Faculty Member at MANIT, and since August 1991, she has been with the Department of Instrumentation and Control Engineering, Netaji Subhas Institute of Technology, University of Delhi, New Delhi, India, where she is working as Professor in the same division since September 2008 and as Dean of Undergraduate Studies. She also worked as Associate Head of the Instrumentation and Control Engineering Division at NSIT, New Delhi, India, from April 2004 to November 2007 and from September 2008 to December 2011. She was Dean of Postgraduate Studies from November 2015 to January 2018. She is Head of the division since April 2016. She is the author of a number of publications in transactions, journals, and conferences in the areas of neural networks, fuzzy logic, control systems, and biometrics. She has given a number of invited talks and tutorials mostly in the areas of fuzzy logic, process control, and neural networks. Her current research interests include neural networks, fuzzy logic, and hybrid methods in modeling, identification, and control of nonlinear systems and biometrics.

She is Reviewer of *IEEE Transactions on Systems, Man and Cybernetics* (SMC), Part-B, *IEEE Transactions on Fuzzy Systems, International Journal of Applied Soft Computing* (Elsevier), *International Journal of Energy, Technology and Policy* (Inderscience).

Her paper titled 'Identification and Control of a Nonlinear System using Neural Networks by Extracting the System Dynamics' was selected by the Institution of Electronics and Telecommunication Engineers for IETE K S Krishnan Memorial Award for the best system-oriented paper. She was also nominated for 'International Engineer of the Year 2008' by International Biographical Center of Cambridge, England, in 2008. Her name appeared in Silver Jubilee edition of 'Marquis Who's Who in the World' in November 2007, 2008, and 2009. Her paper titled 'New Fuzzy Wavelet Neural Networks for System Identification and Control' was the second most downloadable paper of the year 2006 from the list of journals that come under 'ScienceDirect'.

Prof. Yog Raj Sood (SM'10) is a member of DEIS. He obtained his B.E. degree in electrical engineering with 'Honors' and M.E. degree in power system from Punjab Engineering College, Chandigarh (UT), in 1984 and 1987, respectively. He obtained his Ph.D. degree from Indian Institute of Technology Roorkee in 2003. He is Director of the National Institute of Technology Puducherry, Karaikal, India. He joined Regional Engineering College, Kurukshetra, in 1986. Since 2003, he has been working as Professor in the Department of Electrical Engineering, National Institute of Technology Hamirpur, Himachal Pradesh, India. He has published a number of research papers. He has received many awards, prizes, and appreciation letters for his excellence in academic research and administration performance. His research interests are deregulation of power system, power network optimization, condition monitoring of power transformers, high-voltage engineering, nonconventional sources of energy.

 **Aamir Ahmad** is Research Scientist at the Perceiving Systems Department, Max Planck Institute for Intelligent Systems, Tübingen, Germany. He received his Ph.D. degree in electrical and computer engineering from the Institute for Systems and Robotics, Instituto Superior Técnico, Lisbon, Portugal. His main research interests lie in Bayesian sensor fusion and optimization-based state estimation. The core application focus of his research is vision-based perception in robotics. He has worked with several ground robots in the recent past. Currently, he is focusing on aerial robots and active perception in multiple aerial vehicle systems. He has published 5 peer-reviewed journal articles, 2 chapters, 11 conference papers, and 2 workshop papers.

Annual Energy Savings with Multiple DG and D-STATCOM Allocation Using PSO in DNO Operated Distribution Network

Atma Ram Gupta and Ashwani Kumar

Abstract The installation of renewable energy sources based distributed generation (DG) is increasing continuously due to its various techno-economic advantages. Distribution static compensator (D-STATCOM) based reactive power compensation is gaining importance in smart distribution system management. The main high lights of the paper are

- separate and combined placement of DG and D-STATCOM and their size determination using particle swarm optimization (PSO),
- assessment of impact of DG and D-STATCOM on loss reduction and savings in losses,
- determination of annual energy savings and annual cost savings,

Optimal allocation of DG/D-STATCOM results improvement in voltage profiles, reduction in losses, annual energy savings as well as savings in annual cost of energy loss. The obtained results are tested and compared for IEEE-33 bus radial distribution systems.

Keywords Load flow analysis · DG · D-STATCOM · Distribution system · PSO

1 Introduction

Some of the distributed generation (DG) as well as most of the electrical load (lagging) needs reactive power, consequently the power factor of the system decreases which leads to higher flow of currents and responsible for higher active power losses with reduced voltage profile. Reactive power demand is generally fulfilled by capacitor bank, synchronous condenser, phase advancers etc. Distribution static

A. R. Gupta (✉) · A. Kumar
National Institute of Technology, Kurukshetra, Kurukshetra 136119, Haryana, India
e-mail: arguptanitd@gmail.com

A. Kumar
e-mail: ashwa_ks@yahoo.co.in

© Springer Nature Singapore Pte Ltd. 2019
H. Malik et al. (eds.), *Applications of Artificial Intelligence Techniques in Engineering*, Advances in Intelligent Systems and Computing 698,
https://doi.org/10.1007/978-981-13-1819-1_1

1

compensator (D-STATCOM) can provide variable reactive power with better and faster control.

Some of the key challenges in distribution systems are voltage profile enhancement and reduction of line losses (VPEARL) for effective utilization of existing distribution system. Approx 13% of generated power is wasting as line losses as registered in [1]. Distribution flexible alternating current transmission system (D-FACTS) devices can improve the power factor which leads to decline of current index and consequently it leads to VPEARL of the distribution network. In addition, D-FACTS devices offers many other advantages like it improves power quality, increase existing feeder capacity, increase power transfer capability, load balancing and many more. The fundamental ideology and arithmetical model of STATCOM and D-STATCOM are alike, as a result the power flow model of STATCOM provide the impression to be appropriate for load flow analysis with D-STATCOM [2]. Also, the inclusion of renewable energy based DG (wind, solar etc.) is increasing due its various techno-economic as well as environmental benefits [3]. Distribution network operator (DNO) encourages DG developer to install more number of DG in distribution systems (DS). Some researchers also reported that D-FACTS can increase the DG penetration capacity in distribution system [4, 5]. Also, wind and photovoltaic (PV) based DG demands reactive power from the system [6, 7]. So, DG with D-STATCOM can be the best option in today's scenario. In the last five years several authors have analyzed the distribution system with separate DG [8] and D-STATCOM allocation with various sensitivity based method, optimization method, voltage security margin method and cost benefit methods with an objective of VPEARL [9–11]. But, very few researchers have incorporated simultaneous DG as well as D-STATCOM for VPEARL.

In this work impact of combined placement of multiple DG and D-STATCOM using PSO with various cases is studied. The cases considered for obtaining the results are allocation of single D-STATCOM, two D-STATCOM, single DG, two DG, three DG, and four DG and also with simultaneous DG and D-STATCOM. The complete flow chart for LFA used in this paper for base case is given in [11].

The organization of paper is: Sect. 2 describes the steady state model of DG/D-STATCOM as well as their cost components. Section 3 explains various constraint used for optimization. Section 4 consists of the flow chart for the methodology used in this paper for DG/D-STATCOM allocation. Section 5 represents the results and Sect. 5 provides comparison of results with various methods for RDS and conclusions in Sect. 6.

2 Steady State Model and Cost Components

2.1 DG Model

In this study, the DG is considered as an active power source.

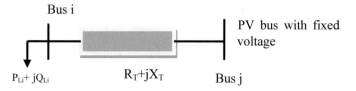

Fig. 1 Steady state model for D-STATCOM

2.2 D-STATCOM Model

In steady state situation, D-STATCOM act in the same way to a shunt reactive power resource that regulates the voltage scale of the node where it is to be placed. If the node i is a load bus of the system with a consumption equivalent to $P_{Li} + jQ_{Li}$, the model of D-STATCOM on bus i can be considered as a new PV bus j that is added to bus i with its active power set to nil. The transformer is modeled by its leakage resistance and reactance; $R_T + jX_T$ and the model are given in Fig. 1.

2.3 Cost of Energy Loss [12]

$$\text{CEL}\,(\$) = (\text{Total Real power Loss}) * (E_C * T) \tag{1}$$

E_c: Energy rate (\$/kWhr), T: Time duration (hr), $E_c = 0.06$ \$/kWhr, $T = 8760$ h.

2.4 Cost of DG [12]

The Cost component of DG for real power

$$C(Pdg) = a * Pdg^2 + b * Pdg + c \,\$/\text{MWh} \tag{2}$$

Cost coefficients are taken as: a = 0, b = 20, c = 0.25

2.5 Cost of D-STATCOM [13]

$$Cost_{D-STATCOM}\,(\$) = Investment\_cost_{D-STATCOM} \frac{[1+B]^{nDST} * B}{[1+B]^{nDST} - 1} \tag{3}$$

$nDST$: The longevity of D-STATCOM, B: is the asset rate of return.

$Investment_{cost} = 50\$/\text{kVAr}$, $B = 0.1$, $nDST = 30$ years.

2.6 Cost Saving

Cost saving (\$) = [Cost of energy loss (\$) before DG and D-STATCOM alloca-
tion]—[Cost of energy loss (\$) after DG and D-STATCOM allocation+ Cost of
D-STATCOM (\$)].

3 Problem Formulation

Problem formulation has been done by conducting LFA of RDS, fixing of the objective function as well as system restriction and finding the best possible allocation of
DG and D-STATCOM using PSO.

The optimization problem is as follows:

Minimize power loss: $Pl = \sum_{i=1}^{n} |I_i^2| R_i$

Voltage constraint: $|V_{min}| < |V| < |V_{max}|$, Current constraint: $|I_i| < |I_{max}|$

$$DG \text{ capacity constraints} : P_{DGimin} \leq P_{DGi} \leq P_{DGimax}$$

$$D - STATCOM \text{ capacity constraints} : Q_{D-STATCOMimin} \leq Q_{D-STATCOMi} \leq Q_{D-STATCOMimax},$$

where, n = number of buses, V is the voltage at any bus, V_{min} and V_{max} are the
minimum and maximum voltage limits respectively. I_i is the current flowing in the
branch, I_{imax} is the maximum current limit on ith branch. R_i is the resistance of the
ith branch, P_{DGimin} and P_{DGimax} is the maximum and minimum capacity of DG connected to ith node, $Q_{D-STATCOMimin}$ and $Q_{D-STATCOMimax}$ is the maximum and minimum
capacity of D-STATCOM connected to ith node of RDS.

4 DG and D-STATCOM Allocation Using PSO

The steps followed for optimal allocation of DG and D-STATCOM using PSO technique [14] is given as a flow chart in Fig. 2. Here first an initial random population
or array of particles with random positions and velocities on a multidimensional
space are generated. Then real power loss is evaluated keeping all the constraint
satisfied which is followed by comparing the objective value of each random particle
with individual best (pbest) and for the current minimization problem with ensuring
the objective value is less than pbest, set this pbest as current pbest and records its

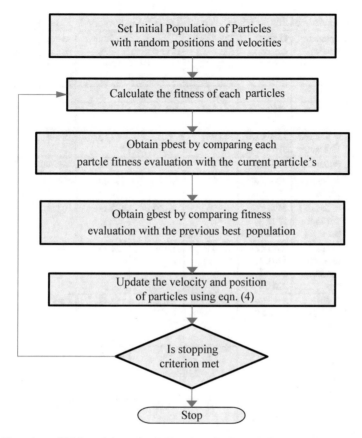

Fig. 2 Flow chart of PSO used for optimal allocation of DG and D-STATCOM

corresponding current position. Found minimum pbest and updated it to their best encounter position and set it as current gbest then update the velocity and position of particles using following equation.

$$v_i = v_i + 2 * rand() * (Pbest - x_i) + 2 * rand() * (Gbest - x_i)$$
$$x_i = x_i + v_i$$

(4)

where x_i = position vector, v_i = velocity vector, This process is repeated till stop criteria is reached.

Table 1 Comparison of Results with DG and D-STATCOM allocation in 33 bus RDS

Device	Optimal location	Optimal size of device	TPL (kW)	TQL (kVAr)	Reduction in TPL (kW)	Annual energy saving (MWh)	Annual cost savings ($)
No device	–	–	210.98	143.02	–	–	–
Single D-STATCOM	@30th bus	1260 kVAr	151.37	103.81	59.61	522.20	24647
Two D-STATCOM	@30th and 13th bus	1260 kVAr, 370 kVAr	142.88	97.42	68.10	596.60	27145
Single DG	@ 6th bus	2590 kW	111.03	81.68	99.95	875.60	52530
Two DG	@ 6th and 15th bus	2590 kW, 470 kW	96.08	68.62	114.90	1006.50	60390
Three DG	@ 6th,15th and 25th bus	2590 kW, 470 kW, 640 kW	88.56	63.02	122.42	1072.40	64340
Four DG	@ 6th, 15th, 25th and 32nd bus	2590 kW, 470 kW, 640 kW, 220 kW	86.44	60.79	124.54	1091.00	65460
Single DG+ Single D-STATCOM	@ 6th bus +@30th bus	2590 kW, 1260 kVAr	58.50	47.15	152.48	1335.70	73457
Two DG+ Single D-STATCOM	@ 6th and 15th bus +@30th bus	2590 kW, 470 kW, 1260 kVAr	45.32	35.36	165.66	1451.20	80387
Two DG+ Two D-STATCOM	@ 6th and 15th bus +@30th and 13th bus	2590 kW, 470 kW, 1260 kVAr, 370 kVAr	39.00	30.56	171.98	1506.50	81745

5 Results with DG and D-STATCOM Allocation

The proposed technique is tested for IEEE 33 bus RDS with 12.66 kV and 100 MVA base values. The line and load data are taken from [15] and the results are obtained using MATLAB programming [16]. The comparison of results with separate as well as simultaneous allocation of DG/D-STATCOM are obtained and recorded in Table 1. TPL variations with iteration after placement of D-STATCOM of 1260 kVAr at 30th and 370 kVAr at 13th bus is shown in Fig. 3. The improvement in voltage profiles for each case are obtained and graphically represented in Fig. 4. Also, TPL variations

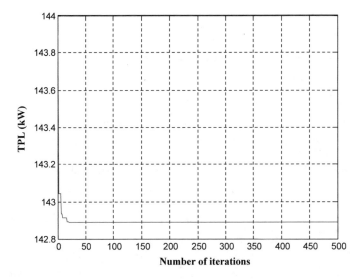

Fig. 3 TPL variations with iteration after placement of D-STATCOM of 1260 kVAr at 30th and 370 kVAr at 13th bus

after placement of DG of 2590 kW at 6th, 470 kW at 15th, 640 kW at 25th and 220 kW at 32nd bus is graphically and Fig. 5. The result shows highest VPEARL with simultaneous multiple allocations of DG and D-STATCOM. Comparison of results with simultaneous DG and D-STATCOM allocation for 33 bus RDS is made in Table 2.

6 Conclusions

Electric utilities are always expected to provide high power quality and reliability in power system. In order to entertain such facilities in the system D-STATCOM devices can be used by the DNO. In this work, steady state analysis of radial distribution system with separate as well as simultaneous allocation of DG and D-STATCOM is studied with an objective of reducing the line losses with improvement in voltage profile. The proposed method of multiple DG and D-STATCOM allocation are also compared with sensitivity as well as existing optimization technique. The percentage of cost saving is highest with simultaneous placement of DG and D-STATCOM. With the installation of D-STATCOM devices the DNO can provide other facilities besides voltage profile improvement and reduction of line losses like power quality improvements, voltage stability and system loading capability enhancement. This study can help DNO to plan distribution system with simultaneous allocation of renewable energy based DG and D-STATCOM device.

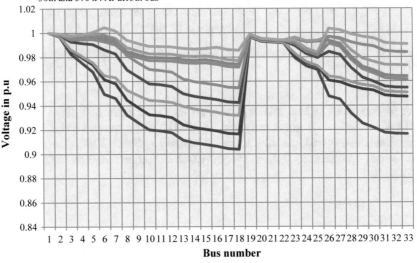

Fig. 4 Variation of voltage of IEEE 33 bus RDS after DG and D-STATCOM allocation using PSO

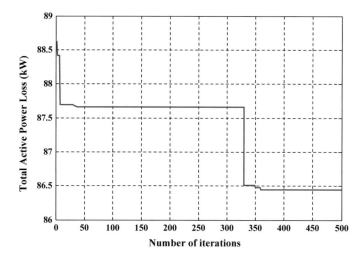

Fig. 5 TPL variation with iteration after placement of DG of 2590 kW at 6th, 470 kW at 15th, 640 kW at 25th and 220 kW at 32nd bus

Table 2 Comparison of results with simultaneous DG and D-STATCOM allocation in 33 bus RDS

Parameters	Devabalaji et al. [17]	Proposed method
	BFOA	PSO
TPL (kW) in base case	210.98	210.98
TPL (kW) with DG and D-STATCOM	70.87	58.50
TPL reduction (%)	66.41	**72.27**
D-STATCOM ratings (kVAr)	1094.60	1260
D-STATCOM locations	@30th bus	@30th bus
DG ratings (kW)	1239.80	2590
DG locations	@10th bus	@6th bus

References

1. A.A. El-fergany, Optimal capacitor allocations using evolutionary algorithms. IET Gener. Transm. Distrib. **7**(6), 593–601 (2013). https://doi.org/10.1049/iet-gtd.2012.0661
2. A. Ghosh., G.F. Ledwich, *Power Quality Enhancement Using Custom Power Devices* (Kluwer Publications, 2002). https://doi.org/10.1007/978-1-4615-1153-3
3. S.N. Singh, Distributed generation in power systems: an overview and key issues, in *24th Indian Engineering Congress* (2009). http://orbit.dtu.dk/files/5202512/24IEC_paper.pdf
4. K.V. Bhadane, M.S. Ballal, R.M. Moharil, Enhancement of distributed generation by using custom power device. J. Electron. Sci. Technol. **13**(3), 246–254 (2015)
5. C.S. Chen, C.H. Lin, W.L. Hsieh, C.T. Hsu, T.T. Ku, Enhancement of PV penetration with DSTATCOM in Taipower distribution system. IEEE Trans. Power Syst. **28**(2), 1560–1567 (2013). https://doi.org/10.1109/TPWRS.2012.2226063
6. E.H. Camm, M.R. Behnke, O. Bolado, M. Bollen, M. Bradt, C. Brooks, W. Dilling et al., Reactive power compensation for wind power plants, in *Power & Energy Society General*

Meeting, PES'09 (IEEE, 2009) pp. 1–7. https://doi.org/10.1109/PES.2009.5275328
7. A. Ellis, R. Nelson, E. Von Engeln, R. Walling, J. MacDowell, L. Casey, E. Seymour et al., Reactive power performance requirements for wind and solar plants, in *Power and Energy Society General Meeting*, (2012) pp. 1–8. https://doi.org/10.1109/PESGM.2012.6345568
8. U. Sultana, A.B. Khairuddin, M.M. Aman, A.S. Mokhtar, N. Zareen, A review of optimum DG placement based on minimization of power losses and voltage stability enhancement of distribution system. Renew. Sustain. Energy Rev. **63**, 363–378 (2016)
9. O.P. Mahela, A.G. Shaik, A review of distribution static compensator. Renew. Sustain. Energy Rev. **50**, 531–546 (2015)
10. A.R. Gupta, Kumar A, Optimal placement of D-STATCOM using sensitivity approaches in mesh distribution system with time variant load models under load growth. Ain Shams Eng. J. (2016). https://doi.org/10.1016/j.asej.2016.05.009
11. A.R. Gupta, A. Kumar, Impact of various load models on D-STATCOM allocation in DNO operated distribution network. J. Proced. Comput. Sci. **125** 862–870 (2018). https://doi.org/1 0.1016/j.procs.2017.12.110
12. V.V.S.N. Murty, A. Kumar, Optimal placement of DG in radial distribution systems based on new voltage stability index under load growth. Int. J. Electr. Energy Power Syst. **69**, 246–256 (2014). https://doi.org/10.1016/j.ijepes.2014.12.080
13. A.R. Gupta, A. Kumar, Reactive power deployment and cost benefit analysis in DNO operated distribution electricity markets with D-STATCOM. Front. Energy (2017). https://doi.org/10.1 007/s11708-017-0456-8.
14. A. Uniyal, A. Kumar, Comparison of optimal DG placement using CSA, GSA, PSO and GA for minimum real power loss in radial distribution system, in *2016 IEEE 6th International Conference on Power Systems (ICPS)* (IEEE, 2016). https://doi.org/10.1109/ICPES.2016.758 4027
15. S.A. Taher, S.A. Afsari, Optimal location and sizing of DSTATCOM in distribution systems by immune algorithm. Electr. Power Energy Syst. **60**, 34–44 (2014). https://doi.org/10.1016/j.ijepes.2014.02.020
16. MATLAB version 7.8. The MATLAB by Mathworks Corporation (2009)
17. K.R. Devabalaji, K. Ravi, Optimal size and siting of multiple DG and DSTATCOM in radial distribution system using bacterial foraging optimization Algorithm. Ain Shams Eng. J. **7**(3), 959–971 (2016). https://doi.org/10.1016/j.asej.2015.07.002

Optimized 2DOF PID for AGC of Multi-area Power System Using Dragonfly Algorithm

Kumaraswamy Simhadri, Banaja Mohanty and U. Mohan Rao

Abstract This paper presents the automatic generation control (AGC) of an interconnected two-area multi-source hydro-thermal power system. The considered system performance is studied and analyzed with proportional-integral (PI), proportional-integral-derivative (PID) and 2 degree of freedom PID(2DOF PID). The gains of the controllers are optimized using dragonfly algorithm (DA). The performance of the DA algorithm is matched with the genetic algorithm (GA) and hybrid firefly and pattern search technique to state its superiority. The comparative results show that 2DOF PID scheme tuned using DA gives better results than the classical controllers.

Keywords AGC · PID · Interconnected · Optimization · Hydro-thermal

1 Introduction

It is always desirable to produce reliable and quality electric power supply form an interconnected power system. Modern power system network is the combination of different controlled areas to get stable operation, the total generation of each controlled area must satisfy the total load demand plus accompanied system losses. The unbalance between generation and load demand due to sudden change in load demand will cause the change in frequency and tie-line power are sustained by the controller, which adjust the speed of the governor to maintain the balance [1]. An extensive literature on the automatic generation control (AGC) problem in conventional power systems is presented in [2, 3].

K. Simhadri (✉) · B. Mohanty
Electrical and Electronics Engineering Department, VSSUT, Burla, India
e-mail: kumar.simhadri@gmail.com

U. Mohan Rao
Electrical and Electronics Engineering Department, LIET, Vizianagaram, India
e-mail: mohan13.nith@gmail.com

© Springer Nature Singapore Pte Ltd. 2019   11
H. Malik et al. (eds.), *Applications of Artificial Intelligence Techniques in Engineering*, Advances in Intelligent Systems and Computing 698,
https://doi.org/10.1007/978-981-13-1819-1_2

The different authors have proposed number of control approaches such as classical controllers [4], fractional order controllers, digital control approaches to solve AGC problems in a multi-area multi-source power system. The evolutionary techniques such as genetic algorithm (GA) [5], differential evolution (DE) [6], particle swarm optimization (PSO) [7], bacterial foraging optimization algorithm (BFOA) [8], teaching learning based optimization (TLBO) technique was proposed to get optimal parameters of classical controller in multi-sources multi-units power system and superiority of the algorithm was compared with differential evolution (DE) [9]. Firefly algorithm (FA) have been used for tuning the controller gains of the classical controllers to find the solution for the AGC problem on multi-area multi-sourcepower system [10].

As studied in the literature various controller structure and various intelligent techniques are incorporated in multi-area multi-source interconnected power system to solve AGC problems. But, still researchers all over the world are trying and proposing new controllers to obtain the system stability. Two-degree freedom controller is a newly developed controller structure has gained much attention in the control community and applied for PID, FOPID for AGC system. The advantages of 2DOF PID and control structure are already mentioned in [11].

Dragonfly algorithm (DA) is recently developed swarm optimization algorithm by Mirjalili [12]. This algorithm is based on the behavior of dragonfly's navigation, searching for food and protection from enemies. The advantages of this algorithm are to meet the global optimum, and provide very good results when compared to other famous algorithms [12]. To the best of author's knowledge, DA is still not applied for multi-area multi-source power system and required to be implemented. The effectiveness of the DA algorithm is established by comparing it with the hFA-PS tuned PI (ITAE), hFA-PS tuned PI (ISE), and PID controller results in terms of settling time of the deviations in frequency and tie-line power of control areas in power system reported in the literature [13].

2 Power System Model and Mathematical Modeling

The investigated two-area interconnected multi-unit's power system consists of thermal and hydro generating units in each area. The thermal area is provided with a single reheat turbine, hydro generating unit is also provided with turbine whose parameters are taken from [13], details are presented in Appendix-A. So as to implement the optimized controller the complete transfer function of test system model is taken from [13].

The classical controllers are not able to handle the nonlinear constraints present in the power system. The proposed two degree of freedom has two mechanisms, one will take care of closed-loop stability and one will take care to shape the closed-loop response also, it attenuates the measurement noise. The 2DOF PID will damp out the system oscillations compared to single degree of freedom PID, the block diagram and mathematical formulation of 2DOF PID for generating control signals is given in [11]. The controllers considered in both the areas are different hence the control gains are also different. To find out frequency error, integral squared error (*ISE*), integral time absolute error (*ITAE*) and integral time squared error (*ITSE*) has been used as an objective function in this present work [13].

$$ISE = \int_{t=0}^{t_{sim}} \left[(\Delta f_1)^2 + (\Delta f_2)^2 + (\Delta P_{tie})^2 \right] dt \qquad (1)$$

$$ITSE = \int_{t=0}^{t_{sim}} t \left[(\Delta f_1)^2 + (\Delta f_2)^2 + (\Delta P_{tie})^2 \right] dt \qquad (2)$$

$$ITAE = \int_{t=0}^{t_{sim}} t \left[(\Delta f_1) + (\Delta f_2) + (\Delta P_{tie}) \right] dt, \qquad (3)$$

where Δf_1 and Δf_2 are the frequency deviations in area1 and area2, ΔP_{tie} tie-line power and t_{sim} is the simulation time to find out objective function value.

3 Dragonfly Algorithm

Mirjalili proposed one of the new heuristic algorithms is the Dragonfly algorithm (DA) [12]. It is developed based on the unique and rare swarming behavior of dragonflies. The suggested DA is found to be popular and powerful algorithms. Dragonflies move somewhere in large number for the purpose of hunting and migration. The first purpose is called feeding or static swarm, while the latter is termed as the dynamic or migratory swarm.

The primary inspiration of the DA initiates from the static and dynamic swarming behaviors of dragonflies in nature. Two essential phases of optimization are exploration and exploitation are designed by modeling the social interaction of dragonflies in navigating, searching for foods, and avoiding enemies when swarming dynamically or statistically. Based on the movements of individuals in the swarm there are five main factors are considered in position updating of individuals in swarms. The five factors are as follows cohesion, alignment, separation, attraction (towards food sources), and distraction (outwards enemies) of individuals in the swarm. Suitable operators are included to the proposed DA algorithm for solving binary and multi objective problems as well. DA algorithm describes different steps involved in DA.

1. *Separation*: The individual search agent should maintain some distance away from the neighboring search agent, can be calculated as follows:

$$S_l = -\sum_{m=1}^{M} K_l - K_m, \qquad (4)$$

where $l \in \{1, 2, 3, 4....N\}$ is the index of present individual search agents, $m \in \{1, 2, 3, 4....M\}$ is the index of neighboring search agents. K_l is the present individual search agent, K_m neighboring individual agents, M is the no of neighboring individuals.

2. *Alignment*: The velocity of individual search agent should match with the velocity of the neighboring search agents, is found using following equation shown in Eq. (5)

$$A_l = \frac{1}{M} \sum_{m=1}^{M} \Delta K_m \tag{5}$$

3. *Cohesion*: The individual search agent will try to reach toward midpoint of mass of neighboring search agent, can be calculated as follows

$$C_l = \left(\frac{1}{M} \sum_{m=1}^{M} K_m \right) - K_l \tag{6}$$

4. *Attraction toward food sources*: The individual agent fly towards the food source, can be calculated as follows.

$$F_l = F_p - K_l, \tag{7}$$

where F_p is the position of food source, $K_l =$ Position of individual agent.

5. *Distraction outward an enemy*: The individual agent fly against the enemy, can be determined as follows.

$$E_l = E_p + K_l \tag{8}$$

where E_p is the position of enemy.

The food source fitness value F_{fit} and position F_p are updated by using the search agent with the best fitness value. The enemy fitness value E_{fit} and location E_p is updated using the search agent with the worst fitness value. At iteration k, the step and location of each search agent are updated as follows

$$\Delta K_t(d + 1) = {}_s S_1(d) + {}_p A_1(d) + {}_q C_1(d) + {}_r F_1(d) + {}_t E_1(d) + w \Delta K_t(d) \tag{9}$$

$$K_t(d + 1) = K_t(d) + \Delta K_t(d + 1), \tag{10}$$

where s, p, q, r and t are the weights of the behaviors, w is the inertia weight, S_1, A_1, C_1, F_1 and E_1 fitness values of search agent. The neighboring search agent M is at least one then according to Eq. (9) the position of the search agent is updated.

If there is no neighboring search agent then according to Eq. (11) the position will be updated.

$$K_t(d + 1) = K_t(d) + \text{Lévy}(d) \tag{11}$$

$$\text{Lévy (d)} = 0.01 \frac{r_1 \sigma}{|r_2|^{\frac{1}{\beta}}} \tag{12}$$

where r_1 and r_2 are random numbers, β is constant and σ can be calculated as follow:

$$\sigma = \left(\frac{\Gamma(1 + \beta) \sin\left(\frac{\Pi\beta}{2}\right)}{\Gamma\beta\left(\frac{1+\beta}{2}\right) 2^{\left(\frac{\beta-1}{2}\right)}} \right)^{\frac{1}{\beta}}, \tag{13}$$

where $\Gamma(x) = (x - 1)!$

4 Results and Discussion

The proposed DA optimized 2DOFPID based AGC is implemented in MATLAB-simulink environment using MATLAB R 2014a and executed on Intel core i3 2.20 GHz processor PC. The parameters of DA are fixed through trial and error method with number of simulation given in Appendix-B.

4.1 Application of DA Optimized 2DOF PID to Multi-area Hydro-Thermal Power System

The simulations are carried out under the same conditions as of [13], using DA tuned PI, PID and 2DOFPID controllers, the different controller gains are given in Table 1. Using the controller gains of DA optimized controllers, the transient performance of the power system is analysed as shown in Fig. 1a–c and the overshoot, undershoot, settling time of the deviations in frequency and tie-line power and improvement in DA optimized controllers are presented in Table 2 along with the GA and hFA-PS optimized controller performance as given in [13].

The considered objective function *ISE* for PI controller with DA technique is improved by 27.7, 13.88, and 2.62% compared to ZN, GA, and hFA-PS. The settling time of Δf_1 is improved by 76.67, 44.47, and 6.11% with DA tuned PI controller compared to ZN, GA, and hFA-PS PI. The settling time of Δf_2 shows an improvement of 67.16%, 50.23, and 16.06% with DA technique as compared to ZN, GA, and hFA-PS PI. The ΔP_{tie} settling time improves by 78.32, 47.15, and 27.17% with DA in comparison with ZN, GA, and hFA-PS PI.

Dragonfly algorithm steps to solve AGC problems:

Step 1	Read the data	Step 18	End if
Step 2	Set size of swarm N	Step 19	If $func(K_l)$ is worse than E_{fit} then
Step 3	Calculate food source fitness F_{fit} and update the position F_p	Step 20	$E_{fit} \leftarrow func(K_l)$
Step 4	Calculate enemy fitness E_{fit} and update position E_p	Step 21	$E_p \leftarrow K_l$
Step 5	Set the number of neighbors M	Step 22	End if
Step 6	Set maximum iteration K_{max}	Step 23	End for
Step 7	Calculate objective function $func(k)$	Step 24	For each search agent K_l
Step 8	for search agent l to N	Step 25	Calculate M
Step 9	Initialize search agent location K_l	Step 26	Calculate S_l, A_l, F_l, C_l, E_l using Eqs. (4)–(7)
Step 10	Initialize search agent step vector ΔK_t	Step 27	If M > 0 then
Step 11	For iteration $k=1$ to K_{max}	Step 28	Update ΔK_t using Eq. (8)
Step 12	Define w, a, f, s, e and c	Step 29	Update K_t using Eq. (9)
Step 13	For each search agent l	Step 30	else
Step 14	Calculate fitness value $func(K_l)$	Step 31	Update K_t using Eq. (10)
Step 15	If $func(K_l)$ is better than F_{fit} then	Step 32	End if
Step 16	$F_{fit} \leftarrow func(K_l)$	Step 33	End for
Step 17	$F_p \leftarrow K_l$	Step 34	End for

Figure 1 Dynamic performance of DA optimized controllers system with 1.5% SLP to the system (a) Δf_1 versus Time (b) Δf_2 versus Time (c) ΔP_{tie} (P.U MW) versus Time. The considered objective function *ITAE* for PI controller with DA technique is improved by 34.09% compared to hFA-PS. The settling time of Δf_1 improved by 14.46% with DA tuned PI controller compared to hFA-PS PI. The settling time of Δf_2 shows an improvement of 8.13% with DA technique as compared to hFA-PS PI. The ΔP_{tie} settling time improves by 18.06% with DA in comparison with hFA-PS PI.

Table 1 PI/PID/2DOF PID Controllers optimum values for multi-area multi-source system with 1.5% SLP at area-1

Controller Parameters	DA PI (ISE)	hFA-PS PI (ISE) [13]	DA PI (ITAE)	hFA-PS PI (ITAE) [13]	DA PID (ITAE)	hFA-PS PID (ITAE) [13]	DA 2DOF PID (ITAE)
K_{p1}	0.0411	0.0476	−0.033	0.0490	1.9338	1.8457	1.999
K_{p2}	−1.8455	−1.9441	−2	−0.7220	−0.1250	−0.4525	−1.999
K_{p3}	1.1451	1.1591	1.579	1.3594	0.1178	1.2922	1.7612
K_{p4}	−0.6148	−0.5823	−2	−1.7002	0.1878	−1.0720	2
K_{i1}	1.3902	1.4093	1.273	0.6533	1.7984	1.6563	2
K_{i2}	−0.2415	−0.2675	0.634	−0.0301	0.3527	0.1378	0.0992
K_{i3}	0.4112	0.4211	−0.070	0.1119	0.1968	1.8748	0.9190
K_{i4}	0.4859	−0.4942	1.407	−0.0827	0.1197	−1.3785	−1.312
K_{d1}	–	–	–	–	0.8326	0.6109	2
K_{d2}	–	–	–	–	−0.7458	0.4120	0.5662
K_{d3}	–	–	–	–	0.2584	0.4041	1.2439
K_{d4}	–	–	–	–	0.1892	0.4541	−1.9857
P_{w1}	–	–	–	–	–	–	1.996
D_{w1}	–	–	–	–	–	–	−1.616
n_1	–	–	–	–	–	–	72.78
P_{w2}	–	–	–	–	–	–	−1.9271
D_{w2}	–	–	–	–	–	–	−1.054
n_2	–	–	–	–	–	–	36.04
P_{w3}	–	–	–	–	–	–	1.911
D_{w3}	–	–	–	–	–	–	1.960
n_3	–	–	–	–	–	–	15
P_{w4}	–	–	–	–	–	–	2
D_{w4}	–	–	–	–	–	–	2
n_4	–	–	–	–	–	–	100

The Δf_1, Δf_2 and ΔP_{tie} settling times are improved by 30.09, 15.38, and 1.78% with DA optimized PID compared to hFA-PS PID. The overshoots of Δf_1, Δf_2 and ΔP_{tie} with DA optimized PID are improved by 35.87, 8.66, and 48.14% in comparison with hFA-PS PID. The undershoots of Δf_1, Δf_2 and ΔP_{tie} with DA optimized PID are improved by 20.89, 23.07, and 9.09% in comparison with hFA-PS PID.

The overshoot (OS) of Δf_1 has an improvement of 52.01 and 25.17% by DA tuned 2DOFPID as compared to hFA-PS and DA optimized PID controllers. An improvement of 100 and 100% is observed in the overshoot of Δf_2 using DA tuned 2DOFPID controller in comparison to hFA-PS and DA optimized PID controllers. The overshoot of ΔP_{tie} describes the performance improvement of DA by 96.66 and

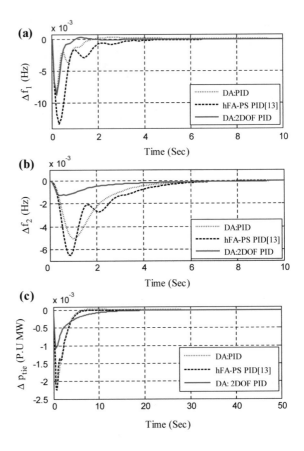

Fig. 1 Dynamic performance of DA optimized controllers system with 1.5% SLP to the system **a** Δf_1 versus Time **b** Δf_2 versus Time, **c** ΔP_{tie} (P.U MW) versus Time

93.57% in comparison to hFA-PS and DA optimized PID controllers. The undershoot(US) of Δf_1 shows an improvement of 34.32 and 16.98% in DA optimized controllers as compared to hFA-PS and DA algorithm. An improvement of 80 and 74% is observed in undershoot of Δf_2 using proposed controller as compared to hFA-PS and DA tuned controllers. The ΔP_{tie} undershoot shows the improvement of 50 and 45% in the performance of DA tuned 2DOFPID controller in comparison to hFA-PS and DA optimized PID controllers.

The settling time(ST) of Δf_1 is found to be improved by 63.52 and 47.82% by DA optimized 2DOF PID as compared to hFA-PS and DA optimized PID controllers. The settling time of Δf_2 shows an improvement of 48.07 and 38.63% in DA technique as compared to hFA-PS and DA. From Fig. 1c, it is shown that the ΔP_{tie} settling time improves by 3.06 and 1.29% using DA optimized controllers as compared to hFA-PS and DA.

The *ISE* objective function value is improved by 69.32 and 61.80% using DA optimized 2DOF PID as compared to hFA-PS and DA tuned PID. The *ITSE* objective function value is improved by 86.80 and 86.59% using DA tuned 2DOF PID

Table 2 Overshoot (OS), undershoot (US), settling time (ST) and with DA comparing to different algorithms

PI Controller ISE						PI Controller ITAE	
Parameters		ZN [14]	GA [14]	hFA-PS [13]	DA	hFA-PS [13]	DA
Δf_1	OS × 10^{-3}	13.6	7.8	12	12.1	3.6	10.9
	US × 10^{-3}	−27.7	−25.6	−26.6	−26.8	−28.8	−27.6
	ST	38.15	16.03	9.48	8.9	6.43	5.5
Δf_2	OS × 10^{-3}	9.4	5.6	8.9	8.9	0.837	5.1
	US × 10^{-3}	−18.7	−16.20	−17.7	−17.8	−19.6	−18.1
	ST	38.98	25.72	15.25	12.8	8.60	7.9
ΔP_{tie}	OS × 10^{-3}	0.74	1.2	0.599	0.64	0.0094	0.51
	US × 10^{-3}	−6.19	−5.7	−6.1	−6.1	−6.8	−6.3
	ST	23.99	9.84	7.15	5.2	5.98	4.9
OBJ	ISE × 10^{-6}	1079	905.8	801	780	805	790
	ITAE × 10^{-3}	1336	625.8	333.8	301	228.5	150.6
	ITSE × 10^{-6}	2890	1238	899.9	800	862.9	800

as compared to hFA-PS and DA tuned PID. The *ITAE* objective function value is improved by 77.70 and 57.73% using DA tuned 2DOF PID as compared to hFA-PS and DA tuned PID. The settling time (ST) of Δf_1, Δf_2 and ΔP_{tie} are found to be improved by 76.47, 50 and 55.81% using DA optimized 2DOF PID as compared to DE optimized 2DOF PID controller [15]. It is observed from the results in Tables 2, 3 and Fig. 1a–c the DA optimized 2DOF PID controller gives less OS, US, ST and objective function values as compared to hFA-PS and DA Optimized PID controllers.

Figure 2 a–c shows the DA optimized 2DOF PID controller gives less OS, US, ST, and objective function values as compared to DE Optimized 2DOF PID controller.

5 Conclusion

In this paper, a Dragonfly algorithm(DA) based 2DOFPID controller is proposed for two-area hydro-thermal interconnected power system. The performance of DA tuned 2DOFPID is found to be superior in comparison with GA, ZN tuned PI, hFA-PS-tuned PID, and DA-tuned PID.

Table 3 Overshoot (OS), undershoot (US), settling time (ST) and improvement with DA comparing to different algorithms

Parameters		hFA-PS PID [13]	DA tuned PID	DE tuned 2DOF PID [15]	DA tuned 2DOF PID	(%) Improvement with			
						DA PID w.r.t hFA-PS PID	DA 2DOF PID w.r.t hFA-PS PID	DA 2DOF PID w.r.t DA PID	DA 2DOF PID w.r.t DE 2DOF PID
Δf_1	$OS \times 10^{-3}$	0.223	0.143	1.4	0.107	35.87	52.01	25.17	92.35
	$US \times 10^{-3}$	−13.4	−10.6	–	−8.8	20.89	34.32	16.98	–
	ST	3.29	2.3	5.1	1.2	30.09	63.52	47.82	76.47
Δf_2	$OS \times 10^{-3}$	0.0127	0.0116	0.1559	0	8.66	100	100	100
	$US \times 10^{-3}$	−6.5	−5.0	–	−1.3	23.07	80	74	100
	ST	5.20	4.4	5.4	2.7	15.38	48.07	38.63	50
ΔP_{tie}	$OS \times 10^{-3}$	0.0675	0.035	0	0.00225	48.14	96.66	93.57	–
	$US \times 10^{-3}$	−2.2	−2.0	–	−1.1	9.09	50	45	–
	ST	3.92	3.85	8.6	3.8	1.78	3.06	1.29	55.81
OBJ	$ISE \times 10^{-6}$	118.4	95.08	122.9	36.32	19.69	69.32	61.80	70.44
	$ITAE \times 10^{-3}$	87	45.9	59.727	19.4	47.24	77.70	57.73	67.51
	$ITSE \times 10^{-6}$	85.7	84.4	81.73	11.31	1.516	86.80	86.59	86.16

Fig. 2 Dynamic performance of the system with DA optimized 2DOF PID controller **a** Δf_1 versus Time. **b** Δf_2 versus Time. **c** ΔP_{tie} versus Time

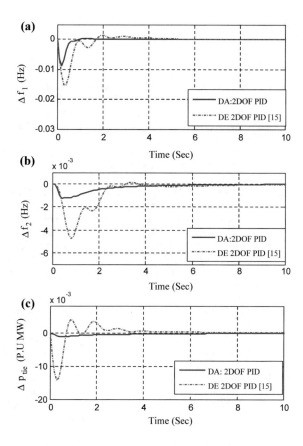

Appendix-A [13]

Parameters of Multi-area Hydro-Thermal Power system are:

$f = 50$ Hz; $B_1 = B_2 = 0.425$ p.uMW/Hz; $R_1 = 2.0$ Hz/p.u; $R_2 x = 2.4$ Hz/p.u; $T_g = 0.08$ s; $T_t = 0.3$ s; $K_p = 100$ Hz/p.u; $T_p = 20$ s; $K_1 = 1.0$; $T_1 = 48.7$ s; $T_w = 1.0$ s; $T_2 = 0.513$ s; $T_r = 5.0$ s; $T_{12} = 0.0707$ p.u; $a_{12} = -1$

Appendix-B [12]

$w = 0.3$, $s = 0.1$, $a = 0.1$, $c = 0.7$, $f = 1$, $e = 1$

References

1. O.I. Elgerd, *Electric energy systems theory an introduction*, 2nd edn. (Tata McGraw Hill, New Delhi, 2000)
2. P. Kumar, D.P. Kothari, Recent philosophies of automatic generation control strategies in power systems. IEEE Trans. Power Syst. **20**(1), 346–357 (2005)
3. S.K. Pandey, R.M. Soumya, N. Kishor, A literature survey on load–frequency control for conventional and distribution generation power systems. Renew. Sust. Energy. Rev. **25**, 318–334 (2013)
4. J. Nanda, S. Mishra, L.C. Saikia, Maiden application of bacterial foraging based optimization technique in multi area automatic generation control. IEEE Trans. Power Syst. **24**, 602–609 (2009)
5. H. Golpira, H. Bevrani, Application of GA optimization for automatic generation control design in an interconnected power system. Energy Convers. Manag. **52**, 2247–2255 (2011)
6. P.K. Hota, B. Mohanty, Automatic generation control of multi-source power generation under deregulated environment. Int. J. Electr. Power Energy Syst. **75**, 205–214 (2016)
7. K. Mohan, C. Anu, K. Satish, Application of PSO technique in multiarea automatic generation control, in *Proceeding of International Conference on Intelligent Communication, Control and Devices, Advances in Intelligent Systems and Computing*, vol. 479. https://doi.org/10.1007/97 8-981-10-1708-7_9
8. E.S. Ali, S.M. Abd-Elazim, Bacteria foraging optimization algorithm based load frequency controller for interconnected power system. Int. J. Electr. Power Energy Syst. **33**, 633–638 (2011)
9. A.K. Barisal, comparative performance analysis of Teaching Learning Based Optimization for automatic load frequency control of multi sources power system. Int. J. Electr. Power Energy Syst. **66**, 67–77 (2015)
10. K. Jagatheesan, B. Anand, S. Samanta, N. Dey, A.S. Amira, V.E. Balas, Design of a proportional-integral-derivative controller for an automatic generation control of multi-area power thermal systems using firefly algorithm. IEEE/CAA J. Autom. Sin
11. M.A. Johnson, H.M. Mohammad, *PID Control: new identification and design methods* (Springer, 2005)
12. S. Mirjalili, Dragonfly algorithm: a new meta-heuristic optimization technique for solving single-objective, discrete, and multi-objective problems. Neural Comput. Appl. **29** (2015)
13. R.K. Sahu, S. Panda, S. Padhan, A hybrid fire fly algorithm and pattern search technique for automatic generation control of multi areapowersystems. Int. J. Electr. Power Energy Syst. **64**, 9–23 (2015)
14. K.R.M.V. Vijaya Chandrakala, S. Balamurugan, K. Sankaranarayanan, Variable structure fuzzy gain scheduling based load frequency controller for multi-source multi area hydro thermal system. Int. J. Electr Power Energy Syst. **53**, 375–381 (2013)
15. R.K. Sahu, S. Panda, U.K. Rout, DE optimized parallel 2-DOF PID controller for load frequency control of power system with governor dead-band nonlinearity. Int. J. Electr. Power Energy Syst. **49**, 19–33 (2013)

Wide Area Monitoring System Using Integer Linear Programming

Hasmat Malik and M. D. Gholam Wahid Reza

Abstract With the enhancement of power system network topology, it is necessary to require the tools and techniques for handling the blackout condition due to any kind of disturbances. When any kind of fault or disturbance happen, protection, and control measures play the vital role to prevent further degradation of the system; restore the system back to a normal state. Continuous technological innovation in information and communication technology, various kinds of sensors and measurement instrument principle in general have promoted the advent of phasor measurement units (PMUs). This paper describes the modeling and testing of phasor measurement unit for different bus system using MATLAB/Simulink and wide area protection using PMU are analyzed. Since PMUs are costly, so optimal placement of PMU has been done using integer linear programming (ILP) in MATLAB solver for different bus systems. The results are tested and validated by using IEEE 7, IEEE 14 and IEEE 30 bus system under zero injection and without zero injection condition.

Keywords Wide area measurement system (WAMS) · Phasor measurement unit (PMU) · Global positioning system (GPS) · Optimal placement

1 Introduction

In the present scenario, the power system network is increasing day-and-night basis as the requirement of the increase population to provide the uninterrupted supply. But due to unpredictable scenario such as storm, heavy rain and/or snow fall, many major problems may occur in the power system. Therefore, blackouts and load shedding are big challenging problems faced by power engineers in the whole world. These can

H. Malik
EE Department, IIT Delhi, New Delhi, India
e-mail: hmalik.iitd@gmail.com

M. D. Gholam Wahid Reza (✉)
Division of ICE, NSIT, New Delhi, India
e-mail: wahidreza16@gmail.com

© Springer Nature Singapore Pte Ltd. 2019
H. Malik et al. (eds.), *Applications of Artificial Intelligence Techniques in Engineering*, Advances in Intelligent Systems and Computing 698,
https://doi.org/10.1007/978-981-13-1819-1_3

23

be overcome by using advanced condition monitoring and controlling the existing power grid. Proper monitoring and controlling of power grid can be done using a PMU. The Phasor Measurement Unit (PMU) is a device that is employed to detect the voltage and current waveform that is synchronized with a clocking signal obtained continuously from the global positioning system (GPS) [1]. Integrating with the GPS receiver the base station is able to receive the synchronous data from each PMU in real time. The location of malfunction circuits or transmission lines can be immediately identified if phase differences between different PMUs are detected [2]. It is neither economical nor necessary to install a PMU at each bus of a WAMS. As a result, the problem of optimal PMU placement (OPP) concerns with where and how many PMUs should be implemented to a power system to achieve full observability at minimum number of PMUs [3].

2 Phasor Measurement Unit

Phasor Measurement Unit has been defined by IEEE as "a device that produces synchronized Phasor, frequency, and rate of change of frequency (ROCOF) estimates from voltage and/or current signals and a time synchronizing signal" [4]. These synchronized measurements are termed synchrophasor. PMU devices are time stamped with GPS. The concept of a PMU was introduced in the 1980s [5]. A research group, led by Prof. Phadke, performed most of the original work at Virginia Tech.

Mathematical description of Synchrophasor:

An AC waveform can be mathematically represented as

$$x(t) = X_m \cos(\omega t + \Phi) \tag{1}$$

where,

X_m Magnitude of Sinusoidal waveform
ω Fundamental frequency
Φ Angular starting Point of the waveform

In phasor notation it can be represented as

$$\bar{x} = \frac{X_m}{\sqrt{2}} < \phi \tag{2}$$

where, $\frac{X_m}{\sqrt{2}} = $ rms magnitude of waveform and $\Phi = $ phase angle.

Fig. 1 Flow chart for PMU
model

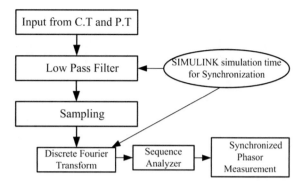

2.1 Simulation of PMU

The flow chart for simulation of a PMU is represented in Fig. 1 which includes seven basic steps. In step1, the current and voltages are measured with the help of CT and PT respectively and then it will pass through the low pass filter as shown in step2. Obtained results from LPF are utilized for sampling purposes so that the sampled signals may be processed through DFT. After this, sequence analyzer is used to analyze the DFT signal and finally, synchronized Phasor measurement unit will use the data. The basic Phasor measurement process is that of estimating a positive sequence (also negative and zero are available), fundamental frequency phasor representation from voltage or current waveforms. The positive sequence phasor can be calculated as follows:

$$V = 1/3(V_a + \alpha V_b + \alpha^2 V_c), \tag{3}$$

where $\alpha = 1\angle 120°$. V_a, V_b and V_c are the DFT phasor coefficients of each of the three phases.

In designing the PMU, the dynamic of the system is considered, as the transient after fault is visible in the figure. Moreover, the 3-phase voltage then measured by V-I measurement block and feed to the PMU block. PMU block is shown in Fig. 2, where the voltage feed into the Phase Locked Loop (PLL) and then sequence analyzer. That is how the three sequence voltage is built. Frequency can be measured through PLL (Fig. 3).

3 Case Study and Results

By using Simulink platform of the MATLAB, a Simulink model of doubly-fed two-bus system is implemented with the help of PMUs as represented in Fig. 4 with the following network parameters as mentioned in Table 1.

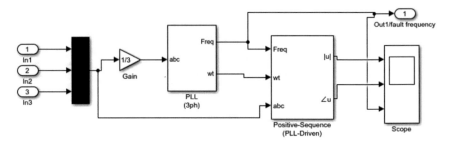

Fig. 2 Simulink model of phasor measurement unit

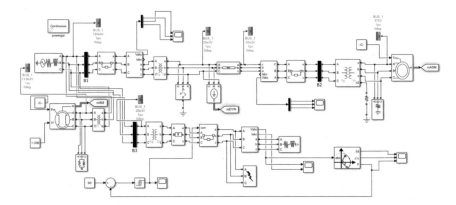

Fig. 3 Simulation model for three bus system with PMU

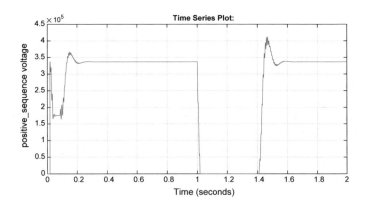

Fig. 4 Positive sequence voltage

The three phase input voltages are measured after each bus. These three phase voltage are given to the PMU and PMU gives us the positive, negative, and zero sequence voltages and also theAngle. Figure 6 illustrates clearing fault with feedback frequency from PMU that is compared with constant frequency 60 Hz and error signal

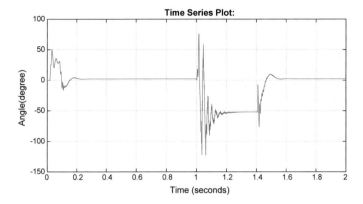

Fig. 5 Positive sequence angle

Table 1 Network parameter

Voltage rating: 120 k
System frequency: 60 Hz
Line constant: Positive sequence resistance: 0.1153; zero sequence resistance: 0.413; (ohm/km) Positive sequence inductance: 1.05e−3; zero sequence inductance: 3.32e−3; (H/km) Positive sequence capacitance: 11.33e−009; zero sequence capacitance: 5.01e−009; (F/km)
Transmission line length: 25 km

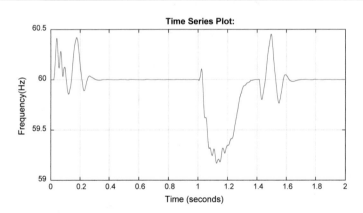

Fig. 6 Frequency during before after and clearing fault

is used to operate relay. The relay is used to switch on and off the three phase breaker. Figures 4 and 5 shows positive sequence voltage and angle before, during and after fault.

4 Optimal Placement of PMU

The objective of PMU placement is to make the entire system observable using a minimum number of PMUs. One of the assumptions about the considered PMUs is that each PMU has enough channels to measure bus voltage and all incident branch current phasors at a given bus.

4.1 Formulation of ILP

Objective function: min $\sum_{i=1}^{n} x_i$ where n = no of buses
    Constraints

$$f(x) = Ax \geq b$$

$$x = [x_1 x_2 \ldots x_n]^T$$

$$x_i \in \{0, 1\},$$

where, x_i is a PMU placement Variable and

$$b = [1\ 1\ 1\ 1 \ldots]^T$$

The entries of binary connectivity matrix A are defined as (Fig. 7);

$$Aij = 1 \text{ if } i = j \text{ or connected}$$

$$\text{Otherwise } = 0$$

Fig. 7 7 bus system

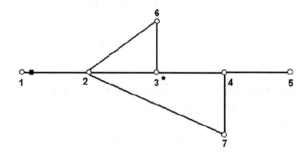

For 7-bus system min $\sum_{i=1}^{7} x_i$

$$
A.x = \begin{array}{l}
x_1 + x_2 \geq\sim 1 \\
x_1 + x_2 + x_3 + x_6 + x_7 \geq 1 \\
x_3 + x_4 + x_5 + x_7 \geq 1 \\
x_3 + x_4 + x_5 + x_7 \geq 1 \\
x_4 + x_5 \geq 1 \\
x_2 + x_3 + x_6 \geq 1 \\
x_2 + x_4 + x_7 \geq 1
\end{array}
$$

The operator "+" serves as the logical "OR" and the use of 1 in the right-hand side of the inequality ensures that at least one of the variables appearing in the sum will be non-zero. Incidence matrix for 7-bus system as;

$$
A.x = \begin{bmatrix}
1 & 1 & 0 & 0 & 0 & 0 & 0 \\
1 & 1 & 1 & 0 & 0 & 1 & 1 \\
0 & 1 & 1 & 1 & 0 & 1 & 0 \\
0 & 0 & 1 & 1 & 1 & 0 & 1 \\
0 & 0 & 0 & 1 & 1 & 0 & 0 \\
0 & 1 & 1 & 0 & 0 & 1 & 0 \\
0 & 1 & 0 & 1 & 0 & 0 & 1
\end{bmatrix};
$$

4.2 Optimal Results

The proposed approach is implemented and tested on distinct power system network models; out of them some are mentioned in Table 2. The obtained results for IEEE 7, 14, and 30 bus systems are represented in Table 2 under two different conditions (PMU with zero-injection bus and PMU without zero-injection bus). Both results (PMU with zero-injection bus and PMU without zero-injection bus) for IEEE 7 bus system are same whereas for IEEE 14 and IEEE30 bus system results are different. For IEEE 14 bus system, optimal numbers of PMUs are selected 3 and 4 for PMU with zero-injection bus and PMU without zero-injection bus system respectively. Similarly, IEEE 30 bus systems, optimal number of PMUs are selected 9 and 10 for PMU with zero-injection bus and PMU without zero-injection bus system respectively. Therefore, PMU with zero-injection bus condition is better than PMU without zero-injection as represented in Table 1.

Table 2 Optimal placement of PMU

IEEE bus system	No of PMU with zero-injection bus	No of PMU without zero-injection bus
IEEE 7 Bus	2(1, 3)	2(1, 3)
IEEE 14 Bus	3(2, 6, 9)	4(2, 6, 7, 9)
IEEE 30 Bus	9(2, 4, 6, 9, 10, 12, 15, 18, 27)	10(2, 4, 6, 9, 10, 12, 15, 18, 25, 27)

5 Conclusions

The PMUs are the core for wide area monitoring system (WAMS) and their performance determines the performance of WAMS for monitoring and control of modern power system. From this work we are able to achieve the Voltage and Currents measurements in Phasor form with accurate Time stamping. For the Optimal placement integer linear programming method is used in this paper. As the objective function and constraints are linear the method is very simple to understand. By optimal placement of PMU we can reduce the overall installation cost as well as make our grid more reliable and smarter.

References

1. A.G. Phadke, Synchronized phasor measurement in power system. IEEE Comput. Appl. Power **6**(2), 10–15(1993)
2. P. Zhang, F. Li, N. Bhatt, Next-generation monitoring, analysis, and control for the future smart control center. IEEE Trans. Smart Grid **1**(1), 186–192 (2010)
3. R.F. Nuqui, A.G. Phadke, Phasor measurement unit placement techniques for Complete and incomplete Obs. IEEE Trans. Power Del. **20**, 2381–2388 (2005)
4. D.R. Gurusinghe, A.D. Rajapakse, K. Narendra, Evaluation of steady-state and dynamic performance of a synchronized phasor measurement unit, in *2012 IEEE Electrical Power and Energy Conference (EPEC)* (2012), pp. 57–62
5. A.G. Phadke, Synchronized phasor measurements—a historical overview, in *IEEE/PES Transmission and Distribution Conference and Exhibition 2002: Asia Pacific, 6–10 Oct 2002*, vol.1 (IEEE, 2002), pp. 476–479
6. B. Xu, A. Abu, Observability analysis and measurement placement for Systems with PMUs, Proc. IEEE Power System Conf. Expo., **2**, 943–946, 2004

Fault Classification and Faulty Phase Selection Using Symmetrical Components of Reactive Power for EHV Transmission Line

Piyush Khadke, Nita Patne and Sahitya Bolisetty

Abstract The scholarly paper describes the method using the symmetrical components of the reactive power for both the classification of the fault as well as for the selection of the faulty phase for single-circuit transmission lines. The evaluation of the different types of faults is done using MATLAB/ SIMULINK software and provides very fast results for the type of the fault and the faulted phase. This method just requires the raw value of the system voltage and does not require any threshold value unlike other methods. The proposed fault detection method is tested for shunt and series compensated systems also which shows its potential for future use. This method was proposed to have the fast and reliable operation of the protective relays.

Keywords Fault classification · Symmetrical components · Reactive power

1 Introduction

Fault in the power system leads to over current, under voltage, unbalance of the phases, failure of the equipments etc. Faults are classified into two types i.e. open circuit faults and short circuit faults. These faults are further classified into symmetrical and unsymmetrical faults. Open circuit faults or series faults occur due to the failure of one or more conductors. The causes of this fault include failures of joints of cables and overhead lines and failure of circuit breaker in one or more phases and due to melting of fuse or conductor. Short circuit faults or shunt faults are due

P. Khadke (✉)
Electrical Engineering Department, Shri Ramdeobaba College of Engineering
and Management, Nagpur, India
e-mail: piyushkhadke88@gmail.com

N. Patne
Electrical Engineering Department, Visvesvaraya National Institute
of Technology Nagpur, Nagpur, India

S. Bolisetty
Nagpur Metro Rail Corporation, Nagpur, India

© Springer Nature Singapore Pte Ltd. 2019
H. Malik et al. (eds.), *Applications of Artificial Intelligence Techniques in Engineering*, Advances in Intelligent Systems and Computing 698,
https://doi.org/10.1007/978-981-13-1819-1_4

31

to the abnormal connection of very low impedance between two points. These are due to the insulation failure between phase conductors or between phase and earth conductors. These are severe faults resulting in the flow of high currents through the equipment; it leads to extreme damage to the equipments.

The symmetrical faults are also called as balanced faults which include line-line-line (L-L-L) and line-line-line ground (L-L-L-G) faults. The occurrence of this fault is 2–5% of total fault system but causes severe damage to the equipments if they occur. Unsymmetrical faults are unbalanced faults as they cause unbalanced currents in the system having different magnitude and unequal phase displacement. It includes open circuit fault with one or two phase open condition and short circuit faults of line-ground (L-G), line-line (L-L), double line to ground (L-L-G). L-G fault is the most common fault and 70–80% of the faults that occur in power system are of this type and less severe as compared to other types of faults. Line-line fault occurs when one live conductor get in contact with other live conductor. There are less severe faults and occurrence ranges from 15–20%. Double line-ground fault occurs when two lines come in contact with each other as well as with the ground; the occurrence range is about 10% of the total power system.

As these faults can cause severe damage to the electrical equipments and transmission lines, different methods are employed to determine the type of fault and hence reduce the damage caused by the faults to the equipments. Various methods for the classification of the faults and the selection of the faulted phase have been proposed earlier. These methods involve different techniques such as depending upon the on the angles between superimposed positive-sequence currents and negative sequence currents, the ratio of differential superimposed voltage to differential superimposed current, ANN (Artificial Neural Network), etc.

One of the interesting methods, proposed by Behnam Mahamedi and Jian Guo Zhu [1] presents a method of the classification of the fault based on the action of the faulted phase for single-circuit transmission lines which depends on the symmetrical components of reactive power. In this method, the result does not hold good if the distances are beyond specific limits. A method proposed by Silva et al. [2], included an application of Haar wavelet transform for fault classification. The energy of the detail coefficients of the phase currents are used for the faulted phase selection. The coefficients of the neutral current are used to distinguish the faults with and without ground involvement. The fault incidence angle variations involved in this method decrease the performance of this method. Another method proposed by Sulee Bunjonjit and Attapol Ngaopittakul [3], used ANN and discrete wavelet transform (DWT) for the classification of fault on single-circuit transmission line. Positive sequence current signals are used in fault measurement and calculation in fault detection. Radial basis function (RBF) neural network, Back propagation neural network (BPNN) and probabilistic neural network (PNN) are compared. However, this method involves huge computation and data analysis. Pillai [4] presented another approach for accurate fault classification in TCSC Compensated Transmission Line using SVM. This method uses three phase current measurement for the fault classification. One support vector machine is trained for classification of fault. It involves local current measurement and that has not been an easy task. A strategy proposed by Jamehbozorg and Shahrtash [5, 6], introduced a method based on decision tree for classification of fault in single and double circuit and transmission lines using the traveling waves initiated by the fault and exerting half cycle discrete Fourier transform (HCDFT). Calculations up to nineteenth harmonic are to be made which is laborious, consequently increasing the computing burden. The method proposed by [7] is depending on ANN and metal oxide varistor (MOV) energy, employing Levenberg–Marquardt training algorithm. The novelty of this scheme is the use of MOV energy signals of fixed series capacitors (FSC) as input to train the ANN. The wavelet transform concept and its value in techniques of classification and feature detection schemes are presented by Youssef [8]. Although the wavelet transform is very effective in detecting transient signals generated by the faults, the wavelet transform may not be adequate to complete characterization. Further to the wavelet transform concept, Bancha Sreewirote [9] uses the Support vector machine (SVM) algorithm and concept of discrete wavelet transform (DWT) for the classification of the faults. Here, the fault detected is analyzed using the high frequency component by DWT and further the fault will be classified using SVM technique using the co-efficient in the DWT as an input to SVM. A. S. Neethu [10] uses the concept of ANN and DWT for finding the location of the fault and done in MATLAB. One more algorithm is developed for finding the type of fault using wavelet transform by carrying out the simulation in MATLAb Simulink. Based on the initial current traveling wave another algorithm for the classification of fault and faulted phase selection has been proposed by Dong et al. [11]. The characteristics of different faults were investigated on the basis of the Karenbauer transform. A high sampling frequency of 400 kHz has been used for evaluation in this method. Mallik [12] uses the modified fuzzy Q learning (MFQL) technique to classify the different types of faults in the transmission line based on the raw values available of currents and voltages at supply side and load side, then are processed using empirical mode decomposition, J48 algorithm and are taken as input to MFQL fault classifier. A method proposed by [13, 14], for the fault classification and detection of SVC (static VAR Compensator) integrated differential relay based double circuit transmission line, is for the compensated transmission line using DWT. This method Supervised machine learning (SML) proposed by [15] for the fault classification uses DWT to extract the data from the current and voltage Waveforms and then the harmony search algorithm is used to identify the optimal parameters of the wavelets. Then the values are simulated in transmission and distribution test simulation models for the result.

In proposed method of this paper, the reactive powers formed by positive as well as negative sequence components are used to determine line to line or phase to phase faults and the actual phases involved in such types of faults Fig. 1. To differentiate phase to phase to earth fault from single phase to earth fault, the ratio of zero sequence reactive power to negative sequence reactive power is used. To select the faulty phase in a ground fault, the absolute value of reactive power formed by negative and zero sequence components, will be employed. In a double phase to earth fault, minimum value for the absolute value of reactive power formed by negative and zero sequence components will arise in one of the faulty phases. In single phase to earth fault, this value will be maximum.

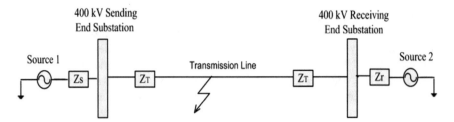

Fig. 1 Power system

The above-mentioned papers propose different methods like ANN, DWT, BPNN, PNN, HCDFT, and other machine learning techniques like MFQL, SML, etc. to classify the fault. The methods used like ANN, DWT, etc., requires stringent and complex computations, and these methods may or may not give the exact results. The machine learning techniques prove to give exact results but needs more time and are also involved with combinational use of different techniques. But the method proposed in this paper is rugged to give the result for any condition of the fault, is accurate, requires less time and even does not involve rigorous computations to arrive at the result.

The scholarly paper is arranged in different sections as with proposed method in which the total description of the method will be explained then comes the simulation part, where the simulation model of the paper and the flow chart for the simulation model is explained and then after comes the result part, where for all types of faults are discussed. And the last section is the conclusion in which the overall description of the paper is explained.

This paper is arranged with: (1) Sect. 2—Proposed Method, (2) Sect. 3—Simulations, (3) Sect. 4—Results and (4) Sect. 5—Conclusion.

The effect of compensation using Static VAR compensator, STATCOM and SSSC in the transmission line is compared using MATLAB/SIMULINK software.

2 Proposed Method

The method used here employs the symmetrical components of reactive power. They are defined as follows:

$$Q_1 = \text{image} \left(V_1 . I_1^* \right) \tag{1}$$

$$Q_2 = \text{image} \left(V_2 . I_2^* \right) \tag{2}$$

$$Q_0 = \text{image} \left(V_0 . I_0^* \right) \tag{3}$$

(*) denotes the conjugated component of current used where Q_1, Q_2, Q_0 are the positive, negative, and zero sequence components of reactive power respectively.

Fig. 2 Flow chart

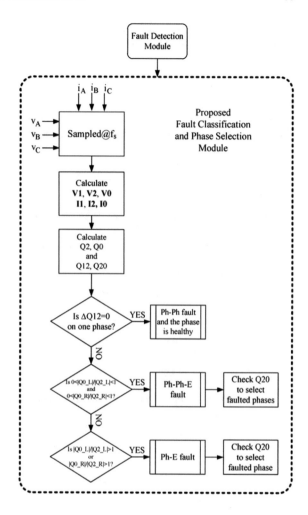

V_1, V_2, V_0 are the positive, negative, and zero sequence components of the voltage observed at relay point (either at sending or receiving end). Similarly I_1, I_2, I_0 are the positive, negative, and zero sequence components of the current observed at the relay point. The type of fault and faulty phase can be found for different power systems as described in following explanation. The program for determining various types of faults is written from the below mentioned flow chart (Fig. 2).

A. *Single Phase to Earth Faults*:

In this fault, the ratio of zero sequence reactive power to negative sequence reactive power is greater than 1 either from sending end or receiving end, i.e.,

$$Q_0/Q_2 > 1 \tag{4}$$

For the selection of faulted phase, $|Q_{20}|$ of different phases are calculated. The value of $|Q_{20}|$ will be maximum on the faulted phase.

$$Where, Q_{20} = image\ (V_2 + V_0) \times (I_2 + I_0)^* \tag{5}$$

B. *Double Phase to Earth Faults*:

Here, the ratio of zero sequence reactive power to negative sequence reactive power must lie between 0 and 1, i.e.,

$$0 < Q_0/Q_2 < 1 \tag{6}$$

For the selection of faulted phase, $|Q_{20}|$ of all the three phases are calculated.

And, $|Q_{20c}|$ will be maximum for ABG (L − L − G) fault,

$|Q_{20a}|$ will be maximum for BCG fault,

$|Q_{20b}|$ will be maximum for CAG fault.

C. *Phase to Phase Faults*:

To distinguish the phase to phase faults, $|Q_{12}|$ of different phases are used.

$$Where, |Q_{12}| = image\ (V_1 + V_2) \times (I_1 + I_2)^* and$$

$|Q_{12c}|$ will be minimum for AB fault

$|Q_{12a}|$ will be minimum for BC fault

$|Q_{12b}|$ will be minimum for CA fault

However, the accuracy of this type of fault decreases as fault resistance increases.
The flow chart in Fig. 2 explains about how the fault is classified, i.e., basically the phase currents and phase voltages are taken as the input samples and from them the zero sequence, positive sequence, and the negative sequence currents and voltages are calculated separately. From that, the reactive power components are also calculated of the entire Zero, negative and positive sequence components. Now, after the calculation of reactive power components difference between the negative reactive power component and positive reactive power component of any one phase is found

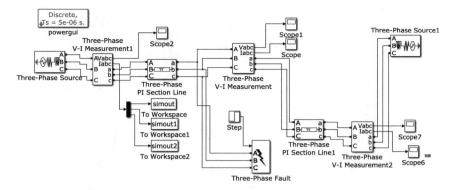

Fig. 3 Simulation diagram for proposed system

and if, the value is equal to zero then it can be described that, the particular phase is healthy and it will be determined to be a phase to phase fault, but if that difference value is found not to be equal to zero, then ratio of zero sequence reactive power to positive sequence reactive power of other phases is to be found and confirmed weather it is between 0 and 1, if yes, then it is confirmed to be phase to phase to earth fault. But, now if the ratio is greater than 1, then it is to be confirmed to be phase to earth fault, and the faulted phases in any case can be known by checking the difference between positive sequence reactive power and zero sequence reactive power.

3   Simulations

In this section, the presented method will be computed using MATLAB/SIMULINK software. From the MATLAB/SIMULINK software we calculate the values of voltages and currents for different fault distances. Considering the power system shown in the (Fig. 1) with the following parameters and (Fig. 3) shows the detail simulation diagram for the proposed system.

For source: $f = 50$ Hz; Voltage level $= 400$ kV; Fault resistances: $R_f = 0 - 150\,\Omega$;

$$Z_{1S} = Z_{2S} = 0.32 + j6.56\,\Omega;$$
$$Z_{0S} = 1.76 + j9.28\,\Omega;$$
$$Z_{1r} = Z_{2r} = 0.48 + j8.8\,\Omega;$$
$$Z_{0r} = 1.76 + j11.68\,\Omega;$$

For Line: Total line length $= 250$ km; $Z_{11} = Z_{21} = 0.021 + j.278\,\Omega$/km; $Z_{01} = 0.302 + j0.905\,\Omega$/km; $C_{11} = 13.170$ nF/km; $C_{01} = 8.396$ nF/km.

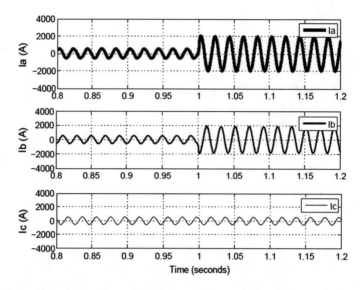

Fig. 4 Figure shows the waveform of the currents in the phases during AB line-line fault

Thus, the calculated values of currents and voltages are used in the program and the type of fault and the phase of the faulty line are produced as output. The method has been extended to run for the compensated systems. In compensated systems, compensation devices are used for compensating the reactive power in the power system. The compensators that are connected in series with the power systems are called as series compensators and that are connected in parallel are called as shunt compensators.

Examples of shunt compensators include static synchronous compensator (STATCOM), Thyristor-controlled reactor (TCR), Static VAR compensator (SVC), Thyristor-switched reactor (TSR), Thyristor-switched capacitor (TSC), Mechanically switched capacitor (MSC). Examples of series compensators includes Static synchronous series compensator (SSSC), Thyristor-controlled series capacitor (TCS), Thyristor-controlled series reactor (TCSR), Thyristor-switched series capacitor (TSSC), Thyristor-switched series reactor (TSSR). SVC is employed in this system.

4   Observations from Simulations

Figure 4 shows plot of per phase currents for line to line fault.

In the preceding Table 1, (Q_0/Q_2) is the ratio of zero sequence reactive power to that of negative sequence. For the Line to ground fault, it will be greater than 1 and for the other two faults, it lies between 0 and 1. $|Q_{20}|$ is the absolute value of

Table 1 Results of reactive power components of L-G and L-L-G faults for a fault distance = 10 km (sending end)

Type of fault	(Q_0/Q_2)	Q_{20a}	Q_{20b}	Q_{20c}
A-G	1.01	4.38e+006	1.09+006	1.10e+006
B-G	1.0090	1.10e+006	4.38e+006	1.09e+006
C-G	1.0085	1.09e+006	1.10e+006	4.38e+006
A-B-G	0.1116	3.983e+006	3.343e+006	8.95e+006
B-C-G	0.1117	8.95e+006	3.98e+006	3.84e+006
C-A-G	0.1114	3.84e+006	8.95e+006	3.98e+006

Table 2 Results of reactive power components of L-G and L-L-G faults for a Fault distance = 10 km (receiving end)

Type of fault	(Q_0/Q_2)	Q_{20a}	Q_{20b}	Q_{20c}
A-G	0.8373	2.42e+005	6.33e+004	5.88e+004
B-G	0.8474	5.9e+004	2.45e+005	6.42e+004
C-G	0.8431	6.518e+004	5.79e+004	2.44e+005
A-B-G	0.0930	2.48e+005	2.32e+005	5.19e+005
B-C-G	0.0937	5.23e+005	2.49e+005	2.33e+005
C-A-G	0.0194	2.93e+005	4.98e+005	3.07e+005

Table 3 Results of reactive power components of L-G and L-L-G faults for a fault distance = 100 km (sending end)

Type of fault	(Q_0/Q_2)	Q_{20a}	Q_{20b}	Q_{20c}
A-G	0.8373	2.42e+005	6.33e+004	5.88e+004
B-G	0.8474	5.9e+004	2.45e+005	6.42e+004
C-G	0.8431	6.518e+004	5.79e+004	2.44e+005
A-B-G	0.0930	2.48e+005	2.32e+005	5.19e+005
B-C-G	0.0937	5.23e+005	2.49e+005	2.33e+005
C-A-G	0.0194	2.93e+005	4.98e+005	3.07e+005

the reactive power between negative sequence and zero sequence components for different phases. For the Line to Ground fault, the value of $|Q_{20}|$ will be maximum for the faulted phase. And for the Double Line to Ground fault it will be maximum for the healthy phase. In the receiving end Table 2, it is observed that the ratio slightly changes in the Line to Ground fault due to the addition of capacitance and reactance, as the fault occurs at a greater distance from the receiving end (Table 3).

Now in the case of Line-to-Line faults, the ratio of the zero sequence to the negative sequence reactive power will not become handy as its ratio is negligible or zero. Hence, the absolute value of the difference between the positive sequence and negative sequence reactive power after the fault and prior to the fault (ΔQ_{12}) are used. Now this value is minimum for the healthy phase and higher for the other two faulty phases. Also it is noted that ΔQ_{12} measured by the receiving end relay is

Table 4 Results of reactive power components of L-L fault for a fault distance = 10 km (sending end)

Type of fault	ΔQ_{12a}	ΔQ_{12b}	ΔQ_{12c}
A-B	4.18e+008	4.45e+008	1.8e+006
B-C	1.09e+006	4.16e+008	4.43e+008
C-A	4.43e+008	4.5e+004	4.19e+008

Table 5 Results of reactive power components of L-L fault for a fault distance = 10 km (receiving end)

Type of fault	ΔQ_{12a}	ΔQ_{12b}	ΔQ_{12c}
A-B	2.398e+007	3.58e+007	1.737e+007
B-C	7.73e+004	2.39e+007	5.31e+007
C-A	3.54e+007	1.59e+005	6.52e+006

almost the same for A-B fault, in all the three phases. So, the accuracy of this fault decreases as the distance increases from either of the ends (Tables 4 and 5).

It is observed that the value of ΔQ_{12} measured by sending end relay for A-B fault, in all the three phases is almost the same to be precise; it is slightly less for the phase C than for the other two phases. But, ΔQ_{12} is less for the healthy phase than the other two phases. It is to be noted that the faulty phase has the higher ΔQ_{12} value than that of the healthy phase but the value is in the same powers in AB fault for phase A and C measured by sending end relay. So, the accuracy of this fault has to be taken care of. But, the measured values of ΔQ_{12} by both the relays is less for the healthy phase and more for the faulty phase in this case as well.

A. *With shunt compensation Static VAR compensator (SVC)*:

The static VAR compensator (SVC) is the set of electrical equipments for furnishing the fast-acting reactive power control on high-voltage transmission networks. SVCs are a part of the Flexible AC transmission system device family, power factor, regulating voltage, harmonics and stabilizing the system. The SVC is kept at the receiving end and the values of the 3-phase reactive and active power is noted for the different cases, is noted. Values in different cases are noted by varying the Susceptance (B_{ref}.) in (pu/Pbase). In Table 6, it is observed that the reactive power improves on increasing the value of 'B', but the active power does not show any appreciable change. Also, it is noted that the reactive power is compensated largely in the chance of Line to Ground faults, for the sending end as the SVC is kept at sending end.

B. *With series compensation Static synchronous series compensator (SSSC)*:

For the series compensation of the line, Static synchronous series compensator (SSSC) is kept at a distance of 10 km from the sending end and the values are tabulated. SSSC has the base power of 500 MVA. DC link nominal voltage is 40 kV. Maximum rate of change of voltage is 2 pu/s. Fault is at a distance of 100 km from the sending end. R_f is 250 Ω. It is observed from Table 7 that active power and reactive

Table 6 Observations with shunt compensation from sending end

Type of fault	B$_{ref}$. (pu/Phase)	P (W)	Q (var)
No fault	0	2.68e+008	−2.55e+007
No fault	1	2.73e+008	−5.32e+007
L-G	0	3.83e+008	−1.15e+007
L-G	0.5	3.87e+008	−1.44e+006
L-G	1	3.92e+008	9.05e+006
L-L-G	0	5.56e+008	7.62e+007
L-L-G	0.5	5.64e+008	7.62e+007
L-L-G	1	5.72e+008	2.86e+007
L-L	0	5.12e+008	1.98e+007
L-L	0.5	5.19e+008	1.22e+007
L-L	1	5.26e+008	2.28e+007

Table 7 Observations with series compensation from sending end

Type of fault	Vqref (pu)	P (W)	Q (var)
No fault	0.05	3.18e+008	−1.18e+007
No fault	0.2	4.64e+008	3.93e+007
L-G	0.05	4.31e+008	3.06e+006
L-G	0.2	5.78e+008	5.85e+007
L-L-G	0.05	6.05e+008	2.14e+007
L-L-G	0.2	7.52e+008	8.17e+007
L-L	0.05	5.60e+008	1.61e+007
L-L	0.2	7.08e+008	7.52e+007

power changes insignificantly at receiving end but reactive power is significant at sending end.

The superiority of the research work in this paper when compared to [1] is that simulation results for compensated i.e. both series and shunt and also for the non-compensated system are provided, where as in the paper [1] only the algorithm is proposed. Also, advantage of this algorithm is its application to any kind of EHV system.

5 Conclusion

The proposed method uses the symmetrical components of reactive power for the classification of the fault and for the selection of the faulted phase. The symmetrical components of fault voltage and current are calculated and as the faulted phase is calculated from the MATLAB program and then the classification of the fault is done using the zero and negative sequence components of the reactive power for the ground faults and differential reactive power are used for the phase faults. The method gives satisfactory results for a wide range of distances, fault resistances and compensation. It is observed that the reactive power and the power transfer capability of the line is increased, when the value of susceptance is increased for a compensated system. This paper gives algorithm and further detailed simulation results for the mathematical model proposed in the Behnam Mahamedi and Jian Guo Zhu [1].

References

1. B. Mahamedi, J.G. Zhu, Fault classification and faulted phase selection based on the symmetrical components of reactive power for single-circuit transmission lines. IEEE Trans. Power Deliv. **28**(4), 2326–2332 (2013)
2. K.M. Silva, K.M.C. Dantas, B.A. Souza, N.S.D. Brito, F.B. Costa, J.A.C.B. Silva, Haar wavelet-based method for fast fault classification in transmission lines, *in Proceedings of the IEEE/Power Engineering Society Transmission & Distribution Conference and Exposition: Latin America*, Aug 2006, p. 15
3. S. Bunjongjit, A. Ngaopitakkul, Selection of proper artificial neural networks for fault classification on single circuit transmission line. Int. J. Innov. Comput. Inform. Control **8**(1A), 361–374 (2012)
4. P. Tripathi, A. Sharma, G.N. Pillai, I. Gupta, Accurate fault classification and section identification scheme in TCSC compensated transmission line using SVM. Int. J. Electr. Comput. Energ. Electr. Commun. Eng. **5**(12), 1800–1806 (2011)
5. A. Jamehbozorg, S.M. Shahrtash, A decision-tree- based method for fault classification in double- circuit transmission lines. IEEE Trans. Power Del. **25**(4), 2184–2189 (2010)
6. H. Simon, *Neural networks: A comprehensive foundation*, 2nd edn. (Prentice-Hall, Inc., Upper Saddle River, New Jersey, 1999)
7. P. Khadke, N. Patne, A. Singh, G. Shinde, A soft computing scheme incorporating ANN and MOV energy in fault detection, classification and distance estimation of EHV transmission line with FSC. SpringerPlus **5**, 1834 (2016). https://doi.org/10.1186/s40064-016-3533-2
8. O.A.S. Youssef, Fault classification based on wavelet transforms. IEEE/PES Transm. Distrib. Conf. Expos. **1**, 531–538 (2001)
9. B. Sreewirote, A. Ngaopitakkul, Classification of fault type on loop-configuration transmission system using support vector machine, in *2017 6th IIAI International Congress on Advanced Applied Informatics (IIAI-AAI)*, Hamamatsu, 2017, pp. 892–896. https://doi.org/10.1109/iiai-aai.2017.203
10. A.S. Neethu, T.S. Angel, Smart fault location and fault classification in transmission line, in *2017 IEEE International Conference on Smart Technologies and Management for Computing, Communication, Controls, Energy and Materials (ICSTM)*, 2–4 August 2017, Chennai, India, pp. 339-343
11. X. Dong, W. Kong, T. Cui, Fault classification and faulted-phase selection based on the initial current traveling wave, IEEE Trans. Power Deliv. **24**(2), 552–559 (2009)
12. H. Malik, R. Sharma, Transmission line fault classification using modified fuzzy Q learning. IET Gener. Transm. Distrib. **11**(16), 4041–4050, 11 9 (2017)
13. S.K. Mishra et al., An Adaline LMS control and DWT approach based differential relaying STATCOM integrated line. Int. J. Control Theor. Appl. **10**, 281–296 (2017)
14. R. Gangadharan, G.N. Pillai, I. Gupta, Fault zone detection on advanced series compensated transmission line using discrete wavelet transform and SVM. Proc. World Acad. Sci. Eng. Tech. **70**, 176–180 (2010)
15. T.S. Abdelgayed, W.G. Morsi, T.S. Sidhu, Fault detection and classification based on co-training of semisupervised machine learning. IEEE Trans. Ind. Electron. **65**(2), 1595–1605 (2018). https://doi.org/10.1109/TIE.2017.2726961

Optimal Bidding Strategy in Deregulated Power Market Using Krill Herd Algorithm

Chandram Karri, Durgam Rajababu and K. Raghuram

Abstract In this article, Krill Herd algorithm (KHA) is implemented for optimal bidding strategy. The bidding coefficients of suppliers and buyers are selected strategically. The code of proposed KHA has been developed in MATLAB. It has been tested on IEEE 30 bus power system. The simulation results are correlated with the existing algorithms. The results proved flexibility of KHA correlated with Particle Swarm Optimization (PSO), Genetic Algorithm (GA) and Monte Carlo simulation in terms of Market Clearing Price (MCP).

Keywords Bidding strategy · Krill herd algorithm · Deregulation

1 Introduction

Deregulation in the electricity industry is increasing expeditiously around the world [1, 2]. Suppliers and consumers build bid strategically to maximize the revenue. Independent system operator (ISO) determines marginal cost price (MCP) from supply and demand curves provided by suppliers and consumers respectively in the deregulated power markets. Optimal bidding strategy (OBS) problem is addressed by many authors in the past [3–12]. Literature survey on OBS is provided in Fig. 1.

It is observed from the literature that several methods have been used to address optimal bidding strategy. Still, there is a scope to increase the quality of solution in terms of profit. This article suggests the KHA for OBS problem.

C. Karri (✉) · D. Rajababu
Department of Electrical and Electronics Engineering, S R Engineering College,
Warangal, Telangana, India
e-mail: dr.chandram_k@srecwarangal.ac.in

D. Rajababu
e-mail: rajababu_d@srecwarangal.ac.in

K. Raghuram
Laqkya Institute of Technology and Sciences, Khammam, India
e-mail: dr_k_raghuram@yahoo.com

© Springer Nature Singapore Pte Ltd. 2019
H. Malik et al. (eds.), *Applications of Artificial Intelligence Techniques in Engineering*, Advances in Intelligent Systems and Computing 698,
https://doi.org/10.1007/978-981-13-1819-1_5

Literature Survey on Optimal Bidding Strategy		
Conventional	Stochastic	Hybrid
Monte Carlo	Genetic Algorithm	PSO combined with SA
Dynamic programming	Evolutionary Algorithm	Fuzzy adaptive GSA
Game theory	Particle Swarn Optimization	Bi level programming and swarn technique

Fig. 1 Literature survey on optimal bidding strategy

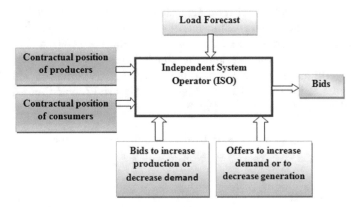

Fig. 2 Structure of modern power system operation

2 Optimal Bidding Strategy

In deregulation, market clearing price (MCP) is determined from supply and demand functions. Due to changes in the MCP, the consumers and suppliers try to perform in bidding action to increase their benefits. The MCP of electricity is determined by crossing point of the respective demand and supply curves. The point of intersection determines the hourly MCP. The structure of operation of modern power system is shown in Fig. 2.

Suppliers bid price is $G_i(P_i) = a_i + b_iP_i$

Consumer load demand price is $W_j(L_j) = c_j - d_jL_j$

To find a set of supplier's outputs and consumer load demands that minimizes the total cost, the mathematical formulation is as follows:

$$a_i + b_iP_i = MCP, \quad i = 1, 2, 3, \ldots n \tag{1}$$

$$c_j - d_jL_j = MCP, \quad j = 1, 2, 3, \ldots m \tag{2}$$

The power equality constraint is provided below:

$$\sum_{i=1}^{n} P_i = Q_{MCP} + \sum_{j=1}^{m} L_j \qquad (3)$$

The generation inequality constraints is

$$P_{i,min} \leq P_i \leq P_{i,max}, i = 1, 2, 3, \dots n \qquad (4)$$

The load inequality constraints is

$$L_{j,min} \leq L_j \leq L_{j,max}, \ j = 1, 2, 3, \dots m \qquad (5)$$

The cumulative forecasted pool demand is as follows:

$$Q_{MCP} = Q_0 - K \times MCP \qquad (6)$$

From the above equations, the expression of MCP can be derived. It is provide in Eq. (7)

$$MCP = \frac{Q_0 + \sum_{i=1}^{n} \frac{a_i}{b_i} + \sum_{j=1}^{m} \frac{c_j}{d_j}}{K + \sum_{i=1}^{n} \frac{1}{b_i} + \sum_{j=1}^{m} \frac{1}{d_j}} \qquad (7)$$

$$P_i = \frac{MCP - a_i}{b_i}, \ i = 1, 2, 3, \dots n \qquad (8)$$

$$L_j = \frac{c_j - MCP}{d_j}, \ i = 1, 2, 3, \dots m \qquad (9)$$

The objective function is

$$Max.F(a_i, b_i) = MCP \times P_i - C_i(P_i) \qquad (10)$$

The generator cost function is

$$C_i(P_i) = f_i P_i^2 + e_i P_i \qquad (11)$$

$$Max.H(c_j, d_j) = B_j(L_j) - MCP * L_j \qquad (12)$$

Consumer and suppliers may set MCP to get maximum profit when they know their bidding co-efficient's. Each GENCO may predict their rivals bidding coefficients using probability distribution function (pdf).

$$pdf_p(x_i, y_i)$$

$$= \frac{1}{2\pi\sigma_i^{(x)}\sigma_i^{(y)}\sqrt{1-\rho_i^2}}$$

$$\times \exp\left\{-\frac{1}{2(1-\rho_i^2)}\left[\left(\frac{x_i - \mu_i^{(x)}}{\sigma_i^{(x)}}\right)^2 - \frac{2\rho_i\left(x_i - \mu_i^{(x)}\right)\left(y_i - \mu_i^{(y)}\right)}{\sigma_i^{(x)}\sigma_i^{(y)}} + \left(\frac{y_i - \mu_i^{(y)}}{\sigma_i^{(y)}}\right)^2\right]\right\}$$

(13)

where, $\mu_i^{(x)}$, $\mu_i^{(y)}$, $\sigma_i^{(x)}$ and $\sigma_i^{(y)}$ are the parameters of the joint normal distribution. ρ_i is the correlation between x_i and y_i. The marginal distributions of x_i and y_i are both normal with mean values $\mu_i^{(x)}$, $\mu_i^{(y)}$, and standard deviations $\sigma_i^{(x)}$ and $\sigma_i^{(y)}$ respectively.

This estimate is conveyed in condensed form as follows:

$$(x_i, y_i)_p \sim N\left\{\begin{bmatrix} \mu_i^{(x)} \\ \mu_i^{(y)} \end{bmatrix}\begin{bmatrix} \sigma_i^{(x2)} & \rho_i\sigma_i^{(x)}\sigma_i^{(y)} \\ \rho_i\sigma_i^{(x)}\sigma_i^{(y)} & \sigma_i^{(y2)} \end{bmatrix}\right\}$$

(14)

Similarly, from pth large consumers point of view, the bidding coefficients of from jth large consumer can be found with the following probability density function:

$$\left(u_j, v_j\right)_p \sim N\left\{\begin{bmatrix} \mu_j^{(u)} \\ \mu_j^{(v)} \end{bmatrix}\begin{bmatrix} \sigma_j^{(u2)} & \rho_j\sigma_j^{(u)}\sigma_j^{(v)} \\ \rho_j\sigma_j^{(u)}\sigma_j^{(v)} & \sigma_j^{(v2)} \end{bmatrix}\right\}$$

(15)

3 Krill Herd Algorithm (KHA)

KHA is a new stochastic search algorithm developed by Gandomi and Alavi [13]. It depends on the herding performance of individual krill. Large number of krill's search for food in a multidimensional area. The location of krill individuals are variables. The space between krill location to food location is considered as an objective function. In the KHA, the individual krill moves to better location by altering other krill individuals. The moments of individual krill is influenced by induction process, foraging activity, and random diffusion.

(a) **Induction process**: The velocity of the ith krill at the mth movements can be expressed as follows:

$$V_i^m = \alpha_i V_i^{max} + \omega_n V_i^{m-1}$$

(16)

$$\alpha_i = \sum_{j=1}^{N_s} \left[\frac{f_i - f_i}{f_{i} - f_i} \times \frac{Z_i - Z_j}{|Z_i - Z_j| + rand(0, 1)} \right] + 2 \left[rand(0, 1) + \frac{i}{i_{max}} \right] f_i^{best} X_i^{best}$$

(17)

The parameter of sensing distance can be used to find the adjoining number of particular krill. The sensing distance is expressed by

$$SD_i = \frac{1}{5n_p} \sum_{j=1}^{n_p} |Z_i - Z_k|$$

(18)

(b) **Foraging action**: The foraging velocity of ith krill at mth action can be represented as

$$V_{fi}^m = 0.02 \left[2 \left(1 - \frac{i}{i_{max}} \right) f_i \frac{\sum_{k=1}^{N_s} \frac{Z_k}{f_k}}{\sum_{k=1}^{N_s} \frac{1}{f_{kj}}} + f_i^{best} X_i^{best} \right] + \omega_x V_{fi}^{m-1}$$

(19)

(c) **Random diffusion**: In KHA, random diffusion process is added to upgrade the population diversity. The diffusion speed of individual krill can be represented as

$$V_{Di}^m = \mu V_D^{max}$$

(20)

(d) **Position update**: In the KHA, the updated position of the ith krill can be expressed as

$$Z_i^{m+1} = Z_i^m + \left(V_i^m + V_{fi}^m + V_{Di}^m \right) P_t \sum_{j=1}^{N_d} \left(u_j - l_j \right)$$

(21)

In order to increase the convergence and to find the better solution, the crossover and mutation operations of DE are combined in this algorithm.

Crossover
Based on the crossover probability, every individual krill interacts with other krill to update the krill position. The jth components of the ith krill can be updated by

$$Z_{ij} = \begin{cases} Z_{k,j} \ if \ rand \le C_{Ri} \\ Z_{i,j} \ if \ rand < C_{Ri} \end{cases} \quad where \ k = 1, 2, 3, \ldots n_p; k \ne i,$$

(22)

where C_{Ri} is the cross-over probability and is calculated by

$$C_{Ri} = 0.2 f_i^{best}$$

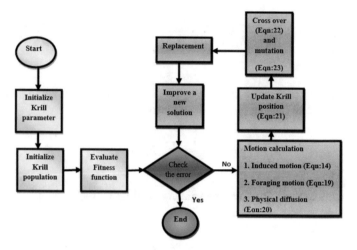

Fig. 3 Flowchart of Krill Herd algorithm

Mutation

In this method, a scalar number F_R scales the difference of randomly selected two vectors $Z_{m,j}$ and $Z_{m,j}$ and the scaled difference is added to the best vector $Z_{best,j}$ hence the mutant vectors $Z_{i,j}^m$ is calculated as

$$Z_{i,j}^m = Z_{best,j} + F_R(Z_{m,j} - Z_{n,j}) \tag{23}$$

Using mutation probability (M_R) the modified value, $Z_{i,j}^{mod}$ is selected from $Z_{i,j}^m$ and $Z_{i,j}$ and it can be mathematically represented as

$$Z_{i,j}^{mod} = \begin{cases} Z_{i,j}^m \ if \ rand \leq M_R \\ Z_{i,j} \ if \ rand < M_R \end{cases} \tag{24}$$

The flowchart of Krill herd algorithm is shown in Fig. 3.

4 Methodology

Complete process of the KHA for solving OBS problem is presented in this section.

Stage 01: Data.
Stage 02: Initialization a and *c* values are randomly generated and fixed.
Stage 03: Maximizing probability density function (pdf) 'b' and *'d'* values are obtained such that the *pdf* function is maximized. MCP is calculated for the above

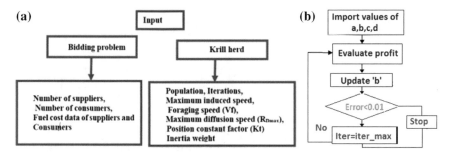

Fig. 4 a data set **b** Pdf maximization using KHA

values. This process is run for some iterations and the best values of '*a*' and '*c*' are chosen. The flowchart of pdf maximization is shown in Fig. 4.

(i)   Calculate MCP using Eq. (7)
(ii) Evaluate suppliers powers using Eq. (8) and consumers using Eq. (9)
(iii) Calculate revenue benefits of suppliers and consumers using Eqs. (10) and (12).

Stage 04: Iterations Starts for KHA.

   (i)   Ranking the krills according to their position
  (ii) Move all krills to the better locations
 (iii) Foraging and Physical diffusion
 (iv) Updating the krill individual position
  (v)   Determine MCP for modified bidding coefficients
 (vi) Calculate the powers and test the error
(vii) Obtain the optimal solution
(viii) End of iterations.

Stage 05: Final Results.

5 Case Study

Code of the KHA is implemented in MATLAB (version 2012 A). Data have been taken from [10]. The numerical values of control parameters are shown in Table 1.

The bidding coefficients are listed in Table 2.

The MCP values for first 200 iterations are shown in Fig. 1. The best value of the MCP is 19.4516 $.

The powers of suppliers and consumers are given in Table 3.

The profit of generator and consumers and total profit are given in Table 4.

Table 1 Numerical values of various control parameters

S.no	Control parameter	Value
1	Qo	300
2	K	5
3	Population	1000
4	Iterations	500
5	Maximum induced speed	0.01
6	Foraging speed (V_f)	0.02
7	Maximum diffusion speed (RD_{max})	0.05
8	Position constant factor (K_t)	100
9	Inertia weight	0.9

Table 2 Bidding coefficients of generators and consumers

Generator	x	Y
1	8.0102	0.0695
2	6.3738	0.1103
3	3.6422	0.2888
4	11.8371	0.0684
5	10.926	00.2025
6	10.926	0.2025
Consumer	u	V
1	36.4219	0.08650
2	30.3516	0.0810

Table 3 Powers of suppliers and consumers are given

S.no	Power (MW)	Profit ($)
G1	164.6366	1864.2
G2	118.6187	945.9
G3	54.7510	488.6
G4	111.3806	766.5
G5	42.0986	307.1
G6	42.0986	307.1
L1	196.2734	529.4492
L2	134.5687	203.3849

Table 4 The profit of generator and consumers and total profit

Profit ($/hr)		Total profit ($/hr)
Generator	Load	
4679.3	732.8	5412.1

6 Conclusions

Optimal bidding strategy (OBS) problem is solved using the KHA in this paper. This algorithm is one of the best algorithms for solving the OBS problem. From the results, it is identified that it is the best method in terms of profit. Real time OBS problem can be solved for better competition under deregulated environment.

References

1. L.L. Lai, *Power System Restructuring and Deregulation-Trading Performance and Information Technology* (Wiley, New York, 2001)
2. C.W. Richter Jr., G.B. Shible, Genetic algorithm evolution of utility bidding strategies for the competitive market place. IEEE Trans. Power Syst. **13**, 256–261 (1998)
3. A.K. David, F. Wen, Strategic bidding in competitive electricity markets: a literature survey. IEEE Power Eng. Soc. Summer Meet. **4**, 2168–2173 (2000)
4. F.S. Wen, A.K. David, Optimal bidding strategies for competitive generators and large consumers. Electr. Power Energy Syst. **23**, 37–43 (2001)
5. A.K. David, Competitive bidding in electricity supply. IEE Proc. Gener. Trans. Distrib. **140**, 421–426 (2002)
6. V.P. Gountis, A.G. Bakirtzis, Bidding strategies for electricity producers in a competitive electricity market place. IEEE Trans. Power Syst. **19**, 356–365 (2004)
7. P. Attaviririyanupap, H. Kita, E. Tanaka, J. Hasegawa, New bidding strategy formulation for day-ahead energy and reserve markets based on evolutionary programming. Electr. Power Energy Syst. **27**, 157–167 (2005)
8. S. Soleymani, Bidding strategy of generation companies using PSO combined with SA method in the pay as bid markets. Electr. Power Energy Syst. **33**, 1272–1278 (2011)
9. A. Azadeh, S.F. Ghaderi, B.P. Nokhandan, M. Sheikhalishahi, A new genetic algorithm approach for optimizing bidding strategy viewpoint of profit maximization of a generation company. Expert Syst. Appl. **39**, 1565–1574 (2012)
10. J.V. Kumar, D.M.V. Kumar, K. Edukondalu, Strategic bidding using fuzzy adaptive gravitational search algorithm in a pool based electricity market. Appl. Soft Comput. **13**, 2445–2455 (2013)
11. Z. Qiu, N. Gui, G. Decininck, Analysis of equilibrium-oriented bidding strategies with inaccurate electricity market models. Electr. Power Energy Syst. **46**, 306–314 (2013)
12. G. Zhang, G.L. Zhang, Y. Gao, J. Lu, Competitive strategic bidding optimization in electricity markets using bilevel programming and swarm technique. IEEE Trans. Industr. Electron. **58**, 2138–2146 (2011). https://doi.org/10.1109/TIE.2010.2055770
13. A.H. Gandomi, A.H. Alavi, Krill herd: a new bio-inspired optimization algorithm. Nonlinear Sci Numer. Simul. **17**(12), 4831–4845 (2012)

A Novel Intelligent Bifurcation Classification Model Based on Artificial Neural Network (ANN)

Hasmat Malik and Tushar Sharma

Abstract This paper presents the classification of various types of bifurcations in the voltage stability analysis of electrical power system based on ANN (Artificial Neural Networks) techniques. The bifurcations being analysed are saddle-node bifurcation and Hopf bifurcation. The bifurcations are carried out by simulating the IEEE 14 bus model under healthy and faulty conditions in PSAT toolbox. The model is created by carrying out the analysis of the data obtained from bifurcations using neural network toolbox in MATLAB.

Keywords Bifurcation · Voltage stability · ANN (Artificial Neural Networks)
PSAT · Saddle-node bifurcation · Hopf bifurcation

1 Introduction

Recently, power systems are operating under more stressed conditions in comparison to those in the past due to factors like restricted transmission expansion due to environmental pressures, electrical consumption increment in heavy load areas, new system loading patterns, etc. These stressed conditions introduced a new type of unstable behaviour characterized by sudden (or slow) voltage drops, sometimes escalating to voltage collapse [1]. Due to highly nonlinear nature of electrical power systems, any variation of any parameter in the system may result in complicated

H. Malik
Department of Electrical Engineering, IIT Delhi, New Delhi, India
e-mail: hmalik.iitd@gmail.com

T. Sharma (✉)
Department of Instrumentation and Control Engineering, NSIT, Sector 3, Dwarka, New Delhi
110078, India
e-mail: tushars49@gmail.com

© Springer Nature Singapore Pte Ltd. 2019   53
H. Malik et al. (eds.), *Applications of Artificial Intelligence Techniques
in Engineering*, Advances in Intelligent Systems and Computing 698,
https://doi.org/10.1007/978-981-13-1819-1_6

behaviour which gives rise to system instability. In any nonlinear dynamical system, the qualitative change in the behaviour of the system with the corresponding change of one or more parameters is due to bifurcations [2]. The objective of this paper is to analyse power system voltage stability using saddle-node bifurcation and Hopf bifurcation by simulating the healthy and faulted conditions in PSAT [3] and designing a classification model from the parameters obtained from bifurcations using ANN techniques in MATLAB.

The paper is organised as follows: Sect. 2 gives a model description along with the techniques used for designing classification model. Section 3 shows the results obtained followed by conclusions in Sect. 4.

2 Model Description

2.1 IEEE 14 Bus Model Simulation

The models shown in Figs. 1 and 2 below are simulated in PSAT (version 2.1.10) both in healthy and faulted conditions.

The model running in faulty condition is shown below. The fault is created at bus 1 along with STATCOM. The fault time is taken 0.11 s with fault clearing time as 1.25 s (Fig. 2).

2.2 Bifurcation Classification Model Formation

The data (eigenvalues complex conjugate pairs) obtained from Hopf bifurcation (HB) and saddle-node bifurcation (SNB) for both healthy and faulty systems is used for designing the classification model using neural network pattern recognition toolbox in MATLAB. The complete details of ANN based classification model implementation is given in [3].

A training input dataset is formed by combining 18 samples of Table 1 data and 51 samples of Table 2 data (including healthy and faulted systems).

A training output dataset is formed by providing pattern recognition keys (0 1) to all the 18 samples and (1 0) to all the 51 samples of the training input dataset. An example is shown as follows (Table 3).

Testing input dataset is formed by combining 6 samples of Tables 1 and 9 samples of Table 2. The testing output dataset is formed similar to training output dataset.

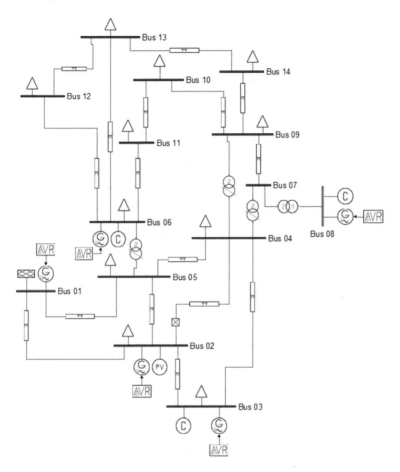

Fig. 1 IEEE 14 bus power system in healthy condition

The number of hidden neurons for training is calculated according the formula:

$$\left\{ \frac{Training\ input + Training\ output}{2} + \sqrt{No.\ of\ samples} \right\}$$
$$\pm 10\% \left\{ \frac{Training\ input + Training\ output}{2} + \sqrt{No.\ of\ samples} \right\} \quad (1)$$

There are two columns of training input and training output respectively with 69 samples. By calculating from the formula, 11 hidden neurons are implemented for training purpose.

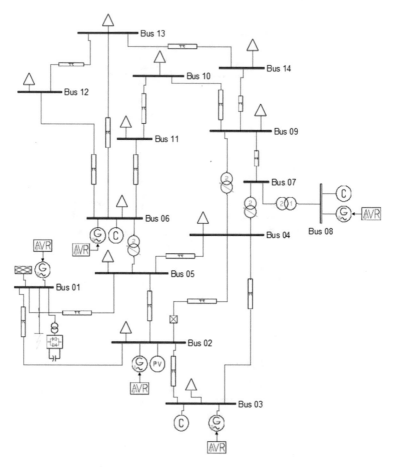

Fig. 2 IEEE 14 bus model in faulty condition

3   Results and Discussion

The following results are obtained from the pattern recognition of the datasets formed as shown in classification model design. The epoch denotes the complete cycle of training, testing and validation. The cross-entropy decreases as the epoch increases with the best validation performance of 0.011883 occurring at 31st epoch (Fig. 3). The gradient at 37th epoch of the training state plot (Fig. 6) is 0.064017. The training and testing phase confusion matrix are shown in (Fig. 5) and (Fig. 7) respectively (Fig. 4 and Table 4).

Table 1 Hopf Bifurcation (HB) data

Point	(Healthy system)	(Faulted system)
1	-47.6508, 8.35113	–39.5241, 5.59175
2	–47.6508, –8.35113	–39.5241, –5.59175
3	–4.70705, 10.0182	–6.81649, 2.22583
4	–4.70705, –10.0812	–6.81649, –2.22583
5	–1.79385, 10.76035	–1.77015, 10.8005
6	–1.79385, –10.76035	–1.77015, –10.8005
7	–1.88475, 9.90326	–1.81676, 9.89028
8	–1.88475, –9.90326	–1.81676, –9.89028
9	–1.3202, 8.95434	–1.3039, 8.96208
10	–1.3202, –8.95434	–1.3039, –8.96208
11	–0.07829, 0.88442	–0.17197, 0.88937
12	–0.07829, –0.88442	–0.17197, –0.88937

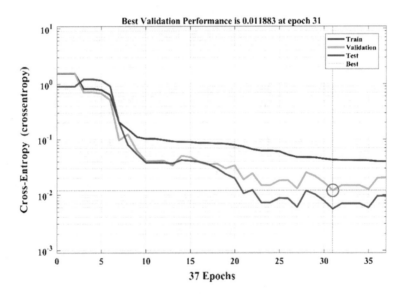

Fig. 3 ANN performance curve

Table 2 Saddle node bifurcation (SNB) data

Point	(Healthy system)	(Faulted system)
1.	−1000, 0	−1000, 0
2.	−1000, 0	−1000, 0
3.	−1000, 0	−1000, 0
4.	−1000. 0	−1000, 0
5.	−1000. 0	−1000, 0
6.	−49.927, 0	−49.9269, 0
7.	−38.3509, 0	−37.4816, 0
8.	−30.0846, 0	−32.0029, 0
9.	−22.5722, 0	−19.5633, 0
10.	−17.07, 0	−17.0849, 0
11.	−16.2665, 0	−17.6304, 0
12.	−8.27399, 0	−8.30219, 0
13.	−8.56499, 0	−8.58127, 0
14.	0, 0	0, 0
15.	−0.18961, 0	−0.18944, 0
16.	−0.77259, 0	−0.77246, 0
17.	−1.04916, 0	−1.0493, 0
18.	−0.87091, 0	−0.87003, 0
19.	−1.01868, 0	−1.01886, 0
20.	−1.0182, 0	−1.10142, 0
21.	−1.0014, 0	−1.00139, 0
22.	−50, 0	−50, 0
23.	−1, 0	−1, 0
24.	−0.57291, 0	−0.57291, 0
25.	−50, 0	−50, 0
26.	−1, 0	−1, 0
27.	−0.57291, 0	−0.57291, 0
28.	−50, 0	−50, 0
29.	−1, 0	−1, 0
30.	−2.652, 0	−2.652, 0

Table 3 Example of training input dataset along with training output dataset

Point	Eigenvalues (Real Part)	Eigenvalues (Imaginary Part)	Pattern recognition key	Pattern recognition key
1.	−1000	0	1	0
2.	−49.927	0	1	0
3.	−4.70705	10.0182	0	1
4.	−4.70705	−10.0182	0	1

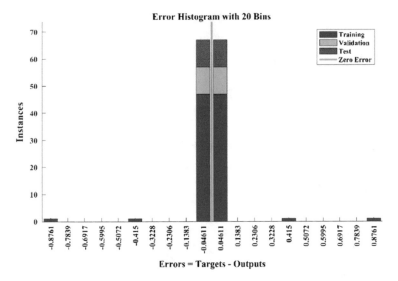

Fig. 4 Error Histogram

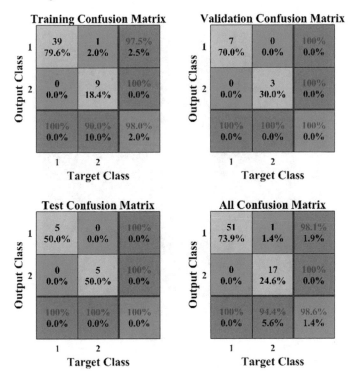

Fig. 5 Training phase confusion matrix

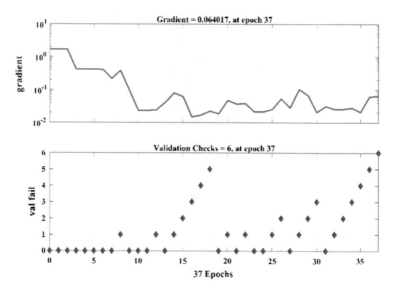

Fig. 6 Training state plot

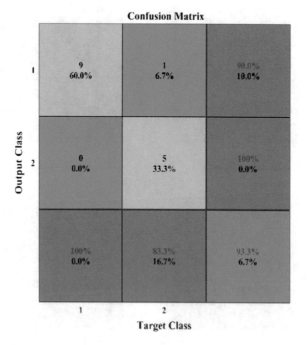

Fig. 7 Testing phase confusion matrix

Table 4 Training and testing phase performance results

Phase	Accuracy (%)	Error (%)
Training phase	98.6	1.4
Testing phase	93.3	6.7

4 Conclusion

As shown in the results, the performance of the system is 93.3% in testing phase in contrast to 98.6% in training phase for an IEEE 14 bus power system. The accuracy of the results can be further increased by the use of more advanced pattern recognition techniques. Moreover, the future work can be proposed on more complex system like IEEE 39 and IEEE 57 bus systems and more practical systems like regional and national grids.

References

1. C. Vournas, T. Van Cutsem, *Voltage stability of electric power systems* (Springer, US, 1998)
2. V. Ajjarapu, B. Lee, Bifurcation theory and its applications to nonlinear dynamical phenomena in an electrical power system, in *IEEE transactions on power systems*, vol. 7(1), Feb 1992
3. F. Milano, *Power system modelling and scripting* (Springer, Berlin, Heidelberg, 2010)

Teaching Learning Based Optimization for Frequency Regulation in Two Area Thermal-Solar Hybrid Power System

D. K. Mishra, T. K. Panigrahi, A. Mohanty and P. K. Ray

Abstract This novel work analyses the model of Automatic generation control having two unequal deregulated power system in Lab VIEW platform. Teaching learning based optimization (TLBO) has been implemented for tuning the gains of the PID controller. Further a comparative study has been given in this work with DE-based PID and TLBO-based PID controller. From the Simulation results implementing approaches of DE-optimized PID and TLBO-optimized PID establishes that the performance of TLBO optimization is better in improving the system performance index. The execution of TLBO-PID enhances performance in comparison with PI, PID controller with regards to stability, overshoot, and damping.

Keywords AGC · CSP · ED · LFC · LabVIEW · PID · TLBO

1 Introduction

The effectiveness of an interconnected power systems are effective because of its fast and easy control action. But it is difficult to achieve controllability and stability when different renewable energy sources are integrated with the power system [1]. Many literatures have discussed the stability of AGC-based power system using different controllers with simple design and smooth control action [2, 3]. There is always a possibility to investigate of functioning of AGC-based hybrid power system with classical controller. The main aim behind all the approach to acquire constant system frequency and tie line power of the AGC [4, 5]. As grid power needs the load and power generation increases progressively to maintain the load demand and genera-

D. K. Mishra · T. K. Panigrahi (✉) · P. K. Ray
Department of Electrical Engineering, IIIT Bhubaneswar, Bhubaneswar 751003, India
e-mail: tapas@iit-bh.ac.in

D. K. Mishra
e-mail: chandanmishra4@gmail.com

A. Mohanty
Department of Electrical Engineering, CET Bhubaneswar, Bhubaneswar 751003, India

© Springer Nature Singapore Pte Ltd. 2019
H. Malik et al. (eds.), *Applications of Artificial Intelligence Techniques in Engineering*, Advances in Intelligent Systems and Computing 698,
https://doi.org/10.1007/978-981-13-1819-1_7

tion. Adjustment of the system can be carried out measuring the system frequency as more power is used when machine in the systems are stimulated. AGC promises with automatic voltage regulation, load frequency control and economic dispatch. During load perturbation, frequency gets changed and governing action comes into picture, which responds quickly. If the change in frequency is more, governor system is incapable of restoring the frequency. During that time supplementary control loop is needed in load frequency control.

In recent times power system have been found to be more complex and unpredictable. Because of its size and more integration of RES [6]. Penetration of RESs is rising due to shortage of conventional power sources and environmental outcome. RES are also abundantly available and they are environment friendly. Further the conversion efficiency in case RES based power system is more which make the power system more reliable and sustainable [7].

2 System Under Study

This work proposes two unequal area (Thermal-CSP) based power plant in which the area-1 comprises of solar plus thermal with reheat an darea-2 having thermal with reheat. The capacity of plants in each area has been fixed tom 2000 MW and GRC is added to the thermal unit at a rate of 3% per minute (Fig. 1).

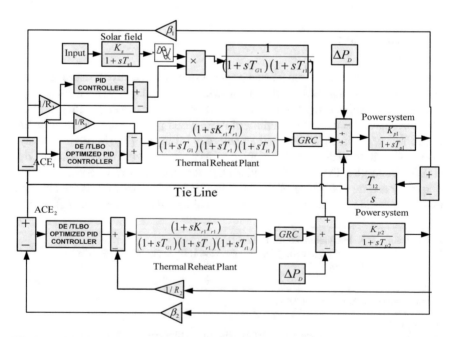

Fig. 1 Transfer function model of two area hybrid power system

TLBO optimization technique has been applied for tuning the gains of the PID controllers. TLBO algorithm has been applied for tuning the gains of the PID controllers. The transient behavior is measured with change in frequency and tie line power. The modeling of the proposed system with transfer function is given by [4, 8].

3 Solar Thermal Power System

Solar thermal technologies, otherwise called concentrating solar power (CSP), utilizes the temperature produced by solar radiation to generate power. CSP technologies are based on different types of axis-based tracking such one axis and two axis different mirrors arrange to focus the solar radiation either line focus based concentrators or point focus based concentrators. The temperature generated in case of concentrators has been used to generate the steam by the intermediate heat exchanging fluid. That can be used for heat energy, steam turbine, thermal engine, etc. Nowadays four different types of CSP have been employed in several applications like parabolic trough, power tower, linear fresnel reflector, and dish sterling design. The best application of CSP technologies is found in parabolic trough. Advance researches has been approved different parts of the world for solar power tower and solar power trough, which are the main part solar based thermal power system [1].

3.1 Control Technique

In present days of control system application, control inputs (U_1 and U_2) of two area power system can be made over linear arrangement of all states of the system. Controller adjusts the set point of the system in order to minimize large disturbance error, overshoot, and different oscillations. In this particular work PID controller gas been utilized to control the frequency deviation of each area in addition to control of tie line power [9].

3.1.1 PID Controller

PID controller stands for summation of Proportional, Integral, and Derivative controller. This controller is extremely important and it is utilized in several industrial applications. Further this controller executes different controlling action without considering the dynamic achievement of the plants [10].

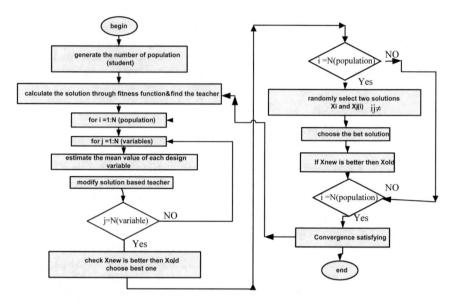

Fig. 2 TLBO Flowchart

3.2 *Teaching Learning Based Optimisation (TLBO)*

TLBO based optimisation technique was first developed by [11]. This method is now very popular and most effective technique and employed in many fields of application. This technique has numerous advantages like lesser time is required for the best solution and gives more stable performance with multiple frequency constraint. TLBO has been divided into two phases such as "teacher" and "learning" phases. Furthermore students acquire knowledge during teaching phase while they communicate with other students in learner phase (Fig. 2).

3.3 *Choice of Objective Function*

The main objective of the AGC is to reduce the Area Control Error (ACE) with a very short period of time. In order to get low value of ACE, the cost function can be defined as

$$f = \int_0^t |\Delta w_1 - \Delta w_2| \cdot t dt, \tag{1}$$

where dt is change in time, Δw_1 and Δw_2 are the frequency deviation in area-I and area-II, respectively. Present study, a step load disturbance of 1% is changed in area-I and start simulation with TLBO optimization technique for 50 numbers to achieve optimal gain margin of PID controller. Two more parameters are put into algorithm, i.e., Number of population and maximum number of iteration.

4 Results and Discussion

In this section, comparative analysis is done and shows the effectiveness of TLBO-PID controller. The optimum value of PID gains has been found out by TLBO optimization technique. The transient behavior of two area power system is measured for 10% step load disturbance in area-I through LabVIEW environment.

Different responses of frequency deviation, tie line power deviation and Area control error are shown in Figs. 3, 4 and 5 with different cases. TLBO-based optimization technique optimizes the PID controller to achieve more stable results and quite robust as compared to other controllers.

4.1 (10% Step Load Change in Area-I)

4.1.1 Two Equal Area (Thermal–Thermal)

In the first case, sequential increase in demand in steps has been applied in the area-I without solar integration. Variation in frequency in area-I (ΔF_1), variation in

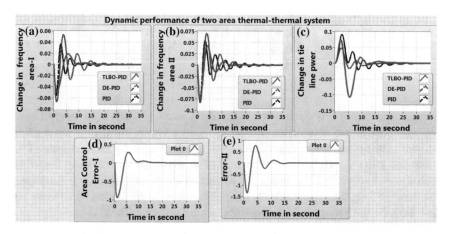

Fig. 3 Dynamic performance of two area system without solar. **a** ΔF_1; **b** ΔF_2; **c** ΔP_{tie}; **d** ACE-I; **e** ACE-II

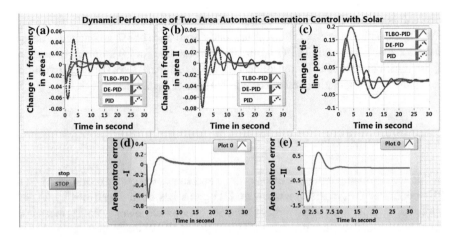

Fig. 4 Dynamic performance of two area system with solar integration. **a** ΔF_1; **b** ΔF_2; **c** ΔP_{tie}; **d** ACE-I; **e** ACE-II

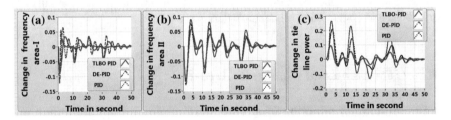

Fig. 5 Dynamic performance of two area system with random loading. **a** ΔF_1; **b** ΔF_2; **c** ΔP_{tie}

Table 1 Performance analysis

Controller	Settling time in s				
	ΔF_1	ΔF_1	ΔP_{tie}	ACE_1	ACE_2
PID	34.31	33.68	32.10	34.86	33.01
DE-PID	28.17	25.89	22.82	29.39	26.10
TLBO-PID	15.62	18.34	18.51	17.31	18.92

frequency in area-II (ΔF_2), variation in tie line power (ΔP_{tie}) and area control error (ACE) are measured and shown in Fig. 4. Three different controllers are applied into the same system and compared as depicted in Fig. 3a–e. Optimum value of PID controller gain (K_P, K_I, K_D) is tuned with TLBO algorithm, which shows the better response. It is observed that using TLBO technique, system stability is increases, decreases settling time and overshot. The performance index dealing with frequency deviation, tie line power and area control error are shown in Table 1.

Table 2 Performance analysis

Controller	Settling time in s				
	ΔF_1	ΔF_1	ΔP_{tie}	ACE-I	ACE-II
TLBO-PID	9.58	12.17	13.34	9.75	11.98
DE-PID	17.61	21.20	24.81	18.69	21.29
PID	28.98	29.34	29.84	29.30	29.80

4.1.2 Thermal-CSP Hybrid

In this section, Concentrating Solar Power (CSP) is added into the area-I which make hybrid system, i.e., Thermal-CSP hybrid system. In the same system, three different controllers such as PI, DE-PID, and TLBO-PID is applied and compared as shown in Fig. 4. As seen the result, it is clear that using CSP in area-I and TLBO optimized PID bring the system more stable, smaller settling, overshoot decreases and zero steady state error. Frequency deviation, tie line power deviation and area control error is shown in Fig. 4a–e. The result demonstrates the impact of CSP on AGC in multi area system and settling time of different controller is shown in Table 2.

4.1.3 Random Loading

Figure 5 shows the random loading of CSP-Thermal hybrid system. The change in frequency in area-I, change in frequency in area-II and tie line deviation is shown in Fig. 5a–c with three different controllers.

4.1.4 Parameter Variation

This section describes the impact on frequency deviation, tie line power by parameter variation and working condition of the PID controller gains in the two area CSP-Thermal hybrid system.

Here the variable parameter is T_{12} (synchronizing coefficient). T_{12} is varied $\pm 20\%$ and $\pm 40\%$ from their nominal value. It is observed that, by variation of T_{12} in the proposed system, a very very small change in frequency deviation in area-I and area-II, but some impact on tie line deviation as shown in Fig. 6a–c.

5 Conclusion

Design of AGC having unequal area has been discussed with TLBO-based algorithm for finding out the optimum value of PID controller. The power balance of proposed power system is maintained with governing action of thermal power plants. The

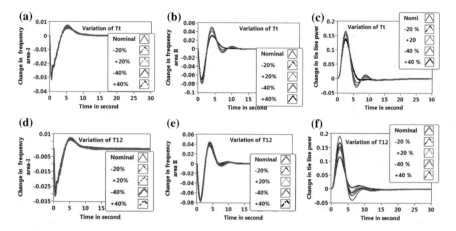

Fig. 6 Robustness analysis with ±20% and ±40% loading parameters

dynamic performance of two area systems is measured with three different controllers such as PID, DE-PID, and TLBO-PID. A TLBO-PID controller result shows more robust and better performance. In this study, also random load is applied into the same model and it shows the performance of the system. At last robustness analysis is done by ±20% and ±40% loading.

References

1. A. Rahman, L.C. Saika et al., AGC of dish striling solar thermal integrated thermal system with BB based optimised three degree of freedom PID controller. IET Renew. Power Gen. 1–10 (2016)
2. H. Shabani, B. Vahidi, M. Ebrahimpur, Roboust PID controller based on imperialist competitive algorithm for load frequency control of power system. ISA Trans. **52**(1), 88–95 (2013)
3. A. Mohanty, D.K. Mishra, K. Mohan, P.K. Ray, *Optimised Fractional Order PID Controller in Automatic Generation Control* (Taylor and Francis Group, 2016), pp. 215–219
4. O.I. Elgard, C. Fosha, Optimium megawatt frequency control of multi area electric energy system. IEEE Trans. Power Syst. **89**(4), 556–563 (1970)
5. H. Bevrani, *Robust Power System Frequency Control* (Springer book, 2009)
6. H. Bevrani, A. Ghosh et al., Renewable energy source and frequency regulation: survey and new perspective. IET Renew. Power Gen. **4**(5), 438–457 (2010)
7. V.S. Reddy, S.C. Kaushik et al., State of the art of solar thermal power plants—a review. Renew. Sustain. Energy Rev. **27**, 258–273 (2013)
8. L. Wang, C.C. Huang, Dynamic stability analysis of a grid connected solar concentrated ocean thermal energy conversion system. IEEE Trans. Sustain. Energy **1**(1), 10–18 (2010)
9. H. Gozde, M.C. Taplamacioglu, Automatic generation control application with craziness based particle swarm optimization in a thermal power system. Int. J. Electr. Power Energy Syst. **33**(8), 1–16 (2011)
10. D.K. Mishra, A. Mohanty, P.K. Ray, Multi area automatic generation control with FOPID and TID. Int. J. Control Theory Appl. **10**(6), 461–470 (2017)

11. R.V. Rao, V.J. Savsani, D.P. Vakharia, Teaching–learning-based optimization: a novel method for constrained mechanical design optimization problems. Comput. Aided Des. **43**(3), 303–315 (2011)

An Approach to Minimize the Transmission Loss and Improves the Voltage Profile of Load Bus Using Interline Power Flow Controller (IPFC)

Abdul Quaiyum Ansari, Mashhood Hasan and Noorul Islam

Abstract The interline power flow controller (IPFC) has two converter/inverters connected back to back with DC link. One of the converter is knows as series inverter which improve voltage quality of the load bus and second inverter known as shunt converter is used to compensate the reactive power of load and minimize the losses of the transmission line. In this work, the impact of the IPFC is seen in IEEE-3 Bus systems. The first one is connected between Load Bus or PQ-Bus and the Generating Bus or PV-Bus to inject voltage at certain angle. And it compensates reactive demand of the load using novel control approach. Moreover, the second converter of the IPFC is connected between Slack Bus and Load Bus to inject the compensating current using the Instantaneous Symmetrical Voltage Component Theory (ISVCT) based control algorithm. And it compensates the reactive demand of the load. The impact assessment of the IPFC on IEEE-3 Bus system is verified using MATPOWER.

Keywords Control approach · Converter · IEEE-3 bus · IPFC · Reactive power

1 Introduction

The restructuring and deregulation of the power system is increased to satisfy the transmission thermal limit. The thermal limit is because of extra burden of active and reactive power of transmission line which is demanded. The restructuring can be enhanced the power flow in transmission system even without increasing the thermal limit. The inclusion of Flexible AC Transmission System (FACTS) controlling devices in restructuring system gives wonderful results in AC power transmission system. It is because of application of power electronics in FACTS devices.

A. Q. Ansari · N. Islam (✉)
Department of Electrical Engineering, Jamia Millia Islamia, New Delhi 110025, India
e-mail: noorul_i@hotmail.com

M. Hasan
Department of Electrical Engineering, Jazan Univesity, Jazan, Kingdom of Saudi Arabia
e-mail: mhasen@jazanu.edu.sa

© Springer Nature Singapore Pte Ltd. 2019
H. Malik et al. (eds.), *Applications of Artificial Intelligence Techniques in Engineering*, Advances in Intelligent Systems and Computing 698,
https://doi.org/10.1007/978-981-13-1819-1_8

The concept of FACTS devices has been exposed in the late 1980s [1]. This is based on power electronics, which incorporates the controller to enhance power stability and controllability of the power system [2]. There are two kinds of methodology behind the power electronics based controller, first one is conventional thyristor switched capacitors (TSC) or thyristor switched reactors (TSR) and second one is self-commutated switching converters. Moreover, compensation flexibility and response time in the self commutated switching converters have been explored better than conventional thyristor switched converters. Since the size of passive elements limit the application of the self-commutated switching converter. Thus, most promising FACTS devices are available static synchronous series compensator (SSSC), static compensator (STATCOM), and unified power flow controller (UPFC) and interline power flow controller (IPFC). These are used in the AC power transmission system to mitigate the reactive power, enhance voltage profile of the bus and minimize the losses of the transmission system.

Optimal location of static var compensator is adjusted by applying continuation approach. Continuation approach improves voltage profile of power network and maximizing system loadability [3].

One of the most important and sensitive issue in power system is power loss reduction. At the planning time, the main objective in power system design is to minimize power loss and improve voltage profile. To optimize this problem, harmony search algorithm is used, which is self adaptive [4]. The UPFC and IPFC [5, 6] is most promising device to be used in power system which offers advantages over SSSC and STATCOM. The use of any one of these FACTS devices is gradually attractive in modern restructuring and deregulation system.

The UPFC has two converter/inverter connected back to back DC link voltage. The converter one is series converter/inverter which control the voltage related issues whereas second converter/inverter is known as shunt converter which control currents of power system. These two converters are connected in same transmission line. Moreover IPFC has same internal configuration of UPFC but the difference is that the IPFC is connected between two lines thus it is known as interline power flow controller. The most of the researcher have used UPFC for optimal power flow control [7] and congestion control [8] in transmission system. In [9–11], has proposed impact assessment of optimal placement of FACTS controller to minimize the real power loss power system. In this paper, authors have proposed a novel approach to mitigate the reactive power and improve voltage profile by injecting phasor sum of two voltages through series converter of the IPFC. The shunt converter of the IPFC is connected between slack bus and PQ bus, here it will compensate reactive power by injecting compensating current.

Figure 1, the practical 3-bus system is given to extract the practical data using power world simulator. This is actual load power flow condition where all the demand and generation are given. In 3-Bus system the Generating Bus where Generator is connected and active power (P) and voltage (V) are given, thus Generating Bus is also known as PV bus. The Slack Bus where another Generator is connected to supply the extra power where voltage and power angle is specified and third Bus is Load

Fig. 1 Practical 3-bus system using power world simulator

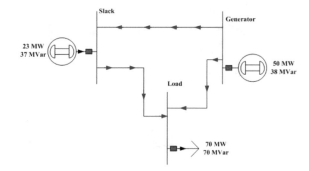

Bus where active and reactive power is demanded, it is also known as PQ-Bus. These data of Fig. 1 are analyzed the effect of IPFC using MATPOWER.

In the next section, proposed configuration is discussed. In Sect. 3, ipfc algorithm (series and shunt converters) is presented. In Sect. 4, Result and discussion has been presented. The last section concludes the proposed work.

2 Proposed Configuration

In Fig. 2, proposed configuration of IEEE-3 Bus system is integrated with IPFC. The IPFC is connected between Generating Bus and Slack Bus. The two converters of IPFC are shunt converter and series converter. It is optimally connected near to Load Bus. The series converter injects voltage in series and with certain angle to minimize transmission losses and compensate reactive power of Generator Bus. Additionally, shunt converter can absorb or injects real power demanded by series converter via DC link voltage (V_{DC}). The V_{DC} could be distributed generator (DG), DC battery and DC capacitor. The series converter can also injects or absorb reactive power of the Slack Bus.

3 Algorithm of IPFC

An IPFC have series and shunt converters. These converters want command to work with power system. The commands have to develop for series and shunt converters of the IPFC. Therefore, in this work, a novel approach is developed for firing the series converter. And a shunt converters is required an instantaneous symmetrical voltage component theory based algorithm to firing the shunt converter of the IPFC.

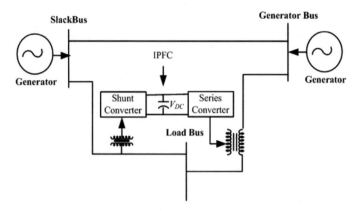

Fig. 2 Proposed configuration of IEEE-3 bus system integrated with IPFC

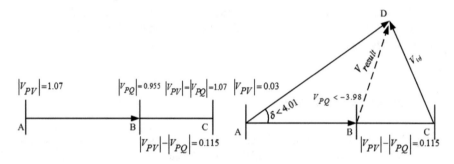

Fig. 3 Phasor sum of series voltage injection

3.1 An Algorithm for Series Converter

The series converter of IPFC is connected between PV-Bus and PQ-Bus which is near to PQ-Bus to optimize the transmission losses. In this case the PV-BUS voltage magnitude and PQ-Bus Voltage magnitude is compared to inject the difference of this voltage magnitude. These voltages are in series with the transmission line voltage and minimize the transmission losses. The mathematical calculations are as follows,

$$V_{PV} - |V_{PQ}| = |V_{sr}|, \tag{1}$$

where $|V_{PV}|$ is the per unit (pu) voltage magnitude of Generating Bus, $|V_{PQ}|$ is the pu Load Bus voltage magnitude and difference of these two is $|V_{sr}|$ in pu which is injected through series converter of the IPFC. It is shown in Fig. 3a. According IEEE-3 Bus data of PV-Bus and PQ-Bus the required series injection voltage magnitude is as follows.

$$1.07(V) - 0.955(V) = 0.11(V)$$

Thus required injection voltages are in series 0.11 (V) pu to minimize the line losses. This is first injection voltage which is reaches up to required load voltage, i.e., 1.07 pu. Now the second voltage is injected through series converter of the IPFC is shown in Fig. 3b. Here, some constraint is required to flow of power between PV-Bus to PQ-Bus. If the phase angle and magnitude of PV-Bus and PQ-Bus are same the power will not flow. And from Fig. 3a, after the injection of voltage in series of the line voltage the voltage magnitude is near about same of PV-Bus and PQ-Bus. Thus, it is mandatory to maintain the phase difference between the Bus. The following parameters are required which is given as,

Phase angel of PV-Bus is θ_1 and phase angle of PQ-Bus is θ_2 and difference of these two are given as,

$$\theta_1 - \theta_2 << \delta \leq 40^0, \tag{2}$$

where angle δ is power angle which can be extracted by (2) and injection voltage from Fig. 3b is evaluated as [8].

$$V_{inj} = |V_{PV}| \times 2Sin\left(\frac{\delta}{2}\right) \tag{3}$$

at last the phase sum of Eqs. (1) and (3) can be calculated as given.

$$|V_{result}| = \sqrt{(-V_{sr})^2 + V_{inj}^2 + 2 \times (-V_{sr}) \times \left(V_{inj}\right) \times \cos \Phi_{sr}}, \tag{4}$$

where, $\phi_{sr} = (90+\delta/2)$. Finally the resultant voltage V_{result} is injected through series converter to improve the voltage profile and to compensate the reactive burden of the PQ-Bus.

The propose flow chart is given in Fig. 4 to calculate the actual magnitude of the injection voltage using series converter of the IPFC.

3.2 An Algorithm for Shunt Converter

The control algorithm of shunt converter is based on reference currents using instantaneous symmetrical component theory.

The purpose of which is to inject reactive current to compensate the load reactive power and maintained DC link voltage at the constant value. To develop the algorithm for shunt converter the step wise calculations are as follows,

Step-1, the zero sequence (v_{sa}^0), positive sequence (v_{sa}^+) and negative sequence voltages (v_{sa}^-) are extracted. This can be evaluated as follows:

Fig. 4 Flow chart to series injection voltage

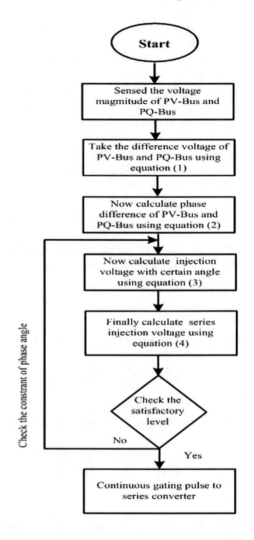

$$
\begin{vmatrix} v_{Sa}^0 \\ v_{Sa}^+ \\ v_{Sa}^- \end{vmatrix} = \begin{vmatrix} 1 & 1 & 1 \\ 1 & a & a^2 \\ 1 & a^2 & a \end{vmatrix} \begin{vmatrix} v_{Sa} \\ v_{Sb} \\ v_{Sc} \end{vmatrix}, \tag{5}
$$

where a is the complex operator $e^{j2\pi/3}$.

Step-2, The required reference current is extracted. Which is as follows,

$$
i_{sabc}^* = \frac{v_{Sabc} - v_{Sabc}^0}{\Delta}, \tag{6}
$$

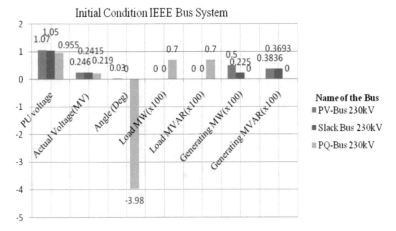

Fig. 5 Initial condition IEEE-3 bus system

where, $v_{Sabc}^0 = \frac{1}{3}\sum_{j=a,b,c} v_{Sj}$ and $\Delta = v_{Sabc}^0 = \frac{1}{3}\sum_{j=a,b,c} v_{Sj}^2 - 3(v_{Sabc}^0)^0$, for balanced sinusoidal signals zero sequence voltage v_{Sabc}^0 is zero, under unbalance and harmonics AC mains it is not zero.

Step-3, The reference currents is subtracted with actual load currents to know the required shunt injection currents. Which is as follows,

$$i_{shabc}^* = i_{labc} - i_{sabc}^*(P_{lavg} + P_{loss}), \tag{7}$$

where P_{lavg} is the average load power and P_{loss} is switching losses and ohmic losses. The P_{loss} is generated by DC link capacitor and taken from PI controller and sum it up with the P_{lavg} in the actual practice.

Step-4 The Eq. (7) is used to gating pulse of the shunt converter to compensate the reactive power of the Slack Bus.

This algorithm of the shunt converter is capable to handle reactive power of the Slack Bus and maintained the DC link voltage.

4 Results and Discussion

IEEE-3 bus system is used to see the impact of IPFC. The data of IEEE-3 bus system is extracted from MATPOWER. All initial condition is shown in Figs. 1, 5 and Table 1. In Table 2 when series controller works where Table 3 where both series and shunt converter of the IPFC work. The impact of series and shunt converter of the IPFC has seen one by one.

Table 1 Initial conditions of IEEE-3 bus system

Name of the bus	PV-bus	Slack bus	PQ-bus
Normal voltage (kV)	230	230	230
PU voltage	1.07	1.05	0.955
Actual voltage	246.103	241.500	219.669
Angle (Deg)	0.03	0.00	−3.98
Load MW	0.0	0.0	70
Load MVAR	0.0	0.0	70
Generating MW	50.00	22.50	00
Generating MVAR	38.36	36.93	00

Table 2 Series converter

Name of the bus	PV-bus	Slack bus	PQ-bus
Normal voltage (kV)	230	230	230
PU voltage	1.07	1.05	1.00
Actual voltage (kV)	246.103	241.500	230
Angle (Deg)	2.01	0.00	−3.98
Load MW	0.0	0.0	70
Load MVAR	0.0	0.0	70
Generating MW	50.00	20.50	00
Generating MVAR	19.36	36.93	00

Table 3 Series converter and shunt converter

Name of the bus	PV-bus	Slack bus	PQ-bus
Normal voltage (kV)	230	230	230
PU voltage	1.07	1.05	0.955
Actual voltage (kV)	246.103	241.500	219.669
Angle (Deg)	2.02	0.00	−3.98
Load MW	0.0	0.0	70
Load MVAR	0.0	0.0	70
Generating MW	50.00	20.50	00
Generating MVAR	19.36	18.93	00

4.1 Performance of Series Converter of IPFC Using Novel Approach

The series converter of IPFC is connected between PV bus and PQ bus. The line losses between PV-bus and PQ bus is because of resistance in transmission line. The losses can be compensated by injecting extra voltage using series converter of the

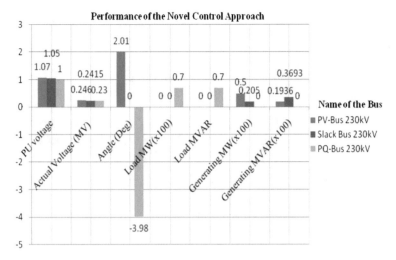

Fig. 6 Series converter performance

IPFC, therefore IPFC is very near to the PQ-bus. While the reactive power can be compensated using active and reactive power of the PV bus. The constraint of the reactive power compensation depends on the power angle constraints which is given in Eq. (2). Table 2 shows that PV bus has supplied 19 kVAR only where 50% of reactive is supplied through series converter of the IPFC. Thus, burden of the PV-Bus is reduced to compensate reactive power of the load and improve voltage profile to minimise the losses. All the data of Table 2 or Fig. 6 shows that the performance of the novel control approach is working satisfactorily.

4.2 Performance of Shunt Converter of the IPFC Using Current Control Algorithm

The shunt converter of IPFC is connected between slack bus and PQ bus. The line losses between slack bus and PQ bus are because of resistance in transmission line. The losses can be compensated by injecting compensating current using shunt converter of the IPFC. The compensating currents compensate the reactive power of Slack Bus. The constraint of the reactive power compensation depends on the injection of shunt compensating current which is given in Eq. (7). The Table 3 shows that Slack Bus has minimum reactive power which maintains the flow of the power between Slack Bus and the PQ bus. All the data of Table 3 and Fig. 7, shows that the performance of the shunt converter is working satisfactorily.

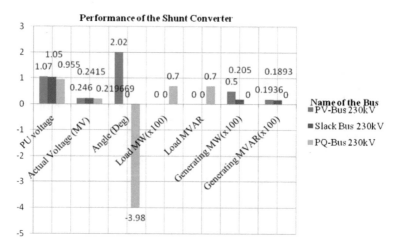

Fig. 7 Series converter and shunt converter performance

5 Conclusions

A novel approach to compensate reactive power of the load using the data of IEEE-3 system has been verified in this paper. The concept of IPFC has been implemented satisfactorily. Two voltage, series voltage and voltage at certain angle are injected at same time using series converter of the IPFC to minimize the losses of the transmission line and mitigate the reactive power of the PV bus correspondingly. Moreover, the shunt converter of the IPFC has been used to compensate the reactive power and the line losses of the Slack Bus. It has been seen that combination of shunt and series converters is able to mitigate the reactive power and minimize the line losses of the IEEE-3 Bus system by using appropriate control algorithm.

References

1. N.G. Hingorani, Introducing custom power. IEEE Spectr. **32**(6), 41–48 (1995)
2. H. Fujita, H. Akagi, The unified power quality conditioner, the integration of series- and shunt-active filters. IEEE Trans. Power Electron. **12**(2), 315–322 (1998)
3. A.K. Mehrdad, S. Soleymani, S.B. Mozafari, S.M. Hosseini, An approach to optimal allocation of SVC in power systems connected to DFIG wind farms based on maximization of voltage stability and system loadability. J. Intell. Fuzzy Syst. **29**(5), 2147–2157 (2015). https://doi.org/10.3233/ifs-151690
4. S. Vahid, Z. Masoud, O. Mahmoud, An intelligent method to reduce losses in distribution networks using self-adaptive harmony search algorithm. J. Intell. Fuzzy Syst. **30**(3), 1393–1402 (2016). https://doi.org/10.3233/ifs-152052
5. H. Mehrjerdi, A. Ghorbani, Adaptive algorithm for transmission line protection in presence of UPFC. Electr. Power Energy Syst. **91**, 10–19 (2017)

6. A. Vinkovic, R. Mihalic, A current-based model of an IPFC for Newton-Raphson power flow. Electr. Power Syst. Res. **79**, 1247–1254 (2009)
7. M. Noroozian, L. Angquist, M. Ghandari, G. Andarson, Use of UPFC for optimal power flow. IEEE Trans. Power Deliv. **12**(4), 1629–1634 (1997)
8. B. Bhattacharyya, S. Kumar, Approach for the solution of transmission congestion with multi-type FACTS devices. IET Gener. Transm. Distrib. **10**(11), 2802–2809 (2016)
9. B. Singha, V. Mukherjee, P. Tiwari, Genetic algorithm optimized impact assessment of optimally placed DGs and FACTS controller with different load models from minimum total real power loss viewpoint. Energy Build. **126**, 194–219 (2016)
10. A.Q. Ansari, B. Singh, M. Hasan, Algorithm for power angle control to improve power quality in distribution system using unified power quality conditioner. IET GTD **9**, 1439–1447 (2015)
11. U. Koteswara Rao, M.K. Mishra, A new control strategy for load compensation in power distribution system, in *International Conference on Power Systems, 2009, ICPS '09*, Kharagpur, 2009, pp. 1–6

A Novel Intelligent Transmission Line Fault Diagnosis Model Based on EEMD and Multiclass PSVM

Hasmat Malik and Deepti Chack

Abstract The performance of a power network is frequently affected by the faults occurring in transmission line and for maintaining healthy operation, fault diagnosis is necessary. In recent times, a significant amount of research work has been directed to address this problem and the techniques such as ANN, WT, FIS, SVM, Decision tree, etc., have already been employed. This study presents an intelligent framework for transmission line fault classification, applying an innovative machine learning algorithm called Proximal Support Vector Machine (PSVM) that requires small training time to solve nonlinear problems and is applicable to high-dimension application. EEMD method has been utilized for raw electric signals' decomposition into Intrinsic Mode Functions (IMFs), which act as input variable in PSVM-based classifier for identification of faults. Simulations and evaluations suggest that the proposed scheme is effective, reliable, and accurate.

Keywords Fault diagnosis · Proximal support vector machine (PSVM)
Ensemble empirical mode decomposition (EEMD)
Intrinsic mode functions (IMFs)

1 Introduction

Disturbances in power quality in the industrial plants and electrical appliances are one of the major issues that is being faced, of late. The performance of a power network is frequently affected by the transmission line faults, which gives rise to disruption

H. Malik
Department of Electrical Engineering, IIT Delhi, New Delhi 110016, India
e-mail: hmalik.iitd@gmail.com

D. Chack (✉)
Department of Instrumentation and Control Engineering, NSIT,
Sector 3, Dwarka, New Delhi 110078, India
e-mail: deepti.chack@yahoo.com

© Springer Nature Singapore Pte Ltd. 2019
H. Malik et al. (eds.), *Applications of Artificial Intelligence Techniques in Engineering*, Advances in Intelligent Systems and Computing 698,
https://doi.org/10.1007/978-981-13-1819-1_9

Table 1 Parameters of proposed model

Parameters	Sending end source	Receiving end source
Source voltage	500 kV L-L, 0°	475 kV L-L, 15°
Z_{S1}	17.177 + 45.5285 jΩ	15.31 + 45.9245 jΩ
Z_{S2}	2.5904 + 14.7328 jΩ	0.37432 + 1.1315 jΩ

in power flow. Therefore, some of the challenges for incessant supply of power are detection, location, and classification of faults occurring in the transmission line.

The transmission line fault may be defined as any abnormal electrical condition that occurs causing the failure of electrical equipment or power outage. Asymmetrical or short circuit faults wherein the three phases of the line current are not affected equally are majorly responsible for the unstability in the power quality. Asymmetrical faults are more complicated to analyze as compared to the symmetrical. Hence, a fast and reliable technique for diagnosis of faults is required so as to provide safe and secured operation of the power system. Proper selection of algorithm is an imperative decision to be made in order to have correct fault analysis.

In the field of fault classification many researchers have come up with various different algorithms [1–11]. The techniques are quoted as Classical algorithms [1] or impedance matching based methods [2], fuzzy logic-based algorithms [3], and artificial neural network-based algorithms. In literature [4] a new combination making the use of EMD and SVM has been presented. Though these neural networks and reinforcement learning techniques bring out satisfactory results, but these approaches are quite slow in their process of learning and can make the system go sluggish. Due to redundancy and complexity, the computational burden gets increased in case of heuristic fuzzy logic approach [5]. Thus, an algorithm that can work superfast and provide greater efficiency is the need of the hour.

The paper is organized as follows. Section 2 presents the proposed research. Results and discussions are given in Sect. 3 followed by conclusion in Sect. 4.

2 Methodology

2.1 Model Description

To depict the suitability of PSVM for fault diagnosis, a 300 km, 500 kV, 50 Hz transmission line has been developed making the use of distributed parameters on MATLAB and the proposed scheme has been tested on it. The simulation of this model produces data sets for ten types of faults and one healthy condition of the electrical line. The system parameters have been presented in the Table 1 (Fig. 1):
Transmission line parameters: $C_0 = 7.4982E - 9$ F/km $C_1 = 1.2097E - 8$ F/km $Z_0 = 1.2097E - 8$ F/km $Z_1 = 0.15483 + 0.36608$ jΩ/km

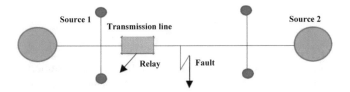

Fig. 1 Simplified transmission line model

2.2 Ensemble Empirical Mode Decomposition (EEMD)

EEMD is an extended and newer version of the frequently used approach called EMD (Empirical Mode Decomposition). The role of both these methods is to disintegrate a complex signal into a series of mono symmetric components called as IMFs (Intrinsic Mode Functions).

However, there is a problem of mode mixing in case of EMD [12, 13] and to overcome it, EEMD has been adopted which is a noise-assisted data analysis method that skillfully removes the mode mixing phenomenon depicted by EMD. It does so by adding a series of variant white noise into the original signal in numerous trials and obtaining an ensemble mean of the produced IMFs in several decompositions as the final result.

The EEMD algorithm [6] is stated as follows-

Step 1: To a given signal $X(t)$ a white noise series $U_n(t)$ is added to produce a new time series in the nth trial giving the equation

$$Y_n(t) = X(t) + U_n(t) \tag{1}$$

for $n = 1, 2, 3 \ldots, N$, where N corresponds to the ensemble number.

Step 2: This new noise-affected signal is decomposed into a set of simpler components called IMFs and a residual making it to be

$$Y_n(t) = \sum_{m=1}^{M-1} IMF_m^{(n)}(t) + r_M^{(n)}(t), \tag{2}$$

$M - 1$ is the total no. of IMFs produced after decomposition of $Y_n(t)$. $IMF_m^{(n)}(t)$ is mth IMF and $r_M^{(n)}(t)$ is the obtained residual in nth trial. To procure an equal number of IMFs for each decomposition, a sifting number of 10 (fixed) has been taken into account.

Step 3: Steps 1 and 2 are repeated for N trials. For every case a distinct white noise is appended to the original signal.

Step 4: To get the final IMF of the EEMD, average of m IMFs is obtained for N trials, represented as

$$IMF_m^{AVG} = \frac{1}{N} \sum_{n=1}^{N} IMF_m^{(n)}(t) \tag{3}$$

The decomposition using EEMD depends on the value of ensemble number (N) and the amplitude (A) of the noise added and the following relation is needed to be satisfied $\beta = A/\sqrt{N}$, where β is the standard deviation of the calculated error. The error is found between sum of all the IMFs obtained from EEMD and the original signal.

In this work, the Ensemble number is chosen to be $N = 100$, while the standard deviation of the noise is 0.2 times the standard deviation of the raw signal.

2.3 Proximal Support Vector Machine (PSVM) [14–16]

Proximal support vector machine is a high-powered classification technique that requires shorter training time in solving nonlinear problems of higher dimensions, invented by Glenn Fung and Mangasarian [7]. PSVM is an improved and more advanced version of the classical Support vector machine (SVM).

Proximal Support Vector Machine classifies the points on the basis of their proximity to one of the parallel planes that are disjoined from one another to maximum possible extent. In PSVM, p data points in R^q characterized by $p \times q$ matrix A and a diagonal matrix D of $p \times p$ dimension of ± 1 labels representing each row of matrix A. The subsequent steps form the whole PSVM algorithm.

Step 1: H is defined using equation, $H = D[A - e]$, e represents a $p \times 1$ vector of ones that calculates u for some positive value of nu.

$$u = \left(nu \left(1 - H \left(\frac{1}{nu} + H'H \right)^{-1} H' \right) \right) e \tag{4}$$

The value of nu is specialized or estimated and it depends on the expert. The range of the values that nu can take lies from $10^6 - 0.01$. Better fitting of the training data is achieved with bigger values of nu.

Step 2: Calculate (w, γ)

$$w = A' * Du; \gamma = -e' * Du; y = u/nu \tag{5}$$

The gap between the hyperplanes is maximized with respect to both the location γ and the orientation w relative to the origin.

Step 3: Classification of new data sample x using

$$\left(x'w - \gamma \right) = \begin{cases} > 0 & then x \in A+; \\ < 0 & then x \in A-; \\ = 0 then\ x \in A + or x \in A-; \end{cases} \tag{6}$$

After training, new data set is tested. Multiclass PSVM classification is performed by making a combination of multiple PSVM classifiers in one by one fashion.

3 Results and Discussions

The presented scheme is used to classify ten different fault types along with one healthy condition of a Power transmission line. The simulated electrical model generates 125,001 samples of data for each case namely, source current, source voltage, load current, load voltage. These data samples are fragmented into smaller signal components known as IMFs using EEMD. Ten distinct PSVM models are designed for each of the cases using 1200 samples. These are then loaded into the PSVM classifier for classifying faults. The highest training and testing accuracy was 94.340% and 90.017% respectively that was for the case of current source Imfs. Plot for Gamma and weights of ten PSVM models for this case has also been produced.

Figures 2, 3, 4, 5, 6 and 7 provide a glimpse of the achieved results on MATLAB for further validation of the performance of our proposed research.

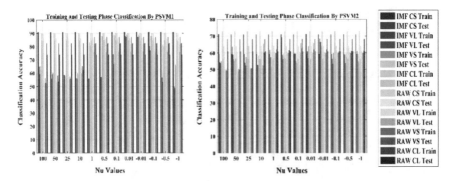

Fig. 2 Bar plots for training and testing by PSVM 1 and PSVM 2

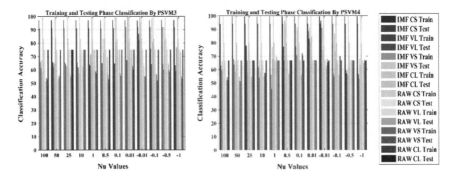

Fig. 3 Bar plots for training and testing by PSVM 3 and PSVM 4

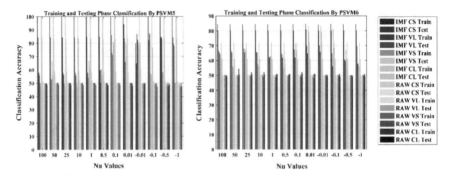

Fig. 4 Bar plots for training and testing by PSVM 5 and PSVM 6

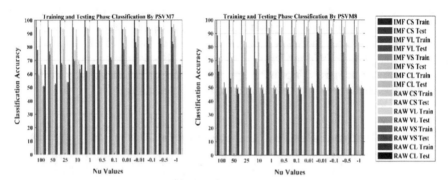

Fig. 5 Bar plots for training and testing by PSVM 7 and PSVM 8

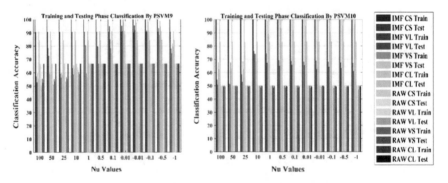

Fig. 6 Bar plots for training and testing by PSVM 9 and PSVM 10

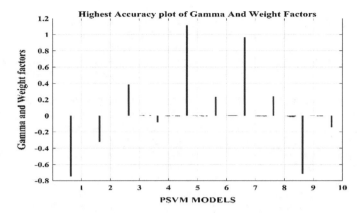

Fig. 7 Gamma and weight factors for PSVM models

4 Conclusion

The presented paper shows the combination of EEMD and PSVM techniques for the classification of faults in an electrical transmission line. The data sets obtained from the simulation of power system model are disintegrated into intrinsic mode functions employing EEMD. Furthermore, the PSVM models are developed, validated and tested for classification of the faults occuring in power transmission line. The results suggest that PSVM proves to be effective and takes much shorter training/testing time.

References

1. R.K. Aggarwal, Y. Aslan, A.T. Johns, New concept in fault location for overhead distribution systems using superimposed components. IET Gener. Transm. Distrib. **144**(3), 309–316 (1997)
2. C.K. Jung, J.B. Lee, X.H. Wang, Y.H. Song, Wavelet based noise cancellation technique for fault location on underground power cables. Electr. Power Syst. Res. **77**(10), 1349–1362 (2007)
3. I. Sadinezhad, M. Joorabian, A new adaptive hybrid neural network and fuzzy logic based fault classification approach for transmission lines protection, in *International Conference on Power and Energy* (2008), pp. 895–900
4. N.R. Babu, B.J. Mohan, Fault classification in power systems using EMD and SVM. Ain Shams Eng. J. (2015). http://dx.doi.org/ https://doi.org/10.1016/j.asej.2015.08.005 (In Press). ISSN 2090-4479
5. O.A.S. Youssef, Combined fuzzy-logic wavelet-based fault classification technique for power system relaying. IEEE Trans. Power Deliv. **19**(2), 582–589 (2004). https://doi.org/10.1109/tp wrd.2004.826386
6. K. Khaldi, M. Turki-HadjAlouane, A.O. Boudraa, A new EMD denoising approach dedicated to voiced speech signals, in *2nd International Conference on Signals, Circuits and Systems, 2008. SCS 2008*, Issue 7–9, Nov 2008, pp. 1–5

7. G. Fung, O.L. Mangasarian, Proximal support vector machine classifiers, in *Proceedings of KDD-2001: Knowledge Discovery and Data Mining*, San Francisco, CA, 2001, pp. 77–86. Accessed on ftp://ftp.cs.wisc.edu/pub/dmi/tech-reports/01-02.pdf

8. H. Malik, R. Sharma, Transmission line fault classification using modified fuzzy Q learning. IET Gener. Transm. Distrib. **11**(16), 4041–4050 (2017). https://doi.org/10.1049/iet-gtd.2017.0331

9. H. Malik, R. Sharma, EMD and ANN based intelligent fault diagnosis model for transmission line. J. Intell. Fuzzy Syst. **32**(4), 3043–3050 (2017). https://doi.org/10.3233/JIFS-169247

10. A. Aggarwal, H. Malik, R. Sharma, Selection of most relevant input parameters using weka for artificial neural network based transmission line fault diagnosis model, in *Proceedings of the International Conference on Nanotechnology for Better Living*, vol. 3, no. 1 (2016), pp. 176. https://doi.org/10.3850/978-981-09-7519-7nbl16-rps-176

11. H. Malik, A. Aggarwal, R. Sharma, Feature extraction using EMD and classification through probabilistic neural network for fault diagnosis of transmission line, in *Proceedings of IEEE ICPEICES-2016* (2016), pp. 1–6. https://doi.org/10.1109/icpeices.2016.7853709

12. H. Malik, S. Mishra, Artificial Neural network and empirical mode decomposition based imbalance fault diagnosis of wind turbine using TurbSim, fast and simulink. IET Renew. Power Gener. **11**(6), 889–902 (2017). https://doi.org/10.1049/iet-rpg.2015.0382

13. H. Malik, S. Mishra, Application of GEP to Investigate the imbalance faults in direct-drive wind turbine using generator current signals. IET Renew. Power Gener. **11**(6), 889–902 (2017). https://doi.org/10.1049/iet-rpg.2016.0689

14. H. Malik, S. Mishra, Fault identification of power transformers using proximal support vector machine (PSVM), in *Proceedings of IEEE International Conference on Power Electronics (IICPE 2014)*, 8–10 Dec 2014, NIT Kurusherta, India. https://doi.org/10.1109/iicpe.2014.7115842

15. A.P. Mittal, H. Malik, V. Talur, S. Rastogi, External fault identification experienced by 3-phase induction motor using PSVM, in *Proceedings* of IEEE International Conference on Power India (PIICON 2014), 5–7 Dec 2014, New Delhi, India. https://doi.org/10.1109/poweri.2014.7117762

16. H. Malik, S. Mishra, Proximal support vector machine (PSVM) based imbalance fault diagnosis of wind turbine using generator current signals. Energy Procedia **90**, 593–603. https://doi.org/10.1016/j.egypro.2016.11.228

Optimal Power and Performance Using Fletcher–Reeves Method in Dynamic Voltage Scaling

Anuradha Pughat, Parul Tiwari and Vidushi Sharma

Abstract Power consumption and cost reduction are always two important performance parameters in determining the optimal functioning of wireless sensor networks. This paper proposes a Fletcher–Reeves algorithm to minimize the power and performance cost of a sensor node using an ultra-low power microcontroller MSP430. The computational results have been compared with the Gradient-Descent algorithm, which shows that the Fletcher–Reeves (FR) algorithm outperforms over Gradient-Descent (GD) algorithm and saves the node power significantly. The simulation results are validated through theoretical results. The parameters such as average wait time, service time, operating frequency, optimized voltage, and cost are used as performance measures. It is worth mentioning that obtained results saves power significantly and reduces the performance cost of DVS when compared with the existing literature.

Keywords Dynamic voltage scaling · Sensor node · Optimization
Power and performance · Cost · Frequency scaling · MSP430 microcontroller
Waiting time · Mathematical modeling

1 Introduction

In the present era, connectivity behaves like ears and mouth of a human being and it provides communication flexibility to sensors in the practical world. The sensor network technique is constantly playing its vital role in communication technology.

A. Pughat · P. Tiwari (✉)
Jaypee Institute of Information Technology, Noida, Noida, Uttar Pradesh, India
e-mail: parultiwari8922@gmail.com

A. Pughat
e-mail: anuradha.pughat@gmail.com

V. Sharma
SoICT, Gautam Buddha University, Noida, Uttar Pradesh, India
e-mail: svidushee@gmail.com

© Springer Nature Singapore Pte Ltd. 2019   93
H. Malik et al. (eds.), *Applications of Artificial Intelligence Techniques
in Engineering*, Advances in Intelligent Systems and Computing 698,
https://doi.org/10.1007/978-981-13-1819-1_10

The ever-increasing demand of WSNs showed the necessity of using energy efficient microcontrollers to control the processing. Thus, it is a good choice to use TI MSP430 microcontroller with Von Neumann architecture that is specially designed for low power applications. It can operate in several modes depending on the active and sleep time of the device. Large numbers of registers contained in its CPU are used as local variables to enhance the efficiency of the device. The overall power consumption can be minimized by utilizing various sleep modes in an efficient way and facilitate the user with the characteristic to shut down the multiple sleep modes when they are not in use.

Some of the major issues in developing sensor network systems and algorithms are energy optimization, quality of service management, service time management, optimal cost estimation, and security, etc. Optimization plays key role to achieve these objectives. It provides an elegant amalgamation of theory and practices. It is a basic and regularly applied task for most engineering applications. It is a decisive step of any optimization to choose a proper algorithm because an efficient optimizer can ensure the existence of optimal solutions.

Dynamic Voltage Scaling (DVS) is frequently used in sensor node networks to obtain significant dynamic power decrement. In DVS the speed and voltage of devices are changed dynamically to achieve the increment in energy efficiency of their operation [1, 2]. Net power consumption of a sensor node can be reduced by applying an appropriate voltage scaling. A quadratic reduction in dynamic power is obtained in [3] by reducing the supply voltage. The objective of DVS is to save energy by reducing the attainment of the network systems without causing an application to miss its specified deadlines. Dynamic voltage scaling has been applied in [4–6] on wireless sensor nodes and a passive voltage scaling (PVS) technique is suggested in [7]. However, this approach is unable to readjust the voltage level and thus this problem was handled in [8] by using an energy efficient step down converter. Dynamic optimizations ensure better flexible options by continuously optimizing a WSN/sensor node during runtime, providing better adaptation to recent challenging application needs and practical situations.

Existing work in the literature on DVS has focused mostly on the suitable voltage scaling techniques not on the optimization methods. Steepest decent optimization [9] with SimEvents is applied to design a dynamic voltage scaling controller. To scale the various parameter for WSNs, a dynamic optimization method is performed in [10]. DVS along with optimization using an energy-response time product trade-off metric with sleep states is proposed in [11]. There is no single and specific algorithm which can be applied to resolve all types of issues related to service time management or energy reduction. Numerous optimization algorithms are available in the literature known as classical and bio-mimetic optimization methods. Classical optimization methods include derivative, or gradient-based and derivative-free algorithms. Bio-mimetic algorithms are generally pattern matrix-based methods which provide random solutions to the related problems, whereas classical optimization methods use a Hessian matrix or gradients [12]. For multiprocessor embedded systems, gradient-based scheduling is discussed in [13].

In [14] energy is conserved for computational cloud using dynamic voltage and frequency scaling (DVFS). Machine learning algorithms/artificial intelligence techniques can also be applied for optimization but these are heavy algorithms and may take more time to execute results for resource constraint based WSNs. Dynamic voltage and frequency scaling based on coarse grained learning for video decoding is proposed in [15]. In this paper the authors proposed workload prediction and online learning based algorithm for DVFS. The approach can be used in WSNs for table-based scaling. Although the steepest descent algorithm is one of the commonly used optimization methods, it suffers with poor rate of convergence if proper scaling of the function is not done. This difficulty can be resolved by using a conjugate set of gradient directions which ensures a quadratic rate of convergence and the optimal solution is achieved in limited number of steps [16]. Some of the conjugate gradient methods are explained in [17, 18]. Fletcher–Reeves conjugate gradient method optimizes the objective function more rapidly. Several versions of this method are available in the literature and used by the researchers in different fields. Modified Fletcher–Reeves is applied for solving mono-domain model [19]. Later, its Modified version is used for solving unconstrained optimization [20]. The smoothing Fletcher–Reeves conjugate gradient method is applied in [21]. Dynamic voltage scaling optimization along with power aware system on chip test is performed in [22]. Energy harvesting for wireless sensor network nodes using dynamic voltage and frequency is performed [23]. Recently a modified hybridization method of Fletcher–Reeves and Polak-Ribière-Polyak for optimization problems is discussed in [24, 25]. In the present paper, a Fletcher–Reeves conjugate gradient method is proposed to optimize the power and performance metric in dynamic voltage scaling. Further the paper is organized as follows.

Section 2 gives the mathematical modeling for DVS on MSP430 using Gradient-Descent and Fletcher–Reeves methods. M/M/1 queuing model has been used to visualize the stochastic nature of event arrival at input of sensor node. Then, Sect. 3 presents the result analysis on the proposed system. Finally, Sect. 4 concludes the paper.

2 Mathematical Modeling

TI MSP430 can be modeled with varying processing rate, voltage and operates with ultra-low power. The operating voltage can vary from 2 to 3.3 V and the frequency in the range of 4.6–7 MHz. While modeling the performance metric, the inter-arrival rate (λ) and service time (μ) is assumed to be exponentially distributed. Thus the cost function $C(\mu)$ will follow the M/M/1 queuing model and is given by Eq. 1.

$$C(\mu) = \frac{\mu}{1 - \lambda\mu} \tag{1}$$

The energy consumption is related quadratically with voltage (V) and shows a linear relation with the number of operations (N) performed. The relation is given in Eq. 2. Where, the operating voltage is $V = \frac{1-\mu c_1}{\mu c_2}$. The constants c_1 and c_2 are the device dependent coefficients and are taken as 1.8 and 0.91 respectively. The performance metric $J(\mu)$ is related to energy consumption $E(\mu)$ and associated cost $C(\mu)$ function as in Eq. 3, Where, the parameter 'w' represent the weight function assigned to the optimization and signifies the price of energy relative to delay in processing. Here, the weight parameter is taken as 100. Using Eqs. 1 and 2 in Eq. 3, the performance metric is obtained, Eq. 4.

$$E(\mu) = NV^2 \tag{2}$$

$$J(\mu) = wE(\mu) + C(\mu), \tag{3}$$

$$J(\mu) = wN\left(\frac{1 - \mu c_1}{\mu c_2}\right)^2 + \frac{\mu}{1 - \lambda\mu} \tag{4}$$

The Conjugate Gradient Fletcher–Reeves method is used as a conjugate direction method. It performs the gradient of the function using first direction search as steepest descent. Since the rate of convergence of this method is quadratic, it improves the convergence characteristics of the gradient-descent method. This property ensures that this method will converge in a finite number of iterations. To get the optimum value of 'μ', the Fletcher–Reeves method is applied and this process is iterated until a desired level of accuracy is achieved, see Algorithm 1. This method is compared with the existed methods, i.e., Gradient-Descent method, refer Algorithm 2.

Algorithm 1: Fletcher–Reeves method

i. Choose an arbitrary initial value as a lower bound for service time say μ_1.
ii. Determine the initial search direction $(S_1) = -\left(\frac{dJ}{d\mu}\right)_{\mu_1}$ according to Steepest descent search.
iii. Find the next point $\mu_2 = \mu_1 + h_1^* S_1$, where h_1^* is the appropriate step size in the direction S_1 that minimizes $J(\mu_2)$.
iv. Find $\nabla J(\mu_i)$ and set $S_i = -\nabla J_i + \frac{|\nabla J_i|^2}{|\nabla J_{i-1}|^2} S_{i-1}$.
v. Calculate the optimum step size h_i^* in the direction S_i and find $\mu_{i+1} = \mu_i + h_1^* S_i^*$.
vi. Stopping criterion is according to the optimum value of μ_{i+1}, if it is optimum, process is to be stop, otherwise set the next value of $i(= i + 1)$ and move to step iv and repeat the process.

Algorithm 2: Gradient-Descent method

i. Start with an arbitrary μ_1.
ii. Find search direction $S_i = -\nabla\{J(\mu_i)\}$.

iii. Determine optimal step length h_i^* and set $\mu_{i+1} = \mu_i + h_i^* S_i$.
iv. Check the new value of μ_{i+1} for optimum solution and if exists, stop, else iterate
   again.

3 Results Analysis

The proposed simulation model for MSP430 processor has been designed using
SimEvents. The dynamic voltage scaling using FR method has been tested for optimal
voltage and cost. Then, the simulation results obtained from the proposed model
(Algorithm 1) have been compared with the existing model (Algorithm 2). Assume
that the processor performs an event in 1×10^6 operations. Then, the model has been
tested for optimality at varying event arrival rate (e.g. 3.88 events/h, 2 events/h, 0.67
events/h). Figure 1 shows the voltage scaling for 10 number of sampled time values.
The operating voltage is scaled down as the number of events at input of processor
queue increases. For an arrival of 0.67 events/h, the FR minimizes its voltage from 3
to 2.1 V. On the other hand, GD minimizes the processor voltage form 3.13 to 2.24 V
for the same event arrival rate.

When compared the FR method with the GD method, it has been found that
the proposed (FR) method potentially work for the minimized operating voltage
at same event arrival rate. After getting stationarity in the simulation results, the

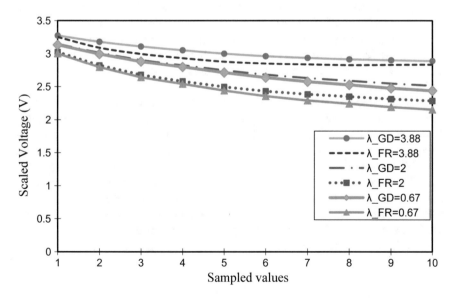

Fig. 1 Comparison of scaled voltage for Gradient-Descent (GD) and proposed Fletcher–Reeves
(FR) methods

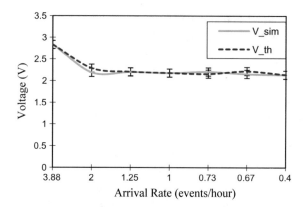

Fig. 2 Error computation for theoretical versus simulated results

SimEvents-based simulation model is checked for its accuracy in scaling voltage. For this purpose, the simulation results are validated with the theoretical results and less than 0.05% error is computed for a wide range of arrival rate. Figure 2 depicts the error bars for theoretical versus simulation results.

The voltage scaling is used to reduce the power consumption by operating the processor at a minimum voltage and desired performance depending on the application requirement. The scaled down voltage costs in terms of the speed of system operation, i.e. its frequency. The frequency of operation can be scaled down with voltage to the point it meets the deadline criteria. Figure 3 presents the optimized voltage and total cost for various arrival rates. The higher arrival rate costs more and requires higher operating voltage. The optimization starts from right hand side and converges to its minimum values. The graph shows a large voltage difference (from 2.1 to 2.89 V) in optimized voltages at minimum ($\lambda = 0.4$) and maximum ($\lambda = 3.88$) arrival rates.

In M/M/1 model, an event arrives to the FIFO queue and waits for its processing. The wait time depends on the system performance, its service time and delay. Now, the average waiting time for processing the events using both the methods have been computed, Fig. 4. For higher event arrival rate, our proposed (FR) method outperforms and gives even lower average wait time for lower arrival rate, for random 30 samples. Further, the same sample size has been chosen to see the power consumption curves for both the methods. We computed the power consumption for MSP430-based sensor node and shown the difference between proposed and existing method, see Fig. 5. It shows that our proposed method optimizes voltage and cost to lower value and consumes lesser power in processing events than GD method based optimization.

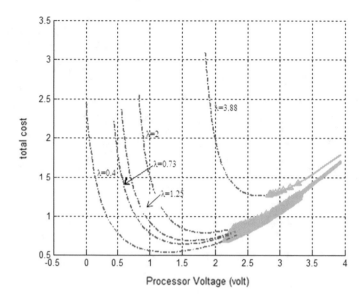

Fig. 3 Optimal voltage versus cost at various arrival rate

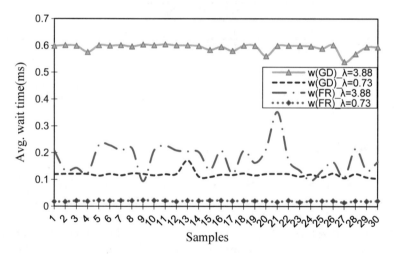

Fig. 4 Compared average wait time at various arrival rates

Finally, Table 1 gives the comparison statements for microcontrollers AT9OS8535 (GD) and MSP430 (GD) using Gradient-Descent method and proposed MSP430 for FR method. As expected, a sensor node using proposed method can operate on lesser voltage and higher performance than previous GD method. Thus, works for least average power consumption among others.

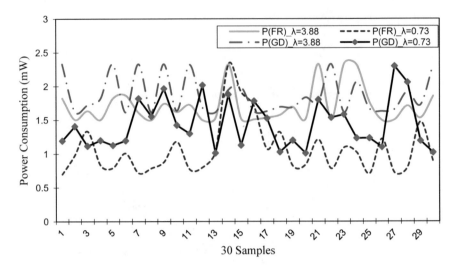

Fig. 5 Comparision of power consumption for GD and proposed FR methods

Table 1 Comparative voltage, service time, frequency and average power

| | t_interarrival = 0.2577, lambda = 3.88 | | | |
	Voltage	μ	$f = 1/\mu$	Avg. power
AT9OS8535 (GD)	3.996	0.1668	5.9952	95.73
MSP430 (GD)	2.982	0.1593	6.2775	55.82
MSP430 (FR)	2.83324	0.16639401	6.009	48.23

4   Conclusion

In the present paper, a Fletcher–Reeves conjugate gradient method is applied to determine the optimum voltage and cost using MSP430 microcontroller for a sensor node. After simulating the results, it has been observed that the proposed optimization method (FR) provides lower operating voltage and consumes less power. Significance of FR method in optimization is shown by comparing the results with those reported in literature using GD method. Our future plans are to implement the proposed method for voltage and cost optimization on Telosb sensor node.

Acknowledgements The authors would like to acknowledge Department of Science and Technology, Ministry of Science and Technology, Govt. of India for supporting financially the project (Grant No.: SR/WOS-A/ET-1043/2014). The authors are very much thankful to Gautam Buddha University, Greater Noida for providing infrastructure, help and support in completion of this research work.

References

1. T. Burd, R. Brodersen, Energy efficient CMOS microprocessor design, in *Proceedings of the 28th Hawaii International Conference on System Sciences*, vol. 1 (1995), pp. 288–297
2. D.P. Pillai, K. Shin, Real-time dynamic voltage scaling for low-power embedded operating systems, in *Proceedings of the Eighteenth ACM Symposium on Operating Systems Principles*, (2001), pp. 89–102
3. D.B. Rane, S.B. Sajgure, M.Y. Dighe, P.S. Gunjal, A method based on dynamic voltage scaling technique for power optimization. Int. J. Adv. Res. Innov. Ideas Educ. **2**(2), 924–928 (2016)
4. W. Tuming, Y. Sijia, W. Hailong, A dynamic voltage scaling algorithm for wireless sensor networks, in *3rd International Conference on Advanced Computer Theory and Engineering (ICACTE)*, vol. 1 (2010), pp. 554–557
5. T. Hamachiyo, Y. Yokota, E. Okubo, A cooperative power-saving technique using dvs and dms based on load prediction in sensor networks, in *Fourth International Conference on Sensor Technologies and Applications (SENSORCOMM)* (2010), pp. 7–12
6. W. Dargie, Dynamic power management in wireless sensor networks: state-of-the-art. IEEE Sens. J. **12**(5), 1518–1528 (2012)
7. Y. Cho, Y. Kim, N. Chang, PVS: passive voltage scaling for wireless sensor networks, in *2007 ACM/IEEE International Symposium on Low Power Electronics and Design (ISLPED)* (2007), pp. 135–140
8. L.B. Hoermann, P.M. Glatz, C. Steger, R. Weiss, Energy efficient supply of wsn nodes using component-aware dynamic voltage scaling, in *Wireless Conference—Sustainable Wireless Technologies (European Wireless)* (2011), pp. 1–8
9. W. Li, C.G. Cassandras, M. Clune, Model-based design of a dynamic voltage scaling controller based on online gradient estimation using SimEvents, in *Proceedings of the 45th IEEE Conference on Decision & Control* (2006), pp. 6088–6092
10. A. Munir, A.G. Ross, S. Lysecky, R. Lysecky, A Lightweight dynamic optimization methodology for wireless sensor networks, in *IEEE 6th International Conference on Wireless and Mobile Computing, Networking and Communications*, 978-1-4244-7742-5/10, (2010), pp. 129–136
11. A. Gandhi, V. Gupta, M.H. Balter, M. Kozuch, Optimality analysis of energy-performance trade-off for server farm management, in *International Symposium on Computer Performance, Modeling, Measurements, and Evaluation* (2010), pp. 1155–1171
12. M.J. Tahk, M.S. Park, H.W. Woo, H.J. Kim, Hessian approximation algorithms for hybrid optimization methods. Eng. Optim. **41**, 609–633 (2009)
13. V. Bharadwaj, L.K. Goh, S. Viswanathan, An energy-aware gradient based scheduling heuristic for heterogeneous multiprocessor embedded systems, in *International Conference on High-Performance Computing* (2007), pp. 331–341
14. A.P. Florence, V. Shanthi, C.B. Simon, Energy conservation using dynamic voltage frequency scaling for computational cloud. Sci. World J. (2016). https://doi.org/10.1155/2016/9328070
15. J. Guo, M. Potkonjak, Coarse-grained learning-based dynamic voltage frequency scaling for video decoding, in *26th International Workshop on Power and Timing Modeling, Optimization and Simulation (PATMOS)* (2016), pp. 84–91
16. X.S. Yang, S. Koziel, *Optimization Algorithms Methods and Algorithms* (Springer, Germany, 2011), pp. 13–31
17. Z. Sun, Y. Tian, H. Li, Two modified three-term type conjugate gradient methods and their global convergence for unconstrained optimization. Math. Probl. Eng. 1–9 (2014)
18. C. Li, Conjugate gradient type method for the nonnegative constraints optimization problems. J. Appl. Math. 1–6 (2013)
19. K.W. Ng, A. Rohanin, Modified Fletcher-Reeves and Dai-Yuan conjugate gradient methods for solving optimal control problem of Monodomain model. Appl. Math. **3**, 864–872 (2012)
20. A. Abashar, M. Mamat, A. Alhawarat, F. Susilawati, Z. Salleh, et al., A modified Fletcher-Reeves conjugate gradient method for unconstrained optimization, in *Proceedings of ISER 41st International Conference, Sydney, Australia* (2016), pp. 5–8. ISBN: 978-93-86291-20-2

21. D. Pang, S. Du, J. Ju, The smoothing Fletcher-Reeves conjugate gradient method for solving finite minimax problems. ScienceAsia **42**, 40–45 (2016)
22. V. Sheshadri, V.D. Agrawal, P. Agrawal, Power-aware SoC test optimization through dynamic voltage and frequency scaling, IEEE (2013). 978-1-4799-0524-9/13
23. X. Li, N. Xie, X. Tian, Dynamic voltage-frequency and workload joint scaling power management for energy harvesting multi-core WSN node SoC. Sensors **17**(2), 1–15 (2017)
24. N. Sindhu, P. Kaelo, M.V. Thuto, A modified quadratic hybridization of Polak-Ribière-Polyak and Fletcher-Reeves conjugate gradient method for unconstrained optimization problems. Int. J. Optim. Control Theor. Appl. **7**(2), 177–185 (2017)
25. S.B. Kafaki, R. Ghanbari, A hybridization of the Polak-Ribiere-Polyak and Fletcher-Reeves conjugate gradient methods. Numer. Algorithms **68**, 481–495 (2015)

Optimal Location and Sizing of DG in Distribution System and Its Cost–Benefit Analysis

Same Ram Ramavat, Shiva Pujan Jaiswal, Nitin Goel and Vivek Shrivastava

Abstract This paper proposed a tactic to find a optimal location of Distributed Generation (DG) and its capacity through Particle Swarm Optimization Technique (PSO). The site and size of DG are determined to achieve the highest benefit at the lowest cost and with the aim that distribution system should be efficient and stable. Cost–Benefit analysis is carried out considering investment cost, operating cost, and maintenance cost. Benefits are quantified on the concept of Present Value Factor (PVF). PVF is based on the inflation rate and interest rate. This factor is used to determine benefit likely to accrue to power distribution network at the current market price of electricity. The developed technique is tested on IEEE 34-bus radial distribution system. Results demonstrate that the PSO based algorithms are capable to decide optimum size and bus number for DG, to minimizing the loss and improves the voltage stability and economic benefits.

Keywords Distributed generation · Optimal location · Optimal size
Cost–benefit · PSO

1 Introduction

Electrical distribution systems are a key element in nation building and play an signif-icant function in development of infrastructure. The planners of electrical distribution structure are to plan a system with high reliability and viable cost. To achieve highest benefit from DG such at minimum cost, low power loss, and voltage stability profile. Two important goals for planners are reduction in cost and stable distribution sys-tem. The interruptions and high loss are two biggest drains on profit and resources in

S. R. Ramavat (✉) · S. P. Jaiswal · N. Goel
Department of Electrical and Electronics Engineering, SET, Sharda University, Greater Noida, India
e-mail: shiva_dei1@rediffmail.com

V. Shrivastava
Department of Electrical Engineering, RTU, Kota, India

© Springer Nature Singapore Pte Ltd. 2019   103
H. Malik et al. (eds.), *Applications of Artificial Intelligence Techniques in Engineering*, Advances in Intelligent Systems and Computing 698,
https://doi.org/10.1007/978-981-13-1819-1_11

distribution system. The outages and poor efficiency of network affect consumers as well as reduce the profit of Utility. The decision of installing DG and its location is a challenge for power system planners and designers. Location and size of DG will increase or decrease energy loss with load variation in feeder circuits of power distribution network. An optimization technique should be employed to find best possible place and its size which is cost effective for both utility and consumers. The optimization technique employs various operating factors and constraints. Several techniques to find optimal size and site of DG are published in different scientific journals. The use of intelligence algorithms, power flow techniques and analytical approach are employed to optimize location of DG to achieve highest economic benefits and lowest energy loss in electricity distribution system. The planners have to design the system to maximize profit by employing considerable resources, attention and suitable technique simultaneously ensuring reduction in power loss. In [1] authors studied and validated the optimal capacity and place of DGs of different technologies, using analytical approach. Minimum energy loss and maximum benefit with a simultaneous improvement in voltage profile was achieved. In [2] sensitivity test has been used to determine optimal locations of DGs. In [3] genetic algorithms is employed to find suitable size and place of DGs with minimum power loss and stability. Ant colony optimization technique has been employed for deciding optimal DG with low power loss and voltage within agreed variation. In [4] authors have utilized optimization technique of Artificial Bee Colony identified optimal site and sizes of DGs which are a happy mix of real and reactive power. In [5] authors determined the sizes of DGs and their operating level keeping various load requirements. The authors have identified various cost functions based on system losses. In [6] authors suggested that planning was carried out keeping in view various scenario of placing DGs at pre-decided locations. These locations are based on assumption that these are beneficial to both consumer and utility. In [7] authors have described the design of grid considering photovoltaic, fuel cells, battery storage and hybrid DG the grid in open market of electricity. Authors in [8] investigated that the optimal placement of distributed generation with loss reduction, improved voltage profile at load buses and overall profit maximization of system. In one of the paper [9], authors presented the optimal location of DGs to maximize profit by reducing real power loss and improving reliability with the help of dynamic programming. The objective was achieved by using different load pattern and model. Single or multi-objective functions can be taken into consideration to get highest advantage from DG. The main objectives which can be considered are voltage profile improvement, real power loss reduction, minimization of reactive power, maximization of the capacity of DG and other economy-related issues. Multi-objective functions can be used for implementation for DG allocation. Various studies have been investigated to evaluate the effect of DG size and location on other parameters such as voltage profile, stability and security of system [10]. While designing scheme for optimal DG allocation, many issues come up for consideration so that all objectives whether single or multi-objective are achieved and all limits parameters remain within permissible range. Various limitations and constraints are taken into consideration or investigated by researchers such as power system stability, reliability, voltage profile, loss reduction, line thermal, short circuit level and profit maximization. In [11] synchronous DG are applied to provide ancillary services. The authors in [12] installed DG in smart grid. Das et al. [13] applied the DG for voltage control of remote buses. Number of research papers also evaluated DG of different technology such as wind power, solar power, fuel cell technology, etc., to achieve happy mix of site and size of DG [14]. In this paper authors developed a PSO-based algorithm to locate optimal DG. A multi-objective function is proposed with installation cost, system loss, and voltage stability as objectives. The developed methodology is analyzed on IEEE 34-bus radial distribution systems. Authors have worked out cost- benefits analysis and proved that DG builds a stable, efficient and economical power distribution system. The present paper is organized into 4 Sections. Section 1 deals with an introduction and general discussion of place-ment of DG through optimum techniques in the research paper. Section 2 deals with formulation of problem and objective function. It deals with factors affecting the cost of installation of DG, operation, and maintenance cost. It elaborates various factors of benefit and quantification of cost–benefit accrued out of the investment. Section 3 deals with various optimization techniques and technique applied herein the paper. Suitability of PSO technique has been discussed. Section 4 deals with the case study of optimal installation of PV Power plant. The actual investment required, benefit accrued and final outcome on objectives of the plant are worked out. The conclusion is discussed in the last section.

2 Problem Formulation

Analysis and investigation are easy on the mathematical model of cost and benefits of DG. In this paper objective is to the prune down the overall generation cost of power, minimum power losses in system along with low voltage profile index of the system [15]. Hence three indices, i.e., index for cost, index of power loss, and voltage profile index are used to make a objective function.

2.1 Objective Function

The objective of the paper is to minimize total cost of generation, reduce system power losses, and improve the voltage profile. For obtaining desired objectives, a multi-objective function is formed by combining TCG, PLI, and VPI. For this linearization problem, weighing factors x, y, and z are used for TCG, PLI and VPI respectively in such a way such that their sum is unity. Hence, multi-objective function can be formulated as under

$$OF = xTCG + yPLI + zVPI \qquad (1)$$

Constraints for the optimization problem are stated below:

The node voltage at each node is restricted within its minimum and maximum limits

$$V_{imin} \leq V_i \leq V_{imax} \tag{2}$$

Active power generation is restricted by its lower and upper limits as

$$P_{DG}^{min} \leq P_{DG} < P_{DG}^{max} \tag{3}$$

Power Loss Index (PLI): Mathematically PLI can -be expressed as Eq. (4).

$$PLI = \frac{Loss\,with\,DG}{Loss\,without\,DG} \tag{4}$$

Voltage Profile Index (VPI): Voltage profile index is given by Eq. (5)

$$VPI = \frac{\sqrt{(1 - V_{imin})^2}}{V_{imin}} \tag{5}$$

Total Cost of generation: Total cost of generation is the present value of the sum of all annual investments such as annual Installation cost, annual maintenance cost, and annual variable maintenance.

Fixed Cost of Installation: The Fixed Cost of Installation includes cost incurred in demand survey, Installation of DG, land cost, tools and plants required for installation, monitoring cost, commissioning cost. Mathematically all these costs can be clubbed into one cost and can be termed as Fixed Cost of Installation per MW.

$$FC = \sum_{j=1}^{n} Cdgj * ICj, \tag{6}$$

where, *FC* is Fixed Cost of Installation, *Cdgj* is the capacity in MW of *j*th DG

ICj is fixed Cost per MW, *j* = No of DG in distribution system.

Variable Operating Cost: The operating cost which is required to produce electricity and is variable, depends on size of DG, cost of fuel, etc., and can be evaluated as per MW of DG

$$VOC = \sum_{j=1}^{n} (Cdgj * OCj) * \Delta T, \tag{7}$$

where, *VOC* is variable operating cost, *OCj* is operating cost per MW of *j*th DG

ΔT is operation time of DG in a year.

Variable Maintenance Cost: Maintenance Cost is a variable component which includes day today maintenance cost, renovation cost carried out annually. Variable Maintenance can be expressed as

$$VMC = \sum_{j=1}^{n} (Cdgj * MCj), \tag{8}$$

where, *VMC* is Variable Maintenance Cost, *MCj* is maintenance cost per MW capacity of *j*th DG.

Total Cost Generation: Total Cost Generation (*TCG*) is expressed as sum of Fixed Cost of Installation (FC), Variable Operating Cost (*VOC*) and Variable Maintenance Cost (*VMC*) in terms of Present worth of *VOC* and *VMC*

$$TCG = FC + PW(VOC) + PW(VMC) \tag{9}$$

3 Optimization Technique

Optimization techniques play a very important for placement of DG. Optimization techniques are classified into the main categories (i) analytical approach [16] (ii) society-based algorithms (iii) nature-inspired techniques. Optimization technique used in this paper is Particle Swarm Optimization Algorithm. Particle Swarm Optimization Technique (PSO) is a evolutionary technique which was introduced by Kennedy and Ebenhart which can investigate continuous variable factors in a dynamic system in a perfect manner In this technique, individual factor called Particle changes his value or position continuously with respect to the timeline [17]. The particle flies in space and gains its position based on its past experience (pbest) and past experience of its neighbor (gbest). The position of Particle so gained is based on the best experience itself and with its neighboring particle. Position of particle can be expressed as process is repeated till the optimum solution is found

$$V_i^n = w \times V_i^{n-1} + c_1 \times R_1 \times \left(P_{besti}^{n-1} - X_i^{n-1}\right) + c_2 \times R_2 \times \left(G_{besti}^{n-1} - X_i^{n-1}\right) \tag{10}$$
$$X_i^n = X_i^{n-1} + V_i^n \tag{11}$$

where:

w	The inertia weight
c_1 and c_2	The acceleration coefficients
R_1 and R_2	Uniformly distributed random numbers between 0 and 1
X_i	Position of ith particle
V_i	Velocity of ith particle
P_{besti}	Best position achieved by the ith particle
G_{besti}	Best position achieved ever

To get the accurate and fast result the following PSO parameters are considered as given in Table 1.

PSO-based Algorithms: The PSO Technique for optimal DG allocation in the distribution network is employed while taking various issues and constraints into consideration. The main aim is the maximization of profit or benefit. Various steps should be taken as per list below

Table 1 PSO parameters

PSO parameter	Value	PSO parameter	Value
Particles number	20	No of iterations	50
Inertia weight	Linearly decreased	Velocity bounds	$\{-3,7\}$
Individual acceleration constant	2.5	R1	0.3
Social acceleration constant	2.0	R2	0.2

Step 1: Insert the bus data, line data, load and generation data etc.

Step 2: The Initial array of the particle be generated with random velocity and random position, i.e., location and size of DG. In the procedure of solving problem, set initial counter of iteration as $j = 0$. All particles should satisfy all constraints.

Step 3: Put all planning data as DG fixed cost, number and number of years planned.

Step 4: Estimate the current index of every particle according to Eq. (1).

Step 5: Compare the value of objective for each particle with its best value. If it found that present value is more than p_{best} value, then let this value be p_{best} and position of the particle should be recorded. Thus the value of all particles should be p_{best}. Now g_{best} value will be overall best value for all particles.

Step 6: Now weight, velocity, and position of all particles should be refreshed using Eqs. (10) and (11).

Step 7: Countdown the iteration number and on reaching proposed maximum number, process as per next step should be carried out such that $n = n + 1$ and process is repeated again and again.

Step 8: Now the solution is obtained as per target to the problem and best optimal size and optimal location of DG are obtained. Optimal size and location yield maximum benefit in a specified time frame.

4   Benefits Accrued from DG

Present Benefits accrued are estimated based on year of planning for which distribution system is likely to operate economically and this depends on the cost of energy in the market. Present Value Factor depends on the rate of interest in banking sector and the inflation rate in the economy of the country. Present Value Factor (π) can be evaluated as under

$$\pi = \sum_{k=1}^{n} (1 + IF)/(1 + IR), \tag{12}$$

where, *IF* is rate of inflation, *IR* is rate of interest, k=useful year of DG operation.

Benefit from Fixed Cost: Distribution utility supplies power to consumers as per demand. If the demand for power is more than the generation, it has to purchase power from open energy trade market. Cost of power in the open market is the major source of benefits accrued to the utility. Lesser the energy drawn from grid more is the benefit. Hence benefit due to saving in power or supplied to grid is evaluated as under

$$BN_1 = \sum_{j=1}^{n} Cdgj * PPg * \Delta T, \tag{13}$$

where, BN_1 is net benefit accrued to utility, PPg is purchase price per KWH of electricity in open market, ΔT is time duration when electricity generated by DG. *Benefit due to Loss Reduction*: One of the aims of optimal DG allocation is the reduction of energy loss which is ultimately revenue gain to power distributor. Maximization of benefits takes place. This is evaluated as under

Benefit due to Loss Reduction

$$BN_2 = \sum_{j=1}^{n} \Delta LRj * PPg * \Delta T, \tag{14}$$

where, ΔLRj is Loss Reduction in real power at jth DG, PPg Power Purchase price per KWH in open market

$$Net\ Benefit = BN_1 + BN_2 \tag{15}$$

Case study

The proposed methodology is tested on 34-bus test Radial Distribution System [18]. It has a totally real and reactive load of 4.636 MW and 2.873 MVAr, respectively. Installation of 3 MW DG at bus number 8 will result in maximum loss reduction, best voltage profile at the lowest cost of generation. Without DG the system loss is 221.72 kW while after installation of DG losses reduce to 167 kW. Figure 1 shows that voltages of every bus are enhanced after integration of DG that is the VPI is improved. Figures 2 and 3 demonstrate the capability of DG and effectiveness of developed algorithm to reduce the active loss as well as reactive loss in power distribution network.

Table 2 demonstrated the effectiveness of the developed method. The selected size and location of DG is the best location because it improves the efficiency, voltage profile, and profit.

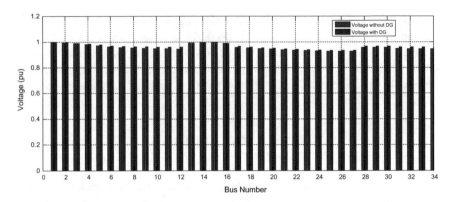

Fig. 1 Voltage profile of IEEE 34-bus system

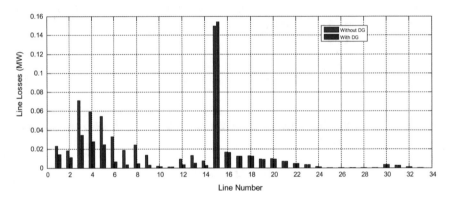

Fig. 2 Active losses in lines of IEEE 34-bus

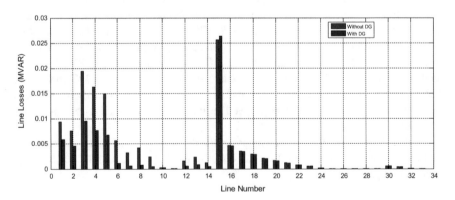

Fig. 3 Reactive losses of lines of IEEE 34-bus

Table 2 Effectiveness of developed algorithm on IEEE 34-bus RDS

Technique	Objective	Size of DG (MW)	Location	Losses (MW)	NEB/Yr (INR)
w/o DG	–	–	–	0.22172	–
PSO	Cost of generation, PLI, and VPI	3	8	0.167	82.96×10^6
PSO [18]	Total losses	0.1996	21	0.20398	11.42×10^6
ILC [19]	Total losses	2.9676	24	0.9374	105.74×10^6

5 Conclusion

In this paper, authors have applied PSO to allocate DG of optimum size DG in power distribution network. It is observed that multiple objectives have been achieved like voltage stability and minimum power loss with the lowest investment for DG installation. The analysis was carried out on 34-bus power distribution network by installing DG in the 8th bus and achieving the reduction in power loss and enhancement of bus voltage profile and profit. The initial investment is recovered very easily considering 15 years as the useful life of DG. Even though tax holiday in some production zone are available which make the proposal further lucrative and benefit of the liberalization can be passed on to consumers. Lifespan will further improve due to less line thermal because of optimum loading of all circuits.

References

1. D. Singh, D. Singh, K. Verma, GA based energy loss minimization approach for optimal sizing & placement of distributed generation. Int. J. Knowl. Intell. Eng. Syst. **12**, 147–156 (2008). https://doi.org/10.3233/kes-2008-12206
2. F.S. Abu-Mouti, M.E. El-Hawary, Heuristic curve-fitted technique for distributed generation optimization in radial distribution feeder systems. Proc. IET Gener. Trans. Distrib. **5**(2), 172–180 (2011). https://doi.org/10.1049/iet-gtd.2009.0739
3. W. Prommee, W. Ongsakul, Optimal multiple distributed generation placement in microgrid system by improved reinitialized social structures particle swarm optimization. Eur. Trans. Electr. Power **21**(1), 489–504 (2011). https://doi.org/10.1002/etep.456
4. F.S. Abu-Mouti, M.E. El-Hawary, Optimal distributed generation allocation and sizing in distribution systems via artificial bee colony algorithm. IEEE Trans. Power Deliv. **26**, 2090–2101 (2010). https://doi.org/10.1109/TPWRD.2011.2158246
5. M. Mohammadi, S.H. Hosseinian, G.B. Gharehpetian, GA-based optimal sizing of micro grid and DG units under pool and hybrid electricity markets. Elect. Power Energy Syst. **35**, 83–92 (2012). https://doi.org/10.1016/j.ijepes.2011.09.015
6. R.K. Singh, S.K. Goswami, Optimal allocation of distributed generations based on nodal pricing for profit, loss reduction and voltage improvement including voltage rise issue. Elect. Power Syst. Res. **36**, 637–644 (2010). https://doi.org/10.1016/j.ijepes.2009.11.021

7. N. Khalesi, N. Rezaei, M.R. Haghifam, DG allocation with application of dynamic programming for loss reduction and reliability improvement. Elect. Power Energy Syst. **33**, 288–295 (2011). https://doi.org/10.1016/j.ijepes.2010.08.024
8. S. Kansal, R. Tyagi, V. Kumar, Cost-Benefit analysis for Optimal DG placement in distribution system. Int. J. Ambient. Eng. (2015). https://doi.org/10.1080/01430750.2015.1031407
9. K. Akbari, E. Rahmani, A. Abbasi, M. Askari, Optimal placement of distributed generation in radial networks considering reliability and cost indices. J. Intell. Fuzzy Syst. **30**(2), 1077–1086 (2016). https://doi.org/10.3233/IFS-151883
10. S. Kansal, V. Kumar, B. Tyagi, Optimal placement of different types of DGs in distribution networks. Elect. Power Energy Syst. **53**, 752–760 (2013). https://doi.org/10.1016/j.ijepes.201 3.05.040
11. Y.J. Kim, J.L. Kirtley, L.K. Norford, Reactive power ancillary coordination with voltage control devices. IEEE Trans. Smart Grid **8**(2), 515–527 (2017). https://doi.org/10.1109/tsg.2015.247 2967
12. R. Hasanpour, B.M. Kalesar, J.B. Noshahr, P. Farhadi, Reconfiguration of smart distribution network considering variation of load and local renewable generation, in *2017 EEEIC/I&CPS Europe*, Milan, 2017, pp. 1–5. https://doi.org/10.1109/eeeic.2017.7977556
13. S. Das, D. Das, A. Patra, Reconfiguration of distribution networks with optimal placement of distributed generations in the presence of remote voltage controlled bus. Renew. Sustain. Energy Rev. **73**, 772–781 (2017). https://doi.org/10.1016/j.rser.2017.01.055
14. R. Khorshidi, F. Shabaninia, State estimation in smart distribution grids considering renewable energy sources. J. Int. Fuzzy Syst. **28**(5), 2169–2178 (2015). https://doi.org/10.3233/IFS-141 499
15. V.S. Bhadoria, N.S. Pal, V. Shrivastava, S.P. Jaiswal, Reliability improvement of distribution system by optimal sitting and sizing of disperse generation. Int. J. Reliab. Qual. Saf. Eng. **24**(6), 1740006 (11 pages) (2017).https://doi.org/10.1142/S021853931740006X
16. S.P. Jaiswal, V. Shrivastava, Allocation of UPFC in distribution system to minimize the losses. IEEE International Conference on Innovative Applications of Computational Intelligence on Power, Energy and Controls with their Impact on Humanity (2014). https://doi.org/10.1109/C IPECH.2014.7018212
17. V.S.Bhadoria, N.S. Pal, V. Shrivastava, *Comparison of Analytical and Heuristic Techniques for Multi-objective Optimization in Power System,* Chapter 13, (IGI Global, Hershey, PA, 2016). https://doi.org/10.4018/978-1-4666-9885
18. S. Devi, M. Geethanjali, Optimal location and sizing determination of distributed generation and DSTATCOM using particle swarm optimization algorithm. Int. J. Electr. Power Energy Syst. **62**, 562–570 (2014). https://doi.org/10.1016/j.ijepes.2014.05.015
19. A.N. Mehdi, A. Ahmadreza, M.S. Zahra, R. Mehdi, DG allocation in distribution network using ILC method. vol. 8. World Appl. Sci. J. 1459–1464 (2011)

IDMA-Based on Chaotic Tent Map Interleaver

Divya Singh and Aasheesh Shukla

Abstract Interleave Division Multiple Access (IDMA) is a popular multiple access scheme and can be considered as one of the powerful NOMA (non-orthogonal multiple access) technique for 5G communication requirements. In IDMA, the interleaver plays a vital role as the means to distinguish the different users in the channel. An "efficient" Interleaver in IDMA scheme must be simple and non-colloidal in nature. So, in this paper, the analysis and designing of a new Tent map based Interleaver (TMI) matrix is proposed for IDMA system. Further, the simulation experiments have been performed among TMI and other famous interleaver designs to validate the bit error rate performance of "TMI". The simulation results explain that Tent map based chaotic (TMI) interleaver performs better in terms of bit error rate and computational complexity.

Keywords CDMA · IDMA · MAI · Chaos theory · TMI

1 Introduction

High speed, bandwidth efficient, and secure wireless communication are few among many, required features for 5G communication system. IDMA can be considered as a powerful code domain NOMA technique and a better candidate for 5G communication. IDMA scheme is also a better alternative for CDMA scheme as it inherits all the advantages of CDMA and also overcome the limitations such as: multiple access interference (MAI), and intersymbol interference (ISI). In IDMA systems the interleavers are used to distinguish the different users and bandwidth expansion is exploited with the help of low rate codes [1]. Further this scheme is also one of the popular multiuser detection techniques.

D. Singh (✉) · A. Shukla
Department of Electronics & Communication, GLA University, Mathura, India
e-mail: divya.rti@gmail.com

A. Shukla
e-mail: aasheesh.shukla@gla.ac.in

© Springer Nature Singapore Pte Ltd. 2019
H. Malik et al. (eds.), *Applications of Artificial Intelligence Techniques in Engineering*, Advances in Intelligent Systems and Computing 698,
https://doi.org/10.1007/978-981-13-1819-1_12

Interleavers are used to spread the information and to protect these bits from error bursts due to multipath propagation and noise sources. Hence, efficient interleaver can increase the throughput of iterative MUD receivers [2]. Interleavers should be easy to generate and not to consume large bandwidth and memory. Interleavers should also be non-colloidal in nature [3, 4]. Many Interleavers were studied and proposed in the literature. Random Interleaver (RI), Nested Interleaver (NI) [3], Shifting Interleavers (SI) [5], Tree-based Interleaver and Helical Interleaver are some popular interleaving patterns [4].

On the other hand, role of chaos in wireless communication has been widely explored since 90s. So, in this paper, a new Tent map based chaotic interleaving matrix is proposed and analyzed in IDMA communication system. The chaotic sequences are random like sequences and the interleavers based on chaos theory offers several advantages over conventional schemes. Such as: (1) Interleavers are non-colloidal in nature, i.e., orthogonal; (2) Computational cost and complexity is very low; (3) Less memory requirement, and (4) Bandwidth efficient [6].

The content of the paper is organized as follows. In Sect. 2, the introduction and role of chaos in communication is discussed. Section 3, defines the IDMA system and algorithms for interleaver generation. Section 4, develops the performance analysis of chaos based IDMA schemes and finally Sect. 5 conclude the paper.

2   Chaos in Communication

Chaos is a very widely spread concept and used in many areas of science to understand the physical behavior of system. The role of chaos in spread spectrum communication has been explored in late 90s. Many researchers proposed in literature that the application chaos theory in spread spectrum communication offers several advantages. The noise like behavior, deterministic nature and dependence on initial conditions are some important and key features which helps in the generation of secure signals or sequences [7, 8]. In the literature there are many chaotic maps available and on the basis of application any chaotic map among Henon map, Tent map or logistic map can be chosen to generate different chaotic sequences. The popular maps are presented in Table 1 along with their state equations.

For our interleaver design we have chosen Tent Map because signals generated by this map have different behaviors but easily summerizable and it is also piecewise linear so, have uniform invariant density and hence allow analytical calculations.

2.1   Tent Map

This is well known that chaotic sequences perform equivalently to random sequences due to its noise like structure. The chaotic sequences express some of the key features such that; these sequences exhibits deterministic behavior and highly dependent on

Table 1 Popular chaotic maps

S. No	Chaotic map	Equation
1	Logistic map	$X_{n+1} = AX_n(1 - X_n)$
2	Tent map	$F_m(x) = \begin{cases} mx_n, & if\ 0 \leq x \leq \frac{1}{2} \\ m(1 - x_n), & if\ 1/2 \leq x \leq 1 \end{cases}$
3	Baker map	$F(x, y) = \begin{cases} (2x, y/2) & 0 < x < 1/2 \\ ((2x - 1), (y + 1/2)) & 1/2 < x < 1 \end{cases}$
4	Henon map	$X_{n+1} = 1 - aX_n^2 + Y_n$ $Y_{n+1} = bX_n$

the initial conditions used for the generation of sequences. Among many available chaotic maps, the Tent map is suitably good for the generation of chaotic sequences and so interleaving sequences. The Tent map can described in the parameterized form (also piecewise) by:

$$F_m(x) = \begin{cases} mx_n, & if\ 0 \leq x \leq \frac{1}{2} \\ m(1 - x_n), & if\ 1/2 \leq x \leq 1 \end{cases} \tag{1}$$

Equation (1) shows the behavior of Tent map, the values $x \in [0, 1]$ and x_0 is the initial value parameter. Bifurcation parameter is represented as "m" which decides the behavior of system and clearly depicts the region of chaos, which gives the parameters for the generation of chaotic sequences. Generally for the Tent map the parameter belongs to $m \in (1, 2]$. With the help of Eq. (1), the Lyapunov exponent can also be calculated to determine the chaotic region of Tent map. Further the value of Lyapunov exponent for the above-said system states can be calculated and written as

$$\lambda = \lim_{n \to \infty} \left\{ \frac{1}{n} \sum_{j=0}^{n-1} \ln|f'(x)| \right\} \tag{2}$$

So after calculation if the value of L.E. (i.e., $\lambda > 0$) is positive then the system exhibits the chaotic behavior. Simply the first derivative of system presented in Eq. (1) can be written as $F'(x) = \begin{cases} m, & if\ 0 \leq x \leq 0.5 \\ -m, & if\ 0.5 \leq x \leq 1 \end{cases}$. Further according to the relation in Eq. (2) the L.E. may calculated as $\lambda = \lim_{n \to \infty} \left\{ \frac{1}{n} \sum_{j=0}^{n-1} \ln|\pm m| \right\}$, which clearly shows that L.E. cannot be simplified t=negative and hence Tent map can exhibit chaotic behavior.

3 System Description

Figure 1 shows the IDMA transmitter and receiver structure with K simultaneous users. The length of transmitted sequence is taken as n and the data vector for n-length and k users can be written as: $d_k = [d_{k1}, d_{k2}, \ldots, d_{ki}, \ldots, d_{kn}]$. The data is further spreaded and encoded chips $c_k = [c_{k1}, c_{k2}, \ldots, c_{kj}]$, where j is the Chip length. In the next block, chips are interleaved by a Tent map based interleaving pattern to produce transmitted chip sequence $x_k = [x_{k1}, x_{k2}, \ldots, x_{kj}]$, which sent to Rayleigh fading wireless channel [1, 9].

After passing from channel filter, the received signal at elementary signal estimator can be written as

$$R_j = \sum_{k=1}^{k} h_k x_{kj} + \phi_j, \tag{3}$$

where channel coefficient is h_k for kth user and for the samples of fading $\{\phi_j\}$.

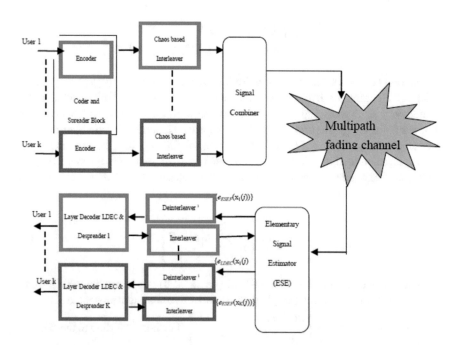

Fig. 1 IDMA communication block for k simultaneous users

3.1 Algorithm: TENT Map Interleaver

As mentioned above the system based on logistic map can be proposed as $F_m(x) = \begin{cases} mx_n, & if\ 0 \le x \le \frac{1}{2} \\ m(1 - x_n), & if\ 1/2 \le x \le 1 \end{cases}$. The first value or initial state of the system starts from X_0 and then the sequence of states may be written as $\{X_1, X_2, \ldots, X_N\}$. All these states toggle between 0 & 1 and N is the length of Interleaver. The bifurcation parameter is denoted by "m".

Tent map algorithm

---

1. Define the required parameters: $1 < m < 2$, chip length $= n$, $l =$ no. of users
2. $X_j^l = l$th $user: 0 < X_j^l < n$, $\xi =$ Foot step, $G_0^k = \lceil X_0^k \rceil$
3. First user interleaver ($\pi^k \equiv G_0^k$), j and n $= 0$
4. If n $<$ N
5. Find X_{j+1}^k, $G_{j+1}^k \equiv \lceil X_{j+1}^k \rceil$
6. If G_{j+1}^k is in the set π^k Increment j by 1 and repeat the main operation
7. Otherwise: $\pi^k \equiv \pi^k \cup F_{j+1}^k$
8. If n $>$ N:
9. $X_{j+1}^k = []$
10. $\Pi^k \equiv \Pi^k \cup \lceil X_{j+1}^k \rceil$
11. End

---

4 Simulation Results Analysis Tent Map Based IDMA

4.1 BER Analysis

In this section, bit error rate of Tent map based interleaver in IDMA system is presented. For simulation experiment in MATLAB, the BPSK modulation is used. The data length is taken as 1024 and 2048 bits and the spreading code length is of 16 chips. The simulation result is shown in Fig. 2 and it clearly depicts that TMI interleaver outperforms the random interleaver for higher values of Eb/N$_0$ [4].

In Fig. 3, a simulation result depicts the performance comparison among RI and logistic map interleaver. With the same simulation parameters (experiment 1) the simulation results shows that TMI performs better among RI and TBI interleaver.

Fig. 2 BER comparison of
TMI and RI

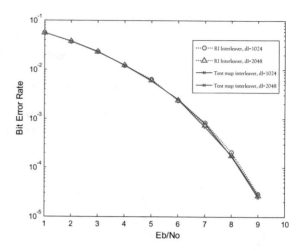

Fig. 3 BER comparison of
TMI, logistic map interleaver
and RI

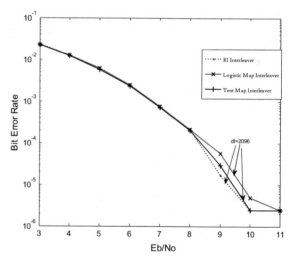

5 Computational Complexity

In the section computational complexity analysis is presented for TMI algorithm.
Here complexity means the number of cycles needed to generate the interleaving
sequences [9, 10]. Table 2 shows that TMI algorithm is most computationally efficient
among the comparisons of orthogonal, Nested and Tree-based interleaver (TBI) of
user k. In fact, the complexity of TMI is O(1) which indicates that computational
complexity of TMI is independent from the number of users.

In Table 3, the order of computational complexity is provided of different algorithms used for interleaver design.

Table 2 Computational complexity of different interleaver algorithm

No. of users	OI	Nested interleaver	TBI	TMI
1	1	1	1	1
2	2	2	1	1
4	4	4	2	1
16	16	16	4	1
50	50	50	5	1

Table 3 Computational complexity for the generation of first interleaver

	OI	Nested interleaver	TBI	TMI
Complexity	$O(n) + O(n^2)$	$O(\log_2(n))$	$O(n^2)$	$O(n^2)$

6 Conclusion

This paper concludes to the new design to generate multiple interleavers regardless the length and number of simultaneous users. This new design is based on chaos theory and use the Tent map and hence named as "Tent map interleaver" (TMI). The simulation results show that the TMI can perform better or same in comparison to other popular interleavers as well as TMI also offers other advantages such as small requirement of memory, implementation complexity and easy to generate.

References

1. L. Ping, L. Liu, K. Wu, W.K. Leung, Interleave-division multiple-access. IEEE Trans. Wirel. Commun. **5**(4), 938–947 (2006)
2. A. Shukla, V.K. Deolia, Performance analysis of modified tent map interleaver in IDMA systems. J. Electr. Eng. (2017)
3. I. Pupeza, A. Kavcic, L. Ping, Efficient generation of interleavers for IDMA, in *IEEE International Conference on Communications, ICC 2006*, vol. 4 (2006), pp. 1508–1513
4. C. Bai et al., Chaos-based underwater communication with arbitrary transducers and bandwidth. Appl. Sci. **8**(2) 2018
5. F.J. Escribano, A. Wagemakers, M.A.F. Sanjuan, Chaos-based turbo systems in fading channels. IEEE Trans. Circuits Syst. **61**(2), 530–541 (2014)
6. A. Shukla, V.K. Deolia, Performance analysis of chaos based interleaver in IDMA system. ICTACT J. Commun. Technol. **7**(4) (2016)
7. B. Akbil, G. Ferre, D. Aboutajdine, Computational complexity and bandwidth resource analysis of NLM interleaver in IDMA system, in *Image Processing and Communications Challenges 4* (Springer, Berlin, Heidelberg, 2013), pp. 241–251
8. A. Shukla, V.K. Deolia, Performance improvement Of IDMA scheme using chaotic map interleavers for future radio communication. ICTACT J. Commun. Technol. **8**(2) (2017)
9. L. Shan, H. Qiang, J. Li, Z. Wang, Chaotic optimization algorithm based on tent map. Control Decis. **20**(2), 179–182 (2005)
10. B. Akbil, G. Ferre, The NLM interleaver design for IDMA system, in *Proceedings of IEEE International Conference on Complex System* (2012), pp. 1–5

Energy Efficient Resource Allocation for LSA Array Based 5G Wireless Communication Systems Using Convex Optimization Techniques

Javaid A. Sheikh, Arshid Iqbal Khan, Mehboob ul-Amin
and G. MohiuddinBhat

Abstract In this paper we consider the problem of sum rate maximization for higher order Multiple Input Multiple Output (MIMO) networks with and without Zero Forcing (ZF) technique. The aim is to formulate the user dependent MIMO channel matrix to maximize the achievable communication sum rate. We devise a method based on convex optimization to obtain a quality to the existing non-convex problem by applying the negative sum rate of the non-convex optimization. Furthermore, the KKT condition is applied to obtain the optimal power allocation channel coefficients. Simulations are performed for 4×4, 8×8, 16×16 and 32×32 MIMO systems. Simulation results prove that ZF technique performs better at higher power levels.

Keywords Large scale antenna array (LSA) · Resource allocation
Convex-optimization · KKT · Zero forcing · Lagrangian function

1 Introduction

The Fifth Generation (5G) of wireless mobile communication systems is a technology with a vision, the vision of providing the highest Quality of Service (QOS) with lowest possible end to end latency [1]. The 5G wireless communication systems will have to feed an unprecedented number of mobile devices, expected to reach 50 billion by the year 2020, satisfying their high volume needs of data and voice traffic [1]. To support such a massive number of connected smart devices and their data needs,

J. A. Sheikh · A. I. Khan (✉) · M. ul-Amin · G. MohiuddinBhat
Department of Electronics and Instrumentation, University of Kashmir, Srinagar, India
e-mail: Khnarshid20@gmail.com

J. A. Sheikh
e-mail: sjavaid_29ku@gmail.com

M. ul-Amin
e-mail: mehboo1197@gmail.com

G. MohiuddinBhat
e-mail: drgmbhat@gmail.com

© Springer Nature Singapore Pte Ltd. 2019
H. Malik et al. (eds.), *Applications of Artificial Intelligence Techniques in Engineering*, Advances in Intelligent Systems and Computing 698,
https://doi.org/10.1007/978-981-13-1819-1_13

the 5G wireless systems are expected to provide a 1000 times increase in capacity enhancement with the same power and spectrum resources currently existing in the performing networks [2]. A close proximity insight into the 5G capabilities and its performance indicators reveals that the current available spectrum resources are insufficient to provide the promised user experienced data rates, connection density, end to end latency, mobility and peak data rates [3] Due to inadequate availability of spectrum, it is recommended to design resource allocation algorithms employing reuse of the spectral resource by multiple communication links [1, 2]. The deployment of 5G communication systems employing the current network resources rely on highly efficient technologies like massive multiple input multiple output (MIMO), cognitive radio (CR) systems, device-to-device communication and low budget link based cooperative and relay communications [3–5].

Massive MIMO, cooperative communication, Cognitive Radio (CR) and compressive sensing are the driving technologies for 5G wireless communication systems to achieve enormously high data rates in order to serve billions of network connected devices and to provide new services like e-Health, Banking, Learning, Vehicle to Vehicle communications (V2V), Device to Device (D2D) [6]. Such explosive data rates cannot be achieved by simply amplifying the transmitter powers, instead such enormous data rates will have to be achieved utilizing the resources of the present networks at a low carrier frequency. It is hence globally accepted fact that the resource allocation (bit/joule of energy) is a central design issue for 5G wirless systems [6, 7]. Radio resource optimization is a key performance indicator in 5G wireless communication systems. Optimization aims at obtaining the maximum value of a parameter subject to certain real time constraints. Convex optimization technique plays a vital role in obtaining optimal parameter values for enhanced energy efficiency, overall cost per bit, optimal power allocation, transmitter capacity by minimizing the negative sum rate. The resource allocation problems are either concave optimization problems or quasi-concave optimization problems and hence do not guarantee optimal solution to the problem. However, by considering the negative sum rate of a concave or quasi concave optimization problem, a less complex closely approaching optimal solution can be obtained. The rest of paper is organized as: Sect. 2 presents related work. The problem formulation is presents in Sect. 3 and its solution in Sect. 4. The System model for rate optimization is presented in Sect. 5. Results are discussed in Sect. 6 followed by concluding remarks in Sect. 7.

2 Related Work

Several works have already been carried out on the efficient radio resource allocation schemes for 5G wireless communication systems. In [2], Xin Liu, Yanan Liu, discussed the non-orthogonal multiple access (NOMA), multiple input and multiple output and relaying technologies to attain much higher data throughput and improved spectral efficiency without the requirement for increased bandwidth and redundant base stations. In [3], the authors presented cooperative and coordinated multi-cell resource allocation methods for 5G ultra reliable low latency connection, considering the typical indoor environment. Moreover [8–10], presents various other resource allocation schemes for 5G networks with application to device to device (D2D) and machine to machine (M2 M) communications. Efficient resource allocation for MIMO and OFDM in 5G is a non-convex optimization problem [11]. A less complex closely approaching optimal solution for such problems can be obtained by considering the negative sum rate of the optimization objective function using the convex optimization routine in MATLAB.

3 Problem Description

Massive MIMO, cooperative communication, etc., are promising technologies for the 5G wireless communication systems. In massive MIMO, a large array of high directive/gain antennas are employed at both the base station and the mobile station multiplexed spatially for highly directed beam-forming to efficiently allocate the channel resources and to support the spectrum reuse. For MIMO- and OFDM-based wireless system architectures, the resource allocation scheme is a non-convex optimization problem and hence do not guarantee the optimal solution of the problem. In this paper we introduce a duality counterpart for obtaining the solution of the non-convex optimization problems by minimizing the negative sum rate of the objective function. The convex optimization techniques form the basis for the efficient resource allocation like energy efficiency, spectrum reuse, bit error rates, etc., the optimal solutions for all the convex optimization problems related to judicious allocation of network resources can be obtained by minimizing the objective function or cost function subject to certain real-time constraints (inequality or equality constraints).

4 Problem Solution

For optimum resource allocation in MIMO and multicarrier (OFDM, FBMC) based 5G wireless communication systems, we formulate mathematically and analytically the various optimization parameters like MIMO rate optimization [12, 13], OFDM rate optimization, optimal MIMO-OFDM power allocation and optimization problems related to effect of multiple antennas in cooperative communication. In this section we first introduce the concept of typical convex optimization problem and then we will analyse various optimization parameters using the convex optimization approach and LaGrange's function. Note that the number of Lagrangian coefficients is always equal to number of the constraints.

4.1 Convex Function and Convex Optimization

If the domain of the optimization function $g(x)$ is a convex set, i.e., x, $y \in$ dom g, the function $g(x)$ is a convex function satisfying the following inequality:

$$g(\alpha x + (1 - \alpha)y) \leq \alpha g(x) + (1 - \alpha)g(y); 0 \leq \alpha \leq 1 \qquad (1)$$

Any convex optimization problem will have the form

$$\text{Minimize } g_0(x)$$
$$\text{Subject to } g_i(x) \leq 0, \ i = 1, 2, \ldots, m \qquad (2)$$
$$h_i(x) = 0, \qquad i = 1, 2, \ldots, n$$

This describes a convex optimization problem for finding the values of the variable x that minimizes $g_0(x) \forall x$ satisfying the constraints $g_i(x) \leq 0$, $i = 1, 2, \ldots, m$ and $h_i(x) = 0$, $i = 1, 2, \ldots, n$. The variable $x \in R^n$ is termed as the optimization variable and the function $g_0 : R^n \rightarrow R$ the optimization function or cost function. The inequalities $g_i(x) \leq 0$ are inequality constraints corresponding to inequality constraints functions $g_i : R^n \rightarrow R$. The equality constraint $h_i(x) = 0$ corresponds to the equality constraint functions $h_i : R^n \rightarrow R$. If there are no constraints, the optimization problem is called unconstrained.

Graphically a convex function represents a chord passing through two points $(x, g(x))$ and $(y, g(y))$ from x to y. An optimization function g is strictly convex, if strict inequality holds i.e. whenever $x \neq y$ and $0 \leq \alpha \leq 1$. If g is concave then $-g$ is convex and g is strictly concave if $-g$ is strictly convex. This is a generalized fact that an optimization functions like a MIMO rate optimization function, OFDM rate optimization function, optimal MIMO-OFDM power allocation functions and optimization problems related to effect of multiple antennas in cooperative communication are concave optimization problems but the negative sum rate of these functions are convex functions. The convex optimization techniques can now be employed to obtain the optimal values of the said functions.

5 System Model for MIMO Rate Optimization

In this section, we will consider a standard MIMO system consisting of 't' transmit antennas at the base station side and 'r' decentralized receive antennas. The MIMO channel can be equivalently modeled as:

$$\bar{Y} = H\bar{X} + \bar{N}. \qquad (3)$$

where $\bar{Y} = [y_1 y_2 y_3 \ldots y_r]$ is the 'r' dimensional receive vector at the MIMO receiver, $\bar{X} = [x_1 x_2 x_3 \ldots x_t]$ is a 't' dimensional transmit vector with each symbol transmitted through each transmit antenna [14]. $\bar{H} = [h_{11}h_{12}h_{13}h_{31}\ldots h_{rt}]$ is the $r \times t$ channel coefficient vector and $\bar{N} = [n_1n_2n_3\ldots n_r]$ is the 'r' dimensional noise vector. The subscripts to the parameters y, x, h, n corresponds to the antenna numbers at transmit and receive sides of the MIMO channel.

The MIMO system introduced represents the parallelization of the MIMO channel with 't' symbols transmitted in parallel and spatially multiplexed. The signal power received at the receiver corresponding to each MIMO channel is given as

$$\sigma_i^2 \left\{ E |\bar{X}|^2 \right\} \tag{4}$$

where σ_i represents the singular values of the channel coefficient matrix $\bar{H}$ [15] of the MIMO channel. The SVD of $\bar{H}$ is given below

$$H = U \sum V^H, \tag{5}$$

where the matrices U, $\sum$ V are $r \times t$, $t \times t$ and $t \times t$ dimensional respectively [16]

The noise power received at the receiver corresponding to each MIMO channel is given by σ_n^2 computed as the value of the covariance of the noise matrix. Therefore the signal to noise ratio at the input of the receiver is given as

$$SNR = \frac{\sigma_i^2 \left\{ E |\bar{X}|^2 \right\}}{\sigma_n^2} \tag{6}$$

From the above SNR expression for the ith channel, the Shannon capacity C_i of the channel can be derived as given below

$$C_i = \log_2 \left(1 + \frac{P_i \sigma_i^2}{\sigma_n^2} \right) \tag{7}$$

The optimal MIMO power allocation problem can now be formulated as

$$\text{maximize.} \sum_{i=1}^{t} \log_2 \left(1 + \frac{P_i \sigma_i^2}{\sigma_n^2} \right) \tag{8}$$

$$subject\ to \sum_{i=1}^{t} P_i \leq P, \tag{9}$$

where P is the total transmit power.

The above optimization problem with the given inequality constraint is a non-convex optimization problem and hence the convex optimization techniques cannot be applied directly to obtain the optimal solution for the MIMO power allocation problem. A non-convex optimization problem is transformed into a convex optimization problem by taking the negative sum rate of the non-convex optimization expression. The optimal MIMO power allocation problem can further be modified and formulated as a convex optimization problem by taking the negative sum rate of. $\sum_{i=1}^{t} \log_2\left(1 + \frac{P_i \sigma_i^2}{\sigma_n^2}\right)$ as under

$$\text{Minimize} \ -\sum_{i=1}^{t} \log_2\left(1 + \frac{P_i \sigma_i^2}{\sigma_n^2}\right) \tag{10}$$

$$\text{subject to} \sum_{i=1}^{t} P_i \leq P \tag{11}$$

To solve the above convex optimization problem a series of steps are followed as under:

Step1: *Finding the Lagrangian cost function* $f\left(\bar{P}, \mu\right)$

It is important to note that the number of Lagrangian multiples is equal to the number of constraints- inequality of equality constraints. The Lagrangian cost function $f\left(\bar{P}, \mu\right)$ for the given optimization problem can be formulated as under

$$f\left(\bar{P}, \mu\right) = \sum_{i=1}^{t} \log_2\left(1 + \frac{P_i \sigma_i^2}{\sigma_n^2}\right) + \mu\left(P - \sum_{i=1}^{t} P_i\right) \tag{12}$$

Step2: *Finding the maxima of the Lagrangian cost function.*

Differentiating the above obtained Lagrangian cost function $f\left(\bar{P}, \mu\right)$ with respect to power associated with the ith MIMO channel P_i and setting the result equal to 0, we get

$$\frac{\partial}{\partial x} f\left(\bar{P}, \mu\right) = 0 \tag{13}$$

$$\Rightarrow \frac{\frac{\sigma_i^2}{\sigma_n^2}}{1 + \frac{P_i \sigma_i^2}{\sigma_n^2}} - \mu = 0 \tag{14}$$

Step3: *Finding the optimal* P_i *using KKT conditions*

The kurush Kuhn tucker (KKT) conditions states that if $\frac{\partial}{\partial x} f\left(\bar{P}, \mu\right) = 0$ then there exist local minima P^* for a unique value of the Lagrangian multiple μ as μ^* subject to $\mu^* \geq 0$. Solving the above differential equation $\frac{\partial}{\partial x} f\left(\bar{P}, \mu\right) = 0$ yields

$$\frac{\frac{\sigma_i^2}{\sigma_n^2}}{1 + \frac{P_i \sigma_i^2}{\sigma_n^2}} - \mu = 0 \tag{15}$$

$$\Rightarrow \frac{\sigma_i^2}{\sigma_n^2} \frac{1}{\mu} = 1 + \frac{P_i \sigma_i^2}{\sigma_n^2} \tag{16}$$

$$\Rightarrow P_i = \left(\frac{1}{\mu} - \frac{\sigma_n^2}{\sigma_i^2}\right)^+ \tag{17}$$

P_i Represents the power associated with the ith MIMO channel, the function $\left(\frac{1}{\mu} - \frac{\sigma_n^2}{\sigma_i^2}\right)^+$ always accounts for the positive value of the channel powers, because channel powers cannot be negative. $P_i = \left(\frac{1}{\mu} - \frac{\sigma_n^2}{\sigma_i^2}\right)^+$ is positive if $\frac{1}{\mu} \geq \frac{\sigma_n^2}{\sigma_i^2}$ and 0 otherwise. P_i Can now be formulated as a piecewise optimization function as under

$$P_i = \begin{cases} \left(\frac{1}{\mu} - \frac{\sigma_n^2}{\sigma_i^2}\right)^+, & x \geq 0 \\ 0, & x < 0 \end{cases} \tag{18}$$

Step4: *Finding the Lagrangian multiplier 'μ' for optimal P_i.*

The power allocated to the ith MIMO channel is directly dependent on user defined function σ_i^2. Increasing the σ_i will increase the power allocated to the ith MIMO channel coefficient. Thus the resulting power allocation will result in the water filling phenomenon subject to the constraint $\sum_{i=1}^{t} P_i \leq P$. Employing the same constraint will yield the value of 'μ' for optimal power allocation to the ith MIMO channel.

$$P_i = \left(\frac{1}{\mu} - \frac{\sigma_n^2}{\sigma_i^2}\right), \forall \frac{1}{\mu} \geq \frac{\sigma_n^2}{\sigma_i^2} \tag{19}$$

Solving the above expression for 'μ' we get

$$\mu = \frac{\sigma_i^2}{P_i \sigma_i^2 + \sigma_n^2} \tag{20}$$

For optimal P_i, the Lagrangian multiplier 'μ' should be minimum subject to the condition $\sum_{i=1}^{t} P_i \sigma_i \leq P$.

6 Results and Discussions

The figures from 1–4 represents the capacity (Mbps) associated with the ith MIMO channel as a function of the corresponding power of the same channel- with Zero-Forcing (ZF) and Without Zero-forcing.

Figure 1 shows capacity (Mbps) versus Power (dB) graph with and without ZF technique for 4 × 4 MIMO system. The figure clearly reveals that with ZF technique capacity increases with the increase in power, resulting in water filling phenomenon. It is observed that as the power increases from 2 to 30 dB capacity increases to 22 Mbps. Thus there is the overall improvement of 50% in capacity.

Figure 2 shows capacity (Mbps) versus Power (dB) graph with and without ZF technique for 8 × 8 MIMO system. The figure reveals that at low power (<5 db) there is no improvement in capacity using ZF technique. However after 5 dB power level ZF technique shows an improvement in capacity enhancement, again resulting in

Fig. 1 Capacity (Mbps) versus power (dB) for 4 × 4 MIMO systems

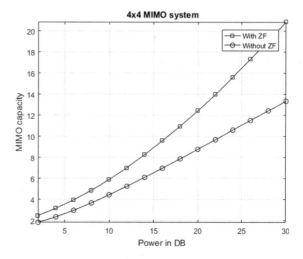

Fig. 2 Capacity (Mbps) versus power (dB) for 8 × 8 MIMO systems

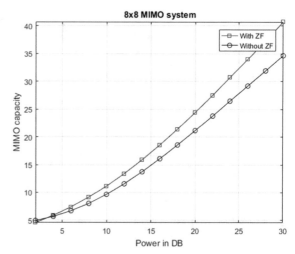

water filling phenomenon. At 30 dB there is an overall increase in 17% as compared to non-ZF.

Figure 3 shows capacity (Mbps) versus Power (dB) graph with and without ZF technique for 16 × 16 MIMO system. The Figure reveals that at low values of power the performances of ZF technique deteriorates as compared to non-ZF technique. After 15 db power level ZF technique again performs better than non-ZF technique. Thus massive MIMO systems will perform better at higher power levels where SINR values will be optimal. This is again shown in Fig. 4 where we again plot capacity vs. power for 32 × 32 MIMO system. ZF technique performs better after 22 dB power levels.

Fig. 3 Capacity (Mbps) versus power (dB) for 16 × 16 MIMO systems

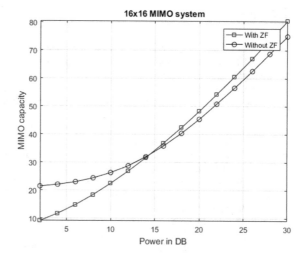

Fig. 4 Capacity (Mbps) versus power (dB) for 32 × 32 MIMO systems

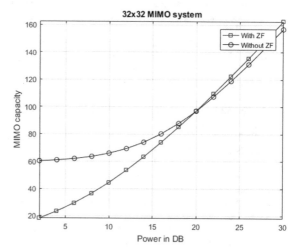

7 Conclusion

In this paper number of iterations has been done to evaluate the performance of proposed ZF technique using convex optimization method. The iteration method was based on applying negative sum rate of non-convex optimization. The sum rate maximization of proposed ZF technique is dealt via CVX toolbox. The optimal values for power allocation coefficient and Lagrangian coefficient obtained in the proposed technique result in water filling phenomenon as depicted from our plots. The proposed ZF technique performs better as compared to non-ZF at higher power levels for higher MIMO order configurations due to lesser interference and noise limited environment.

References

1. R. Tang, J. Zhao, H. Qu, Z. Zhang, User-centric joint admission control and resource allocation for 5G D2D extreme mobile broadband: a sequential convex programming approach. IEEE Commun. Lett. **21**(7), (2017)
2. L. Dai, B. Wang, Y. Yuan, S. Han, C. I, Z. Wang, Non-orthogonal multiple access for 5G: solutions, challenges, opportunities, and future research trends. IEEE Commun. Mag. **53**(9), 74–81 (2015)
3. Z. Ding, Z. Yang, P. Fan, H.V. Poor, On the systems with randomly deployed users. IEEE Sig. Process. Lett. **21**(12), 1501–1505 (2014)
4. J.G. Andrews, S. Buzzi, W. Choi, S.V. Hanly, A. Lozano, A.C.K. Soong, J.C. Zhang, What will 5G be? IEEE J. Sel. Areas Commun. **32**(6), 1065–1082 (2014)
5. A. Zappone, E. Jorswieck, Energy efficiency in wireless networks via fractional programming theory. Found. Trends Commun. Info. Theor. **11**(3–4), 185–396 (2015)
6. S. Andreev, A. Pyattaev, K. Johnsson, O. Galinina, Y. Koucheryavy, Cellular traffic offloading onto network-assisted device-to-device connections. IEEE Commun. Mag. **52**(4), 20–31 (2014). https://doi.org/10.1109/MCOM.2014.6807943
7. H.-W. Tseng, Y.-J. Yu, B.-S. Wu1, C.-F. Kuo, P.-S. Chen1, A resource allocation scheme for device-to-device communication over ultra-dense 5G cellular networks, in *IEEE International Conference on Applied System Innovation IEEE-ICASI* (2017). https://doi.org/10.1109/icasi.2017.7988351
8. X. Liu, X. Wang, Highly efficient 3D resource allocation techniques in 5G for NOMA enabled massive MIMO and relaying systems. IEEE J. Sel. Areas Commun. **35**(12), 2785–2797 (2017)
9. G.D. Swetha, G.R. Murthy, Fair resource allocation for D2D communication in mmWave 5G networks. IEEE (2017). 978-1-5386-2077-9/17/$31.00©2017
10. P. Zhao, L. Feng, P. Yu, W. Li, X. Qiu, A social-aware resource allocation for 5G device-to-device multicast communication. IEEE 2169-3536 (c) (2017). http://www.ieee.org/publications_standards/publications/rights/index.html
11. L. Sharnagat, H. Dakhore, Method of resource allocation in OFDMA using convex optimization. IEEE. (2015). 978-1-4799-1797-6/15$31.00©2015
12. S. Shi, M. Schubert, H. Boche, Rate optimization for multiuser MIMO systems with linear processing. IEEE Trans. Sig. Process. **56**(8), 4020–4030 (2008)
13. J.A. Sheikh et al., Towards green capacity in massive mimo based 4G-LTE a cell using beam forming vector based sectored relay planning. Wirel. Pers. Commun. https://doi.org/10.1007/s11277-017-4809-8
14. H.V. Nguyen, V.-D. Nguyen, H.M. Kim, O.-S. Shin, A convex optimization for sum rate maximization in a MIMO cognitive radio network, in *Eighth International Conference on Ubiquitous and Future Networks (ICUFN)* (2016). https://doi.org/10.1109/icufn.2016.7537081
15. J. Zhu, R. Schober, V.K. Bhargava, Linear precoding of data and artificial noise in secure massive MIMO systems. IEEE Trans. Wirel. Commun. **15**(3), 2245–2261 (2016)
16. H. Yang, B.-C. Seet, S.F. Hasan, P.H.J. Chong, M.Y. Chung, Radio resource allocation for D2D-enabled massive machine communication in the 5G era, 2016. 978-1-5090-4065-0/16$31.00©2016

Vocal Mood Recognition: Text Dependent Sequential and Parallel Approach

Gaurav Agarwal, Vikas Maheshkar, Sushila Maheshkar and Sachi Gupta

Abstract The speech is the most famous and essential method of communication among persons. The communication between human being and machine is called human PC interface. Speech has the capability of being the most essential method of communication with PC. Speech is also an acoustically rich flag that gives impressive individual data about talkers. The statement of emotions in speech sounds and relating capacities to visualize such emotions are both essential parts of human–machine communication. Discoveries from ponders trying to portray the acoustic properties of emotions present in vocal. The speech shows that discourse acoustics give an outside signal to the level of nonspecific excitement related with passionate procedures and, to a lesser degree, the relative charm of experienced feelings. A brief overview on bifurcation of speech, either into single or series of words with persistent or unconstrained speech is also taken into consideration within this paper. This paper represents the methodology that converts the vocal signals into corresponding text. As soon as the signal is converted into text, then the same will be analyzed with the sequential as well as with the parallel sum algorithm works for the parallel random access machine. Using both algorithms, the mood of the speech will be identified as romantic, sad or rock. Comparative studies of sequential and parallel approaches

G. Agarwal (✉) · S. Maheshkar
Department of Computer Science and Engineering, Indian Institute
of Technology (ISM), Dhanbad, Jharkhand, India
e-mail: gaurav13shaurya@gmail.com

S. Maheshkar
e-mail: sushila_maheshkar@yahoo.com

V. Maheshkar
Division of Information Technology, Netaji Subhas Institute of Technology,
Dwarka, New Delhi, India
e-mail: vikas.maheshkar@gmail.com

S. Gupta
Department of Computer Science and Engineering, Raj Kumar Goel Institute
of Technology, Ghaziabad, India
e-mail: sachiagarwal@rediffmail.com

© Springer Nature Singapore Pte Ltd. 2019   131
H. Malik et al. (eds.), *Applications of Artificial Intelligence Techniques
in Engineering*, Advances in Intelligent Systems and Computing 698,
https://doi.org/10.1007/978-981-13-1819-1_14

are also discussed in this paper. This paper concludes that the time complexity of the sequential executed algorithm can be reduced to a particular extent by using the parallel approaches.

Keywords Voice · Speech · Emotion · Mood · Parallel random access machine Word error rate · Word recognition rate

1   Introduction

The speech stream is an exceptionally intricate and variable flag that is most specifically considered by breaking down its acoustic properties, or sound examples [1]. We know from ordinary experience that talkers give data about their emotional states through the acoustic properties of their speech [2, 3]. For example, a large number of people have encountered talking in an unwittingly boisterous voice when feeling joyous, talking in a uniquely sharp voice when welcoming a sexually attractive individual, or chatting with checked vocal tremor while giving an open speech [4, 5]. Thusly, an audience member is apparently proficient at making exact assessments of emotional states even without visual signals, as routinely happens in the course of phone discussions. The speech is essential method of correspondence among persons and furthermore the most characteristics and productive type of trading data among human in speech [6, 7]. Vocal recognition can be characterized as the way toward changing over speech flag to an arrangement of words by implies algorithm executed as a PC program. Speech handling is one of the energizing regions of flag handling. The objective of speech acknowledgment range is to created procedure and framework to produce for speech contribution to machine [8, 9]. In light of major progressed in statically displaying of speech, programmed speech acknowledgment today finds broad application in errand that require human being machine interface, for example, automatic call answering.

1.1   Overview of Speech Recognition

A speech acknowledgment framework comprises of four pieces: Feature extraction, acoustic demonstrating, pronunciation displaying, and decoder [10]. The procedure of speech acknowledgment starts with a speaker making an expression which comprises the sound waves [11, 12]. These sound waves are then caught by a receiver and changed over into electrical signs [13, 14]. These electrical signs are then changed over into computerized frame to make them justifiable by the speech framework. Speech flag is then changed over into discrete grouping of highlight vectors, which is accepted to contain just the pertinent data about given articulation that is critical for its right acknowledgment [15]. A vital property of highlight extraction is the concealment of data insignificant for amend arrangement, for example, data about

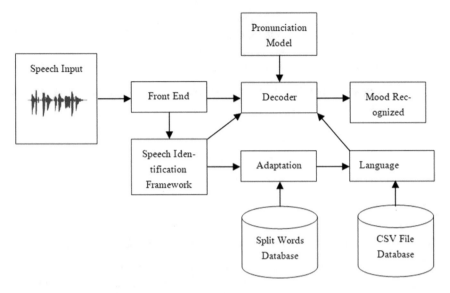

Fig. 1 Speech recognition system

speaker (e.g., basic frequency) and data about transmission channel (e.g., normal for a mouthpiece). At long last acknowledgment, segment finds the best match in the learning base, for the approaching element vectors. Some of the time, however, the data passed on by these element vectors might be corresponded and less separating which may back off the further preparing. Highlight extraction techniques like Mel frequency cepstral coefficient (MFCC) give some approach to get uncorrelated vectors by methods for discrete cosine transforms (DCT) as shown in Fig 1.

1.2 Word Error Count

Word error count is a conventional matrix for the evaluation of a machine interpretation or vocal recognition framework. The universal trouble in performance evaluation lies in the approach that the superficial word string can have an alternate length from the reference word arrangement [16]. This issue is illuminated by first adjusting the superficial word succession with the reference (talked) word grouping then utilizing dynamic sequence arrangement.

Word error count would then be able to be registered as

$$\text{Word Error Count (WEC)} = \frac{I + S + D}{N}$$

where

I   is the number of new words inserted
S   is the number of words in substitution
D   is the number of wrongly recognized words
N   is the total number of words in reference.

Word recognition rate (WRR) can also be used during the evaluation process of the vocal recognition system.

$$WRR = 1 - WEC$$
$$= \frac{N - I - S - D}{N} = \frac{CR - I}{N}$$

where

WEC   is word error count
CR   is N(S + D), the number of correctly recognized words.

This paper represents an algorithm approach for vocal mood recognition, by analyzing on sequential as well as on parallel platform. Section 2 describes the proposed algorithms approach in detail, along with result obtained, and Sect. 3 is devoted to result analysis. Finally, Sect. 4 elucidates the conclusion derived from the proposed work.

2   Proposed Work for Vocal Mood Recognition

Speech recognition consequently aspires to recognize the passionate condition of a person from his or her vocal words [16]. Depends on rigorous testing, an automatic procedure for generating the speech signal, extracting a few components which contain expressive data from the speaker's voice, and applying suitable prototype identification methods to identify mood states of the person. In this paper, a method is proposed which bifurcates the speech into isolated words with the help of Sphinx or Google API. Vocal mood can be recognized either in offline or online mode. For offline recognition of the mood of speech, suggested algorithm uses Sphinx and for online recognition Google API will play a vital role. With the help of the proposed Algorithm 1 and 2, Rock, Sad, or Romantic mood will be recognized. The information flow for the proposed work is as shown in Fig. 2.

2.1   Algorithm for Vocal Mood Recognition (Sequential)

Algorithm for the vocal mood recognition will work in two phases. During phase 1 an input from the user will be taken in .wav format. Now the file containing the speech signal will be split into single word text either by online or offline mode. After

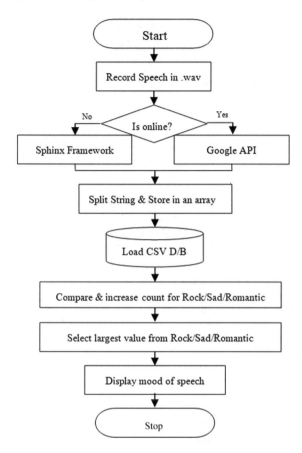

Fig. 2 Information flow for the proposed work

converting the file into text, it will be split into split string and will be read in word by word format. In phase 2 first, we initialize all mood that can be recognized by zero. After loading the database in comma-separated values (.CSV) word by word will be read from the split string and compare against the words available in the database. Here we are emphasizing the recognition of mood for the English language vocal input. The proposed algorithm can be used for the recognition of other language mood if we have the database related to that particular language. Moreover, if the two moods end up with the similar value for the initialize variable then this problem will be solved with the help of bifurcation of the input string. Sequential version of vocal mood recognition is shown in Algorithm 1. Here, we are recognizing three basic moods of speech that are romantic, sad, and rock.

Algorithm 1 Vocal Mood Recognition **(sequential)**

Input: Input from User in Vocal Form (.wav Format)
Output: Mood of the words spoken by user like Romantic / Sad / Rock
1: audio = r.record   //user will speak what so ever he wants
2: aa = r.recognize(audio)   //used to store the speech in text format
3: audio1 = audio.split(" ")
4: romantic = 0
5: sad = 0
6: rock = 0
7: Load the database file in CSV format
8: for i = 1 to row do
9:        for j = 1 to n (word in audio1) do
10:            if (audio1[j] = = row[i])
11:                 romantic = romantic + int (row ['rom'])
12:                 rock = rock + int (row ['rock'])
13:                 sad = sad + int (row ['sad'])
14:          end if
15:      end for
16: end for
17: if romantic > rock && romantic > sad
18:        display "Romantic"
19: else if sad > rock && sad > romantic
20:        display "Sad"
21: else if rock > romantic && rock > sad
22:        display "Rock"
23: else if romantic = = sad || romantic = = rock || sad = = rock
24:        audio1 = audio1 / 2
25:        repeat steps from 4 to 22
26: else
27:        display "Mixed mood identified"
28: end if

2.2 Algorithm for Vocal Mood Recognition (Parallel)

According to Amdhal's law [17], parallel random access machine (PRAM) has less time complexity than an optimal random access machine (RAM). This is so because parallelism has been used. The PRAM algorithm for the below-discussed methodology will be as shown in Algorithm 2 and the time complexity of the same will be as shown in Table 2. Actually PRAM algorithm works in two phases:

Phase 1.   Required number of processors will be activated. And
Phase 2.   Activated processors perform computation in parallel.

Algorithm 2 Vocal Mood Recognition **(Parallel)**

Input: Input from User in Vocal Form (.wav Format)
Output: Mood of the words spoken by user like Romantic / Sad / Rock
Assumptions: N, A [0.....(N-1)], j
1: audio = r.record   //user will speak what so ever he wants
2: aa = r.recognize(audio)   //used to store the speech in text format
3: audio1 = audio.split(" ")
4: Begin   // Parallel Execution Starts
5:     Spawn (P_0, P_1, ..., $P_{n/2-1}$)
6:     for all Pj where $0 \leq j \leq (n/2-1)$
7:         romantic = 0
8:         sad = 0
9:         rock = 0
10:        Load the database file in CSV format
11:        for k=0 to[(log n) - 1]do
12:           if ((j mod 2^k)=0 and $2j+2^k < n$)
13:              A[2j]=A[2j]+A[$2j+2^k$]   // Processor number for step14
14:              Call Vocal Mood Recognition **(Sequential)** // Steps 8 to 28
15:           end if
16:        end for
17:    end for
18: end

---

3   Result Analysis

The dataset used in this study has been created by own self by the author in his parent institute IIT Dhanbad, India. The vocal mood recognition system is built using Python 2.7.13 software. The sampling frequency of the vocal signal is by default 44.1 kHz. In phase 1 of vocal mood recognition algorithm, six different speakers are asked to vocalize the word 20 times from the given list of words and the recognition of the words is noted as result. The recognition percentage of the word is given in Table 1 and efficiency is represented by the chart is shown in Fig 3.

The evaluation of vocal recognition frameworks is typically determined by two factors that are precision and speed. Precision might be measured on the basis of performance precision which is normally reviewed with word error count (WEC) as discussed in Sect. 1.2. Here, we consider that speaker will speak out either word or sentence in English language. The spoken taxonomy will be converted to text then to split string. Split string will be compared with the comma-separated values stored in the database. The mood having the maximum integer value at last will be considered as the mood of the vocal. Suppose user speaks a particular sentence for the duration of 10 s like Hello How are you baby I love you Juliet I am here to marry you. Now hereafter the spoken sentence will be converted into the string of words. Here string of words means, each and every spoken word will be stored at a particular index later on if required the index can also be used for the analysis. The vocal input that converted to text and comma-separated values of the same will be as follows:

Table 1 Word recognition percentage of various speakers

Word	S 1 Child (5–10)	S 2 Child (10–15)	S 3 Adult (Male)	S 4 Adult (Female)	S 5 Old (Male)	S 6 Old (Female)
ROMEO	90	99	97	100	100	100
JULIET	95	100	98	100	100	100
ROCK	98	100	100	100	100	100
HURT	97	100	98	100	100	100
LOVE	100	100	100	100	100	100
PRINCE	98	100	100	100	100	100
MARRY	94	100	92	100	100	100
BABY	100	100	98	100	100	100
RING	100	100	99	100	100	100
ROCK	98	100	98	100	100	100
Average (%)	**97**	**99.9**	**98**	**100**	**100**	**100**

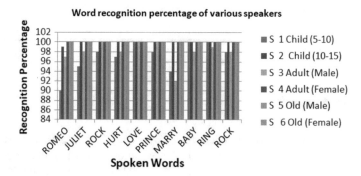

Fig. 3 Efficiency representation of word recognition system

[u' hello, u' how, u' are, u' you, u' baby, u' I, u' love, u' you, u' Juliet, u' am, u' here, u' to, u' marry, u' you].

After the process, the input through the proposed algorithm output generated will be as shown in Table 2. Now, the addition of all individual columns took place and the column have the maximum value will be considered as a mood of the vocal. In the sample shown in Table 2, the individual columns values are 4, 2, and 0 for romantic, sad, and rock, respectively. So, the vocal input is of mood "Romantic". During the analysis if two or more than two moods come up with the same final integer value then that problem will be considered as the problem of ambiguity. Ambiguity is resolved with the bifurcation of string in two equal halves. And the same process is repeated till the input is clearly associated with a particular mood. As mentioned in step 7–16 of an Algorithm 1, split string will be compared with the comma-separated file and the time complexity of an algorithm will be defined as

Table 2 Result generated after matching split string with CSV

Word	Romantic	Sad	Rock
hello	0	0	0
how	0	0	0
are	0	0	0
you	0	0	0
baby	1	1	0
i	0	0	0
love	1	1	0
you	0	0	0
juliet	1	0	0
am	0	0	0
here	0	0	0
to	0	0	0
marry	1	0	0
you	0	0	0

Table 3 Time complexity based on number of words

S. No.	Number of words in split string	Number of words in CSV	Time complexity (TC)
1	n	1	$O(n)$
2	1	N	$O(N)$
3	n/N	N	$O(n \times N)/O(N^2)$

$$Time\ Complexity(TC) = O(n * N)$$

where

n   Number of words in split string,
N   Number of words in CSV.

From Table 3 one can easily analyze if the number of words either in split string or in comma-separated value increases the overall execution time of an algorithm goes on increasing. To have the minimum execution time, the parallel version of the proposed algorithm is developed. By using the concept of parallel sum algorithm as shown in Algorithm 2, time complexity is $O(\log_2 N)$ (Table 4).

Now, consider the input string as "how r u baby I love", after process, the input through output of the proposed algorithms generated will be as shown in Table 5.

Now, the addition of all individual columns took place and the column have the maximum value will be considered as a mood of the vocal. In the sample shown in Table 5, the individual columns values are 2, 2 and 0 for romantic, sad, and rock, respectively. So, the vocal input can be of either mood "Romantic" or mood "Sad". As mentioned in Algorithm 2, Step 23 to Step 28 the case of ambiguity has been

Table 4 Time complexity based on parallel sum algorithm

S. No.	Number of words in split string	Number of words in CSV	Time complexity (TC)
1	N	1	$O(\log_2 N)$
2	1	N	$O(1)$
3	N	N	$O(\log_2 N)$

Table 5 Result generated after matching split string with CSV

Word	Romantic	Sad	Rock
how	0	0	0
are	0	0	0
you	0	0	0
baby	1	1	0
i	0	0	0
love	1	1	0

Table 6 Result generated after matching split string with CSV

Word	Romantic	Sad	Rock
baby	1	1	0
i	0	0	0
love	1	1	0

resolved with the recursion. Now the total count of the input string words took place. Here, the total number of input words is 7. Input string will be bifurcated into the two halves and the two halves will be separately processed and the two individual input. The first half will be "Hello how are you" and the other one will be "baby I love". When both halves processed, then the values generated will be as shown in Table 6.

The same process will be repeated again as the count for the romantic and sad mood are again same. So the one half will accommodate "baby I" and the other one accomplished with "love". In this example after recursively applying the procedure count for the romantic is not differing with the sad mood. Hence the input vocal is of mixed mood.

4 Conclusion

In this work, vocal mood of the speech is recognized using the sequential and parallel algorithms models of vocal mood recognition. Speech accuracy is also measured on the basis of different speaker's voice. 97 to 100% accuracy has obtained using the Google and Sphinx API of variable age group starts from 7 and goes up to 60. After successful conversion of speech to text mood of the speech is recognized. Experimental result shows the robustness of the experimental setup. The setup classifies speech into Romantic, Sad or Rock. It is also come into light that if we run the same procedure using the sequential approach than the time complexity will be $O(1)$ to $O(N^2)$. But if the same will be executed on the parallel approach the complexity will lie in $O(1)$ to $O(\log_2 N)$ range. Ambiguity has been resolved using the split methodology; the input stream will be split into two equal halves till the variable for romantic, sad, or rock have greater value or an input stream is an array of single word.

Acknowledgements The author would like to thank the Department of Computer Science and Engineering, Indian Institute of Technology (ISM), Dhanbad, Jharkhand, India for providing a platform and all necessary requirement for generating the dataset used in this study. This dataset has been created by the author in the lab. The authors are also thankful to the organizer to provide a nice platform for presenting the research.

References

1. P.B. Dasgupta, Detection and analysis of human emotions through voice and speech pattern processing. Int. J. Comput. Trends Technol. **52**(1) (2017)
2. R.B. Lanjewar, D.S. Chaudhari, Speech emotion recognition: a review. Int. J. Innov. Technol. Explor. Eng. (IJITEE) **2**(4) (2013)
3. S. Shinde, S. Pande, A survey on: emotion recognition with respect to database and various recognition techniques. Int. J. Comput. Appl. **58**(3) (2012)
4. M. Sarode, D.G. Bhalke, Automatic music mood recognition using support vector regression. Int. J. Comput. Appl. **163**(5) (2017)
5. Mathieu Barthet, Gyorgy Fazekas, Mark Sandler, *Music emotion recognition: from content to content based models* (Springer, Berlin, Heidelberg, 2013)
6. S.B. Davis, P. Mermelstein, Comparison of parametric representations for monosyllabic word recognition in continuously spoken sentences. IEEE Trans. Acoust. Speech Signal Process. **28**(4) (1980)
7. S.G. Koolagudi, K. Sreenivasa Rao, Emotion recognition from speech: a review. Int. J. Speech Technol. (2012)
8. D.D. Joshi, M.B. Zalte, Speech emotion recognition: a review. IOSR J. Electron. Commun. Eng. (IOSR-JECE) **4**(4) (2013)
9. S. Swamy, K.V. Ramakrishnan, An efficient speech recognition system. Comput. Sci. Eng.: Int. J. (CSEIJ) **3**(4) (2013)
10. S. Arora, M. Goel, Survey paper on scheduling in Hadoop. Int. J. Adv. Res. Comput. Sci. Softw. Eng. **4**(5) (2014)
11. S.K. Gaikwad, B.W. Gawali, P. Yannawar, A review on speech recognition technique. Int. J. Comput. Appl. **10**(3) (2010)
12. S. Sharma, R.S. Jadon, Mood based music classification. Int. J. Innov. Sci. Eng. Technol. (IJISET) **1**(6) (2014)
13. S. Karpagavalli, E. Chandra, A review on automatic speech recognition architecture and approaches. Int. J. Signal Process. Image Process. Pattern Recogn. **9**(4) (2016)
14. Y.-H. Cho, H. Lim, D.-W. Kim, I.-K. Lee, Music emotion recognition using chord progressions, in *IEEE International Conference on Systems, Man, and Cybernetics SMC* (2016). https://doi.org/10.1109/smc.2016.7844628

15. W.M. Campbell, D.E. Sturim, D.A. Reynolds, Support vector machines using GMM super vectors for speaker verification. IEEE Signal Process. Lett. (2015)
16. M. Vyas, A Gaussian mixture model based speech recognition system using Matlab. Int. J. Speech Image Process. (SIPIJ) **4**(4) (2013)
17. H. Shen, F. Pétrot, Using Amdahl's law for performance analysis of many-core SoC architectures based on functionally asymmetric processors, in 24th International Conference Camo, Italy, ARCS 2011, LNCS 6566 (Springer, Berlin, Heidelberg, 2011), pp. 38–49

Wideband Slotted Circular Monopole Antenna with Saturn-Shaped Notch in Ground Plane

Karishma Sharma, Dharmendra Kumar Upadhyay, Harish Parthasarthy and Rohit Gurjar

Abstract This paper presents a wideband slotted circular monopole antenna with Saturn-shaped notch in ground plane. The proposed design is compared to be better than a square monopole antenna with inverted T-shaped notch. Step-by-step analysis of this structure is done to obtain the largest impedance bandwidth. Parametric analysis gives the optimized dimensions of this structure for a bandwidth of ~15 GHz, cross polar level ~22 dB with an acceptable gain value over the whole band of frequencies. The proposed circular monopole antenna is simple and compact in size which covers C, X, Ku, K band and is suitable for WiFi devices, wireless computer networks, satellite transmissions and many other wireless communication system applications.

Keywords Circular monopole antenna · Partial ground · Saturn-shaped notch
Slotted antenna · Wideband · Wireless communication systems

1 Introduction

Microstrip patch antennas (MPA) are simple in shape, easy and inexpensive to fabricate using printed-circuit technology, mechanically robust and very versatile in terms of radiation and impedance characteristics [1]. These are extensively used in wireless communication systems. The basic design of such an antenna consists of a metal patch placed on a grounded substrate. This radiating patch may take many different configurations such as rectangular, circular, or any arbitrary shape [1, 2]. One of the major operational disadvantage of these antennas is the narrow frequency bandwidth.

K. Sharma (✉) · D. K. Upadhyay · H. Parthasarthy · R. Gurjar
Division of Electronics and Communication Engineering, Netaji Subhas Institute
of Technology, Delhi University, Dwarka, New Delhi, India
e-mail: krrissh_16@yahoo.co.in

R. Gurjar
e-mail: gurjarnsit@gmail.com

© Springer Nature Singapore Pte Ltd. 2019   143
H. Malik et al. (eds.), *Applications of Artificial Intelligence Techniques
in Engineering*, Advances in Intelligent Systems and Computing 698,
https://doi.org/10.1007/978-981-13-1819-1_15

Ultra-wideband (UWB) is a technology for transmitting information over a large bandwidth, and the demand of increased bandwidth in commercial, military, and medical applications is increasing [3]. Therefore there is a need of UWB antennas which are low cost, resistant to jamming, have large bandwidth, compact size etc. for such wireless communication applications [4].

Large bandwidth in MPA may be achieved by many different ways such as, increasing the substrate thickness, decreasing the permittivity of dielectric material, change in feeding technique may help in adjusting the design parameters to provide more bandwidth, modification in the radiating patch, truncating the ground plane, introduction of slots in patch or ground, stacking of elements, using coplanar parasitic elements, etc. [5–9]. Also, an efficient antenna design for desired bandwidth may be obtained using an artificial neural network (ANN) [10–12].

In this paper, a new design of a slotted printed circular monopole antenna is proposed. This antenna has a truncated ground plane with a Saturn-shaped defect, with rectangular and arc slots in the radiating patch for wide bandwidth. The antenna is derived from a square monopole antenna for UWB applications in the frequency range of 3.12–12.73 GHz [9]. The proposed structure resonates in the C, X, Ku, K bands and may be used for satellite communication, in WiFi devices, weather radar systems, wireless computer networks, etc. [13–16].

The rest of the paper is organized as follows: Sect. 2 explains the evolution of proposed structure design step-by-step. Section 3 gives the impedance and radiation results of the proposed design. This section also shows the comparison with the existing design [9] from where it has evolved, along with step-by-step results due to modification for increase in bandwidth. Parametric analysis is also given to achieve the optimized dimensions for best results. The paper is concluded in Sect. 4.

2 Structure Design

The proposed antenna structure evolved from the UWB square monopole antenna [9]. The new structure had a slotted circular shaped monopole antenna with a notched partial ground. The step-by-step procedure to obtain such a structure is explained in this section.

2.1 Configuration of Printed Circular Monopole with Partial Ground

A circular monopole wideband microstrip patch antenna (MPA) is designed. The structure evolved from a small square monopole antenna for UWB applications [9]. The shape of the square patch shown in Fig. 1a is changed to a circular shape with its diameter equal to the side of the square $W = 10$ mm. Then, the radius of the

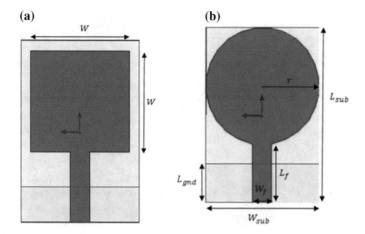

Fig. 1 Printed monopole antenna. **a** Square shaped patch; **b** Circular shaped patch

circular shaped patch is optimized to give the best results. The circular monopole is printed on the FR4 epoxy substrate material with relative permittivity $\varepsilon_r = 4.4$ having the value of dielectric loss tangent of $\delta = 0.02$. The substrate is of size $18 \times 12 \times 1.6 \, \text{mm}^3 \, (L_{sub} \times W_{sub} \times h)$.

A conducting partial ground plane is placed at the bottom of the substrate with length L_{gnd} and width W_{gnd} as 4 and 12 mm respectively. On the other side of the substrate the circular shaped patch having radius $r = 6$ mm is fed with a microstrip line feed of length $L_f = 6.1$ mm and width $W_f = 2$ mm. Figure 1b shows the antenna structure for this circular printed monopole with partial ground.

2.2 Introduction of Notch in the Ground Plane

In order to improve the bandwidth of this antenna an inverted T-shaped notch is cut in the ground plane as shown in Fig. 2a. The size of the inverted T-shape is taken same as given in [9] for its comparison. The dimensions of this notch is then adjusted for better impedance characteristics. These dimensions are $l_{t_1} = 0.5$ mm, $w_{t_1} = 2$ mm, $l_{t_2} = 0.5$ mm, $w_{t_2} = 6$ mm. Later the shape of this inverted T-shaped notch is changed by embedding an arc to form a Saturn-shaped notch. The modification in the shape of the notch further improves the bandwidth of the structure. The dimensions of this notch on the ground, shown in Fig. 2b are optimized as arc length $l_r = 0.75\pi$ mm, arc radius $r_s = 1.5$ mm. The ring of the Saturn-shaped notch has same dimensions as T-shape from which it has evolved and these values are $l_{t_2} = 0.5$ mm, $w_{t_2} = 6$ mm.

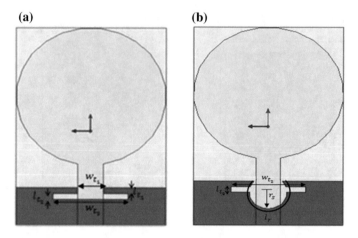

Fig. 2 Notch in the ground plane. **a** T-shaped notch; **b** Saturn-shaped notch

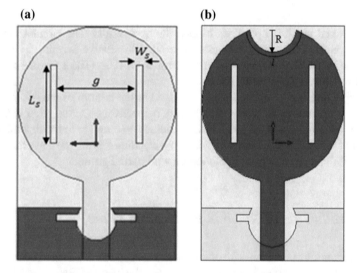

Fig. 3 **a** Rectangular slots cut in the radiating patch of printed antenna; **b** Proposed slotted circular monopole antenna

2.3 Slot and Arc Cuts in Patch Shape

Modification in the shape of the patch by creating slots may also improve the impedance and radiation characteristics [17]. The circular printed monopole patch modified by creating two rectangular slots of same dimensions having length $L_s = 6$ mm and width $W_s = 0.5$ mm, shown in Fig. 3a. The gap between the two rectangular slots is $g = 6$ mm. Later an arc is cut on the top of the patch with arc length $l = 4\pi$ mm and arc radius $R = 2$ mm. The creation of these slots and removal of arc,

Table 1 Dimensions of ground and substrate of the proposed monopole antenna

Permittivity of the dielectric material, ε_r	Substrate length, L_{sub} (mm)	Substrate width, W_{sub} (mm)	Height of the substrate, h (mm)	Partial ground length, L_{gnd} (mm)	Partial ground width, W_{gnd} (mm)
4.4	18	12	1.6	4	12

Table 2 Dimensions of the printed patch, feedline, slots in the proposed circular monopole antenna

Radius of the patch, r (mm)	Length of the feed-line, L_f (mm)	Width of the feed-line, W_f (mm)	Ring length, w_{t_2} (mm)	Ring width, l_{t_2} (mm)	Incomplete circle radius of Saturn slot, r_s (mm)	Arc length of Saturn shape l_r (mm)	Patch slot length, L_s (mm)	Patch slot width, W_s (mm)	Gap between the two rectangular slots, g (mm)	Length of arc cut on patch, l (mm)	Radius of arc cut on patch, R (mm)
6	6.1	2	6	0.5	1.5	$0.75\,\pi$	6	0.5	6	4π	2

improves the bandwidth further. The final proposed structure is shown in Fig. 3b. Tables 1 and 2 gives the final dimensions of the proposed wideband printed microstrip patch antenna (MPA).

3 Results and Discussion

The proposed antenna structure is simulated on Ansoft HFSS software [18]. Figure 4a compares the return loss of square monopole antenna with circular monopole antenna. It may be observed that the square monopole resonates at 8.71 GHz with an impedance bandwidth of 5.28 GHz, whereas circular monopole resonates at 10.23 GHz with bandwidth of 6.99 GHz. Therefore, the change in the

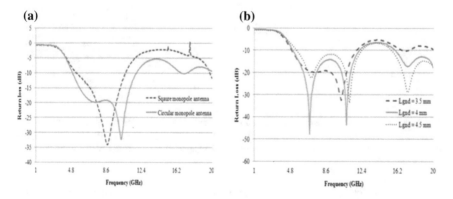

Fig. 4 **a** Comparison of simulated return loss of square and circular monopole antenna; **b** Simulated return loss of a circular monopole antenna with different values of L_{gnd}

Table 3 Comparison of simulated results for different ground length L_{gnd} of a circular monopole antenna

Ground length (mm)	Frequency range (GHz)	Impedance bandwidth (GHz)	Cross polar level (dB)
3.5	4.93–11.82	6.89	24.37
4	5.03–12.17	7.14	34.78
4.5	5.45–12.35	6.90	23.58

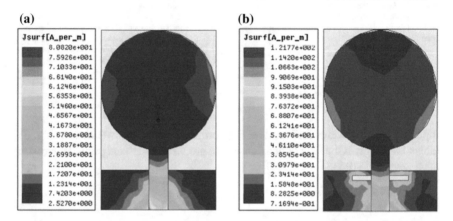

Fig. 5 Simulated current distribution for circular monopole antenna. **a** without inverted T-shaped notch; **b** with inverted T-shaped notch in ground plane

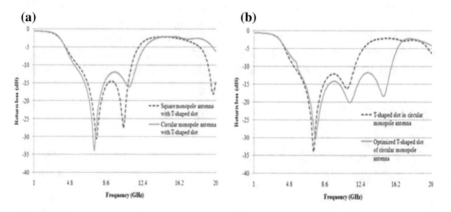

Fig. 6 **a** Comparison of simulated return loss for square and circular monopole antenna with inverted T-shaped slot of dimensions as in [9]; **b** Comparison of simulated return loss for a circular monopole antenna with optimized dimensions of T-shaped slot in ground plane with the other dimensions [9]

shape of the patch has increased the bandwidth. Also, there is a slight increase in the cross polar level from ~21 to ~24 dB. Further to achieve better results the ground length is varied and best results are obtained at $L_{gnd} = 4$ mm. The compared results

Table 4 Comparison of simulated results for monopole antennas with T-shaped notch in ground plane

	Resonant frequency (GHz)	Frequency range (GHz)	Impedance bandwidth (GHz)	Gain (dB)	Cross polar level (dB)
Square monopole antenna with dimensions of T-shaped slot as in [9]	7.57	5.56–11.26	5.70	1.67	18.58
Circular monopole antenna with T-shaped slot [9]	7.38	5.26–11.71	6.45	3.63	26.00
Optimized T-shaped slot of circular monopole antenna	7.54	5.64–15.68	10.04	3.27	27.35

are shown in Fig. 4b. It is observed that, increase in length first increases the bandwidth along with the cross polar level up to $L_{gnd} = 4$ mm, then these values decrease. These observations are tabulated in Table 3.

An inverted T-shaped slot is created in the ground plane as given in [9] which modifies the capacitance between the ground plane and patch hence modifying the impedance characteristics. Figure 5 shows the current distribution between the patch and partial ground, with and without T-shaped slot. It may be observed that, the current near the inverted T-shaped slot and at the circumference of the patch increases which may thus increase the field magnitude and hence the energy to improve the radiation and impedance characteristics. In Fig. 6a return loss for T-shaped slot in the partial ground of monopole antenna with square and circular shaped patch is compared. This T-shaped slot has dimensions of $w_{t_1} = 1$ mm, $l_{t_1} = 1$ mm, $w_{t_2} = 6$ mm, $l_{t_2} = 0.5$ mm [9]. The square monopole with T-shaped slot resonates at 7.57 GHz with an impedance bandwidth of 5.7 GHz. The broadside gain value of this antenna is 1.67 dB with cross polar level of $\sim$ 19dB. Whereas an improvement in bandwidth, gain and cross polar level may be observed for a circular monopole patch with same dimensions of T-shaped slot. It resonates at 7.38 GHz with a bandwidth of 6.45 GHz, having gain and cross polar level values 3.63 dB and $\sim$ 26 dB respectively. Later the dimensions of this T-shaped slot is optimized as given in Fig. 2a, which improves the impedance matching for a larger frequency range and hence increases the bandwidth to 10.04 GHz as shown in Fig. 6b. The comparison results are given in Table 4. The circular monopole antenna with T-shaped slot is observed to be better in terms of impedance and radiation results. Then, for optimized dimensions of T-shaped slot in

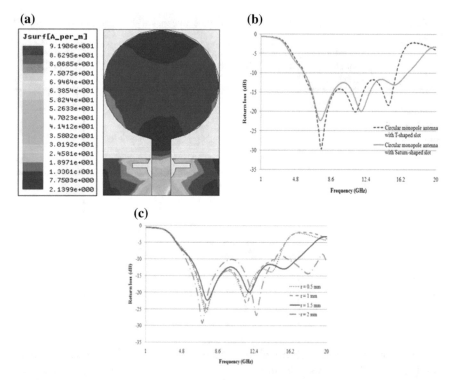

Fig. 7 **a** Simulated current distribution for circular monopole antenna with Saturn-shaped notch in the ground plane; **b** Comparison of simulated return loss for circular monopole antenna with T-shaped and Saturn-shaped slot; **c** Simulated return loss a circular monopole antenna with Saturn-shaped notch for different values of arc radius r_s

circular monopole the value of gain slightly decreases but there is an improvement in cross polar level with large increase in bandwidth.

Further modifications are done in the partial ground by removing an arc at T-shaped slot to form a Saturn type shape with a ring. Removal of this arc increases the current near the slot which increases the capacitance between the patch and the ground to improve the impedance characteristics as shown in Fig. 7a. This improvement in impedance bandwidth from the stage of T-shaped slot to modified Saturn-shaped slot may be observed from Fig. 7b. The resonant frequency is seen to shift slightly to 7.46 GHz with decrease in S_{11} value, but this value is less than -10 dB for a larger frequency range of 5.71–16.92 GHz. The improved bandwidth value is 11.21 GHz with slight decrease in gain and cross polar level, to 3.03 dB and ~ 21 dB respectively. These values are for the optimized arc radius in the Saturn-shaped slot. Figure 7c shows the parametric analysis of return loss for the different radii of the arc. It is observed that best results for bandwidth are obtained when the arc radius $r_s = 1.5$ mm.

(a)   **(b)**

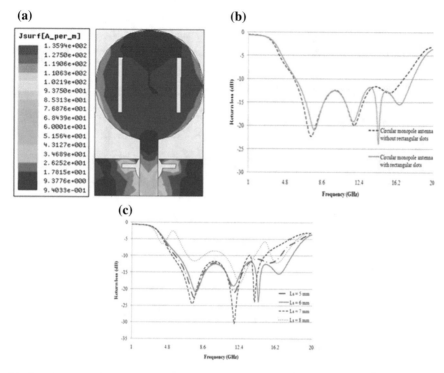

Fig. 8 a Simulated current distribution for circular monopole antenna with Saturn-shaped notch after creation of rectangular slots in patch area; **b** Comparison of simulated return loss for circular monopole antenna with and without rectangular slots in radiating patch; **c** Simulated return loss a circular monopole antenna with Saturn-shaped notch and rectangular slots in patch with different values of slot length L_s

Later modifications are done in the patch itself. Two rectangular slots are cut in the circular patch which increases the current near the slot as shown in Fig. 8a. These slots provide the current a longer path to travel within the patch to affect the resonance. This shifts the resonance towards right at 7.80 GHz with an impedance bandwidth of 12.03 GHz (5.83–17.86 GHz), shown in Fig. 8b. There is a decrement in gain value to 1.38 dB with similar cross polar level of $\sim$ 21 dB. The length of the rectangular slot is varied from 5–8 mm to obtain the best results in bandwidth. Figure 8c shows this comparison and it is observed that, at $L_s = 6$ mm more bandwidth is obtained.

Finally an arc is cut on top of the radiating circular patch having rectangular slots. Figure 9a, b shows that the removal of arc further improves the bandwidth of the monopole antenna as the current flows to the end of the patch, near the arc with improved radiations. The current is more near the edges which denotes increased fringing and increases the antenna perimeter to improve the bandwidth. The frequency range of this proposed structure is 5.52–20.31 GHz (bandwidth = 14.79 GHz, fractional bandwidth = 191.83%). Gain value of this antenna is 2.04 dB. The cross polar level of the proposed structure is also good with a value of 21.61 dB. The

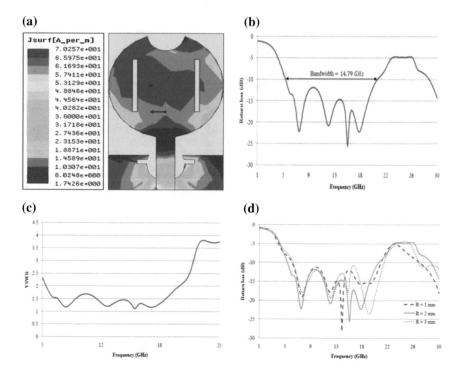

Fig. 9 **a** Simulated current distribution for proposed slotted circular monopole antenna with Saturn-shaped notch; **b** Simulated return loss of proposed antenna structure; **c** Simulated VSWR w.r.t. frequency for proposed antenna; **d** Simulated return loss a slotted circular monopole antenna with Saturn-shaped notch with different values of top arc radius R

Table 5 Simulated impedance and radiation results of proposed antenna structure

Resonant frequency (GHz)	Frequency range (GHz)	Impedance bandwidth (GHz)	Gain (dB)	Cross polar level (dB)	VSWR
7.71	5.52–20.31	14.79	2.04	21.61	1.09

VSWR is less than 2 for the whole band of frequencies as shown in the Fig. 9c. The radius of the arc which is removed from the top is varied from 1–3 mm to obtain the best arc dimension which gives the largest bandwidth with optimum gain and cross polar level as shown in Fig. 9d. Table 5 gives the values of impedance and radiation characteristics obtained for the proposed antenna structure.

The antenna impedance and radiation parameters may further be enhanced with neural network models. In such a model an exhausted set of antenna designs with different sizes, shapes, slot dimensions or any other design parameter, may be taken as inputs to train the processors. These processors extract the required information and stores this knowledge. The desired performance parameter shall be given as an external input to the optimization network, which may help to predict the best suitable design according to the desirable outputs and applications. The artificial neural network (ANN) model thus generalize, simplify, increase accuracy, decrease time of the parametric analysis of antenna, to obtain the desired performance results. Such a model hence may be used in the present work to achieve wideband for a slotted circular monopole antenna.

4 Conclusion

A wideband monopole antenna with slotted circular patch and notched partial ground is proposed to obtain fractional bandwidth of 192%. The $-10\,\text{dB}$ return loss is achieved for a band of frequencies from 5.52 to 2.31 GHz, i.e., a bandwidth of 14.79 GHz. Such an antenna may be used in wireless communication system applications like WiFi devices, wireless computer networks etc. The new design evolved from a small square monopole antenna with T-shaped notch in ground and has an increase of 5.18 GHz in bandwidth with a shift in resonance towards right. Parametric analysis gives the optimized dimensions of the structure for the best impedance and radiation results. The beauty of the proposed monopole antenna is that it has wideband which covers C, X, Ku, and K bands with good cross polar levels, and agreeable gain value.

References

1. C.A. Balanis, *Antenna Theory—Analysis and Design* (Wiley, New York, 1997)
2. K. Sharma, D.K. Upadhyay, H. Parthasarathy, Perturbation theory-based field analysis of arbitrary-shaped microstrip patch antenna. Int. J. Microw. Wirel. Propag. **79**, 1–11 (2017). https://doi.org/10.1017/s1759078717000368
3. H. Schantz, *The Art and Science of Ultra Wideband Antennas* (Artech House, Norwood, MA, 2005)
4. I. Oppermann, M. Hamalainen, J. Iinatti, *UWB Theory and Applications* (Wiley, New York, 2004)
5. P. Kakaria, R. Nema, Review and survey of compact and broadband microstrip patch antenna, in *IEEE International Conference on Advances in Engineering and Technology Research (ICAETR-2014)* (IEEE Press, Unnao, India, 2015). https://doi.org/10.1109/icaetr.2014.7012 846
6. M. Karamanoglu, M. Abbak, S. Simsek, A planar ultra-wideband monopole antenna with half-circular parasitic patches, in *IEEE Microwave Symposium (MMS-2013)* (IEEE Press, Saida, Lebanon, 2013). https://doi.org/10.1109/mms.2013.6663077
7. O. Ahmed, A. Sebak, A printed monopole antenna with two steps and a circular slot for UWB applications. IEEE Antennas Wirel. Propag. Lett. **7**, 411–413 (2008). https://doi.org/10.1109/LAWP.2008.2001026
8. M.K. Khandelwal, B.K. Kanaujia, S. Dwari, S. Kumar, A.K. Gautam, Bandwidth enhancement and cross-polarization suppression in ultrawideband microstrip antenna with defected ground plane. Mircow. Opt. Technol. Lett. **56**(9), 2141–2146 (2014). https://doi.org/10.1002/mop.28 499

9. M. Ojaroudi, C. Ghobadi, J. Nourinia, Small square loop monopole antenna with inverted t-shaped notch in the ground plane for UWB applications. IEEE Antennas Wirel. Propag. Lett. **8**, 728–731 (2009). https://doi.org/10.1109/LAWP.2009.2025972
10. B. Banerjee, A Self-Organizing auto-associative network for the generalized physical design of microstrip patches. IEEE Trans. Antennas Propag. **51**(6), 1301–1306 (2003). https://doi.or g/10.1109/TAP.2003.812266
11. D.K. Neog, S.S. Pattnaik, D.C. Panda, S. Devi, B. Khuntia, M. Dutta, Design of a widebland microstrip antenna and the use of artificial neural networks in parameter calculation. IEEE Antennas Propag. Mag. **47**(3), 60–65 (2005). https://doi.org/10.1109/MAP.2005.1532541
12. T. Khan, A. De, M. Uddin, Prediction of slot-size and inserted air-gap for improving the performance of rectangular microstrip antennas using artificial neural networks. IEEE Antennas Wirel. Propag. Lett. **12**, 1367–1371 (2013). https://doi.org/10.1109/LAWP.2013.2285381
13. W. Jiang, W. Che, A novel UWB antenna with dual notched bands for WiMAX and WLAN applications. IEEE Antennas Wirel. Propag. Lett. **11**, 293–296 (2012). https://doi.org/10.110 9/LAWP.2012.2190490
14. M. Ojaroudi, G. Ghanbari, N. Ojaroudi, C. Ghobadi, Small square monopole antenna for UWB applications with variable frequency band-notch function. IEEE Antennas Wirel. Propag. Lett. **8**, 1061–1064 (2009). https://doi.org/10.1109/LAWP.2009.2030697
15. T. Li, C. Zhu, X. Cao, J. Gao, Bandwidth enhancement of compact monopole antenna with triple band rejections. Electron. Lett. **52**(1), 8–10 (2016). https://doi.org/10.1049/el.20 15.2301
16. Y. Dong, T. Itoh, Planar ultra-wideband antennas in Ku- and K-Band for pattern or polarization diversity applications. IEEE Trans. Antennas Propag. **60**(6), 2886–2895 (2012). https://doi.or g/10.1109/TAP.2012.2194680
17. K. Sharma, D.K. Upadhyay, H. Parthasarathy, Modified circular-shaped microstrip patch antenna, in *IEEE International Conference on Computational Intelligence & Communication Technology (CICT)* (IEEE Press, Ghaziabad, India, 2015), pp. 397–399. https://doi.org/1 0.1109/cict.2015.47
18. Ansoft Simulation Software High Frequency Structure Simulator (HFSS). ver. 13 (Ansoft Corporation, Canonsburg, PA, 2005)

An 8-Shaped Self-affine Fractal MIMO Antenna for UWB Applications

Rohit Gurjar, Dharmendra Kumar Upadhyay, Binod K. Kanaujia
and Karishma Sharma

Abstract A compact 8-shaped self-affine fractal multiple-input-multiple-output (MIMO) antenna for ultra-wideband (UWB) communications presented. In order to achieve miniaturization and wideband phenomena, self-affine fractal geometry is used in the antenna design. The 8-shaped self-affine fractal geometry has been designed through iterated function system (IFS) up to the second iteration. The antenna consists of two 8-shaped self-affine fractal monopole antenna elements, a ground stub and vertical slots cut on the ground stub to enhance isolation and increase impedance bandwidth. The proposed antenna has compact dimensions of 26×40 mm^2 and impedance bandwidth ($S_{11} < -10$ dB) from 3 to 10.6 GHz with an isolation better than 19 dB over entire UWB. The various diversity performance parameters are also determined such as ECC, MEG, TARC and CCL, and all such parameters are within their allowable limits. These all results show that the proposed 8-shaped self-affine fractal MIMO antenna suitable for UWB applications.

Keywords Self-affine antenna · Fractal antenna · 8-shaped antenna
Multiple-input multiple-output antenna

R. Gurjar (✉) · D. K. Upadhyay · K. Sharma
Division of Electronics and Communications Engineering,
Netaji Subhas Institute of Technology, New Delhi, India
e-mail: gurjarnsit@gmail.com

D. K. Upadhyay
e-mail: upadhyay_d@rediffmail.com

K. Sharma
e-mail: krrissh_16@yahoo.co.in

B. K. Kanaujia
School of Computational and Integrative Sciences,
Jawaharlal Nehru University, New Delhi, Delhi, India
e-mail: bkkanaujia@yahoo.co.in

© Springer Nature Singapore Pte Ltd. 2019
H. Malik et al. (eds.), *Applications of Artificial Intelligence Techniques in Engineering*, Advances in Intelligent Systems and Computing 698,
https://doi.org/10.1007/978-981-13-1819-1_16

1 Introduction

Ultra-wideband (UWB) technology allows high data transmission at a low power in wide frequency band [1]. This advantage made industries and academic researchers to show interest in UWB technology. The UWB technology has many applications such are cancer sensing, high data rate communication and radar imaging, etc. But a major problem of multipath fading occurs in UWB system due to diffraction and reflection between two communicating antennas [2]. Multiple-input multiple-output (MIMO) technology may overcome this problem. MIMO system contains multiple antennas to receive and transmit electromagnetic waves with different fading properties, which improve channel capacity and system reliability [3]. However, the characteristics of antenna degrade because of mutual coupling when placed in a smaller area. Whereas, the large-sized MIMO antenna structures create difficulty in integration into small mobiles devices. The fractal geometry miniaturizes the structure and provides wideband property because of its self-affine and self-similar properties [4]. The desired bandwidth may be obtained using an artificial neural network (ANN) [5].

The word fractal means irregular or broken fragments and was coined by Benoit Mandelbrot. Fractal gives the opportunity to design small, broadband and efficient antennas in restricted space. Fractals are abundant in nature, with a few examples of nature like coastlines, trees, mountain ranges and ferns, etc. [6]. A fractal structure contains inherent self-similar or self-affine properties. A self-affine geometry [7] is obtained by the contraction of an image by different vertical and horizontal scaling factors. The fractal structure provides increment in electrical path length [8]. Various fractal geometries is used for the design of UWB antennas, for example, Sierpinski triangle, Koch snowflake and hexagonal shaped structure. In this paper, Sierpinski carpet type an 8-shaped self-affine fractal structure is designed using iterated function system (IFS) for UWB applications.

The remaining paper is arranged in the following way: Sect. 2 shows the geometry of 8-shaped self-affine fractal MIMO antenna and explains the construction of 8-shaped self-affine fractal structure using iterated function system. Section 3 explains the effects of ground stub and ground slots on MIMO antenna performance. Section 4 shows the S-parameters, radiation patterns, and realized gain of the proposed MIMO antenna. Diversity performance parameters are determined in Sect. 5. Section 6 concludes the proposed paper.

2 Antenna Configuration

The geometry of an 8-shaped self-affine fractal MIMO antenna is shown in Fig. 1. Overall dimensions of 8-shaped self-affine fractal MIMO antenna are 40 mm × 26 mm × 0.8 mm and designed on a Rogers substrate, RO4350B, with loss tangent of 0.004 and dielectric constant of 3.5.

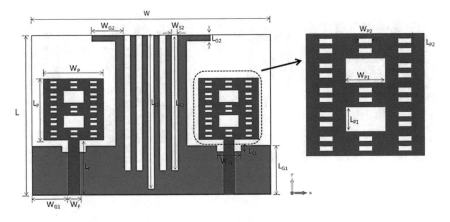

Fig. 1 Geometry of proposed 8-shaped self-affine fractal MIMO antenna (dark blue: upper side, light blue: bottom side)

Table 1 Dimensions of an 8-shaped self-affine fractal MIMO antenna (Unit. mm)

L	L_P	L_{P1}	L_{P2}	L_F	L_{G1}	L_{G2}	L_{S1}	L_{S2}	L_{S3}
26	10	2	0.4	9	8	1	1	22	25
W	W_P	W_{P1}	W_{P2}	W_F	W_{G1}	W_{G2}	W_{S1}	W_{S2}	
40	10	3.33	1.11	1.8	6.1	5.5	4	1	

The two fractal radiators have identical dimensions of $L_P \times W_P$ that are printed on the upper surface of substrate. Each radiator is fed by a microstrip line of 50 Ω impedance with $L_F \times W_F$ dimension. The ground plane with dimensions of $L_{G1} \times W$ is printed on the back surface of substrate. Vertical ground stub is placed between monopole elements to improve impedance matching and increase its impedance bandwidth. Five vertical slots are cut on ground stub of antenna structure for improving isolation. The dimensions of the proposed 8-shaped self-affine fractal MIMO antenna are given in Table 1.

In this paper, an 8-shaped self-affine fractal geometry [6, 9] is constructed from zero iteration structure, as shown in Fig. 2a by scaling a square with different scaling factors, scaling factor of 3 in the horizontal direction, and scaling factor of 5 in the vertical direction. This scaling gives fifteen rectangles, out of which the central two rectangles removed to make 8-shapes structure as shown in Fig. 2b, this is called the first iteration. Then repeat the same scaling process on each of the remaining 15 rectangles, which gives a structure as shown in Fig. 2c, this is called second iteration. The following design method is called iterated function system (IFS).

The initial two iterations of an 8-shaped self-affine fractal structure are shown in Fig. 2. The 8-shaped self-affine fractal MIMO antenna is designed in the EM simulation tool CST and its results are verified by comparing with HFSS results.

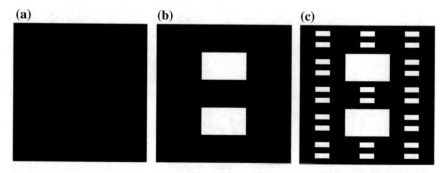

Fig. 2 Iterations of the 8-shaped self-affine fractal geometry

3 Analysis of MIMO Antenna

Here, the ground stub and ground slots effects on MIMO antenna performance are studied.

3.1 Effect of Ground Stub

The 8-shaped self-affine fractal MIMO antennas without ground stub and with ground stub are shown in Fig. 3. The ground stub enhances isolation and improves impedance matching as reflecting radiation. Simulated S-parameters of 8-shaped self-affine fractal MIMO antennas without ground stub and with ground stub are shown in Fig. 4. Antenna without using ground stub has a low cutoff frequency (for $S_{11} < -10$ dB) of about 4.3 GHz (which is more than needed UWB 3.1 GHz low cutoff frequency) and mutual coupling (S_{21}) is above -15 dB (for MIMO antennas, it should be $S_{21} < -15$ dB) between the two ports. It is clearly visible in Fig. 4, that a resonance is generated at both lower and higher frequency sides by adding ground stub in the antenna. Mutual coupling (S_{21}) is suppressed significantly across entire UWB band.

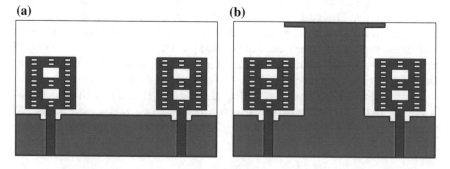

Fig. 3 8-shaped self-affine fractal MIMO antenna **a** without a ground stub **b** with a ground stub

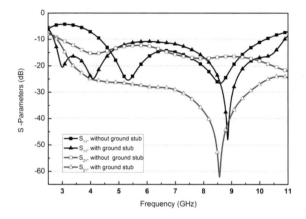

Fig. 4 S-parameters of 8-shaped self-affine fractal MIMO antennas with and without a ground stub

3.2 *Effect of Ground Slots*

For improving isolation, five vertical slots are cut on inserted ground stub as shown in Fig. 5. Figure 6 shows a comparison between S-parameters of 8-shaped self-affine fractal MIMO antennas with and without ground slots. The mutual coupling (S_{21}) is improved near the 3 GHz frequency of desired frequency band.

The current distribution is also considered in the study to analyze the effects of the ground stub and ground slots in isolation. The current distribution of 8-shaped self-affine fractal MIMO antennas without any stub, with ground stub and ground slots at frequency 4.4 GHz are shown in Fig. 7. When no ground stub is used in MIMO antenna, there is strong coupling current between two ports. Figure 7 shows that when a ground stub and ground slots are used in MIMO antenna then more current concentrates near stub and slot region and less current near the other port. So, mutual coupling (S_{21}) is significantly improved.

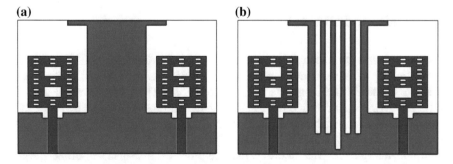

Fig. 5 8-shaped self-affine fractal MIMO antenna **a** without ground slot **b** with ground slots

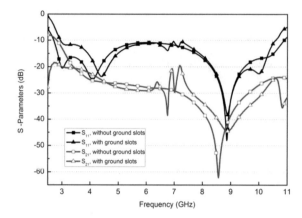

Fig. 6 S-parameters of 8-shaped self-affine fractal MIMO antennas with and without ground slots

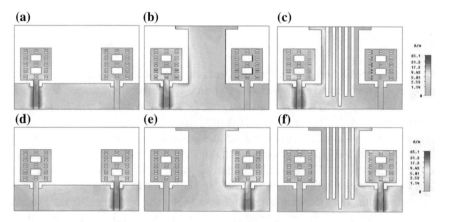

Fig. 7 Surface current distribution at 4.4 GHz when port 1 is excited (**a** without ground stub, **b** with ground stub, **c** with ground slots) and when port 2 is excited (**d** without ground stub, **e** with ground stub, **f** with ground slots)

4   Results and Discussion

Figure 8 shows S-parameters of proposed 8-shaped self-affine fractal MIMO antenna, which holds good agreements in both CST and HFSS. Figure 8a shows that 8-shaped self-affine fractal MIMO antenna has impedance bandwidth ranges from 3 to 10.6 GHz (for $S_{11} < -10$ dB) and impedance bandwidth ranges from 2.7 to 10.8 GHz (for $S_{11} < -6$ dB), which is allowed for MIMO antennas [10]. Figure 8b shows that mutual coupling of 8-shaped self-affine fractal MIMO antenna is below -19 dB for entire operating band. This indicates that the proposed 8-shaped self-affine fractal MIMO antenna is preferable throughout the FCC UWB.

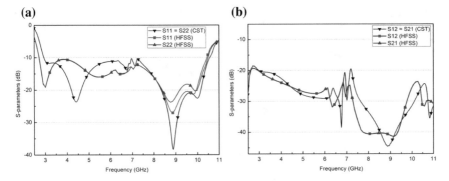

Fig. 8 Comparison between CST and HFSS simulation S-parameters of 8-shaped self-affine fractal MIMO antenna **a** S_{11} and S_{22}, **b** S_{21} and S_{12}

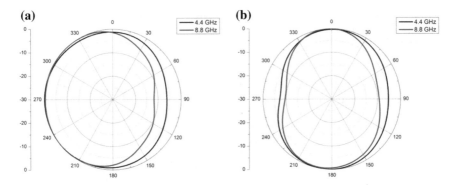

Fig. 9 2-D radiation patterns **a** E-plane and **b** H-plane simulated at frequencies of 4.4 GHz (black color: simulated with port 1) and 8.8 GHz (red color: simulated with port 1)

The radiation pattern of the proposed 8-shaped self-affine fractal MIMO antenna has been evaluated by EM solver CST studio suite. Figure 9 shows the 2-D radiation patterns at 4.4 and 8.8 GHz frequencies.

Figure 10 shows realized gain of an 8-shaped self-affine fractal MIMO antenna. The realized gains of CST and HFSS are comparable. Realized gain is ranging from 1.07 to 5.98 dBi in operating frequency band.

5 Diversity Performance

To corroborate the ability of proposed 8-shaped self-affine fractal MIMO antenna, it is absolutely needed to determine various MIMO diversity performance parameters. These MIMO performance parameters are evaluated by the EM simulation tool CST and these results are verified with HFSS simulation software results.

Fig. 10 Comparison
between CST and HFSS
simulation realized gains of
the proposed MIMO antenna

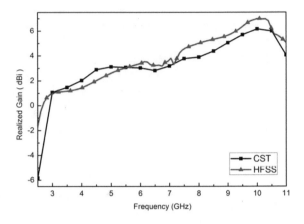

Fig. 11 Comparison
between CST and HFSS
simulation ECC of proposed
MIMO antenna

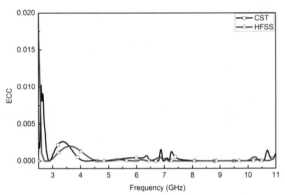

    Figure 11 displays the envelope correlation coefficient (ECC) variation of the proposed 8-shaped self-affine fractal MIMO antenna. The parameter ECC describes the amount of correlation between the communication channels. ECC is determined by using S-parameters for uniform propagation environment as given in [11]. For any practical application, parameter ECC value should be lower than 0.05 [11]. In the proposed 8-shaped self-affine fractal MIMO antenna, parameter ECC is even less than 0.01 for the operating bandwidth. It shows good diversity performance of the 8-shaped self-affine fractal MIMO antenna.

    Mean effective gain (MEG) is also another important diversity performance parameter. It explains the gain performance of an antenna considering the environmental effects. The MEG of port-1 and port-2 are determined while considering the radiation efficiency by using equations as given in [12]. The difference of MEG_1 and MEG_2 should lie in between 0 and -3 dB for considerable MIMO antenna system [12]. Figure 12 shows calculated MEG plots of the proposed 8-shaped self-affine fractal MIMO antenna. The difference between MEG of two ports is within the acceptable limit.

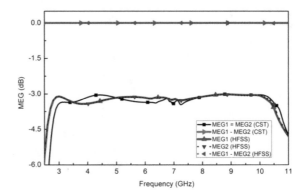

Fig. 12 Comparison between CST and HFSS simulation MEG of the proposed MIMO antenna

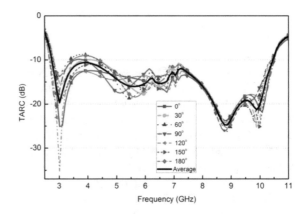

Fig. 13 Simulated TARC of proposed 8-shaped self-affine fractal MIMO antenna

Total active reflection coefficient (TARC) is also essential diversity performance parameter. Resonant frequencies and impedance bandwidth of MIMO antennas are determined by TARC curves. Two ports MIMO antenna system's TARC is determined by the using S-parameters as given in [13]. TARC plot of the proposed 8-shaped self-affine fractal MIMO antenna is shown in Fig. 13. The TARC curves show stable characteristics for desired bandwidth. It is due to good isolation between two ports of 8-shaped self-affine fractal MIMO antenna.

Channel capacity loss (CCL) is also another essential MIMO diversity performance parameter. It gives facts about the maximum attainable limit of the signal transmission rate. It is calculated by using formula as given in [14]. Figure 14 shows CCL of 8-shaped self-affine fractal MIMO antenna. The CCL should be lesser than 0.4 b/s/Hz for MIMO antenna system [14]. From Fig. 14, it is observed that CCL lies within an acceptable limit of channel capacity loss for the entire bandwidth.

Fig. 14 Comparison
between CST and HFSS
simulation CCL of the
proposed MIMO antenna

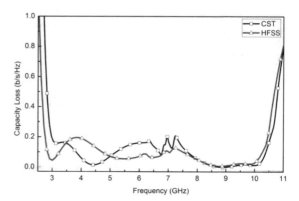

6 Conclusion

An 8-shaped self-affine fractal MIMO antenna with a compact dimension of 26 ×
40 mm^2 has been designed for UWB applications. Application of fractal geometries
in antenna engineering helps in achieving miniaturization and wideband properties.
The two 8-shaped self-affine fractal geometry monopole elements are utilized to
achieve UWB operations from 3 to 10.6 GHz. The total area of monopole elements
of MIMO antenna is reduced by 24.88% at second iteration from the zero iteration
square shaped monopole elements of MIMO antenna. A ground stub with slots is
used to reduce mutual coupling of less than −19 dB between input ports. CCL and
ECC are less than 0.2 b/s/Hz and 0.01, respectively, over the entire desired bandwidth.
Other diversity performance parameters are also within acceptable limits of MIMO
antennas. All results indicate that the proposed 8-shaped self-affine fractal MIMO
antenna is convenient for UWB applications.

References

1. L. Liu, S.W. Cheung, Compact MIMO antenna for portable devices in UWB applications. IEEE
   Trans. Antennas Propag. **61**(8), 4257–4264 (2013)
2. S. Tripathi, A. Mohan, S. Yadav, A compact octagonal fractal UWB MIMO antenna with
   WLAN band-rejection. Microw. Opt. Technol. Lett. **57**(8), 1919–1925 (2015)
3. I.M. Ben, L. Talbi, M. Nedil, K. Hettak, MIMO-UWB channel characterization within an
   underground mine gallery. IEEE Trans. Antennas Propag. **60**, 4866–4874 (2012)
4. M.R. Hashemi, M.M. Sadeghi, V.M. Moghtadai, Space-filling patch antennas with CPW feed.
   Prog. Electromagnet. Res. Symp. (2006)
5. D.K. Neog, S.S. Pattnaik, D.C. Panda, S. Devi, B. Khuntia, M. Dutta, Design of a widebland
   microstrip antenna and the use of artificial neural networks in parameter calculation. IEEE
   Antennas Propag. Mag. **47**(3), 60–65 (2005). https://doi.org/10.1109/MAP.2005.1532541
6. R. Gurjar, S. Dwivedi, S. Thakur, M. Jain, A self-affine 8-shaped fractal multiband antenna for
   wireless applications. Int. J. Elect. Commun. Eng. Technol. **4**(2), 103–108 (2013)

7. H.O. Peitgen, H. Jurgens, D. Saupe, *Chaos and Fractals: New Frontiers in Science* (Springer-Verlag, New York, 1992)
8. D.H. Werner, R.L. Haupt, P.L. Werner, Fractal antenna engineering: the theory and design of fractal antenna arrays. IEEE Antennas Propag. Mag. **41**(5), 37–58 (1999)
9. R. Gurjar, R. Singh, S. Kumar, Elliptically slotted self-affine 8-Shaped fractal multiband antenna, in *Proceedings of the International Conference on Recent Cognizance in Wireless Communication & Image Processing* (Springer, 2016), pp. 783
10. M.S. Sharawi, M.U. Khan, A.B. Numan, D.N. Aloi, A CSRR loaded MIMO antenna system for ISM band operation. IEEE Trans. Antennas Propag. **61**(8), 4263–4274 (2013)
11. H.S. Singh, G.K. Pandey, P.K. Bharti, M.K. Meshram, Design and performance investigation of a low profile MIMO/Diversity antenna for WLAN/WiMAX/HIPERLAN applications with high isolation. Int. J. RF Microw. **25**, 510–521 (2015)
12. J. Nasir, M.H. Jamaluddin, M. Khalily, M.R. Kamarudin, I. Ullahnad, R. Selvaraju, A reduced size dual port MIMO DRA with high isolation for 4G applications. Int. J. RF Microw. **25**, 495–501 (2015)
13. S. Su, C. Lee, F. Chang, Printed MIMO-antenna system using neutralization-line technique for wireless USB-dongle applications. IEEE Trans. Antennas Propag. **60**, 456–463 (2012)
14. Y.K. Choukike, S.K. Sharma, S.K. Behera, Hybrid fractal shape planer monopole antenna covering multiband wireless communications with MIMO implementation for handheld mobile devices. IEEE Trans. Antennas Propag. **62**, 1483–1488 (2014)

Design and Implementation of Conductor-to-Dielectric Lateral Sliding TENG Mode for Low Power Electronics

Khushboo and Puneet Azad

Abstract The triboelectric nanogenertaor is a type of generator which is used to convert all type of mechanical energies that are available but wasted in our daily life such as human motion, walking, mechanical triggering, vibration and rotating tire, etc., in electrical energy. This paper mainly focuses on the working model of lateral sliding mode of triboelectric energy harvesting technique based on conductor-to-dielectric method. In this technique copper and aluminum are selected as two electrode and ptfe as dielectric, these two electrodes having distinct charges are used to generate energy by repeated back and forth motion through a mechanical setup. We have experimentally analyzed the charging behavior of different capacitor. The maximum voltage is found to be of 3.3 V across 4.7 μF capacitor. The maximum power is found to be 0.257 μW across 7.8 MΩ resistor and 4.7 μF capacitor. The maximum energy computed is 95 μJ/cm^{-3} across 33 μF capacitor after 420 s. Such type of energy can be used to drive low power electronic devices, biomedical devices and sensor networks, etc.

Keywords Energy harvesting · Lateral sliding mode · Conductor to Dielectric TENG materials · AC-DC conversion · Storage

1 Introduction

Various kinds of energy sources are available in today's environment including Solar, Thermal (heat), Mechanical, RF, and so on. These energy sources are used to produce energy, stored, and drive low power portable and wearable devices. This complete

Khushboo (✉)
USICT, Guru Gobind Singh Indraprastha University, New Delhi, India
e-mail: kkhushboo_2008@yahoo.com

P. Azad
Department of Electronics and Communication Engineering, Maharaja Surajmal
Institute of Technology, New Delhi, India
e-mail: puneet.azad@msit.in

© Springer Nature Singapore Pte Ltd. 2019
H. Malik et al. (eds.), *Applications of Artificial Intelligence Techniques in Engineering*, Advances in Intelligent Systems and Computing 698,
https://doi.org/10.1007/978-981-13-1819-1_17

Fig. 1 Block diagram of
energy harvesting model

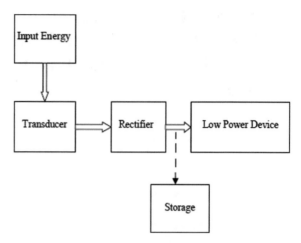

process is called as energy harvesting system. The general block diagram of energy harvesting model is divided into five blocks as shown in Fig. 1.

1. Input Energy: Various free energy sources are available as input energy like solar, mechanical, and thermal. We can opt one of the free energy sources as input energy or can select combination of any two as hybrid energy harvesting [1–4]. We have chosen mechanical energy source for our designed model.
2. Transducer: Transducer works to convert one form of energy into another form. Mechanical energy is converted into electrical energy as we designed the model using mechanical energy source.
3. Rectifier: The first block transducer delivered AC output voltage, so full wave bridge rectifier (FWR) is required to convert this AC input voltage into DC output voltage.
4. Storage: Storage is an optional block, it depends on the requirement of the application. Storage could be possible using rechargeable batteries or super capacitor.
5. Low power device: This type of model can be used to drive low power electronic devices.

In 2012 Zhong Lin Wang invented the first triboelectric nanogenerator, also known as TENG. This generator works on the basic principle of electrostatic and contact electrification effect. This effect states that when two different materials of opposite charges come into contact through a mechanical friction, they get electrically charged and a potential drop is created between these two. The four main types of triboelectric nanogenerator are as follows- vertical contact-separation mode, lateral/contact-sliding mode, single electrode mode and freestanding triboelectric layer mode. We have chosen to work with second mode of TENG that is lateral/contact-sliding mode. It is the simplest type of TENG, which is made with two different materials.

This paper reports a lateral sliding mode triboelectric nanogenerator (TENG) on conductor-to-dielectric method. In this principle a set of two distinct materials are used, one has positive charge such as Aluminum and second has negative charge

Fig. 2 Lateral sliding TENG
mode with the combination
of AL—PTFE—CU

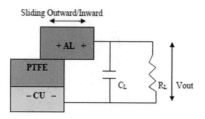

such as Copper with one dielectric as PTFE (Polytetrafluoroethylene/Teflon). Two distinct materials are rubbed to each other in lateral direction and a potential drop is created hence a current starts flowing through the circuit. Figure 2. shows the actual mechanism of lateral sliding electrode mode [5–7].

2 Architecture of Lateral Sliding Mode

We have presented our work on double-electrode lateral sliding TENG mode. The architecture of the lateral sliding mode is shown in Fig. 3. This mode works on the principle of triboelectrification and electrostatic induction including different charge materials used to generate low power by repeated back and forth motion through a mechanical setup. Lateral sliding mode TENG is also called as attached-electrode sliding mode TENG.

Two types of lateral sliding modes are:

1. Dielectric to dielectric
2. Conductor to Dielectric

The architecture is divided into three main sections:

1. TENG source
2. Full wave bridge rectifier
3. Load/Storage

The first block of TENG source represents the Conductor-to-Dielectric mode in which Metal 1 (Aluminum) behave as positive charge, Metal 2 (Copper) behave as negative charge and Dielectric (PTFE) behave as insulator as this will not drain out the charge for a maximum amount of time. Metal 2 with dielectric is kept fixed while

Fig. 3 Architecture of
lateral sliding mode

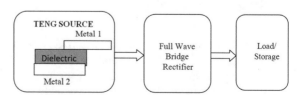

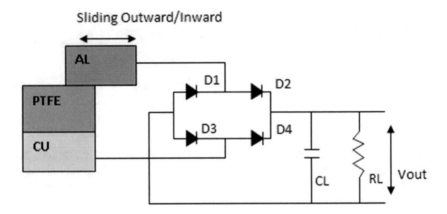

Fig. 4 Schematic of lateral sliding TENG model

the metal 1 sliding inward and outward in continuous manner to keep generating AC input voltage [8, 9].

Full wave bridge rectifier is the second block used for AC to DC conversion. The third block could be a storage device like capacitor/super—capacitor/battery or directly powered a low power electronic device. Figure 4 shows the schematic diagram of the circuit. With the repetitive back and forth motion of the triboelectric materials a periodic AC input voltage is generated which is then rectified and stored in different electrolytic capacitor (4.7, 33, 47, and 100 μF) for practically purpose. Output voltage across different C_L with their energy was obtained.

Equation (1) provides the output current across R_L.

$$Isc = C\frac{\partial v}{\partial t} + V\frac{\partial c}{\partial t} \tag{1}$$

C: System capacitance
V: System voltage

First half of the above equation represents the changes in the voltage across the two materials and second half represents the changes in the system capacitance across the two materials.

3 Simulation Results

The graph shows the correlation between vout (v) versus time (sec) across the different load capacitor C_L (4.7, 33, 47 and 100 μF) in Fig. 5. We have stored the rectified output voltage in different values of load capacitors C_L without load resistor R_L. This rectified output voltage initially increases quickly but with the increased time it starts

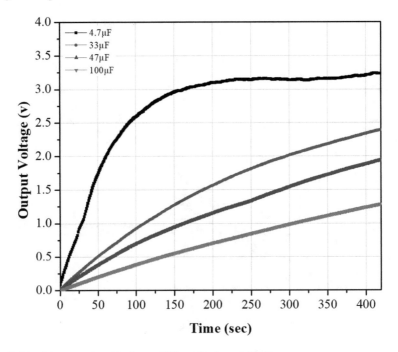

Fig. 5 Vout (v) versus time (s) plots at different load capacitor C_L

getting saturated. After comparing the charging voltage among themselves we have found the maximum voltage of 3.3 V across 4.7 μF capacitor. It is also observed that small value capacitor, i.e., 4.7 and 33 μF charges quickly as compared to large value capacitor i.e. 47 and 100 μF. Large value capacitor takes more time to get fully charged.

Energy across different value of capacitors is calculated using the equation number (2). The amount of energy stored by 4.7 μF capacitor is very less that is 23 μJ/cm^{-3} as it is compared with other capacitors. This is due to less amount of charge is stored by small value capacitor. The maximum energy i.e. 95 μJ/cm^{-3} is found across 33 μF capacitor in 420 s shown in Fig. 6 [10–13]. Equation (2) provides the stored energy density across C_L:

$$E = \frac{1}{2}C_L V^2 \tag{2}$$

$$P_{out} = \frac{V^2}{R_L} \text{ or } VI \tag{3}$$

Table 1 describes the power output across different value of load resistor (MΩ) and load capacitor (μF) for the time period of 420 s [14–16]. It is to be calculated using equation number (3). After analyzed the complete data of power output we have found the maximum power across 7.8 MΩ load resistor than 1 MΩ for all the

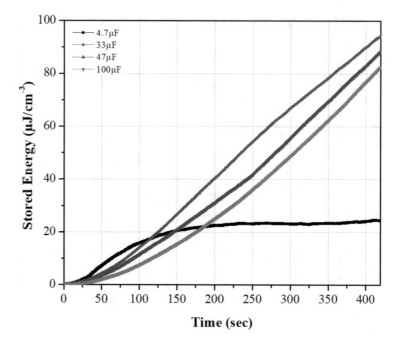

Fig. 6 Stored energy versus time (s) plots at different C_L

Table 1 Maximum power output (μW) (after 420 s) at various R_L (MΩ) and C_L (μF)

Maximum power output (μW)					
Capacitance (μF)	1 MΩ	7.8 MΩ	10.4 MΩ	13 MΩ	20.80 MΩ
4.7	0.107	0.257	0.182	0.201	0.182
33	0.115	0.191	0.180	0.138	0.117
47	0.118	0.169	0.132	0.123	0.082
100	0.117	0.083	0.071	0.060	0.038

capacitor except for case of 100 μF capacitor. The reason is that 100 μF capacitor charged slowly with a small amount of voltage in 420 s. Simultaneously power output start decreasing above and beyond the 10.4 MΩ load resistor as the electric current decreases at high load resistor R_L [17–19].

Maximum power of the model is obtained by connecting five different load resistor R_L (1, 7.8, 10.4, 13, and 20.8 MΩ) in parallel with load capacitor C_L Fig. 7 presents the output voltage at no load and at different load resistor R_L (1, 7.8, 10.4, 13, and 20.8 MΩ). This plot reveals that large value resistor provides more output voltage as compared with small value capacitor. We noticed that 7.8 MΩ load resistor providing maximum power output of 0.257 μW across all the load capacitor except 100 μF load capacitor. The reason for this is that the flow of more current through higher value resistor and the electrodes gets more charged, resulting in higher output voltage.

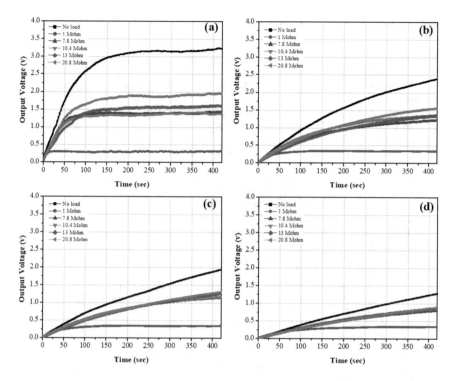

Fig. 7 Vout (v) versus time (s) across different load capacitor **a** 4.7μF **b** 33 μF **c** 47 μF **d** 100 μF

4 Conclusion

The working operation of lateral sliding TENG mode based on conductor-to-dielectric technique has been discussed. In the present study, Copper, Aluminum, and PTFE are used as triboelectric materials to generate energy and simulation results have been presented. It is observed that a maximum of 3.3 V is obtained across 4.7 μF capacitor. The maximum power of 0.257 μW is found across 7.8 MΩ resistor and 4.7 μF capacitor. It is also demonstrated that maximum energy of 95 μJ/cm^{-3} across a 33 μF capacitor is obtained in 420 s. Such materials can be used to generate lot of energy suitable for low power electronics.

References

1. M.D. Emilio, Microelectronic Circuit Design for Energy Harvesting Systems. Springer International Publishing AG (2017)
2. S. Priya, D.J. Inman, *Energy Harvesting Technologies* (Springer Science Business Media, LLC, 2009)
3. D. Briand, E. Yeatman, S. Roundy, Micro energy harvesting. Wiley-VCH, (2015)
4. L.B. Kong, T. Li, H.H. Hng, F. Boey, T. Zhang, S. Li, *Waste Energy Harvesting* (Springer-Verlag, Berlin Heidelberg, 2014)
5. Z.L. Wang, L. Lin, J. Chen, S. Niu, Y. Zi, Triboelectric Nanogenerators. Springer International Publishing Switzerland (2016)
6. Z.L. Wang, J. Chen, L. Lin, Progress in triboelectric nanogenerators as a new energy technology and self-powered sensors. Energy Environ. Sci. **8**, 2250–2282 (2015)
7. J.H. Lee, J. Kim, T.Y. Kim, M.S.A. Hossain, S.W. Kim, J.H. Kim, All-in-one energy harvesting and storage devices. J. Mater. Chem. A **4**, 7983–7999 (2016)
8. P. Azad, R. Vaish, Portable triboelectric based wind energy harvester for low power applications. Eur. Phys. J. Plus **132**, 253 (2017)
9. Khushboo, P. Azad, Triboelectric nanogenerator based on vertical contact separation mode for energy harvesting. in IEEE, ICCCA 2017, (IEEE, India, 2016), pp. 1499–1502
10. M. Vaish, M. Sharma, R. Vaish, V.S. Chauhan, Experimental study on waste heat energy harvesting using lead zirconate titanate (PZT-5H) pyroelectric ceramics. Wiley, Energy Technol. **3**, 768–773 (2015)
11. Y. Mao, D. Genga, E. Liang, X. Wang, Single-electrode triboelectric nanogenerator for scavenging friction energy from rolling tires. Nano Energy **15**, 227–234 (2015)
12. M. Vaish, M. Sharma, R. Vaish, V. Chauhan, Electrical energy generation from hot/cold air using pyroelectric ceramics. Integrated ferroelectrics, J. Integr. Ferroelectr., Taylor & Francis **167**(1), 90–97 (2015)
13. Z.L. Wang, Triboelectric nanogenerators as new energy technology and self-powered sensors—Principles, problems and perspectives. Energy Environ. Sci., Faraday Discussions **176**, 447–458 (2014)
14. G. Zhu, J. Chen, Y. Liu, P. Bai, Y.S. Zhou, Q. Jing, C. Pan, Z.L. Wang, Linear-grating triboelectric generator based on sliding electrification. ACS publication, Nano Letters **13**, 2282–2289 (2013)
15. Y. Zil, J. Wang, S. Wang, S. Li, Z. Wen, H. Guo, Z.L. Wang, Effective energy storage from a triboelectric nanogenerator. Nat. commun. **7**, (2016)
16. F. RuFan, Z.Q. Tian, Z.L. Wang, Flexible triboelectricgenerator. Nano Energy, **1** (Elsevier Ltd., 2012), pp. 328–334
17. S. Niu, Y. Liu, Y.S. Zhou, S. Wang, L. Lin, Z.L. Wang, Optimization of triboelectric nanogenerator charging systems for efficient energy harvesting and storage. IEEE transaction on electron devices **62**(2), 641–647 (2015)
18. A. Proto, M. Penhaker, S. Conforto, M. Schmid, Review nanogenerators for human body energy harvesting, trends in biotechnology. Elsevier Ltd. **35**(7), 610–624 (2017)
19. F.K. Shaikh, S. Zeadally, Energy harvesting in wireless sensor networks: A comprehensive review, renewable and sustainable energy reviews. Elsevier Ltd. **55**, 1041–1054 (2016)

Task-Space Pose Decomposition Motion Control of a Mobile Manipulator

M. Muralidharan and S. Mohan

Abstract This paper presents a task-space pose decomposition motion control scheme for a mobile manipulator to simplify the complexity of the conventional task-space pose control. The proposed scheme is decomposing the task-space pose vector (positions and orientation of the end effector) into a position vector and an orientation vector. Further the decomposed position vector (end effector position) is controlled through the task-space control and the orientation vector is controlled by using a joint-space control. The efficacy of the proposed control scheme is demonstrated numerically on a planar mobile manipulator (consists of a mobile platform and two link revolute (2R) serial planar manipulator) and the proposed scheme is compared with the conventional task-space pose control scheme.

Keywords Mobile manipulation · Task-space control · Motion control
Proportional-derivative control and planar mobile manipulator

1 Introduction

Mobile manipulator consists of a manipulator arm mounted on a mobile base. Mobile manipulators had a strong coupled dynamics between the mobile platforms and the arm. Over the last few years, the motion control of mobile manipulators has been driving interest from the researchers and industries [1]. In the recent past, many researchers have developed various motion control schemes for the mobile manipulators [2–5, 6–8].

M. Muralidharan (✉) · S. Mohan
Centre for Robotics and Control, Mechanical Engineering, Indian Institute
of Technology (IIT) Indore, Indore 452552, India
e-mail: phd1601103001@iiti.ac.in

S. Mohan
e-mail: santhakumar@iiti.ac.in

© Springer Nature Singapore Pte Ltd. 2019
H. Malik et al. (eds.), *Applications of Artificial Intelligence Techniques
in Engineering*, Advances in Intelligent Systems and Computing 698,
https://doi.org/10.1007/978-981-13-1819-1_18

In general, finding the inverse kinematic solution of a serial configuration is complex. Especially kinematically redundant systems like mobile manipulators, the inverse kinematics is more complex than regular configurations because of the existence of multiple solutions and finding the closed-form solutions are not possible in general to these configurations. The task-space and joint space is mapped with the help of the Jacobian matrix, which do the mapping between the end effector velocities and the joint angle velocities. In mobile manipulators, finding the inverse of the Jacobian matrix is not possible because, the Jacobian matrix is not a square matrix. Researchers mostly used pseudo inverse technique to solve this problem. From the literature, it is found that the complete task-space control of both position and orientation is complex. In case of spatial mobile manipulators, finding the combined Jacobian of the end effector position and orientations is complex. Further taking the pseudo inverse of Jacobian matrix is complex in which, mixed units of the position and orientations are present. So far, task-space position control only reported [9, 10, 11–13]. In this paper a task-space pose decomposition scheme is presented to simplify the control of both position and orientation in task-space.The paper is structured as follows. The dynamic modeling of the mobile manipulator is derived in Sect. 2. In Sect. 3, design of proposed task-space pose decomposition scheme is discussed. Performance analysis of the proposed scheme is presented in Sect. 4. Finally, Sect. 5 consists of the conclusion.

2   Dynamic Modeling of a Mobile Manipulator

The mathematical model of a planar mobile manipulator is derived to study the dynamic behavior of the proposed control scheme. The conceptual design of the mobile manipulator is shown in Fig. 1. The generalized dynamic model of the mobile manipulator in a matrix and vector form can be expressed as follows:

$$\mathbf{M(q)\ddot{q} + n(q, \dot{q}) = \tau} \tag{1}$$

where, $M(q) \in \Re^{n \times n}$ is the inertia matrix of the mobile manipulator, $n(q, \dot{q}) \in \Re^n$ is the vector of other effects (which includes restoring, frictional, centripetal and Coriolis effects). $q \in \Re^n$, $\dot{q} \in \Re^n$ and $\ddot{q} \in \Re^n$ are the joint position, velocity, and acceleration vector respectively. $\tau \in \Re^n$ is the generalized vector of input torque.

3   Task-Space Pose Decomposition Motion Control Scheme

The proposed scheme can be applied only if the last three joints of the manipulator arm are revolute joints and having mutually perpendicular joint axes. In order to demonstrate and study the performance of proposed control scheme, a proportional-

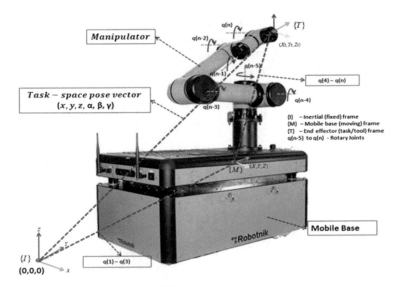

Fig. 1 Conceptual design of a mobile manipulator

derivative (PD) controller is adopted for the motion control. Motion control of the majority of the industrial manipulators are controlled using PD control law because of its simple structure. This section describes the design of proposed task-space pose decomposition scheme for a mobile manipulator. The connection between task-space variables and the configuration-space variables given as

$$\mu = \text{fun}(\mathbf{q}) \tag{2}$$

where, μ is the task-space pose vector and is given as

$$\mu = \begin{bmatrix} p_\mu \\ o_\mu \end{bmatrix} = \begin{bmatrix} \left(\mu_x\ \mu_y\ \mu_z \right)^T \\ \left(\alpha\ \beta\ \gamma \right)^T \end{bmatrix} \tag{3}$$

p_μ is the task-space position vector and o_μ is the task-space orientation vector (generally, these orientations are represented as Euler angles with an axis convention of ZYX). fun(q) is the vector function representing the connection between the configuration-space variables to the task-space variables. i.e., forward kinematics of the system. Differentiating (2) with respect to time gives the velocity relationship as

$$\dot{\mu} = \mathbf{J}(\mathbf{q})\dot{\mathbf{q}} \tag{4}$$

where, $J(q)$ is the Jacobian matrix which maps the task-space velocities and joint-space velocities. The proposed task-space decomposition scheme decomposes the pose vector of task-space coordinates μ into ρ and o. Where ρ is the function of task-space pose vector. o is the function of task-space orientation vector.

$$\begin{bmatrix} \rho \\ o \end{bmatrix} = \begin{bmatrix} fun_1(\mathbf{p}_\mu, \mathbf{o}_\mu) \\ fun_2(\mathbf{o}_\mu) \end{bmatrix} \tag{5}$$

In the proposed scheme, the configuration-space position vector $\mathbf{q}$ is decomposed as

$$\mathbf{q} = \begin{bmatrix} \eta \\ o \end{bmatrix} = \begin{bmatrix} \left(q_1\ q_2\ \cdots\ q_{n-3} \right)^{\mathrm{T}} \\ \left(q_{n-2}\ q_{n-1}\ q_n \right)^{\mathrm{T}} \end{bmatrix} \tag{6}$$

where, η is the decomposed generalized joint-space vector coordinates to find the position and o is the decomposed generalized joint-space vector coordinates of last three joint of the manipulator to find the orientation. The actual position vector is can be found as follows.

$$\rho = fun_3(\eta) \tag{7}$$

$fun_3(\eta)$ is the vector function representing the task-space pose decomposed position. The position error $\tilde{\rho}$ and orientation error $\tilde{o}$ is obtained as follows

$$\tilde{\rho} = \rho_d - \rho \tag{8}$$
$$\tilde{o} = o_d - o \tag{9}$$

where, ρ_d is the desired position and ρ is the actual position. o_d is the desired orientation o is the actual orientation. The PD control law is given as

$$\mathbf{f}_\rho = k_{p1}\tilde{\rho} + k_{d1}\dot{\tilde{\rho}} \tag{10}$$

$$\tau_o = k_{p2}\tilde{o} + k_{d2}\dot{\tilde{Q}} \tag{11}$$

where f_ρ is the task-space force vector obtained by the task-space control of positions using a PD control and τ_o is the vector of input torques obtained by the joint-space control of orientations using a PD control. The task-space force vector f_ρ is transformed to the vector of the joint-space torques τ_η and given as

$$\tau_\eta = \mathbf{J}_\rho(\eta)^{\mathrm{T}} \mathbf{f}_\rho \tag{12}$$

Fig. 2 Block diagram of the proposed task-space pose decomposition scheme

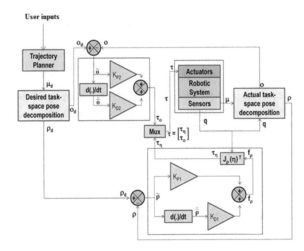

Where, the $J_\rho(\eta)$ is the Jacobian matrix obtained from the vector $\eta = \left(q_1 \ q_2 \ \cdots \ q_{n-3} \right)^T$. The vector of generalized joint torques τ is given as

$$\tau = \begin{bmatrix} \tau_\eta \\ \tau_o \end{bmatrix} = \begin{bmatrix} J_\rho(\eta)^T f_\rho \\ \tau_o \end{bmatrix} \tag{13}$$

The corresponding block diagram of the proposed scheme is presented in Fig. 2. The flow of the block diagram starts from the user inputs which are given to trajectory planner based on the end effector task. The trajectory planner gives the desired time trajectory of the task-space pose vector to the desired task-space pose decomposition block. This block decomposes the task-space pose vector into vector of pose decomposed position ρ_d and vector of pose decomposed orientation o_d.

Through sensors, the task-space and joint-space feedbacks are given to the actual task-space decomposition block. This block gives the position vector ρ and orientation vector o. The position error and orientation error is obtained from a summing point. Further, the PD control law is adopted. From the task-space control of position, force vector f_ρ is obtained. From the joint-space control of orientation, torque vector τ_o is obtained. The force vector is transformed to corresponding joint-space torque vector τ_η through Jacobian matrix $J_\rho(\eta)$. Finally the multiplexer block combines the different torques and gives the combined torque vector τ. The torque is further given to the actuator of the mobile manipulator. The proposed scheme is compared with the conventional task-space control scheme. The PD control for the conventional task-space control scheme is given as

$$\mathbf{f}_\mu = k_p \widetilde{\boldsymbol{\mu}} + k_d \dot{\widetilde{\boldsymbol{\mu}}} \tag{14}$$

where f_μ is the vector of task-space forces. $\tilde{\mu}$ is the vector of task-space errors and is given as

$$\tilde{\mu} = \mu_d - \mu \tag{15}$$

where, μ_d is the desired task-space vector and μ is the actual task-space vector. The task-space force vector f_μ is transformed as a joint-space torque vector τ through the Jacobian matrix is given as

$$\tau = \mathbf{J(q)}^T \mathbf{f}_\mu \tag{16}$$

4 Performance Analysis

4.1 System Description

To validate the proposed scheme, a PD control on a planar mobile manipulator is adopted. The conceptual design of planar mobile manipulator is shown in Fig. 3.

The task-space pose vector of the planar mobile manipulator is $\mu = \begin{bmatrix} x & y & \varphi \end{bmatrix}^T$. x and y are the end effector positions and φ is the end effector orientation. The vector of joint-space position variables of the planar mobile manipulator is $q = \begin{bmatrix} x_v & y_v & \theta_v & \theta_2 \end{bmatrix}^T$. Where, x_v and y_v are the centre coordinate points of the platform, θ_v is the Platform angle, θ_1 is the angle of link1 of the manipulator, θ_2 is the angle of link2 of the manipulator. The 5DOF planar mobile manipulator is a kinematically redundant system where the joint-space variables (n) are greater than the task-space variables (m). In the proposed scheme, the 5DOF planar mobile manip-

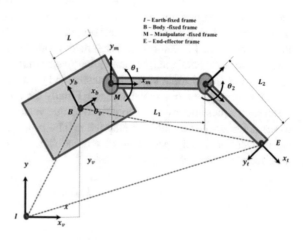

Fig. 3 Conceptual design of a planar mobile manipulator

Fig. 4 Desired trajectory in
task-space

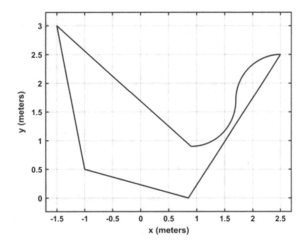

ulator is decomposed into 4DOF for position control and last 1DOF for orientation control. The proposed scheme is valid only if the last joint is revolute joint in planar case.

4.2 Task Description

The task preferred here requires the mobile manipulator's end effector to start at a customized primary position and orientation and come back to the same pose after passing through a defined trajectory. The trajectory loop consists of an upward ramp and a concave up arc and concave down arc and an upward ramp and a downward ramp in vertical direction and succeeded by a downward ramp as shown in Fig. 4.

4.3 Results and Discussions

The numerical simulation in MATLAB/SIMULINK is performed to validate the performance of the proposed scheme on a planar mobile manipulator. The physical parameters considered for the mobile manipulator simulation is as follows. Mass of the mobile platform (mv)=177 kg, Mass of link1 (m1)=10.60 kg, Mass of link2 (m2)=9.20 kg, Distance between base frame of the arm and fixed frame of mobile platform (av)=0.3 m, Length of link1 of the manipulator (a1)=0.3 m, Length of link2 of the manipulator (a2)=0.25 m, Moment of Inertia of mobile platform (Izv)= 10.15 kg m^2, Moment of Inertia of Link1 (Iz1)=0.17 kg m^2, Moment of Inertia of Link2 (Iz2)=0.12 kg m^2. The forward kinematic equations of the planar mobile manipulator are derived as follows.

Table 1 Comparison of proposed scheme based onIAE

Scheme	Proposed Scheme				Conventional scheme		
Controller Gain	$k_{p1} = 25$	$k_{d1} = 10$	$k_{p2} = 25$	$k_{d2} = 10$	$k_p = 25$		$k_d = 10$
IAE	x	y	φ		x	y	φ
	0.0385 m	0.0566 m	1.1506 rad		0.0364 m	0.0855 m	0.0878 rad

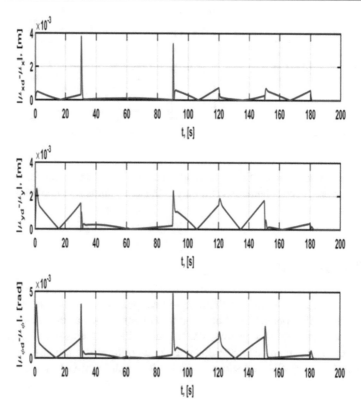

Fig. 5 Errors in conventional scheme

$$x_t = x_v + av * \cos(\theta_v) + a1 * \cos(\theta_v + \theta_1) + a2/2 * \cos(\theta_v + \theta_1 + \theta_2) \quad (17)$$

$$y_t = y_v + av * \sin(\theta_v) + a1 * \sin(\theta_v + \theta_1) + a2/2 * \sin(\theta_v + \theta_1 + \theta_2) \quad (18)$$

Integral absolute error (IAE) is taken as a performance index to compare the proposed scheme with the conventional scheme.

$$IAE = \int_0^T |e(t)| dt \quad (19)$$

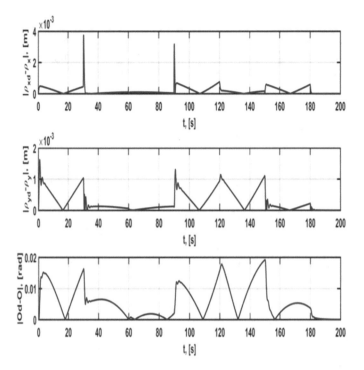

Fig. 6 Errors in proposed scheme

where, $|e(t)|$ is the vector of absolute error of task-space variables. Table 1 shows the comparison of proposed scheme with conventional scheme based on IAE. The same controller gain values are considered for the comparison. The controller gains are chosen in such a way that, the system behaves as a critically damped system. The main objective of this work is to check the effectiveness of the proposed scheme with the conventional scheme.

Further, to reduce the orientation error of the joint-space control in the proposed scheme, the control gain values are taken as $k_{p2} = 18 \times 18$ and $k_{d2} = 2 \times 18$ in which, IAE value for orientation is reduced as 0.0898 rad.

From Figs. 5 and 6, it is observed that the absolute error of task-space position in the xdirection of the proposed scheme is almost equal to the conventional scheme values. The peak values of task-space position errors in both x and y directions are higher in the conventional scheme. The absolute error of task-space position in y direction is small in the proposed scheme as compared to conventional scheme. The absolute error of the orientation in the proposed scheme is higher as compared to the conventional scheme. In case of spatial mobile manipulators, finding the combined Jacobian of the end effector position and orientations is complex. Further taking the pseudo inverse of Jacobian matrix is complex in which, mixed units of the position and orientations are present. The proposed scheme simplifies this complexity. The proposed scheme can be easily extended to a spatial mobile manipulator where the conventional task-space control scheme is more complex.

5 Conclusions

This paper presents a new approach to solve the complexity of the complete task-space control of both position and orientation of the mobile manipulator. The simulation results are presented. Simulation results show that the proposed scheme is as effective as conventional scheme in planar case. It can be extended further to spatial mobile manipulator where the complete task-space control of both position and orientation is more complex. Further, experimental validation of the proposed scheme for planar mobile manipulator with different motion control schemes are under progress.

References

1. V. Andaluz, F. Roberti, J. M. Toibero, R. Carelli, Adaptive unified motion control of mobile manipulators. Control Eng. Pract. 1337–1352 (2012) https://doi.org/10.1016/j.conengprac.2012.07.008
2. S. Lin, A. Goldenberg, Neural-network control of mobile manipulators. IEEE Trans. Neural Netw. **12**, 1121–1133 (2001). https://doi.org/10.1109/72.950141
3. S. Lin, A. Goldenberg, Robust damping control of mobile manipulators. IEEE Trans. Syst. Man Cybern. B Cybern. **32**, 126–132 (2002). https://doi.org/10.1109/3477.979968
4. Y. Liu, Yangmin Li, Dynamic modeling and adaptive neural-fuzzy control for nonholonomic mobile manipulators moving on a slope. Int. J. Control Autom. Syst. **4**(2), 197–203 (2006)
5. D. Xu, D. Zhao, J. Yi, X. Tan, Trajectory tracking control of omnidirectional wheeled mobile manipulators: robust neural network-based sliding mode approach. IEEE Trans. Syst. Man Cybern. B Cybern. **39**, 788–799 (2008). https://doi.org/10.1109/TSMCB.2008.2009464
6. V. Andaluz, R. Carelli, L. Salinas, J. Toibero, F. Roberti, Visual control with adaptive dynamical compensation for 3D target tracking by mobile manipulators. Mechatronics **22**, 491–502 (2011). https://doi.org/10.1016/j.mechatronics.2011.09.013
7. N. Chen, F. Song, G. Li, X. Sun, C. Ai, An adaptive sliding mode backstepping control for the mobile manipulator with nonholonomic constraints. Commun. Nonlinear Sci. Numer. Simul. **18**, 2885–2899 (2013). https://doi.org/10.1016/j.cnsns.2013.02.002
8. X. Wu, Y. Wang, X. Dang, Robust adaptive sliding-mode control of condenser-cleaning mobile manipulator using fuzzy wavelet neural network. Fuzzy Sets Syst. **235**, 62–82 (2013). https://doi.org/10.1016/j.fss.2013.07.009
9. M. Boukattaya, T. Damak, M. Jallouli, Robust adaptive control for mobile manipulators. Int. J. Autom. Comput. 8–13 (2011). https://doi.org/10.1007/s11633-010-0548-y
10. M. Galicki, Control of mobile manipulators in a task space. IEEE Trans. Autom. Control **57**, 2962–2967 (2012). https://doi.org/10.1109/TAC.2012.2195935

11. M. Galicki, Robust task space trajectory tracking control of robotic manipulators. Int. J. Appl. Mech. Eng. **21**(3), 547–568 (2016). https://doi.org/10.1515/ijame-2016-0033
12. S. Mishra, P. Londhe, S. Mohan, S.K. Vishwakarama, B. Patre, Robust task-space motion control of a mobile manipulator using a nonlinear control with an uncertainty estimator. Comput. Electr. Eng. 1–12 (2017). https://doi.org/10.1016/j.compeleceng.2017.12.018
13. S. Zhou, Yazhini C. Pradeep, Ming Zhu, Kendrick Amezquita-Semprun, Peter Chen, Motion control of a nonholonomic mobile manipulator in Task space. Asian J. Control **20**(5), 1–10 (2018). https://doi.org/10.1002/asjc.1694

Structural Triboelectric Nanogenerators for Self-powered Wearable Devices

Sudha R. Karbari

Abstract Self-energized sensors operate on the energy generated from the nanostructures in the sensors. The simulated Nanostructures are constructed with the polymers like PDMS and PMMA, gold nanoparticles is sputtered between the polymer layers to increase the electrons movement between the layers. The structure operates on the principle of triboelectrification. The structure is simulated and analyzed for different dimensions of the structure in steps of 1 cm $\times$ 1 cm varying from 2 cm $\times$ 2 cm to 5 cm $\times$ 5 cm. The output parameters like stress, displacement, electric potential, and capacitance are analyzed for the structural triboelectric nanogenerators impregnated into the device. The output voltages from the simulations give us an analytical measure of conversion from mechanical energy to electrical energy. The potential induced by 5 cm $\times$ 5 cm design are best suitable for any Wearable applications. The optimized design of 5 cm $\times$ 5 cm dimension is simulated for different eigen frequencies (22.073, 43.977 and 44.252 kHz) offers 8 V potential with displacement and capacitance value of 3.5 μm and 1.46 fF respectively at the eigen frequency of 43.977 kHz. The complete device with its complete operation and design can be integrated to the bottom of the shoe sole to make it self-powered for LECHAL GPS based tracking device.

Keywords STENG (Structural triboelectric Nanogenerators) · TENG (Triboelectric Nanogenerators) · PDMS (polydimethylsiloxane) · PMMA (Polymethyl methacrylate)

1 Introduction

The triboelectric effect exhibits in a material that becomes electrically charged after it contacts different material through friction, where a chemical bond is formed between some parts of the two functionized surfaces, called adhesion and charges a move from

S. R. Karbari (✉)
Department of ECE, R. V. College of Engineering, Bengaluru, India
e-mail: sudhark@rvce.edu.in

© Springer Nature Singapore Pte Ltd. 2019
H. Malik et al. (eds.), *Applications of Artificial Intelligence Techniques in Engineering*, Advances in Intelligent Systems and Computing 698,
https://doi.org/10.1007/978-981-13-1819-1_19

one material to the other to equalize their electrochemical potential attained through the process. The transferred charges can be electrons or may be ions/molecules. When separated from a bonded atoms have the tendency to retain electrons, and some a tendency to give them away the electrons, producing triboelectric effect with respect to the charges on the surfaces. The existence of charges in a tribolectric from on dielectric surfaces can be the force for driving electrons in an electrode of the device to flow in towards the respective layer in order to balance an electric potential drop created. mechanical energy. Many mechanisms of energy harvesting techniques have been developed that are based on various mechanisms including piezoelectric effect, electrostatic effect, and electromagnetic induction. The triboelectric nanogenerators (TENGs) are an excellent approach to convert mechanical strength into electrical energy, which is dependent on the coupling between triboelectrification and electrostatic induction. The TENG has the potential of harvesting mechanical energies such as vibrations, rotation, wind, human motion and water wave energy. Structural nanogenerators are the nanogenerators that are sonicated and impregnated are having functional properties and hence are called as structural nanogenerators (STENG).

As the structural nanogenerator device harvests an open circuit voltage in the range of 1200 V/m^2, the current pulse generated by the human foot fall when a contact is made is used to glow the commercial LED's and can be integrated to the bottom of the shoe sole to make it self-powered for GPS based tracking device for giving one detailed route guidance at every turn during the trekking and adventurous activity.

When materials are brought into contact through friction, a difference in potential is created by the distance of separation between the two material surfaces. Electrostatic induction phenomenon is an electricity generating process such that electrons in one electrode would flow to the other electrode through the external load in order to balance the potential difference. As for TENGs, they realize the conversion efficiency of mechanical energy into usable electric energy by the integration of triboelectrification with electrostatic inductions [1].

In triboelectric nanogenerators, two sheets are rubbed together that act as two electrodes. One electrode donates while the other electrode accepts electrons produced. After rubbing these two electrodes, they are separated instantly generating a gap between them that separates the electrons on PMMA and generate positive and negative charges. This would be so sensitive that it can detect as low as 13 mPa pressure. Hence it would be able to detect even a mass equivalent of small feather or water droplet and produce small amount of current that touches the surface of generator. Moreover, as this device can be transparent up to 75% so the nanogenerators can be used in touch screens that will generate their own power on touching, using the vibration produced [2].

Here Sect. 1.1 includes the history of the nanogenerators, and the construction of the structure to induce the triboelectric effect by the use of the triboelectric phenomenon. Section 2 consists of the description of the proposed model and the materials used in the construction of the model along with the material properties. Section 4 deals with the results of simulation for different Eigen frequencies. In Sect. 5, the comparison of different dimensions of the structure is done.

1.1 History of Triboelectric Nanogenerators

When the floors are engineered with piezoelectric-based technology, the electrical power produced by the pressure of human footsteps is converted to an electric charge by piezoelectric transducers, then store. These sensors make use of the piezoelectric effect, in which materials have the ability and tendency to generate an electrical potential for strain applied to them. These devices have a drawback with high impedance and can pick up the stray voltages in the connecting wires. The voltage generated so is less and hence the efficiency of the device is less.

Two different layers are used for the generation of electricity. When two dielectric materials are applied by an external stimulus, the two materials are in contact with each other. There is a surface charge transfer then occurs on two contacting surfaces due to triboelectric effect [2]. There are four modes of operation of the TENG that can be applied in wide range of application. The complete area power density reaches 1200 W/m^2, volume density reaches 490 kW/m^3, and an energy conversion efficiency of 50–85% [3].

The self-powered sensors for detecting water or ethanol in gaseous form or in liquid form, can be powered with the triboelectric nanogenerators. The triboelectric nanogenerators are made of polyamide film or polytetrafluoroethylene film. The performance of the active sensors has been understood in reference to the level of wettability of solid polymer surfaces. This new approach for powering the sensors could be advantageous of simple fabrication, low-cost and easy application [4].

2 Description for Proposed Model

Here, we simulated and analyzed triboelectric generators based on the principle of surface charge transfer that is surface modification which is simplified and yielding high power output with low input in terms of pressure enabled by surface modification (Fig. 1).

The system for triboelectric energy harvesting consists of two substrates each made of structural polymers PMMA and PDMS [5]. The bottom substrate simulated is fabricated using spin coating process with PMMA acting as a top surface with thin deposited gold layer using sputtering for the charge transfer between the layers of polymers. The dimensions chosen for the base substrate is 2 cm × 2 cm with thickness of 0.25 cm. The top substrate comprises of PMMA and PDMS with a thin coating of gold layer sandwiched in between the two layers. The nanoparticles of metals can also be used as nanogenerators instead of the gold layer for both the substrates as replacements are having structural functioning such as their dimensions and so

Fig. 1 Description of the model with all the layers for simulation [5]

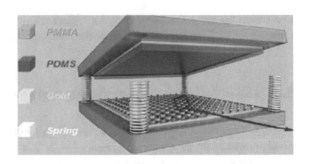

called as structurally programmed nanogenerators and hence Structural Triboelectric Nanogenerators STENG. The two layers are further separated by the two substrates are connected by four springs at the four corners forming anchors, leaving a narrow spacing between the contact electrode and the PDMS [6]. The anchor boundaries of the structure are fixed in COMSOL Multiphysics to achieve the effect of the spring for contact and henceforth relaxation of layers to induce the triboelectric effect.

The electric energy generation process can be explained by the coupling between triboelectric effect and electrostatic effect [7]. Here we have successfully simulated the results for the electrostatic effect by solving the electrostatics effect using COMSOL multiphysics.

2.1 Material Properties

The polymer materials PMMA and PDMS are thermoplastic and suitable for the energy harvesting applications (Table 1).

The values of various material parameters in terms of Coefficient of thermal expansion, heat capacity, poisons ratio, young's modulus, Thermal conductivity, relative permittivity and density. The material gold has excellent conductivity of

Table 1 Material properties of STENG [7]

Properties	PDMS	GOLD	PMM A
Density	970 (kg/m^3)	19,300 (kg/m^3)	1190 (kg/m^3)
Relative permittivity	2.75	1.62	3.0
Thermal conductivity	0.16 (W/m K)	317 (W/m K)	0.17 (W/m K)
Young's modulus	750 kPa	$70 e^9$ Pa	$3 e^9$ Pa
Poisson's ratio	0.50	0.44	0.40
Heat capacity	1460 (J/kg K)	129 (J/kg K)	1420 (J/kg K)
Coefficient of thermal expansion	$9 e^{-4}$ (1/K)	$14.2 e^{-6}$ (1/K)	$70 e^{-6}$ (1/K)

317 W/m and henceforth chosen as the nanogenerators layer for the simulation. Gold nanolayer has a highest Young's modulus with a value of 70 e^9 Pa. The design of the structure includes PMMA with 0.25 cm, PDMS with 0.01 cm and a gold layer with 0.99 mm.

Here various values of stress, strain, and the resonant frequency of the various designs for parametric analysis for energy harvesting applications using triboelectric nanogenerators is considered. Here the materials chosen and the structure chosen are very crucial as they play a major role in the device functioning for the specific property which is triboelectric.

3   Experimental Results

The parametric optimization is performed by varying the dimensions of the structural polymers along with the sandwiched gold layer from 9 cm^2 surface area to a wide range of 25 cm^2.

3.1   Design 1 of STENG

The simulated structure is resonant at a frequency of 1.44 e^{5Hz}–1.052 e^{5Hz}. The total displacement for the Eigen frequency of 144.2, 146.3 and 105.2 kHz are 9 e^{-7} cm, 7 e^{-7} cm, and 5 e^{-7} cm. The shape of the modal frequency is along the Z direction as the bottom layer is fixed and the boundary of the top layers is fixed. As the top substrate comes in contact with the bottom substrate there is charge transfer along the surface because of the surface modification and hence forth contributing to an electric potential. The displacement for an eigen frequency of 30 kHz is 2 e^{-8} cm is the lowest displacement that can be experienced by the structure constituting the top and bottom substrates. The total dimensions of design 1 chosen for simulation is 2 cm × 2 cm.

Figure 2a shows the multislice view of the structure in COMSOL graphics window. This option is enabled so as to view along the x direction with the displacement values. The number of planes chosen for the multislice operation is five. The electric potential that is obtained by the surface modification due to the charge carriers is also illustrated and the value is 4.8 V maximum for a frequency of 105.2 kHz.

The values of different Eigen modes versus displacement in cm for the structure built in the graphics window using the geometry options available for the triboelectric nanogenerators. The resonant frequencies of the structures are 105 kHz to 146 kHz with values of displacement from 5 e^{-7} cm to 9 e^{-7} cm with 7 e^{-7} cm in an alternate step in between.

Figure 3a shows plot of displacement versus arc length for three eigen frequency. The value of displacement varies along the arc length with the maximum of 9 e^{-7} cm for 144.2 kHz and 7 e^{-7} cm for 146.3 kHz, and 5 e^{-7} for 105.2 kHz respec-

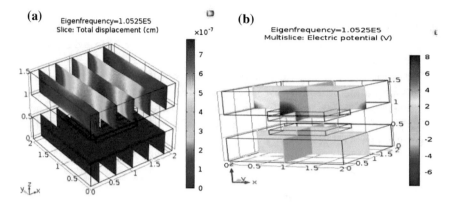

Fig. 2 a Electric potential of STENG 1. **b** Total displacement of STENG 1

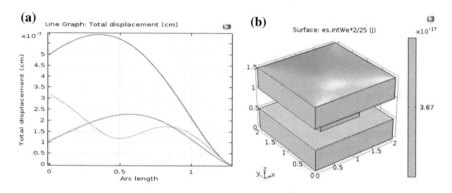

Fig. 3 a Total displacement of STENG1. **b** 3D plot of the capacitance versus arc length

tively. The capacitance value need to be calculated as that is a factor which conveys the charge storing capacity for any energy harvesting application. As gold is sandwiched between the two-polymer substrate and henceforth improving the capacitance contributing to surface charge modification. For the given design and specification, the value of capacitance is $3.67\,e^{-16}$ F.

3.2 Design 2 of STENG

The simulated structure is resonant at a frequency of 53.639–99.852 kHz. It can be inferred that the maximum displacement occurs at 53.639 kHz with a value of 1.2 e^{-6} cm. The total dimensions of design 2 chosen for simulation is 3 cm × 3 cm.

The total displacement in cm for an Eigen frequency of 99.132 kHz with a displacement value of $7\,e^{-7}$ the displacement occurring at 99.852 kHz along the Z

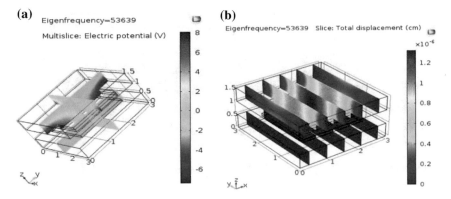

Fig. 4 **a** Electric potential of STENG 2. **b** Total displacement of STENG 2

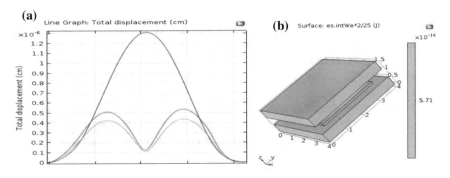

Fig. 5 **a** Total displacement of STENG 2. **b** 3D plot of the capacitance versus arc length

direction and value of the displacement as measured is $6\,e^{-7}$ cm. Figure 4a above shows the multislice view of the structure in COMSOL graphics window. This option is enabled so as to view along the Y direction with the displacement values. Also, the electric potential that is obtained by the surface modification because of the charge carriers is also illustrated and the value is 5.6 V maximum for a frequency of 53.639 kHz.

Figure 5a shows plot of displacement versus arc length for three Eigen frequency. The value of displacement varies along the arc length. The capacitance value need to be calculated as that is a factor which conveys the charge storing capacity for any energy harvesting application. As gold is sandwiched between the two-polymer substrate and henceforth improving the capacitance contributing to surface charge modification for the given design and specification the value of capacitance is $5.71\,e^{-16}$ F.

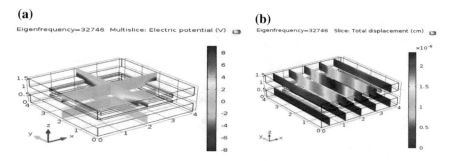

Fig. 6 a Multislice electric potential of STENG 3. b Total displacement of STENG 3

3.3 Design 3 of STENG

The simulated structure was resonant at a frequency of 32.746–64.528 kHz. It can be inferred that the maximum displacement occurs at 32.746 kHz with a value of 2 e^{-6} cm. The displacement occurring at 32.746, 64.081 and 64.528 kHz are 2 e^{-6}, 1.2 e^{-6} and 1 e^{-6} respectively. The total dimensions of design 3 chosen for simulation is 4 cm × 4 cm.

Figure 6a shows the multislice view of the structure in COMSOL graphics window. This option is enabled so as to view along the Y direction with the displacement values. Also, the electric potential that is obtained by the surface modification because of the charge carriers is also illustrated and the value is 6.8 V maximum for a frequency of 32.746 kHz.

Figure 7a shows plot of displacement versus arc length for three eigen frequency. The value of displacement varies along arc length. The capacitance value need to be calculated as that is a factor which conveys the charge storing capacity for any energy harvesting application as gold is sandwiched between the two-polymer substrate and henceforth improving the capacitance contributing to surface charge modification. For the given design and specification, the value of capacitance is 8.99 e^{-15} F (Fig. 8).

3.4 Design 4 of STENG

The simulated structure was resonant at a frequency of 22.073–44.252 kHz. The total displacement in cm for an eigen frequency of 22.073 and 43.977 kHz and 44.252 kHz are 3.5 e^{-6} cm and 1.8 e^{-6} cm and 1.4 e^{-6} cm.

Similarly, the displacement verses the eigen frequency for an eigen frequency of 30 kHz is 4 e^{-8} cm. The total dimensions of design 4 chosen for simulation is 5 cm × 5 cm.

Also the electric potential that is obtained by the surface modification because of charge carriers are also illustrated and the value is 8 V maximum for a frequency

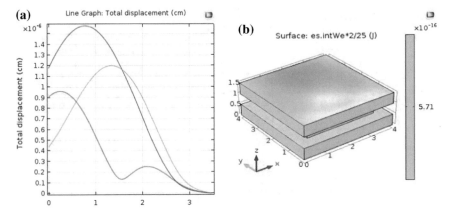

Fig. 7 a Total displacement of STENG 3. **b** 3D plot of the capacitance versus arc length

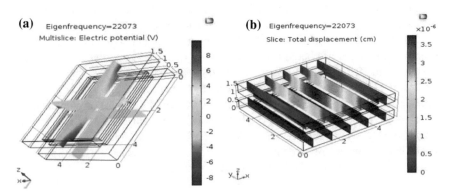

Fig. 8 a Multislice electric potential of STENG 4. **b** Total displacement of STENG 4

of 22.073 MHz. The values of the displacement varies along the arc length with maximum of 3.5 e^{-6} for 22.073 kHz and 1.8 e^{-6} for 42.977 and 1.4 e^{-6} for 44.252 respectively. The capacitance value of $1.46 \times e^{-15}$ F.

4 Comparison of the Proposed Designs

The design 4 has the maximum capacitance value of 1.46 e^{-15} F and the highest displacement of 3.5 e^{-6} cm. The frequency of the swing is very high when compared to the other dimension. At the resonant frequency of 22.073 kHz the structure experiences maximum displacement (Table 2).

Table 2 Comparison of designs for various parameters

Design	Capacitance	Electric Potential (V)	Stress (J/m^3)	Displacement (cm)	Resonant frequency (kHZ)	Energy (J/m^3)
	$3.67\,e^{-16}$	4.8	$1.8\,e^3$	$2\,e^{-8}$	105.2	$8\,e^{-6}$
2	$5.71\,e^{-16}$	5.6	$2.5\,e^3$	$6\,e^{-7}$	99.852	$1.6\,e^{-5}$
3	$8.99\,e^{-15}$	6.8	$3\,e^3$	$2\,e^{-6}$	32.746	$6\,e^{-4}$
4	$1.46\,e^{-15}$	8	$5\,e^3$	$3.5\,e^{-6}$	22.073	$1.2\,e^{-3}$

5 Conclusion and Discussions

In this work we have compared the four different designs with different dimensions and the optimization of different parameters in terms of stress, resonant frequency, capacitance, displacement, electric potential. The spacing between the top surface and bottom surface is chosen to be larger than the polymer thickness to ensure effective generation of charge carries and hence contributing to the surface charge modification. Here instead to using the springs and fixing it at the edges we have made of the COMSOL Multiphysic to obtain the same effect. The values of electric potential, stress, displacement and capacitance induced by the use of gold nanoparticles for 5 cm × 5 cm are 8 V, 5 kJ/m^3, 3.5 e^{-6} cm and 1.46 e^{-15} F. Form the dimensional analysis it can be concluded that 5 cm × 5 cm STENG gives the best results in terms of specification as discussed. The STENG can be fabricated and used in shoe sole for powering the LECHAL GPS tracking.

References

1. Z.L. Wang, J. Chen, L. Lin, Progress in triboelectric nanogenerators as a new energy technology and self-powered sensors. Energ. Environ. Sci. **8**(8), 2250–2282 (2015). https://doi.org/10.1039/c5ee01532d
2. Y. Su, J. Chen, Z. Wu, Y. Jiang, Low temperature dependence of triboelectric effect for energy harvesting and self-powered active sensing. Appl. Phys. Lett. **106**(1), Article ID 013114 (2015). https://doi.org/10.1063/1.4905553
3. B.K. Yun, J.W. Kim, H.S. Kim et al., Base-treated polydimethylsiloxane surfaces as enhanced triboelectric nanogenerators. Nano Energy **15**, 523–529 (2015). https://doi.org/10.1063/1.4945052
4. Y. Wu, Q. Jing, J. Chen et al., A self-powered angle measurement sensor based on triboelectric nanogenerator. Adv. Funct. Mater. **25**(14), 2166–2174 (2015). https://doi.org/10.1021/nn500695q
5. G. Zhu, Z.-H. Lin, Q. Jing et al., Toward large-scale energy harvesting by a nanoparticle-enhanced triboelectric nanogenerator. Nano Lett. **13**(2), 847–853 (2013). https://doi.org/10.1021/nl4001053

6. N. Zhang, C. Tao, X. Fan, J. Chen, Progress in triboelectric nanogenerators as self-powered smart sensors. J. Mater. Res. **32**(5), 1628–1646 (2017). https://doi.org/10.1021/nn506832w
7. Y. Yang, H. Zhang, Z.-H. Lin et al., Human skin based triboelectric nanogenerators for harvesting biomechanical energy and as self-powered active tactile sensor system. ACS Nano **7**(10), 9213–9222 (2013). https://doi.org/10.1021/nn403838y

Analyzing the Optimal Scenario for Energy-Efficient Communication in Underwater Wireless Sensor Network

Sheena Kohli and Partha Pratim Bhattacharya

Abstract Underwater Wireless Sensor Networks comprise of a spread of acoustic sensors deployed in a geographical area of interest under the sea or ocean. The nodes on capturing data from the particular region send it to a central processing point via cluster heads or/and gateway nodes acting as relay nodes. The performance of networks depends upon different parameters and scenarios. The paper aims at identifying the optimal scenario for acoustic networks that can be built up for conserving the energy of the network or utilizing the energy of nodes in an efficient way. Optimal scenario in the network here refers to finding the best position of sink or the base station, location of gateway nodes and topology of the sensors, with the aim of saving energy.

1 Introduction

The concept of wireless sensor technology that involves a sea or an ocean as the area of observation having sensors deployed is known as Underwater Wireless Sensor Network (UWSN). It focuses on connecting a number of tiny acoustic nodes, capable of sensing, detecting or measuring the different parameters of the various ambient activities from the water body. Figure 1 illustrates the arrangement of an Underwater Wireless Senor Network. They are being widely utilized in modern times in different areas of marine research [1, 2]. All the sensor nodes report to Base Station (BS) or sink, controlling unit of network. Besides the normal nodes deployed in the network, gateway nodes aid in streamlining the communication of data from nodes to the sink [3].

This work gives an analysis of optimal scenarios for energy-efficient communication in UWSNs. It has been justified where to position the Base Station and the

S. Kohli (✉) · P. P. Bhattacharya
Mody University of Science and Technology, Lakshmangarh, Rajasthan, India
e-mail: sheena7kohli@gmail.com

P. P. Bhattacharya
e-mail: hereispartha@gmail.com

© Springer Nature Singapore Pte Ltd. 2019
H. Malik et al. (eds.), *Applications of Artificial Intelligence Techniques in Engineering*, Advances in Intelligent Systems and Computing 698,
https://doi.org/10.1007/978-981-13-1819-1_20

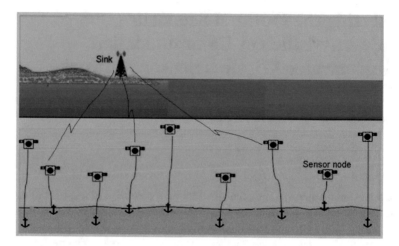

Fig. 1 Structure of underwater wireless sensor network

gateway nodes in a network dealing with clustered- and region-based routing. Further, the topology of the network has also been observed. Simulations have been done in Matrix Laboratory (MATLAB) [4]. The paper includes the following sections: design methodology followed, routing technique used, energy model, optimal scenarios developed with their analysis and finally the conclusion of the work carried out. Related work has been done in this area but the scenarios considered here have not been implemented in terms of UWSNs. Analysis has been done for the terrestrial Wireless Sensor Networks. For UWSNs, authors in [5] proposed an energy dissipation model of sensor node on the basis of energy consumption and optimal clustering. The study in [6], addresses the energy efficiency of error correction methods for UWSNs. The work stated in [7] analyzes the energy required to transfer an information bit over a multi-hop underwater acoustic network considering different parameters and tradeoffs.

2 Design Methodology

2.1 Node Deployment

The nodes are deployed or positioned in the network area depending upon the applications generally. The sensors or nodes can be deployed either in an unplanned, i.e., random fashion or in a planned way, like square or triangular grid [8].

2.2 Base Station Position

The Base Station is the chief point of the network where all the data from network is collected finally. The position and number of BS may affect the performance of the network.

2.3 Gateway Nodes Placement

Gateway nodes basically act as entry/exit level nodes in a Wireless Sensor Network. The gateway nodes are programmed to communicate with a sink or Base Station located outside the network field [9]. A gateway is a mingling platform between the application and the wireless nodes in network [10].

3 Routing

Routing is the way of deciding which path shall be followed by the data for reaching from source to destination in the network.

3.1 Clustered-Based Routing [11]

This category of routing involves sensors organized in groups or clusters which periodically produce information from the proximity being monitored. Data collected from sensors are sent to the cluster head first and then it is forwarded to the Base Station. Cluster heads help in eliminating the data redundancy [12]. Heinzelman et al. [13] introduced a clustering protocol called Low Energy Adaptive Clustered Hierarchy (LEACH) that uses a random technique of selecting cluster heads to distribute energy consumption over all nodes.

3.2 Region-Based Routing [14]

The network is divided into specific regions or segments. Each region has a cluster head selected or a leader for all the sensor nodes. All the cluster heads from different regions or zones send the data to the gateway node which is nearer to them. The gateways will further send the data to the Base Station. This forms the multi-hop communication environment.

4 Energy Model Followed

The energy model used here is inspired from [15]:

$$E = Er + Et \tag{1}$$

$$Er = ERX^*CM^*cluster^*CM/h^*B \tag{2}$$

$$Et = ETX^*(CM + DM) + Efs^*(CM + DM)^*min\_dis^*min\_dis^*(CM + DM)/h^*B \tag{3}$$

The transmission energy Et and reception energy Er sum up to total energy E.

ERX is electronic energy for reception; ETX is electronic energy for transmission; cluster is the number of cluster heads formed; CM is control packet size; DM is data packet size; Efs is energy dissipated; min_dis is the distance between node and its corresponding cluster head; B is bandwidth available; h is bandwidth efficiency of modulation in bps/Hz, given by the equation:

$$h = \log_2(1 + SNR) \tag{4}$$

Signal-to-Noise Ratio (SNR) is signal strength relative to the background noise [16].

5 Optimal Scenario Analysis

5.1 Simulation Parameters and Values

The simulation parameters with values are given in Table 1.

5.2 Results

For the analysis of the same, we have considered three design parameters as follows.

5.2.1 Node Deployment

Sensors can generally be placed in an area of interest either deterministically (i.e. planned based) or randomly [17]. Both involve the clustering approach in paper. The sensor nodes are randomly spread over the given area. Figure 2 is a pictorial representation of a random distribution for 100 nodes. Sensors are placed in the network field in a predefined uniform strategy as shown in Fig. 3.

Table 1 Simulation parameters and values

Parameter	Values
Network size	200 m × 200 m
Number of nodes	100
Cluster head probability	0.05
Fusion rate	0.6
Initial node energy	0.5 J
Control packet size	500 bits
Data packet size	4000 bits
Energy dissipation	10 * 0.000000000001 J
Energy for transmission	50 * 0.000000001 J
Energy for reception	50 * 0.000000001 J
Energy for data aggregation	5 * 0.000000001 J
Bandwidth	4 kHz
Signal to Noise Ratio	20

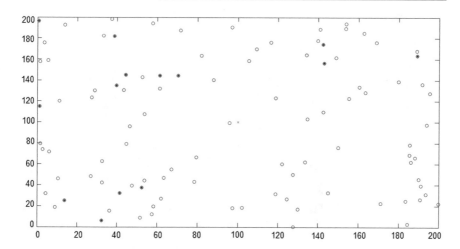

Fig. 2 Random placement

The energy consumption in the network for transmission and reception of data has been analyzed for both the cases. It has been observed that the planned deployment consumes more energy as compared to the random one. The results are depicted in Fig. 4.

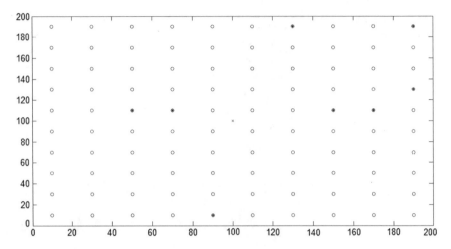

Fig. 3 Planned placement

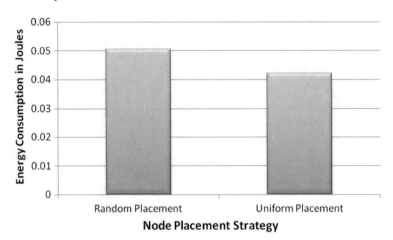

Fig. 4 Energy consumption in different node placement strategies

5.2.2 Base Station Position

Here, in the network area of 200 × 200, BS is placed at the center of the field at 100, 100 as shown in Fig. 5. BS is placed at the edge of the field at 100, 200 as shown in Fig. 6.

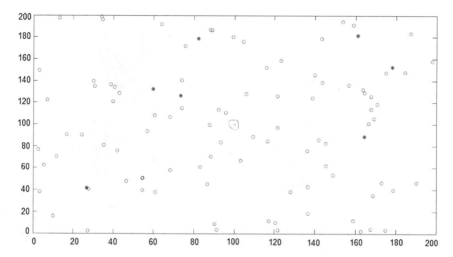

Fig. 5 Base station located at center of network

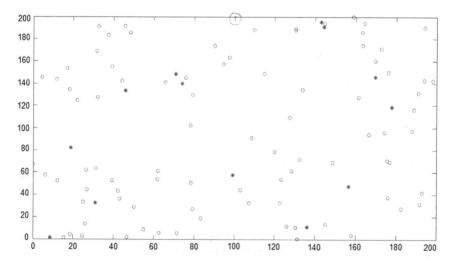

Fig. 6 Base station located at edge of network

While when the BS is at one of the edge of the network, the distance from the nodes situated at the opposite side of the field becomes more, hence resulting in more energy consumption. The same observation can be seen in Fig. 7.

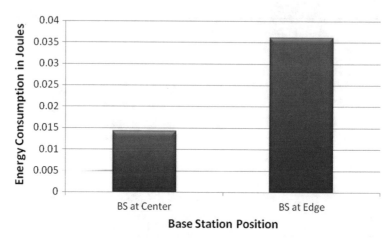

Fig. 7 Energy consumption in different base station positions

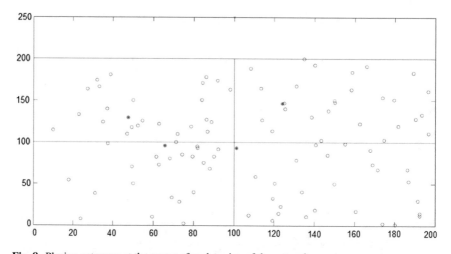

Fig. 8 Placing gateways at the center of each region of the network

5.2.3 Gateway Nodes Placement

The four scenarios analyzed are as shown in Figs. 8, 9, 10 and 11. In each of the scenarios, the nodes or sensors are represented by hollow circles, solid circles show the cluster heads and red hallow circles depict the gateway nodes.

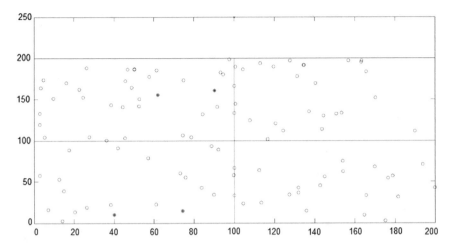

Fig. 9 Placing gateways vertically in a line at the border of regions in the middle of the network

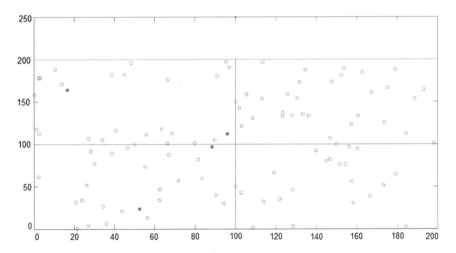

Fig. 10 Placing gateways in the rhombus pattern at the border of region in the network

The energy consumption in the network will be more when gateway nodes are placed in a rhombus pattern that is each gateway node is on each line of the region. The optimal scenario is where the gateways are located in the center of each region. Rest two cases lie in between these options in terms of energy consumption of the network. Results are depicted in Fig. 12.

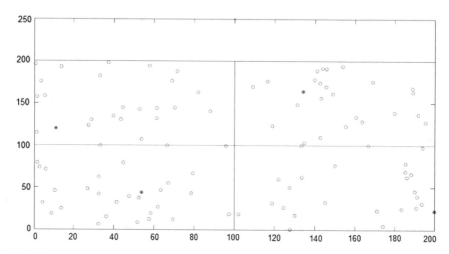

Fig. 11 Placing gateways horizontally in a line at the border of the regions in the middle of the network

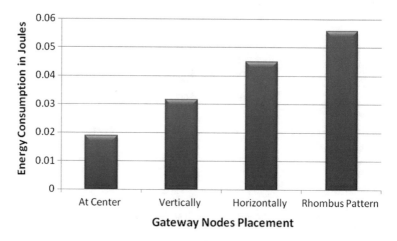

Fig. 12 Energy consumption in different gateway nodes placement

6   Conclusion

The study of the paper concludes that the optimal scenarios for energy-efficient communication in Underwater Wireless Sensor Networks are observed in terms of the node deployment strategy, Base Station position and gateway nodes placement. The planned or the uniform distribution of nodes is better as compared to the random deployment. In terms of distance, the Base Station positioned at center saves energy in transmission and reception and hence prolongs the network lifetime. For the use of gateway nodes, it is optimal to place them at the center of each region for improving the network lifetime. The same scenarios may be implemented for a specific application of Underwater Wireless Sensor Networks.

References

1. I.F. Akyildiz, D. Pompili, T. Melodia, Underwater acoustic sensor networks: research challenges. Ad Hoc Netw. **3**(3), 257–279 (2005)
2. M. Ayaz, A. Abdullah, Underwater wireless sensor networks: routing issues and future challenges, in *Proceedings of the 7th International Conference on Advances in Mobile Computing and Multimedia* (ACM, 2009), pp. 370–375
3. N. Xu, A survey of sensor network applications. IEEE Commun. Mag. **40**(8), 102–114 (2002)
4. D.J. Higham, N.J. Higham, *MATLAB guide*, Society for Industrial and Applied Mathematics, (2005)
5. S. Yadav, V. Kumar, Optimal clustering in underwater wireless sensor networks: acoustic, EM and FSO communication compliant technique, IEEE Access 5: 12761–12776 (2017)
6. M.S. Leeson, S. Patel, Energy efficiency of coding schemes for underwater wireless sensor networks, in *Technological Breakthroughs in Modern Wireless Sensor Applications* (IGI Global, 2015), pp. 27–55
7. F.A. De Souza, B.S. Chang, G. Brante, R.D. Souza, M.E. Pellenz, F. Rosas, Optimizing the number of hops and retransmissions for energy efficient multi-hop underwater acoustic communications IEEE Sens. J. **16**(10): 3927–3938 (2016)
8. A. Salhieh, J. Weinmann, M. Kochhal, L. Schwiebert, Power efficient topologies for wireless sensor networks, in *International Conference on Parallel Processing, 2001* (IEEE, 2001), pp. 156–163
9. S. Taruna, R. Kumawat, G.N. Purohit, Multi-hop clustering protocol using gateway nodes in wireless sensor network, Inter. J. Wireless Mobile Netw.,**4**(4): 169 (2012)
10. J. Cecílio, P. Furtado, *Wireless Sensors in Heterogeneous Networked Systems: Configuration and Operation Middleware* (Springer, 2014)
11. G. Chen, C. Li, M. Ye, W. Jie, An unequal cluster-based routing protocol in wireless sensor networks. Wireless Netw. **15**(2), 193–207 (2009)
12. L. Krishnamachari, D. Estrin, S. Wicker, The impact of data aggregation in wireless sensor networks, in *Proceedings 22nd International Conference on Distributed Computing Systems Workshops, 2002* (IEEE, 2002), pp. 575–578
13. W.R. Heinzelman, A. Chandrakasan, H. Balakrishnan, Energy-efficient communication protocol for wireless microsensor networks, in *Proceedings of the 33rd Annual Hawaii International Conference on System Sciences, 2000* (IEEE, 2000), 10 pp.
14. Y. Liu, H. Xuhui, M.J. Lee, Tarek N. Saadawi, A region-based routing protocol for wireless mobile ad hoc networks. IEEE Network **18**(4), 12–17 (2004)
15. T.V. Padmavathy, V. Gayathri, V. Indumathi, G. Karthika, Network lifetime extension based on network coding technique in underwater acoustic sensor networks. Inter. J. Distrib. Parallel Sys. **3**(3), 85 (2012)
16. M. Felamban, B. Shihada, K. Jamshaid, Optimal node placement in underwater wireless sensor networks, in *2013 IEEE 27th International Conference on Advanced Information Networking and Applications (AINA)* (IEEE, 2013), pp. 492–499
17. M.B. Abbasy, J.P. Ulate, Cluster-based performance analysis of sensor distribution strategies on a wireless sensor network, in *The Fifth International Conference on Sensor Technologies and Applications Sensorcomm, 2011*

Optimal Design of IIR Filter Using Dragonfly Algorithm

Sandeep Singh, Alaknanda Ashok, Manjeet Kumar, Garima and Tarun Kumar Rawat

Abstract The field of infinite impulse response (IIR) filter design mainly focused on the proper selection of filter parameters from the numerous possible combination. This filter design problem is based on determining the optimal set of parameters for unknown model such that its closely matches with the parameters of the benchmark filter. Many researchers have designed IIR filters using gradient based techniques like least mean square (LMS) method, etc. But, these gradient-based techniques have drawback of getting trapped into local solutions. To overcome this problem, evolutionary optimization techniques are used, which give global solutions. This paper utilizes a novel optimization technique known as dragonfly algorithm (DA) for the computation of the parameters of unknown IIR filter. Two benchmark functions are considered to prove the efficacy of the DA for IIR filter design problem. The results obtained using DA are compared with three existing algorithms namely, cat

S. Singh · Garima
Department of Electronics and Communication Engineering,
Maharaja Surajmal Institute of Technology, New Delhi, Delhi, India
e-mail: er.sandeep85@gmail.com

Garima
e-mail: garima@msit.in

S. Singh · A. Ashok
Department of Electronics and Communication Engineering,
Uttarakhand Technical University, Dehradun, Uttrakhand, India
e-mail: alakn@rediff.com

M. Kumar (✉)
Department of Electronics and Communication Engineering, Bennett University,
Greater Noida, Uttar Pradesh, India
e-mail: manjeetchhillar@gmail.com

T. K. Rawat
Department of Electronics and Communication Engineering,
Netaji Subhas Institute of Technology, New Delhi, India
e-mail: tarundsp@gmail.com

© Springer Nature Singapore Pte Ltd. 2019
H. Malik et al. (eds.), *Applications of Artificial Intelligence Techniques
in Engineering*, Advances in Intelligent Systems and Computing 698,
https://doi.org/10.1007/978-981-13-1819-1_21

211

swarm optimization (CSO), particle swarm optimization (PSO), and bat algorithm (BA). The obtained results verify that the performance of DA-based IIR filter design is superior than that achieved by PSO, CSO, and BA.

Keywords IIR filter · Dragonfly algorithm · Bat algorithm
Cat swarm optimization · Particle swarm optimization

1 Introduction

Adaptive filters play tremendous role in the modeling characteristics of a system that varies with time. These filters were employed in many real-world applications like signal processing, ECG detection, prediction, system identification, various applications of image processing, etc. [1–5]. An IIR filter models an unknown system parameters more accurately compared to FIR filters. Also, an IIR filter requires less number of coefficients in comparison with FIR filter, for achieving the desired level of performance [6]. This is because an IIR filter's current output is not only dependent on past and present input but also on the preceding output, whereas, in case of FIR model output is dependent only on present and past input. Here, the IIR filter design problem is considered as an error optimization problem. This error is the difference between the output of the actual system and output of the unknown model. Both, the unknown model and the adaptive IIR filter are excited with the same input. The prime goal is to get the minimum value of the objective function so that optimal set of filter coefficients can be obtained. Earlier gradient based search algorithms such as Quasi-Newton technique, least mean square (LMS), and its different versions were used for designing IIR filters. But, these techniques have the drawback of getting trapped into local solution due to the multimodal and non-quadratic nature of the objective function with respect to the filter parameters [7, 8]. Also, stability issues are there in higher order system as the poles are kept away from the unit circle.

In order to overcome these shortcomings, several researches rely on metaheurstic algorithms, which utilizes natural evolution techniques. These algorithms are nature inspired and population-based search algorithms. Different algorithms utilized by the researchers are genetic algorithm (GA) [9], particle swarm optimization (PSO) [10], artificial bee colony (ABC) [11], cat swarm optimization (CSO) [12], flower pollination algorithm (FPA) [13], seeker optimization algorithm (SOA) [14], gravitational search algorithm (GSA) [15], bat algorithms(BA) [16], and many more [17–22].

Linear and nonlinear system coefficients are estimated by Yao and Sethares using GA [9]. Chen and Luk have reported that optimum results can be obtained using PSO in the designing process of the digital IIR filter[10]. N. Karaboga utilized the behavior of bees in obtaining the digital IIR filter coefficients [11]. G Panda et al. considered CSO technique in [12] and proved that it is superior than GA and PSO in optimizing the system coefficients of the unknown IIR system. Singh et al. presented that flower pollination algorithm estimates the best values of the adaptive IIR system parameters compared to GA and PSO [13]. Dai et al. reported SOA based

IIR filter design which is superior than different versions of differential evolution (DE) and PSO based design [14]. Rashedi et al. proved that GSA is better technique for designing linear and nonlinear filter [15]. Kumar et al. solved the problem of system identification by utilizing BA algorithm [16]. In this work, a new technique called dragonfly algorithm (DA) [23] is employed for IIR system identification problem. A performance comparison is carried out using BA, CSO, and PSO in terms of coefficient optimization and minimum fitness value.

The rest of the paper is organized as: Sect. 2 discusses the problem of system identification along with block diagram. Dragonfly algorithm is introduced along with flow chart in Sect. 3. Simulated results and comparison of DA with other algorithms are described in Sect. 4. Section 5 concludes the paper.

2 Adaptive IIR System Model

This section reviews the basic concepts of IIR system identification. System identification problem involves the diversification of the coefficients of adaptive IIR filter by utilizing a novel algorithm, until it matches with the coefficients of the unknown IIR filter, when similar input signals are applied to the considered unknown system and adaptive filter. Figure 1 depicts the block diagram representation of adaptive IIR system. Here, $d(k)$ and $y(k)$ are the output of the unknown IIR filter and adaptive

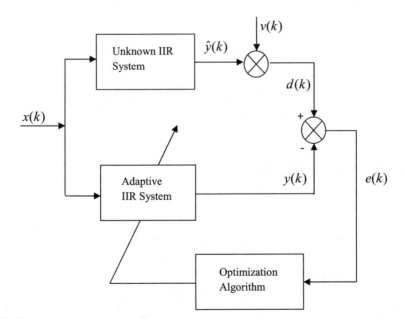

Fig. 1 Adaptive IIR model block diagram

filter, respectively, when $x(k)$ is applied as the input signal to both the system. Error signal, $e(k)$, represents the dissimilarity in the outputs of the two systems.

The adaptive IIR system, depicting the input–output relationship is governed by the following equation:

$$y(k) + \sum_{j=1}^{n} a_j y(k-j) = \sum_{j=0}^{m} b_j x(k-j) \tag{1}$$

The corresponding transfer function in Z-domain is given by

$$H(z) = \frac{\sum_{j=0}^{m} b_j z^{-j}}{1 + \sum_{j=1}^{n} a_j z^{-j}} \tag{2}$$

where, b_j and a_j are the coefficients of numerator and denominator, respectively. For mathematical convenience, the value of coefficient a_0 is taken as 1.

The main task in this work is to find the optimal set of filter coefficients and optimized value of MSE. Objective function MSE is given by

$$J_{b_j, a_j} = \frac{1}{P} \sum_{j=1}^{P} (d(k) - y(k))^2 \tag{3}$$

where P is total number of samples used for coefficient computation.

3 Dragonfly Algorithm (DA)

A novel metaheuristic for system identification problem is being proposed by S. Mirjalili in 2016. It is based on swarming behavior of dragonflies.

3.1 Behavior of Fireflies

Lifecycles of dragonfly comprise of two principle stages: nymph and adult. Dragonflies spend major portion of their life in nymph afterwards they undergo metamorphism to become an adult [23]. One of the amazing facts about dragonflies is their rare and unique swarming behavior. The main two purposes of their swarming behavior are hunting and relocation. The former is known as static (feeding) swarm, and the latter one is known as dynamic (migratory) swarm. Sudden change in flying path and local movements differentiate static swarm from the dynamic swarm. Whereas in case of dynamic swarm, large number of dragonflies fly in one direction over long distances to predate [10]. The main inspiration behind the DA algorithm is these swarming behaviors of dragonflies. These swarming behaviors of dragonflies are same as that of exploration and exploitation phases of metaheuristic optimization. In static swarm, sub-swarms are formed and dragonflies fly over numerous short areas, which is just like the main intent of the exploration phase. In dynamic swarm, dragonflies fly along one direction and in larger swarms which is advantageous in case of exploitation phase.

3.2 Dragonfly Algorithm

Different stages of dragonfly algorithm is shown in Fig. 2. According to Reynolds, the following three prime principles are followed in the behavior of swarm [10]:

- Separation, which is static collision avoidance among neighborhood individuals.
- Alignment, which shows velocity matching among neighborhood individuals.
- Cohesion, which indicates the aptness of neighborhood individuals towards the center of the mass.

The prime goal of a swarm is its existence, therefore dragonflies are fascinated towards the food resource and should distract the opponents outwardly. Considering it, five factors mainly affect the position updating of an individual in a swarm. The separation is expressed as [23]

$$S_i = - \sum_{j=1}^{N} U - U_j \tag{4}$$

where U shows the current location of an individual and U_i is the location of jth neighboring individual, whereas N represents the number of neighboring dragonflies. Alignment is expressed as

$$S_i = \frac{\sum_{j=1}^{N} V_j}{N} \tag{5}$$

where V_j denotes the velocity of jth neighboring individual. The following equation represents the calculation of cohesion:

$$C_i = \frac{\sum_{j=1}^{N} U_j}{N} - U \tag{6}$$

Attraction towards the food is given by

$$F_i = U^+ - U \tag{7}$$

where U^+ represents the location of food resource. Obstraction of an opponent outwardly is given by

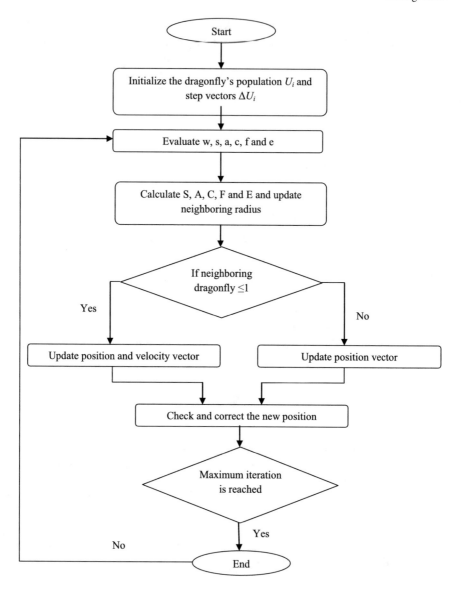

Fig. 2 Flowchart of dragonfly algorithm

$$D_i = U^- + U \tag{8}$$

X^- represents the location of the enemy. These five corrective patterns when combined together determine the behavior of the dragonflies. In order to amend the location of false dragonflies and to track their movements, step $?U$ and position (U) vectors are considered. As the DA algorithm is based on PSO algorithm, step vector resembles with velocity vector is PSO. Step vector which is responsible for directing dragonflies is calculated as

$$\Delta U_{t+1} = (sS_i + aA_i + cC_i + fF_i + dD_i) + x\Delta U_t \qquad (9)$$

where S_i signifies the separation of ith individual, s indicates separation weight, and a and c denote alignment weight and cohesion weight, respectively. Whereas e and f show enemy and food factor respectively, x is the inertial weight, and t represents the iteration counter. After the calculation of step vector, the position vector is expressed as

$$U_{t+1} = U_t + \Delta U_{t+1} \qquad (10)$$

where t denotes the current iteration. During optimization, different exploitative and explorative behavior can be achieved with alignment, food, cohesion, separation, and enemy factors (a, f, c, s and d). Neighborhood of dragonflies is quite important, so within a certain radius of false dragonflies, a neighborhood is assumed. In dynamic swarm dragonflies align themselves while flying with proper maintenance of separation and cohesion. Whereas in case of static swarm, low alignments are there due to high prey attacks. Thus, while examining the search space, low cohesion and high alignment are assigned to dragonflies. The radii of neighborhoods are in direct relation with the number of iteration while considering the transition between exploration and exploitation phase. During optimization, exploration and exploitation can be balanced by adaptively changing the swarming parameters (s, a, c, f, d and x). While progressing the exploration to the exploitation stage of the search space, it is required that dragonflies changes their weights adaptively. As the optimization further proceeds, it is assumed that in order to adjust the flying path dragonflies come across more dragonflies. In this way, convergence of dragonflies is achieved. Best and worst case solution of the whole swarm are used to choose the food source and enemy source. This results in concentration towards the favorable areas of search space and outward divergence from the non-favorable areas of search space. In order to improve the stochastic behavior, randomness, and exploration of the false dragonflies, they are supposed to fly using a random walk (Lévy flight) around the search space when no neighboring solutions are there. Therefore, the position of dragonflies is updated as follows:

$$U_{t+1} = U_t + Levy(d) \times U_t \qquad (11)$$

where d denotes the dimension of position vectors. The Lévy flight is expressed as

$$Levy(u) = 0.01 \times \frac{r_1 \times \sigma}{|r_2|^{\frac{1}{\beta}}} \qquad (12)$$

where r_1, r_2 represent the random numbers in [0, 1], and β is a constant.

Table 1 Control factors of PSO, CSO, BAT, and DA for IIR system identification

Parameters	Symbol	PSO	CSO	BAT	DA
Initial population	n_i	25	25	25	25
Maximum iterations	N_i	300	300	300	300
Tolerance	ϵ	10^{-5}	10^{-5}	10^{-5}	10^{-5}
Lower bound	L_{min}	-2	-2	-2	-2
Upper bound	L_{max}	2	2	2	2
Cognitive constant	C_1	2	–	–	–
Social constant	C_2	2	–	–	–
Initial velocity	v_i^{min}	0.01	–	–	–
Final velocity	v_i^{max}	1.0	–	–	–
Lower frequency (for PSO)	w_{min}	0.2	–	–	–
Higher frequency (for PSO)	w_{max}	1.0	–	–	–
Seeking memory pool	SMP	–	5.0	–	–
Counts of dimension to change	CDC	–	0.6	–	–
Seeking range of selected dimension	SRD	–	2.0	–	–
Stopping criteria	–	Best solution	Best solution	Best solution	Best solution

4 Simulation Results

This section presents the simulated results to prove the efficiency of the utilized technique for the system identification problem. For this, two benchmark functions are taken and for each system under consideration, b_i and a_i are taken as the coefficients of numerator and denominator, respectively. Here, each unknown system is simulated using two cases: (i) same order as that of the adaptive IIR system (ii) adaptive IIR system with reduced order. MSE is taken as the performance measure of the employed technique. The results achieved using DA are compared in terms optimized coefficients and MSE value with the three existing algorithms PSO, CSO, and BA. The control parameter's value for all the four algorithms are reported in Table 1.

Example 1 In this example, a second-order system is considered whose transfer function is represented by

$$H_p(z) = \frac{0.05 - 0.4z^{-1}}{1 - 1.1314z^{-1} + 0.25z^{-2}} \tag{13}$$

Efforts are made to model the unknown IIR system with the applied algorithm using reduced order and same order system. These two cases are described as

Table 2 Optimized parameters of second-order IIR filter modeled using same order in Example 1

Algorithms	Coefficients			
	b_0	b_1	a_1	a_2
Actual value	0.0500	−0.4000	1.1314	−0.2500
DA	0.0500	−0.4000	1.1314	−0.2500
BAT	0.0501	−0.4002	1.1306	−0.2497
CSO	0.0493	−0.4021	1.1248	−0.2433
PSO	0.0536	−0.4184	1.0876	−0.2077

Table 3 Statistical analysis of MSE values of second-order filter designed using second order in Example 1

Algorithm	MSE				
	Best	Worst	Average	Median	SD
DA	6.3993×10^{-09}	8.1012×10^{-08}	2.6262×10^{-07}	1.0565×10^{-07}	3.8349×10^{-07}
BAT	2.1569×10^{-05}	2.2014×10^{-05}	2.1815×10^{-05}	2.1501×10^{-05}	2.3365×10^{-07}
CSO	6.3639×10^{-05}	6.4629×10^{-05}	6.3849×10^{-05}	6.3806×10^{-05}	2.8905×10^{-07}
PSO	1.0116×10^{-04}	2.7405×10^{-04}	1.5491×10^{-04}	1.4519×10^{-05}	5.1800×10^{-05}

Case 1: Same Order This case models second-order system is modeled using second-order unknown system with the system function given by

$$H_s(z) = \frac{b_0 - b_1 z^{-1}}{1 - a_1 z^{-1} - a_2 z^{-2}} \tag{14}$$

Here, the main objective for the system identification is to find the optimized value of the numerator and denominator coefficients b_0, b_1 and a_1, a_2, respectively. The obtained value of the coefficients of the second-order system is reported in Table 2. It is evident from the Table 2 that the applied algorithm predicts the optimal values of the coefficients near to the actual value. Statistical results are obtained using all the four algorithms in terms of MSE and are summarized in Table 3. The best MSE values reported are 6.3993×10^{-09}, 2.1569×10^{-05}, 6.3639×10^{-05} and 1.0116×10^{-04} for DA, BA, CSO, and PSO, respectively. Based on these results, it can be concluded that DA is superior than other applied algorithms.

Case 2: Reduced Order Here, modeling of second-order system is done using first order unknown IIR system with system function given as

$$H_r(z) = \frac{b_0}{1 - a_1 z^{-1}} \tag{15}$$

In the reduced order case, MSE and convergence behavior are considered as the performance criteria for the applied algorithm. MSE value is computed in terms of best, worst, average, median, and standard deviation and is reported in Table 4. Minimum value of MSE observed for DA, BA, CSO, and PSO is 1.8089×10^{-04}, 7.9178×10^{-03}, 1.7515×10^{-02} and 1.7515×10^{-02}, respectively. It is clear from the Table 4 that DA gives the best result among all the four algorithms for the system identification problem.

Example 2 This example considers a third-order adaptive IIR system, whose transfer function is given by

$$H_p(z) = \frac{-0.2 - 0.4z^{-1} + 0.5z^{-2}}{1 - 0.6z^{-1} + 0.25z^{-2} - 0.2z^{-3}} \tag{16}$$

This third-order system can be approximated using a same order as that of the adaptive IIR system and reduced order system.

Case 1: Same order The transfer function of the same order unknown system is given by

$$H_m(z) = \frac{-b_0 - b_1 z^{-1} + b_2 z^{-2}}{1 - a_1 z^{-1} + a_2 z^{-2} - a_3 z^{-3}} \tag{17}$$

The optimal values of the numerator and denominator coefficients b_0, b_1, b_2, a_1, a_2, a_3, of the third-order system are evaluated and listed in Table 5. It is clear from Table 5 that DA estimates the coefficients value near to the actual values among all the four algorithms. The MSE values are calculated in the form of best, worst, average, median, and standard deviation and are stated in Table 6. The best values of MSE observed for DA, BA, CSO, and PSO are 6.2073×10^{-10}, 2.3037×10^{-05}, 6.3520×10^{-05} and 6.3520×10^{-05}, respectively. It can be seen from the results that DA outperform BA, CSO, and PSO in the IIR system identification problem. The comparison of the convergence behavior of DA, BA, CSO, and PSO is depicted in in Fig. 3. It is observed that least fitness value is retained by the DA in comparison with BA, CSO, and PSO. Based on these observations, it can be concluded that DA approach delivers the least MSE and optimal set of system coefficients for system identification problem.

Table 4 Statistical analysis of MSE values of second-order filter designed using first order in Example 1

Algorithm	MSE				
	Best	Worst	Average	Median	SD
DA	1.8089×10^{-04}	2.6790×10^{-03}	9.9829×10^{-04}	5.6664×10^{-04}	1.1000×10^{-03}
BAT	7.9178×10^{-03}	7.9178×10^{-03}	7.9178×10^{-03}	7.9178×10^{-03}	6.9831×10^{-19}
CSO	1.7515×10^{-02}	1.7515×10^{-02}	1.7515×10^{-02}	1.7515×10^{-02}	4.9100×10^{-18}
PSO	1.7515×10^{-02}	5.5840×10^{-02}	3.8807×10^{-02}	5.5840×10^{-02}	2.0199×10^{-02}

Table 5 Optimized parameters of third-order IIR filter modeled using same order in Example 2

Algorithms	coefficients					
	b_0	b_1	b_2	a_1	a_2	a_3
actual value	−0.2000	−0.4000	0.5000	−0.6000	0.2500	−0.2000
DA	−0.2001	−0.3990	0.5002	−0.6000	0.2500	−0.2000
BA	−0.2066	−0.3996	0.4994	0.5983	−0.2497	0.1991
CSO	−0.2050	−0.3927	0.5038	−0.6077	0.2519	−0.2031
PSO	−0.2105	−0.3778	0.4670	0.6123	−0.3134	0.2249

Fig. 3 Convergence profile for Example 2 using DA, BAT, CSO and PSO

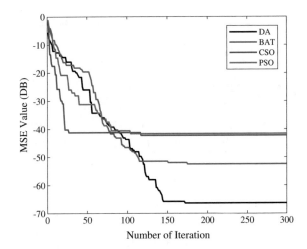

Case 2: Reduced order Here, the third-order system is approximated using second-order unknown IIR system has the following transfer function:

$$H_m(z) = \frac{-b_0 - b_1 z^{-1}}{1 - a_1 z^{-1} + a_2 z^{-2}} \tag{18}$$

To prove the efficiency of the applied algorithm in reduced order case, MSE and convergence profile are taken as the performance measure. The statistical values of MSE are observed and is reported in Table 7. The best value obtained for DA, BA, CSO, and PSO are 2.5603×10^{-04}, 8.3264×10^{-04}, 1.3938×10^{-03} and 1.3938×10^{-03}, respectively. These results proved that the employed algorithms are superior in finding of minimum MSE value as compared to other existing algorithms (Table 7).

Table 6 Statistical analysis of MSE values of third-order filter designed using third-order in Example 2

Algorithm	MSE				
	Best	Worst	Average	Median	SD
DA	6.2073×10^{-10}	7.0786×10^{-09}	9.1240×10^{-09}	2.6890×10^{-09}	3.7012×10^{-09}
BA	2.3037×10^{-05}	2.3045×10^{-05}	2.3039×10^{-05}	2.3482×10^{-05}	2.9143×10^{-12}
CSO	6.3520×10^{-05}	6.3520×10^{-05}	6.3520×10^{-05}	6.3520×10^{-05}	1.6872×10^{-18}
PSO	6.3520×10^{-05}	6.3520×10^{-05}	6.3520×10^{-05}	6.3520×10^{-05}	1.4767×10^{-10}

Table 7 Statistical analysis of MSE values of third-order filter designed using second order in Example 2

Algorithm	MSE				
	Best	Worst	Average	Median	SD
DA	2.5603×10^{-04}	7.6763×10^{-04}	3.4003×10^{-04}	3.5993×10^{-04}	5.0592×10^{-08}
BAT	8.3264×10^{-04}	8.3264×10^{-04}	8.3264×10^{-04}	8.3264×10^{-04}	4.2975×10^{-20}
CSO	1.3938×10^{-03}	1.3938×10^{-03}	1.3938×10^{-03}	1.3938×10^{-03}	1.0842×10^{-19}
PSO	1.3938×10^{-03}	1.3938×10^{-03}	1.3938×10^{-03}	1.3938×10^{-03}	2.9692×10^{-19}

5 Conclusions

In this work, a new nature inspired algorithm, DA, has been utilized for estimating the unknown IIR filter parameters. The performance of the applied algorithm is tested by taking two benchmark functions from the literature and comparing the results with BA, CSO, and PSO. MSE and estimated coefficients value are taken as the performance metrics. On the basis of these observation, it is concluded that the applied dragonfly algorithm estimates the system coefficients near to the actual value and finds the minimum MSE value. The simulated results imitate the effectiveness of the applied algorithm compared with BA, CSO, and PSO. This algorithm can be applied to nonlinear and complex system identification problems in future.

References

1. A. Kumar, D. Berwal, Y. Kumar, Design of high-performance ECG detector for implantable cardiac pacemaker systems using biorthogonal wavelet transform. Circuits Syst. Signal Process. (2018). https://doi.org/10.1007/s00034-018-0754-3
2. A. Kumar, R. Komaragiri, M. Kumar, From pacemaker to wearable: techniques for ECG detection systems. J. Med. Syst. **42**(2) (2018)
3. J. Lin, C. Chen, Parameter estimation of chaotic systems by an oppositional seeker optimization algorithm. Nonlinear Dyn. **76**(1), 509–517 (2014)

4. M.R. Soltanpour, M.H. Khooban, A particle swarm optimization approach for fuzzy sliding mode control for tracking the robot manipulator. Nonlinear Dyn. **74**(1–2), 467–478 (2013)
5. R. Zhang, R. Lu, A. Xue, F. Gao, Predictive functional control for linear systems under partial actuator faults and application on an injection molding batch process. Ind. Eng. Chem. Res. **53**(2), 723–731 (2014)
6. J.J. Shynk, Adaptive IIR filtering. IEEE ASSP Mag. **6**(2), 4–21 (1989)
7. B. Widrow, S.D. Strearns, *Adaptive Signal Processing* (Prentice-Hall, Englewood Cliffs, 1985)
8. Huiyi Hu, Rui Ding, Least squares based iterative identification algorithms for input nonlinear controlled autoregressive systems based on the auxiliary model. Nonlinear Dyn. **76**(1), 777–784 (2014)
9. L. Yao, W.A. Sethares, Nonlinear parameter estimation via the genetic algorithm. IEEE Trans. Signal Process. **42**(4), 927–935 (1994)
10. S. Chen, B.L. Luk, Digital IIR filter design using particle swarm optimisation. Int. J. Modell. Identif. Control **9**(4), 327–335 (2010)
11. N. Karaboga, A new design method based on artificial bee colony algorithm for digital IIR filters. J. Franklin Inst. **346**(4), 328–348
12. G. Panda, P.M. Pradhan, B. Majhi, IIR system identification using cat swarm optimization. Expert Syst. Appl. **38**(10), 12671–12683 (2011)
13. S. Singh, A. Ashok, T.K. Rawat, M. Kumar, Optimal IIR system identification using flower pollination algorithm, in *IEEE International Conference Power Electronics, Intelligent Control and Energy Systems (ICPEICES)* (2016), pp. 1–6
14. C. Dai, W. Chen, Y. Zhu, Seeker optimization algorithm for digital IIR filter design. IEEE Trans. Ind. Electron. **57**(5), 1710–1718 (2010)
15. E. Rashedi, H. Nezamabadi-Pour, S. Saryazdi, Filter modeling using gravitational search algorithm. Eng. Appl. Artif. Intell. **24**(1), 117–122 (2011)
16. M. Kumar, A. Aggarwal, T.K. Rawat, Bat algorithm: application to adaptive infinite impulse response system identification. Arab. J. Sci. Eng. **41**(9), 3587–3604 (2016)
17. M. Kumar, T.K. Rawat, Optimal fractional delay-IIR filter design using cuckoo search algorithm. ISA Trans. **59**, 39–54 (2015)
18. M. Kumar, T.K. Rawat, Optimal design of FIR fractional order differentiator using cuckoo search algorithm. Expert Syst. Appl. **42**(7), 3433–3449 (2015)
19. R. Barsainya, T.K. Rawat, M. Kumar, Optimal design of minimum multiplier fractional order differentiator based on lattice wave digital filter. ISA Trans. **66C**, 404–413 (2017)
20. A. Aggarwal, M. Kumar, T.K. Rawat, D.K. Upadhyay, Optimal design of 2-D FIR digital differentiator using L_1-norm based cuckoo-search algorithm. Multidimens. Syst. Signal Process. **28**(4), 1569–1587 (2017)
21. M. Kumar, A. Aggarwal, T.K. Rawat, H. Parthasarathy, Optimal nonlinear system identification using fractional delay second-order volterra system. IEEE/CAA J. Autom. Sinica (99), 1–17 (2017). https://doi.org/10.1109/JAS.2016.7510184
22. A. Aggarwal, M. Kumar, T.K. Rawat, D.K. Upadhyay, Optimal design of 2-D FIR filters with quadrantally symmetric properties using fractional derivative constraints. Circuits Syst. Signal Process. **35**(6), 2213–2257 (2016)
23. S. Mirjalili, Dragonfly algorithm: a new meta-heuristic optimization technique for solving single-objective, discrete, and multi-objective problems. Neural Comput. Appl. **27**(4), 1053–1073 (2016)

Friends Recommender System Based on Status (StatusFRS) for Users of Overlapping Communities in Directed Signed Social Networks

Nancy Girdhar and K. K. Bharadwaj

Abstract The affluence of signed social networks (SSNs) has attracted the sight of most of the researchers to explore and examine these networks. Besides the notion of friendship, signed social networks also deals with the idea of antagonism among the users in the network. The two fundamental theories of these networks are social balance theory and status theory. Based on the idea of social status of individuals, status theory is suitable for directed signed social networks (DSSNs). Most of the work dedicated to friends recommender system (FRS) is based on social balance theory, grounded on the concept of Friend-Of-A-Friend, which limits to undirected signed social networks and thus, overlooks the impact of direction-link information. In this paper, a friends recommender system based on status (StatusFRS) is proposed to have improved and meaningful friends recommendations. Our contribution is threefold. Initially, by employing genetic algorithm, signed overlapping communities are formed. Further, in order to recommend relevant friends, status of each node in overlapping communities is computed. Finally, on the basis of status, a recommended list of friends is generated for each user. Experiments are performed on real world dataset of Epinions to evaluate the performance of the proposed model.

Keywords Friends recommendation · Genetic algorithm · Status theory
Social balance theory · Overlapping communities
Directed signed social networks

N. Girdhar (✉) · K. K. Bharadwaj
School of Computer and Systems Sciences, Jawaharlal Nehru
University, New Delhi 110067, India
e-mail: nancy.gr1991@gmail.com

K. K. Bharadwaj
e-mail: kbharadwaj@gmail.com

© Springer Nature Singapore Pte Ltd. 2019
H. Malik et al. (eds.), *Applications of Artificial Intelligence Techniques in Engineering*, Advances in Intelligent Systems and Computing 698,
https://doi.org/10.1007/978-981-13-1819-1_22

1 Introduction

In recent years, fast and emerging growth of online social networks have gleaned attention of many researchers to analyze and explore these networks. Online networks function as a platform for users of these social media to get connected with their friends, relatives and acquaintances. Many of these sites have friends recommendation as a feature to suggest users to connect with other users based on homophily [4, 5] and heterophily [9]. By connecting users, these sites allow users to share and receive information which leads to information spread over the network. Also, profile of a user on social networking site help other users to have insight about him/her. For instance a Facebook user A which is not already connected with user B in whom he/her is interested in can consult his/her Facebook network to see whether he/she have any mutual connections who could help him/her to arrange a meeting to introduce him/her. Also, social sites serve as a great platform for marketing and advertising. LinkedIn which is a business- and employment-oriented social networking site, where users from diverse domains create professionals profiles and connect disparate users who might pass on recommendation to help them in job finding. Thus, these social sites act as 'bridge' to link people from diverse multiple domains to share experience, ideas and skills.

For friends recommendations in SSNs, previous work from the literature [7] primarily focus on making recommendations based on social balance theory [4] which is grounded on the concept of Friend-Of-A-Friend but suitable for undirected signed networks only. Hence, while dealing with directed SSNs, researchers typically ignore the direction of links, which mitigates the significance of the information about the direction of creation of links. On the other hand, status theory [3, 4] based on the concept of social or economic status of users takes into account direction of ties thus, suitable for DSSNs. Therefore, in our proposed method StatusFRS, we have exploited status theory to generate recommended friends list for each user in DSSN. The main contribution of our work is summarized as:

- To explore and examine these complex SSNs, we have partition the network in directed signed overlapping communities by employing genetic algorithm.
- Next, we have computed status of nodes in overlapping communities to consider *link density, sign link information, asymmetric, directed and overlapping nature of links* in DSSN.
- Finally, we have proposed new model, StatusFRS based on social status of nodes for generating list of recommended friends for the users of DSSN.

The rest of the article is structured as: Sect. 2 presents the brief of related work on SSNs. Description of the proposed model is provided in Sect. 3. Section 4, elaborates the experiments performed on real world dataset. Finally, Sect. 5 concludes our work and presents some future work directions.

2 Related Work

Traditional approaches available in literature to analyze signed social networks (SSNs) basically deal by considering all links as positive ignoring the antagonistic behavior among users existing in the network as negative links [4, 5]. Recently, researchers have begun to examine these networks by considering both positive and negative links [4, 5, 8], though the field of friends recommendation still remain unexplored. Most of the existing link recommendation algorithm in SSNs are based on common attributes and link structure of the network [1]. Generally, these bespoke recommendations aim to qualify the criteria of FOAF and interest similarity [4] which may not be useful for apportioning and ascertaining suitable content that users actual require. Leskovec et al. [5] proposed a machine learning approach to predict positive and negative links. An inductive learning approach is also proposed [6] for predicting friends and foes in SSNs. In all these techniques, the SSN is considered as undirected network, therefore information about direction of link is simply overlooked in link prediction analysis and thus, in the decision making process of recommending links.

   Another most prevalent theory to investigate SSNs is status theory [4, 5], which is based on the social and/or economic status of users that considers direction of links, unlike social balance theory. Thus, it is more suitable for predicting sign of links in real networks which in most cases, are directed. A transitive node similarity based framework (FriendTNS) in signed networks is proposed to distinguish the trustworthy users from vindictive ones [7]. An unsupervised and semi-supervised link prediction model is presented [10] by capturing behavioral and social interactions of the users in the network. Although, a plethora of approaches are proposed for the analysis of SSNs; yet very less amount of work is dedicated for examining DSSNs. Thus, to have eloquent recommendation of friends list, our proposed model StatusFRS subsume information about direction of links, sign of link and links itself.

3 Proposed Model to Recommend Friends in a DSSN

3.1 Overlapping Community Formation in DSSN

3.1.1 Genetic Representation and Generating Initial Population

An initial population of N individuals is generated randomly. Each chromosome is matrix of size $n*k$ where, n and k are the number of users and communities respectively (Fig. 1). Each user can belong to multiple communities at a time and each community must have at least one user.

Fig. 1 Representation of encoded chromosome

	C_1	.	C_{k-1}	C_k
User$_1$	1	0	1	1
.	0	1	1	1
User$_{n-1}$	1	0	1	0
User$_n$	0	0	1	1

$M=$

3.1.2 Fitness Function

To measure the fitness of each chromosome overlapping signed modularity function, Q_{ov} is deployed [3], Eq. (1), that maximizes positive and minimizes negative intra-community links respectively.

$$\text{Maximize:} Q_{ov} = \frac{1}{2d^+ + 2d^-} \sum i, j \frac{1}{o_i o_j} \left[A_{i,j} - \left(\frac{d_i^+ d_j^+}{2d^+} - \frac{d_j^- d_j^-}{2d^-} \right) \right] \delta(c_i, c_j),$$

(1)

where, c_i is the ith community, d_i^+ and d_i^- are the positive and negative degrees of node i respectively.

3.1.3 Problem Specific Genetic Operators

Selection For selection purpose we have employed *roulette wheel selection*, which selects the chromosome according to their fitness value. Each member of the population is assigned a roulette wheel whose size is according to its fitness value.

Crossover For crossover, we have modified the uniform crossover operator to suit our problem details of which are given as follows:

Modified Uniform Crossover Operator In this crossover, two parents P1 and P2 are taken, and a binary mask of size n is randomly generated (Fig. 2). To generate the first offspring, first consider the mask. If the value is 1, user from first parent is selected and if mask value is equal to 0, user from second parent is taken. To generate the second offspring, consider the mask again if the value is 0, user from first parent is selected and if mask value is 1, user from second parent is taken.

Mutation To introduce diversity in population, the real valued uniform mutation operator is tailored to fit in our application.

Modified Mutation Operator In this, any two random communities of a chromosome are chosen, such that, each of them must have more than one user. Then, two users are taken randomly from these two communities and their membership values are swapped as shown in Fig. 3.

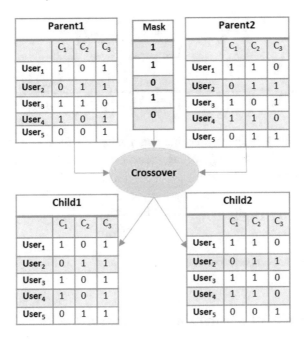

Fig. 2 Modified crossover operator

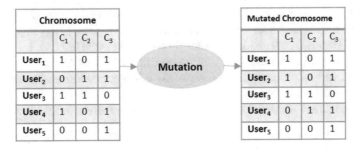

Fig. 3 Modified mutation operator

3.2 Computation of Status

Let $w_{(k)}^+$, $w_{(k)}^-$ are positive and negative weights of node k and $P_{(i,k)}$, $N_{(i,k)}$ are positive and negative contribution of node k in ith community computed as:

$$w_{(k)}^+ = \rho * \left(d_{in_{(k)}}^+ + d_{out_{(k)}}^- \right) \tag{2}$$

$$w_{(k)}^- = \left(\sigma * \left(d_{out_{(k)}}^+ + d_{in_{(k)}}^- \right) \right) * \sum_{\forall i} N_{i,k} \tag{3}$$

$$P_{(i,k)} = \frac{\left(d^+_{in(i,k)} + d^-_{out(i,k)}\right)}{\sum_{\forall i} d^+_{in(i,k)} + d^-_{out(i,k)}} \tag{4}$$

$$N_{(i,k)} = \frac{\left(\left(d^+_{out(k)} + d^-_{in(i)}\right)\right) - \left(\left(d^+_{out(i,k)} + d^-_{in(i,k)}\right)\right)}{\sum_{\forall i} \left(\left(d^+_{out(k)} + d^-_{in(k)}\right)\right) - \left(\left(d^+_{out(i,k)} + d^-_{in(i,k)}\right)\right)} \tag{5}$$

For each node k in ith overlapping community in DSSN with $d^+_{in(k)}$, $d^+_{out(k)}$ as positive in and out degrees, d^-_{in}, d^-_{out} as negative in and out degrees respectively, the status $S_{(i,k)}$ can be computed as

$$S_{i,k} = \frac{w^+_{(k)} * P_{(i,k)} - w^-_{(k)} * N_{(i,k)}}{\left|w^+_{(k)} - w^-_{(k)}\right|}, \tag{6}$$

where, ρ and σ are constant parameters.

3.3 Formation of Recommended Friends List

The basic notion of our proposed approach for recommending friends is based on status theory. According to this theory [4], if User A (sender node) connects positively with User B (receiver node) then, User A believes that User B has higher status. Further, if User B (sender node) is positively connected with User C (receiver node), then, transitively User A tends to make positive link with User C as shown in Fig. 4. This forms the basis for our proposed model to form direct recommended and indirect recommended friends lists.

The model for forming recommended friends lists involves following steps:

Step 1 For a given overlapping community c, for each user A, make set of users with higher status value, let say, $HS_{(A,c)}$.

Step 2 For each user B in $HS_{(A,c)}$ check if user A is already connected or not. If user A and user B are not connected, then add B in direct recommended friends list $DRFL_{(A,c)}$ of A, otherwise go to Step 3.

Fig. 4 Dotted green line depicts direct and indirect friend recommended based on status theory

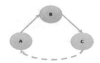

(a) Direct recommended friend

(b) Indirect recommended friend

Step 3   If user A is positively connected with user B, then find set of users which are positively connected and having higher status than user B which are in same community of user A, let say $HS^P_{(B,c)}$.

Step 4   Again check for each user C in $HS^P_{(B,c)}$, if A is connected or not. If user A and user Care not connected then add user C in indirect recommended friends list of A, i.e., $IRFL_{(A,c)}$, otherwise if A is already connected do nothing.

Step 5   Final, recommended friends list RFL_A for user A will be given as

$$RFL_A = \sum_{A=1}^{n} \sum_{c=1}^{k} DRFL_{(A,c)} \cup IRFL_{(A,c)} \tag{7}$$

4 Dataset and Experimental Results

The experiments are conducted on the real world dataset of Epinions [3] which is a directed signed social network. The genetic parameters used for our implementation is given in Table 1. The stagnation in fitness value for five successive generations is used to terminate the GA process.

Illustration with an Example
Due to page constraint, and for the sake of simplicity, here we have shown the working of our model with an illustrative example on randomly generated directed signed social network G, having 10 nodes as shown in Fig. 5. Dotted green and red directed lines represent positive and negative links respectively.

Initially, genetic algorithm is used to discover overlapping communities $\{c_1, c_2,..., c_k\}$ in the network. Let us consider three overlapping communities are detected, say c_1 = {user1, user5, user6, user7, user9, user10}; c_2 = {user1, user2, user3, user5, user8, user9}; and c_3 = {user1, user2, user4, user7, user8, user9}.

After obtaining set of overlapping communities, we have calculated status value for nodes in each overlapping community using Eq. (6) and formed a recommended friends list based on the status of nodes in each community. To demonstrate the working, let say status of users in c_1 = {user1, user5, user6, user7, user9, user10} is S(user1)=0.5, S(user5)=0.37, S(user6)=0.41, S(user7)=0.43, S(user9)=0.31, S(user10)=0.18. As it can be seen easily, user1 has the highest status in c_1 and user10 has the lowest status. Now, based on status of each user, direct and indirect friends are computed as follows:

Table 1 Genetic algorithm parameters

Parameter	Value
Population size	500
Crossover rate (P_c)	0.9
Mutation rate CP_m)	0.1

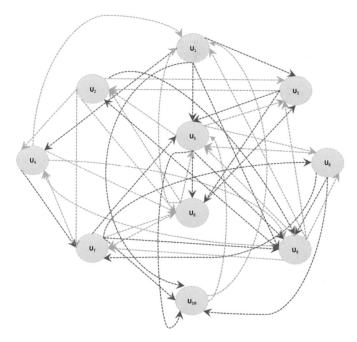

Fig. 5 A directed signed social network

Computation of Recommended Friends List

(i) *Computation of Direct Recommended Friends List (DRFL)*: After comparing
    status of each user with other existing users in the same community, we have
    formed set $HS_{(A,c)}$ of users with higher status value than the user. For user5 in
    c_1, user1, user6 and user7 are having higher status than user5, thus, $HS_{(5,2)} =$
    {user1, user6, user7} and $DRFL_{(5,2)} =$ {user1, user7}, since, user6 is already
    connected with user5.

(ii) *Computation of Indirect Recommended Friends List (IRFL)*: As user6 is pos-
    itively connected with user5 in c_1; $HS_{(6,1)} =$ {user1, user7}. Since user6 has
    positive and negative links with user1 and user7 respectively, thus, $HS^P_{(6,1)} =$
    {user1}and hence, $IRFL_{(5,1)} =$ {user1}. Similarly, for other communities also,
    DRFL and IRFL list for user5 is computed and then using Eq. (7), total rec-
    ommended friends list is prepared for user5, i.e., $RFL_5 =$ {user1, user4, user7,
    user8}. Correspondingly, recommended friends list for all the users are com-
    puted as shown in Table 2.

From the Table 2 following conclusions can be drawn:

- RFL for user1 is { }, indicating that user1 has the highest status in all the communities, thus, no friends recommendation as other users are having lower status
  than user1.

Table 2 Details of direct and indirect recommended friends list computed for all the users in the directed signed social network G

User ID	Community 1		Community 2		Community 3		Recommended friends list
	DRF	IRF	DRF	IRF	DRF	IRF	
User 1	{}	{}	{}	{}	{}	{}	{}
User 2	{1}	{}	{1}	{}	{1}	{}	{1}
User 3	{1, 7, 9}	{4, 7}	{1, 9}	{4, 7}	{1, 9}	{4, 7}	{1, 4, 7, 9}
User 4	{1, 2, 3, 4}	{3, 5, 6}	{2, 3}	{2, 3, 5, 6}	{1, 3, 9}	{3, 7}	{1, 2, 3, 5, 6, 7, 9}
User 5	{1, 4, 7}	{4, 7}	{1, 4}	{4, 7}	{1, 4}	{4, 7}	{1, 4, 7}
User 6	{1, 2, 3, 4}	{2, 3}	{1, 2, 3, 4, 8}	{2, 3}	{1, 2, 3, 4, 8}	{2, 3}	{1, 2, 3, 4, 8}
User 7	{1}	{}	{1, 2, 3}	{1, 2}	{1, 2, 3}	{1, 2}	{1, 2, 3}
User 8	{4, 6, 9}	{4}	{}	{4}	{9}	{4}	{4, 6, 9}
User 9	{}	{8}	{8}	{}	{}	{}	{8}
User 10	{2, 3, 4, 6, 7, 8, 9}	{2, 3, 6}	{2, 3, 4, 6, 7, 8, 9}	{2, 3, 6}	{2, 3, 4, 6, 7, 8, 9}	{2, 3, 6}	{2, 3, 4, 6, 7, 8, 9}

- DRFL for user9 is { } for all the communities, signifying that in each community user9 has the highest status than other nodes in the communities.
- For $c = 1$; DRFL for user2 is {1}, indicating that user2 and user1 both lies in community 1. Also, user2 and user1 are not connected and user1 is having higher status than user2. Thus, user1 is included in DRFL user2.
- For $c = 2$; IRFL for user8 is {4}, indicates that there is no other user in community 2 that has higher status and not connected with user9.
- For $c = 3$; for user9 DRFL and IRFL both are { }, representing that there is there is no node which has higher status and not connected with user9, hence DRFL is { }. Also, positively connected nodes with user9, either does not have any other positive links, or else user9 is already connected with them, so there is no friend left to recommend, thus, IRFL of user9 is { }.

5 Conclusion and Future Work

In the present work, StatusFRS is developed where social status theory is considered to simulate the process of recommending friends in directed signed social networks. The proposed algorithm takes into account the direction of link formed along with the type of relationship (positive or negative) besides the link itself. Thus, our proposed approach is well suitable for both directed and undirected signed social networks. As for further work, we would like to find influential nodes with high social status in signed social networks [9]. Another interesting area will be finding nodes of creation of social circles in directed signed social ego networks [2]. Furthermore, we will focus to study link analysis in dynamic signed networks by deploying the social theories [1].

References

1. V. Agarwal, K.K. Bharadwaj, Friends recommendations in dynamic social networks, in *Encyclopedia of Social Network Analysis and Mining*, 2014, pp. 553–562
2. V. Agarwal, K.K. Bharadwaj, Predicting the dynamics of social circles in ego networks using pattern analysis and GA K-means clustering. Wiley Interdiscip. Rev. Data Min. Knowl. Discov. **5**(3), 113–141 (2015)
3. N. Girdhar, K.K. Bharadwaj, Signed social networks: a survey, in *Proceedings of International Conference* on *Advances in Computing and Data Sciences*, 2016, pp. 326–335
4. J. Leskovec, D. Huttenlocher, J. Kleinberg, Signed networks in social media, in *Proceedings of the SIGCHI conference on Human Factors in Computing Systems*, 2010a, pp. 1361–1370
5. J. Leskovec, D. Huttenlocher, J. Kleinberg, Predicting positive and negative links in online social networks, in *Proceedings of the International Conference on World Wide Web*, 2010b, pp. 641–650
6. A. Patidar, V. Agarwal, K.K. Bharadwaj, Predicting friends and foes in signed networks using inductive inference and social balance theory, in *Proceedings of the 2012 International Conference on Advances in Social Networks Analysis and Mining*, 2012, pp. 384–388
7. P. Symeonidis, E. Tiakas, Transitive node similarity: predicting and recommending links in signed social networks. WWW **17**(4), 743–776 (2014)
8. J. Tang, S. Chang, C. Aggarwal, H. Liu, Negative link prediction in social media, in *Proceedings of the ACM International Conference on Web Search and Data Mining*, 2015, pp. 87–96
9. S. Wang, F. Wang, Y. Chen, C. Liu, Z. Li, X. Zhang, Exploiting social circle broadness for influential spreaders identification in social networks. WWW **18**(3), 681–705 (2015)
10. S.H. Yang, A.J. Smola, B. Long, H. Zha, Y. Chang, Friend or frenemy? predicting signed ties in social networks, in *Proceedings of the International ACM SIGIR Conference on Research and Development in Information Retrieval*, 2012, pp. 555–564

10-Min Ahead Forecasting of Wind Speed for Power Generation Using Nonlinear Autoregressive Neural Network

Amit Kumar Yadav and Hasmat Malik

Abstract This paper present wind speed (WS) forecasting at 10 min ahead using nonlinear autoregressive neural network. For this 45000 measured time series data are utilized. Mean absolute percentage error and correlation are found to be 9.45% and 92% respectively, showing effectiveness and accuracy of the proposed nonlinear autoregressive neural network forecasting model of WS for power generation.

Keywords Forecasting · Nonlinear autoregressive neural network · Wind speed

1 Introduction

Since last two decades wind power (WP) generation is growing at accelerated rates due to global warming and various incentives from government. By 2020 about 12% of world electricity demands is to be supplied by wind [1]. Therefore control generation of WP are necessary for wind farm. This escalate WP prediction for wind farm. WP is a function of WS which depends on various meteorological variables. Therefore forecasting of WS is a severe task. Despite the difficulties, different approaches are given in study [2–9, 10–14]. In recent work Ramasamy et al. [15] used MLP with LM algorithm for prediction of WS for different cities Himachal Pradesh India. ANN model uses inputs as altitude, air pressure, temperature and solar radiation. MAPE and R-value are 4.55% and 0.98% respectively. This work present Nonlinear Autoregressive Neural Network to forecast WS at 10 min ahead.

A. K. Yadav · H. Malik
Electrical and Electronics Engineering Department, NIT Sikkim, Barfung Block, Ravangla 737139, South Sikkim, India
e-mail: amit1986.529@rediffmail.com

H. Malik (✉)
Electrical Engineering Department, IIT Delhi, Hauz Khas, New Delhi 110016, India
e-mail: hmalik.iitd@gmail.com

© Springer Nature Singapore Pte Ltd. 2019
H. Malik et al. (eds.), *Applications of Artificial Intelligence Techniques in Engineering*, Advances in Intelligent Systems and Computing 698,
https://doi.org/10.1007/978-981-13-1819-1_23

Fig. 1 CEEE, NIT
Hamirpur meteorological
station

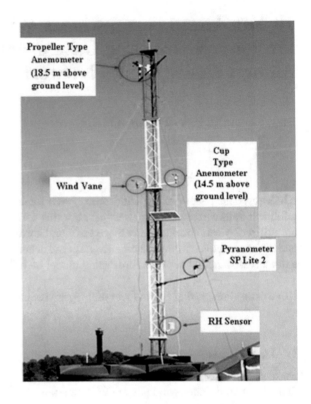

2 Methodology

2.1 Data Measurement

To develop network, time series data of WS for CEEE NIT-H (lat: 31.68 °N, long: 78.52 °E, altitude of 775 m above mean sea level) are used (Fig. 1). The daily variation of WS for year 2012 is shown in Fig. 2 and its normalized value is shown in Fig. 3. Out of 45000 normalized value of WS 44634 value are used for training and 366 values are used for testing the model.

2.2 Nonlinear Autoregressive Neural Network (NANN) Based Forecasting

Forecasting is a type of dynamic filtering (DF) in which historical values of one or more time series are utilized to forecast future values. DF include tapped delay used for forecasting and nonlinear filtering. The developed NANN in this study is shown in Fig. 4.

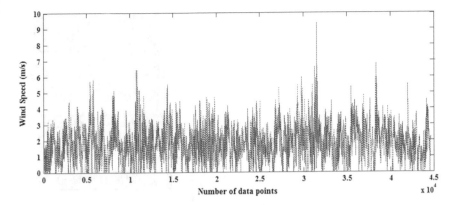

Fig. 2 Time series value of wind speed

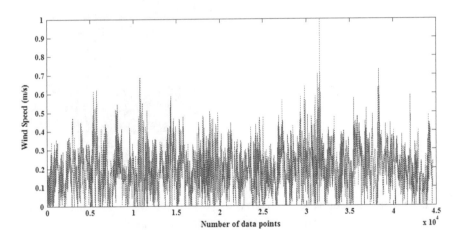

Fig. 3 Normalized value of wind speed

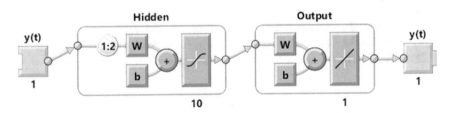

Fig. 4 Nonlinear autoregressive neural network

Fig. 5 Algorithm

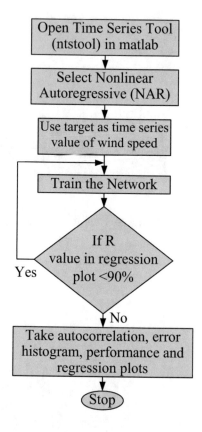

The implementation of NANN is shown in Fig. 5. The WS prediction accuracy is determined by mean absolute error (MAPE) is calculated by Eq. (1):

$$\text{MAPE} = \left(\frac{1}{n} \sum_{i=1}^{n} \left| \frac{WS_{i(NANN)} - WS_{i(measured)}}{WS_{i(measured)}} \right| \right) \times 100, \quad (1)$$

where n is the total number of WS samples used for training and testing, $WS_{i(measured)}$ is measured WS for i day and $WS_{i(NANN)}$ is predicted WS for i day.

The MAPE $\leq 10\%$ means high prediction accuracy given by Lewis [16].

3 Results and Discussion

The NANN model performance plot presents that mean square error (MSE) decreases with increase in number of epochs (Fig. 6). The epoch denotes complete cycle of training, testing and validation. The validation and test set error have similar charac-

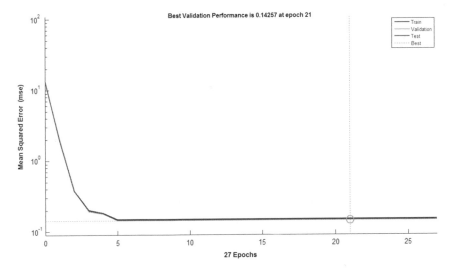

Fig. 6 Performance plot

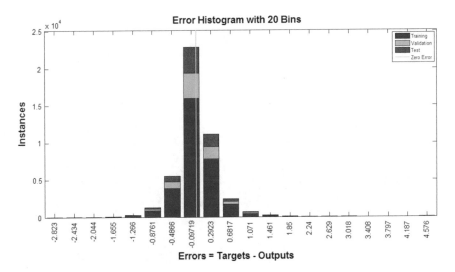

Fig. 7 Error histogram plot

teristics and no major over fitting happens near epoch 4. The correlation coefficient (R-value) shows closeness between outputs and target. R-value close to 1 measures a strong relationship. The R-value is more than 0.92 as shown in Figs. 11, 12, 13 and 14; proving NANN model developed with nstool predicts WS close to measured values (Fig. 10). The error histogram, autocorrelation and fit plots are shown in Figs. 7, 8 and 9 showing authentication of NANN for forecasting of WS. The MAPE is found to be 9.45% showing high forecasting accuracy as per Lewis criteria.

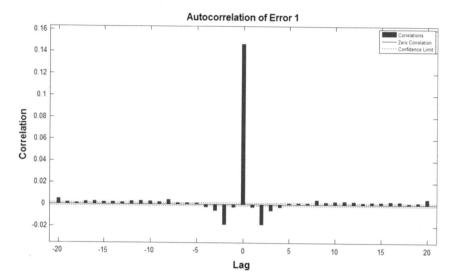

Fig. 8 Autocorrelation plot

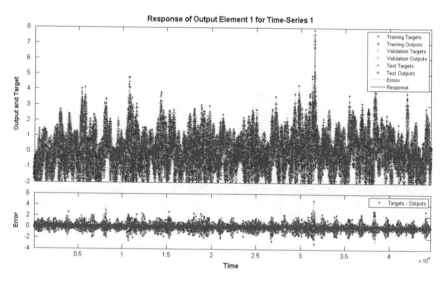

Fig. 9 Fit plot

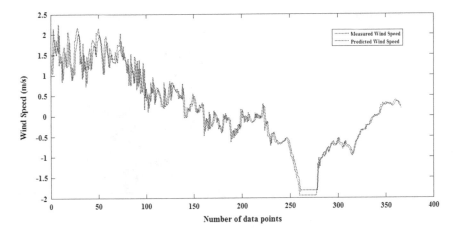

Fig. 10 Comparison between measured and forecasted WS

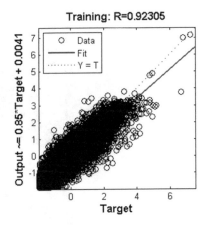

Fig. 11 Regression plot for training

Fig. 12 Regression plot for validation

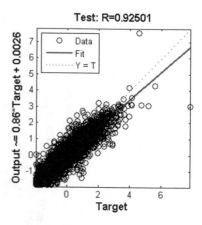

Fig. 13 Regression plot for testing

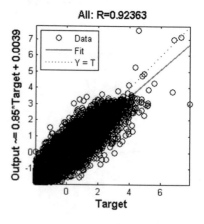

Fig. 14 Overall regression plot

4   Conclusion

The results of this study shows accuracy and effectiveness of the proposed new technique for WS forecasting. The proposed model, i.e., Nonlinear Autoregressive Neural Network forecast wind speeds for ten minute ahead with mean absolute percentage error of 9.45%. Moreover, the results presents that NANN models produce less error and higher coefficients of correlation.

Acknowledgements The authors would like to thanks persons associated with TEQIP III, HOD EEED NIT Sikkim and Director NIT Sikkim for providing funds to present this paper.

References

1. European Wind Energy Association. Wind force 12, 2002. http://www.ewea.org/doc/WindFo rce12.pdf
2. E. İzgi, A. Öztopal, B. Yerli, M.K. Kaymak, A.D. Şahin, in *Determination of the Representatives Time Horizons for Short-Term Wind Power Prediction by using Artificial Neural Networks. Energy Sources, Part A: Recovery, Utilization, and Environmental Effects*, vol. 36, 2014, pp. 1800–1809
3. H. Xiaojuan, Y. Xiyun, L. Juncheng, Short-Time wind speed prediction for wind farm based on improved neural network, in *8th World Congress on Intelligent Control and Automation*, 2010, pp. 5186–5190
4. S.S. Sanz, A. Perez-Bellido, E. Ortiz-Garcia, A. Portilla-Figueras, L. Prieto, D. Paredes, F. Correoso, Short-term wind speed prediction by hybridizing global and mesoscale forecasting models with artificial neural networks, in *International Conference on Hybrid Intelligent Systems*, 2008, 608–612
5. W. Zhang, J. Wang, J. Wang, Z. Zhao, M. Tian, Short-term wind speed forecasting based on a hybrid model. Appl. Soft Comput. **13**, 3225–3233 (2013)
6. A. Daraeepour, D.P. Echeverri, Day-ahead wind speed prediction by a neural network-based model, in *Innovative Smart Grid Technologies Conference*, 2014, pp. 1–5
7. Andrew Kusiak, Wenyan Li, Estimation of wind speed: a data-driven approach. J. Wind Eng. Ind. Aerodyn. **98**, 559–567 (2010)
8. Mehmet Bilgili, Besir Sahin, Abdulkadir Yasar, Application of artificial neural networks for the wind speed prediction of target station using reference stations data. Renew Energy **32**, 2350–2360 (2007)
9. Mohamed A. Mohandes, Shafiqur Rehman, Talal O. Halawani, A neural networks approach for wind speed prediction. Renew. Energy **13**, 345–354 (1998)
10. A. Parul, H. Malik, R. Sharma, Wind speed forecasting model for northern-western region of India using decision tree and multi layer perceptron neural network approach. Interdiscip. Environ. Rev. **19**(1), 13–30 (2018)
11. M.A. Savita, N.S. Pal, H. Malik, Wind speed and power prediction of prominent wind power potential states in India using GRNN, in *Proceedings of IEEE ICPEICES-2016*, 2016, pp. 1–6
12. H. Malik, Savita, Application of artificial neural network for long term wind speed prediction, in *Proceedings of IEEE CASP-2016*, 2016, pp. 217–222, 9–11
13. A. Azeem, G. Kumar, H. Malik, Artificial neural network based intelligent model for wind power assessment in India, in *Proceedings of IEEE PIICON-2016*, 2016, pp. 1–6, 25–27

14. A. Azeem, G. Kumar H. Malik, Application of waikato environment for knowledge analysis based artificial neural network models for wind speed forecasting, in *Proceeding of IEEE PIICON-2016*, 2016, pp. 1–6, 25–27

15. P. Ramasamy, S.S. Chandel, A.K. Yadav, Wind speed prediction in the mountainous region of India using an artificial neural network model. Renew. Energy **80**, 338–347 (2015)

16. C.D. Lewis, in *International and Business Forecasting Methods*. Butterworths, London, 1982

Predictive Control of Energy Management System for Fuel Cell Assisted Photo Voltaic Hybrid Power System

Kurukuru Varaha Satya Bharath and Mohammed Ali Khan

Abstract Distributed generation systems also known as hybrid power systems which involve renewable energy sources are extensively used due to their efficiency and green interface. Considering the varying environmental conditions, these systems are prone to many disadvantages and limitations. In order to overcome these constraints, intelligent techniques which can achieve steady process and power balance are to be implemented. This paper provides an intelligent control using fuzzy inference system and energy management algorithm for Fuel cell assisted PV Battery system. The supervisory control was implemented to achieve utmost feasible efficiency despite varying conditions such as irradiance and Hydrogen levels. With Levelized cost being adapted, an efficient energy management system attributes for even power distribution throughout the day can be implemented. Our thought process was demonstrated, and final software interface was simulated using MATLAB/Simulink to obtain results which confirm the effectiveness of the developed system.

Keywords PVFC hybrid system · Fuzzy logic controller · Inference systems
Energy management · MPPT · Fuel cell

1 Introduction

Today Energy Production and supply is based on Fossil Fuel and Nuclear power plants which are non-sustainable and Nonrenewable. It has major disadvantages and environmental problem. Nuclear power is also based on limited source which having high risk of disposal process of uranium. Renewable energy sources utilization led the world to growth based on environmental and human life [1]. As acknowledged

K. V. S. Bharath (✉) · M. A. Khan
Faculty of Engineering and Technology, Department of Electrical & Electronics Engineering,
Jamia Millia Islamia, New Delhi 110025, India
e-mail: kvsb272@gmail.com

M. A. Khan
e-mail: mak1791@gmail.com

© Springer Nature Singapore Pte Ltd. 2019
H. Malik et al. (eds.), *Applications of Artificial Intelligence Techniques in Engineering*, Advances in Intelligent Systems and Computing 698,
https://doi.org/10.1007/978-981-13-1819-1_24

by Ramos [2] the world's coal stocks are depleting and development of entities which were presuming for greener energy resolutions, were directed to explore new energy generation techniques. To deal with the situation, researchers studied upon various approaches for green technology and proposed the microgrid systems which comprise of several alternative energy sources. These alternate sources deal with solar cells, wind turbines, micro-turbines, fuel cells, batteries, and other storage devices [3, 4]. These sources were unified through the foremost power grid and through the assistance of power electronic devices they were connected to the low voltage distribution network [5, 6]. The interesting facts regarding microgrids are that they involve alternative energy sources which offer high efficiency and less environmental issues when compared to the conventional generation systems. Furthermore, losses due to transmission were eliminated proportionally as they were commissioned at the supply site conferring them more cleansed and efficient [7, 8]. Irrespective of all the advantages, microgrids are still unpredictable as a whole system due to multiple sources [9]. To confront such scenario, simulating the system and inspecting to develop an applicable regulatory is the core for microgrid research [10].

Irrespective of all the research that's being carried out to study the benefits and feasibility of microgrids, it is also difficult to simulate the performance of microgrids at a given power system level due to the absence of specific simulators which include alternative sources. In addition to the aforementioned problem associated with unpredictability, there were some issues with the power electronics especially in the case of simulating the inverters [11, 12].

Dealing with the above-mentioned issues, this study aims to simulate a microgrid system which resides a PV and FC system. The long-term objective of the research is to develop a highly practical, precise model of proposed system, which provides a complete considerate of how the hybrid systems behave. Section 2 validates the various conventional Hybrid Power System topologies. Section 3 elaborates the different generating systems, storages and controlling units developed for grid implemented Hybrid Power Systems. Section 4 deals with energy management system and predictive control of storage devices associated with the generating units. We validate the assumed developments, schematics, simulation, and assessment outcomes in Sect. 5 to approve the efficiency of the system and regulatory controls using MATLAB/Simulink.

2 Conventional Topologies

In the recent past Hybrid Power System drew way more attention due to increase in fuel prices and constraints with transmission systems. To overcome the difficulties and understand the perfect combination of efficient and quality power various configurations of Hybrid Power Systems have been proposed as part of Grid integrated systems, or as standalone systems. Through our research, the established FC prototypical was accustomed using investigational data and the model performance was improvised to depict the real-time behavior of the system accurately. For more pre-

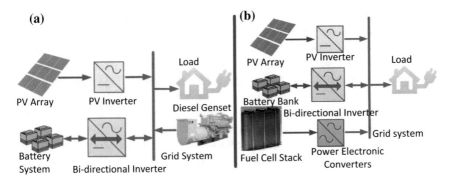

Fig. 1 **a** PV/battery/diesel HPS. **b** Photovoltaic-battery-fuel cell hybrid system

cision, an intelligent controller was implemented for supervisory control and battery management system to balance amongst the power sources and mandate for several times of the day.

2.1 PV/Battery/Diesel HPS

To survive the most terrible weather conditions, a standalone PV system requires significant financial investment and perfect design that is considered enough. When compared to the conventional systems, an integrated hybrid PV system offers cost-effective operation. By connecting a battery bank and diesel generator (Fig. 1a), we can augment the structure strategy and advance its accessibility. At conditions when PV generator fails to balance the power system or at times of certain weather conditions, the backup systems can contribute to ensuring a proper balance of power systems. In addition to balancing the systems, they also recharge the batteries [13]. The advantages are, Accessibility for spares and consumables, Batteries in the present day have improvised a lot and achieved a necessary level of sophistication, and Wide availability of hydrogen batteries with advanced technological capabilities to consumers. The disadvantages are, batteries should be well charged to achieve a good electricity release yield, a poor yield proportionate of 40% is offered by diesel generators for the power required and Constraints with Batteries life when compared to other components.

2.2 Photovoltaic-Battery-Fuel Cell Hybrid System

A fuel cell system which is used as backup systems replaces the conventional diesel generator in the system (Fig. 1b). When the load exceeds the generation limits, the batteries reach the minimum allowable charging level, and thus the backup plays a significant role in balancing the system. Both the Photovoltaic-Battery–Diesel hybrid system and PVFC hybrid system have same advantages regarding the size of PV generation and availability of batteries. Design parameters, size and the operating techniques of hybrid system will be affected due to preeminent disparities that endure between a diesel generator and a fuel cell. Constraints like startup time and output processing were also affected by the change in the system which was in favor of diesel generator initially but improvise in favor of fuel cell significantly [14]. The main disadvantage with the diesel generators is that they should be operated at rated power as much as possible, but the feasible fuel cell capacity can be determined according to the load profile.

High conversion efficiency is the significant advantage of the fuel cell when compared to diesel or petrol generator. In a situation where a 1 kW diesel generator accomplishes overall productivities amongst 8–15% [15], a parallel fuel cell arrangement can accomplish up to 50% productivity when operated with H_2 and O_2. In contradiction, the technical properties of a fuel cell such as low noise level and clean exhaust gases, makes them appeal for standalone power systems especially when pure hydrogen is used as a fuel.

3 Grid-Connected PV and Fuel Cell Power Generating Model

3.1 PEMFC Model

This stack model utilized has a nominal voltage of 45 Vdc and the nominal power of 1.5 kW. The converter is simulated (for test purpose) with a time constant of 1 s. Throughout the initial state, we observed that hydrogen utilization is constant at 99.56% which is measured using a flow rate regulator. After a certain period, to observe the stack voltage variations, we will bypass the flow rate regulator which increases the rate of fuel to a value of 300 lpm which affects the stack efficiency, the fuel consumption, and the air consumption.

3.2 Photovoltaic System Model

This section deals with a detailed model of grid-connected PV system through a DC–DC boost converter and a three-phase three-level VSC. Maximum Power Point Tracking (MPPT) is executed in the boost converter using a Simulink model using the intelligent technique. PV array delivers a determined output of 800 W/m² sun irradiance. 5-kHz DC-DC boost converter increases the PV natural voltage to 500 V DC. Switching duty cycle is optimized by an MPPT controller which uses the intelligent technique. Duty cycle is automatically varied to extract maximum power generate the obligatory voltage. 3-level 3-phase VSC converts the 500 V DC link voltage to 260 V AC conducting unity power factor.

PV Array parameters consent you to select amid numerous array sorts of the NREL System Advisor Model [16]. They also consent you to plot the I–V and PV characteristics for one module and the whole array. The system consists of two inputs that permit you changeable sun irradiance and temperature. The irradiance and temperature outlines are demarcated by a Signal Builder block which is associated to the PV array inputs.

3.3 Fuzzy Inference System

Issues regarding MPPT control, tracking of temperature changes, and nonlinearity of PV system regarding its dependence on operating voltage array. To mend such points, we propose an intelligent control system based on Knowledge-based techniques like fuzzy. By the application of intelligent techniques to boost converter, we can track maximum power point considering all the constraints. Sudden disturbances in MPP due to atmospheric changes are conceivably traced absolutely to such regulatory [16]. The major advantage of the technique is that it depends only on PV array output power irrespective of irradiation or temperature, which reduces the design complexity hence reducing the number and cost of equipment.

Considering various advantages with fuzzy inference system for nonlinear networks, we can also integrate it with the energy management system. The energy management system deals with a battery unit which acts as backup and activates mainly when there is a need for balancing between the load and generation. Such type of integrated control is also known as supervisory control.

In generally fuzzy system works by linguistic variables and membership functions utilizing various platforms for different weights available in the GUI. By all the available inputs and required outputs, a significant set of rules were obtained. The developed set of rules was considered as the heart of the knowledge-based system, and the entire control strategy works on these set of rules. Error and change in error obtained from Generation and Load of the system are the primary inputs for the control system. Triangular and Trapezoidal membership functions with constants like Low, Medium, and High were formulated both for the input and output GUI.

4 Energy Management System (EMS)

The primary goal of energy management system is to maximize the output of renewable energy systems by implementing various control algorithms and strategies. The primary objective of the energy management mainly deals with long-term planning and goals.

4.1 Battery SOC and Predictive Control

Due to indeterministic nature of Renewable energy generation and changing load conditions, it is difficult to determine the power balance shortly. Considering a steady state, after the demand was met by the generation from PVFC system, the remaining power is directed towards the charging of batteries, resulting in ramping operation of the fuel cell. The Ramping operation of the fuel cell depends on the high state of charge of the battery unit whereas the low state of charge is also necessary to accommodate for surplus PV energy in future. To minimize the amount of PV energy dumped due to the battery overcharge it is necessary for energy management system to determine the accurate SOC level. This research develops a predictive controller strategy which is implemented with the help of Fuzzy Inference system to provide a balance between the power generated and load power. Energy management between the generation and demand is carried out for a T_p minutes for enabling prediction of available power which tend to surplus T_p minutes for the future.

For a X_i discrete data series, Smoothing series value at period t is given by

$$\widehat{Y}(t) = \alpha x(t) + (1 - \alpha)\widehat{Y}(t - 1) \tag{1}$$

$$\widehat{Y}(t) = \alpha \left[\begin{array}{c} x(t) + (1 - \alpha)x(t - 1) + \\ (1 - \alpha)^2 x(t - 2) + (1 - \alpha)^3 x(t - 3) + \cdots \end{array} \right] \tag{2}$$

Smoothing parameter tends to be α, when (-1) is the one data point for smoothed output. Net surplus power is illustrated in Eq. (3), where P_G and P_L tend to be power output of PVFC and actual power of the load respectively.

$$P_{su} = P_G - P_L \tag{3}$$

For smoothing out the fluctuation in case of power surplus, an expression similar to (1) can be presented using previous power measurement for a discrete series data.

$$\widehat{P}_{su}(t) = \alpha P_{su}(t) + (1 - \alpha)\widehat{P}_{su}(t - 1) \tag{4}$$

For forecasting surplus power one sample point $\widehat{P}_{su}(t)$ can be considered implicitly. Longer prediction time is desired for allowing longer charging and discharging time of the battery. The process is carried out to smoothen the power constant over the next T_P minutes which is utilized for sampling and holding (S/H). Therefore, the new expressed power tends to be the surplus power which is the step wise constant power as given by expression in Eq. (5). ZOH is the zero-order holding operation which tends to hold the smoothed sample for the surplus power series constant until the next TP minutes.

$$\widehat{P}_{su,new}(t) = ZOH\left\{\widehat{P}_{su}\right\} \tag{5}$$

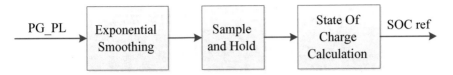

Fig. 2 SOC setpoint generation algorithm

Optimum state charge setpoint is calculated with the help of surplus power which as calculated in Eq. (5) over the period of T_p.

Based on expected surplus power which is computed in Eq. (5) optimum state charge is calculated over T_P minutes. SOC setpoint can be determined as follows for battery bank consisting total capacity Q [Ah] and average voltage $V_{B,AVG}$ [V] as following:

$$SOC_{ref}(t) = SOC_{high} - \frac{\widehat{P}_{su,new} \times T_P}{Q \times V_{B,AVG}} \tag{6}$$

State of charge higher predefined value SOC_{high} is preferred to be below the maximum value of the state of charge for permitting enough space to be allocated for unintentional power which is in surplus because of the forecasting error for charging the battery continuously. In case of zero expected surplus power SOC_{ref} acts as SOC_{high}. Figure 2 illustrates the generation algorithm for SOC_{ref}.

4.2 PVFC System and Batteries Adaptive Difference

Once the setpoint of desired state of charge (SOC_{ref}) is determined, it is used to calculate the additional power being discharged from the battery for state of charge higher than the setpoint. An additional discharge power from the battery means a reduced power coming from the fuel cell. If the state of charge of the battery is less than (SOC_{ref}), the battery is charged with an additional power. The additional battery power calculation is setup to be adaptive. Despite charging or discharging the battery at the rated power, a fraction of a desired high power that is dependent on the battery state of charge is used. For this, two weighting factors given in (7) and (8) are used to determine the fractional (of the desired high power) battery power which is calculated as (9) where PB is a given desired high power. The output power of the fuel cell also varies adaptively and is calculated by subtracting or adding the additional battery power to the fuel cell reference power depending on whether the battery is discharging or charging.

$$\omega_D(t) = \frac{SOC - SOC_{ref}}{SOC_{max} - SOC_{ref}}, SOC \geq SOC_{ref} \tag{7}$$

$$\omega_C(t) = \frac{SOC - SOC_{ref}}{SOC_{ref} - SOC_{Min}}, SOC \geq SOC_{ref} \qquad (8)$$

$$P_{ada} = \omega \times P_B, \omega\epsilon\{\omega_D, \omega_C\} \qquad (9)$$

From (7–9), it can be seen that the battery discharges or charges with battery power equal to a desired high power PB when the battery state of charge is at maximum or minimum allowable level and with a linearly decreasing power otherwise. Note that here charging power is assumed negative ($-1 \leq \omega C \leq 0$) while discharging power is positive ($0 \leq \omega D \leq 1$).

5 Simulation and Results

In the previous section, advantages, disadvantages and control strategies required for design of an efficient model were described. The proposed model consists of two power systems which produce the approximate power up to 6 kW. The dc bus is associated to the dc to dc converter through the load which stabilizes the input of the dc load. In addition to it, the battery management system is also introduced to support the dc load in case of any power shortage. The battery is connected to the dc bus bar with two circuit breakers. These breakers are controlled using the fuzzy logic controller.

In the simulation development, the objective is to perceive the proposed system behavior beneath diverse operating conditions. The solar radiation, panel temperature, and consumer load profiles are all employed to depict the performance of the anticipated hybrid system model, as depicted in Fig. 3.

Furthermore, conferring to the proposed EMS algorithm, Load is determining the functionality of the Battery System, during peak load demand periods. Difference between the power generated and load ($P_G - P_L$) is Negative (i.e., demand is greater than a generation), the remaining energy is satisfied by the Battery System (Discharging). Difference between the power generated and load ($P_G - P_L$) is Positive (i.e., Generation is greater than Demand), at this instant generation is fed to the Battery System by SOC of the system (Charging). If the rate of change of error ($-\Delta(P_G - P_L)$) is high, at this time, the battery bank discharge current is very high, and the battery bank terminal voltage drops significantly.

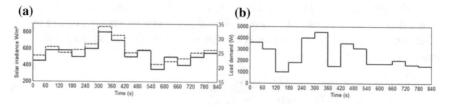

Fig. 3 **a** Solar irradiance and temperature profiles of PV system, **b** Consumer Load profile

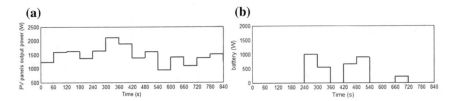

Fig. 4 **a** PV panel output power. **b** Power satisfied by battery

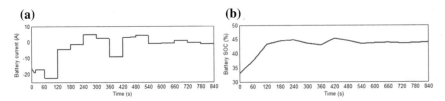

Fig. 5 **a** Rate of change of battery current. **b** Battery state of charge

Therefore, for validation purpose, a test case has been simulated. The rated power of the PVFC system is 6 kW.

Initially, let us consider that the State of Charge of the battery is 33%. Here, when the load demand is low, it is satisfied by the PV system and PEMFC stack, and the extra power will be used to charge the battery (SOC goes high). In this scenario, it is observed that the PEMFC stack power is very high. On the other hand, when the SOC is above 40.5%, and the load demand is high then the energy is satisfied by the battery bank and PVFC system, as shown in Fig. 4b.

As depicted in Fig. 5a, batteries power is met by introducing a current charge at the terminal of the battery bank. Current is transferred from battery bank to the load end when the battery current is positive. Recharging of the battery takes place in case of adverse battery current. Battery SOC is illustrated in Fig. 5b.

Hybrid system model robustness is depicted in the result. Excellent performance was exhibited by hybrid topology under various operating conditions, and SOC was maintained between 40 and 80%. The proposed system results in shorter settling time, low overshoot and zero steady state error when it is compared with conventional systems for managing energy.

6 Conclusion

This paper presents a highly practical, precise model of photovoltaic fuel cell system, which provides a complete idea about how the hybrid systems behave. The proposed standalone FC/PV/Battery Hybrid Power System was simulated using Simulink. The results depict the performance of the system with extreme geographical conditions. It is observed that, a fine-tuned intelligent controller like fuzzy would improve the performance of the system by improving its efficiency. The developed systems also deal with energy management and distribution along with supervisory control without disturbing system performance. Implementation of Battery management system (BMS) has depicted to be highly effective. The research depicted that the developed technique can improve the battery mean state of charge by 20% compared with the conventional techniques. It is also observed that with the help of predictive control, substantial improvements are obtained which adheres to 20–30% increase in PV energy utilization and reduction in fuel usage of Fuel Cell by 25%. Overall results depict a near optimal solution on how to manage different power systems and energy devices.

References

1. J.A. Gow, C.D. Manning, Development of a photovoltaic array model for use in power electronics simulation studies. IEE Proc. Electr. Power Appl. **146**(2), 193–200 (1999)
2. J.A. Ramos, I. Zamora, J.J. Campayo, Modeling of photovoltaic module, in *International Conference on Renewable Energies and Power Quality (ICREPQ'10)*, Granada, Spain, vol. 1, 2010
3. M.G. Villalva, J.R. Gazoli, E. Ruppert, Modeling and circuit-based simulation of photovoltaic arrays. Braz. J. Power Electron. **14**(1), 35–45 (2009)
4. W. DeSoto, Improvement and validation of a model for photovoltaic array performance. M.Sc. thesis, Mechanical Engineering, University of Wisconsin, 2004
5. N. Sharma et al., Modeling and simulation of photovoltaic cell using Matlab/Simulink. Int. J. Sci. Res. Publ. **3**(7) (2013). ISSN 2250-3153
6. G. Walker, Evaluating MPPT converter topologies using a Matlab PV model. J. Electr. Electron. Eng. **21**(1), 49–56 (2001)
7. F. Gonzaez-Longatt, Model of photovoltaic in MatlabTM, in *2nd Latin American Student Congress of Electrical Engineering and Computer Science (II CIBELEC)*, 2006. [Puerto la Cruz. Venezuela]
8. A. Oi, Design and simulation of photovoltaic water pumping system. Master thesis, San Luis Obispo, California Polytechnic State University, 2005
9. A. Luque, S. Hegedus, *Handbook of Photovoltaic Science and Engineering* (Wiley, Hoboken, 2003)
10. A.S. Golder, Photovoltaic generator modeling for large scale distribution system studies. Master Thesis, Drexel University, 2006
11. Z. Jiang, X. Yu, Hybrid DC- and AC-linked microgrids: towards integration of distributed energy resources, in *Proceedings of IEEE Energy 2030 Conference*, 2007, pp. 1–8
12. G. Bayrak, M. Cebeci, Modelling 3.6 Kw installed power PV generator with Matlab Simulink. J. Inst. Sci. Technol. Erciyes Univ. **28**(2), 198–204 (2012)
13. T. Somasak, U. Boonbumrung, S. Tanchareon, N. Jeenkaokam, C. Jovacate, PV- diesel stand alone hybrid system at a royal project research station: observations on power quality. Technical Digest of the International PVSEC-14, Thailand, 2004
14. O. Ulleberg, Stand-alone power systems for the future: optimal design, operation and control of solar-hydrogen energy systems. Ph.D. Dissertation, Norwegian University, Trondheim, 1998
15. J. Benz, B. Ortiz, W. Roth et al., Fuel cells in photovoltaic hybrid systems for stand-alone power supplies, in *2nd European PV-Hybrid and Mini-Grid Conference*, Kassel, Germany, 2003, pp. 232–239
16. https://sam.nrel.gov/. Accessed 8 Mar 2017

Optimal Active–Reactive Power Dispatch Considering Stochastic Behavior of Wind, Solar and Small-Hydro Generation

Jigar Sarda and Kartik Pandya

Abstract Generations from several sources in an electrical network are to be optimally scheduled for economical and efficient operation of the network. Optimal Power Flow (OPF) basically performs an intelligent power flow and optimizes the system operation condition by optimal determination of control variables. The objective of this paper is minimize the total fuel cost of the traditional generators plus the expected cost of an uncertainty cost function for renewable generators while satisfying all operational constraints. The model considers reserve cost for overestimation and penalty cost for underestimation of intermittent renewable sources. In this paper Weibull, Lognormal and Gumbel distributions are used for the wind speed, solar irradiance and river flow respectively. For achieving optimal solution efficiently, it requires a robust and effective solution technique. In this paper, results of the Cuckoo search algorithm (CSA) and Flower pollination algorithm (FPA) are compared to dealing with such type of optimal active–reactive power dispatch problems on IEEE-57 bus system.

Keywords AC optimal power flow · Renewable energy sources
Weibull probability distribution function · Lognormal probability distribution
function · Gumbel probability distribution function

1 Introduction

The electric power industry has lived a significant expansion and growth over the course of the past two decades. The penetration level of renewable sources are increased due to their pollution-free generation techniques and continuous availability. In some days, however a single renewable energy source system cannot

J. Sarda (✉) · K. Pandya
Department of Electrical Engineering, Charusat University, Changa, India
e-mail: jigarsarda.ee@charusat.ac.in

K. Pandya
e-mail: kartikpandya.ee@charusat.ac.in

© Springer Nature Singapore Pte Ltd. 2019
H. Malik et al. (eds.), *Applications of Artificial Intelligence Techniques in Engineering*, Advances in Intelligent Systems and Computing 698,
https://doi.org/10.1007/978-981-13-1819-1_25

provide a continuous source of energy due to low availability. Therefore operation of large-scale renewable stations connected to the grid has been used into the future pattern [1]. On the other hand, the discontinuous nature of wind speed, solar irradiation and flow of water cause difficulties in incorporating uncertainties into the power system. Optimal power flow including wind generation has been described in detail [2]. The inherent uncertain nature of wind speed should be incorporated in OPF for its realistic solution. Due to this fact, [3] characterized uncertain nature of wind speed and consider underestimation cost and overestimation cost of wind generation. Many artificial intelligent techniques are used for solar irradiance forecasting, but for economic dispatch probability distribution function is used [4]. The main objective of economic dispatch is to minimize fuel cost of power generation while fulfilling constraints. In [5] OPF calculation with more complex objectives of multi-fuel options and considers valve-point effect in thermal generators in applying backtracking search optimization algorithm. In [6] multi objective combined economic emission dispatch is converted to single objective economic dispatch using penalty factor and ABC algorithm is used to solve the problem. In [7] an economic performance evaluation method for hydroelectric generating units is proposed. In [8] pumped hydro storage is introduced as an alternate form of storage for a similar standalone hybrid system consisting of a solar PV, a wind turbine and a diesel generator. In [9] SPEA algorithm is used for solving economic dispatch considering single solar and a wind farm is done. Generally speaking, Optimal Power Flow has not been solved for a combination of wind, solar and hydro generator with inclusion of renewable generation cost for overestimation and penalty cost for underestimation of intermittent renewable sources. The paper is organized as follows: Sect. 2 discusses the wind energy stochastic model. Section 3 presents the PV stochastic model. In Sect. 4, the formulation of OPF problem incorporating wind, solar and small-hydro generation cost is done. In Sect. 5, describes the case studies; afterwards, the obtained results are presented for each case study. Finally, the conclusion remarks will come in Sect. 6.

2 Wind Energy Stochastic Model

Wind energy has inherent variances and hence it has been expressed by distribution functions. With the help of probability distribution function behavior of wind velocity at a given site can be specified. Among the probability density functions that have been proposed for wind speed of most locations, the Weibull function has been the most acceptable distribution [10]. The PDF for Weibull distribution is given by:

$$fw(w) = \left(\frac{sh}{sc}\right) * \left(\frac{w}{sc}\right)^{sh-1} * e^{-\left(\frac{w}{sc}\right)^{sk}}, \quad 0 < w < \infty \tag{1}$$

where, w-wind speed (m/s), sc-scale factor and sk-shape factor.

The power output of the wind energy conversion system (WECS) may be given as:

$$W = 0; \quad 0 \le w \le w_i$$
$$W = W_r * \left(\frac{w - w_i}{w_r - w_i} \right); \quad w_i \le w \le w_r$$
$$W = W_r; \quad w_r \le w \le w_0 \tag{2}$$

where W = Output power of WECS; W_r = rated power of WECS; w_i = cut-in speed; w_r = rated speed; w_0 = cut-out speed.

The probability of wind speed being smaller than cut in speed (i.e. $w < w_i$) and larger than cut-out speed (i.e. $w > w_o$) is expressed as follows:

$$F_w(w = 0) = 1 - \exp\left(-\left(\frac{w_i}{sc}\right)^{sk}\right) + \exp\left(-\left(\frac{w_o}{sc}\right)^{sk}\right) \tag{3}$$

The probability of wind speed between rated output speed and cut-out speed is given by:

$$F_w(w = w_r) = \exp\left(-\left(\frac{w_r}{sc}\right)^{sk}\right) + \exp\left(-\left(\frac{w_o}{sc}\right)^{sk}\right) \tag{4}$$

The probability of wind speed between cut in speed and rated output speed is given by:

$$f_w(w) = \left(\frac{sk(W_r - W_i)}{sc * w_r} \right) * \left(\frac{W_i W_r + w * (W_r - W_i)^{sk-1}}{sc * w_r} \right)$$
$$* \exp\left(\frac{W_i W_r + w * (W_r - W_i)^{sk}}{sc * w_r} \right) \tag{5}$$

3 PV Stochastic Model

The solar irradiation to energy conversion function of the PV generator or power output from PV cell is given by [11],

$$P_s(G) = P_{sr} * \left(\frac{G^2}{G_{std} * R_c} \right), \quad 0 < G < R_c$$
$$= P_{sr} * \left(\frac{G}{G_{std}} \right), \quad G > R_c \tag{6}$$

where it is noted that PV cell temperature is neglected and G = solar irradiation forecast in W/m², G_{std} = solar irradiation in the standard environment set as 1000 W/m², R_c = a certain irradiation point set as 150 W/m², P_{sr} = equivalent rated power output of the PV generator.

4 Problem Formulation

The objective of renewable-thermal optimal power flow formulation is to minimize the expected cost of operation with all system constraints satisfied. The different cost components includes; (i) the fuel cost of thermal generator, (ii) expected penalty cost for not utilizing available power given by any of the renewable energy source due to network congestion, (iii) expected reserve cost due to the shortage of power given by any of the renewable energy source. The condition of penalty cost and reserve cost are explained below:

Underestimated Condition
The scheduled power (Psi) in the renewable generator i is lesser than the available real power (Pai), and there will be a cost for underestimate given by: Cu = cu * (Pai − Psi) due to the total available power that is not used in the system (only it is used Psi). It would be a kind of power wasted, but in the real life could be considered as a power directed to an energy storage system with a related cost for using the system (cu).

Overestimated Condition
The scheduled power (Psi) in the renewable generator i is bigger than the available real power (Pai), and there will be a cost for overestimate given by: Co = co * (Psi − Pai) due to the total available power is not enough to get the power to be scheduled in the system (Psi). In this case the network operator must turn on or request more power to another energy source with a related cost (co).

4.1 Operational Cost Function for Wind Power Plant

The operating cost of wind power plant, $oc(v_{ij})$, includes three parts as below:

$$oc(v_{ij}) = c_{vi}(v_{ij}) + c_{p,vi}(V_{ij,av} - v_{ij}) + c_{r,vi}(v_{ij} - V_{ij,av})$$ (7)

The first part is the direct cost involved with wind power from ith plant is modeled as a function of scheduled power is defined as:

$$c_{vi}(v_{ij}) = d_i * f_v(v) * (v_{ij})$$ (8)

where d_i is the cost coefficient output of ith wind farm.

The second part is the penalty cost for the ith wind power plant is defined as:

$$C_{p,vi} = k_{P,vi} * \left(V_{ij,av} - v_{ij}\right) = k_{P,vi} * \int_{vi}^{v_{ri}} \left(v - v_{ij}\right)f_v(v)dv \qquad (9)$$

where $k_{P,vi}$ = coefficient of penalty cost, $V_{ij,av}$ = actual or available wind power. v_{ri} = maximum output of ith wind farm.

The third part is the reserve cost for the ith wind power plant is defined as:

$$C_{r,vi} = k_{r,vi} * \left(v_{ij} - V_{ij,av}\right) = k_{r,vi} * \int_{0}^{v_i} \left(v_{ij} - v\right)f_v(v)dv, \qquad (10)$$

where $k_{r,vi}$ = coefficient of reserve cost.

4.2 Operational Cost Function for PV Power Plant

The operating cost of PV power plant, $oc(pv_{ij})$, includes three parts as below:

$$oc(sp_{ij}) = k_{spi}\left(sp_{ij}\right) + k_{p,spi}\left(SP_{i,av} - sp_{ij}\right) + k_{r,spi}\left(sp_{ij} - SP_{i,av}\right) \qquad (11)$$

The first part is the direct cost involved with PV power from ith plant is modeled as a function of scheduled power is defined as

$$k_{psi}\left(sp_{ij}\right) = h_i * f_{sp}(sp) * (sp_i), \qquad (12)$$

where h_i = cost coefficient, sp_i = scheduled power output.

The second part is the penalty cost for the ith PV power plant is defined as

$$k_{p,spi} = c_{P,spi} * \left(SP_{ij,av} - sp_{ij}\right) = c_{P,spi} * \int_{spi}^{sp(c_{tmax})} \left(sp - sp_{ij}\right)f_{sp}(sp)dsp, \qquad (13)$$

where $c_{P,spi}$ = coefficient of penalty cost, $SP_{ij,av}$ = available PV power, $sp(c_{tmax})$ = maximum output power.

The third part is the reserve cost for the ith PV power plant is defined as:

$$k_{r,spi} = c_{r,spi} * \left(sp_{ij} - SP_{ij,av}\right) = c_{r,spi} * \int_{0}^{sp_i} \left(sp_{ij} - sp\right)f_{sp}(sp)dsp \qquad (14)$$

where $c_{r,psi}$ = coefficient of reserve cost.

5   Simulation Result and Discussion

In this work we have consider IEEE 57 bus system is consider. The IEEE 57 bus system has seven generators, in this paper three of them are considered renewable generators in the buses 2, 6, and 9; respectively. Three different cases of stochastic scenario are simulated in this paper for cuckoo search algorithm (CSA) and flower pollination algorithm (FPA).

Objective Function

$$F(P_G) = \sum_{i=1}^{N} a_i + b_i P_{Gi} + c_i P_{Gi}^2 + oc(w_{ij}) + oc(ps_{ij}) \qquad (15)$$

Constraint

There are three types of constraints:

(i)   Power flow constraints: the constraints are associated to nodal balance of power,

(ii) Constraints penalized in the fitness function: nodal voltages for load buses (42 + 42) allowable transmission line power flows (80), reactive power capability of generator (7 + 7) and maximum active power output of slack generator (1)

(iii) Minimum and maximum levels of optimization variable.

Optimization Variables

31, comprising 13 continuous variables associated to generator active power outputs (6) and generator bus voltage set-points (7), 15 discrete variables associated to stepwise adjustable on-load transformers' tap positions, and 3 binary variables associated to switchable shunt compensation devices.

Considered contingencies (N − 1 conditions): outages at branches 8 and 50.

Number of function evaluations: 50,000.

(1)   **Stochastic OPF for IEEE-57 bus system considering wind generator (Case 1)**

For the case 1 of this paper, it is considered that the three renewable generators are wind generators. It is well known that the wind speed probability distribution follows a Weibull distribution [12].

(2)   **Stochastic OPF for IEEE-57 bus system considering Wind and Solar generators (Case 2)**

For the case 2 of this paper, it is considered that the two renewable generators are wind generator and other solar generator. It is well known that in several parts of the world the solar irradiance probability distribution follows a lognormal distribution [11].

(3)   **Stochastic OPF for IEEE-57 bus system considering Wind, Solar and Small-Hydro generators (Case 3)**

Table 1 Objective function and fitness function values for case-1

Trial	CASE-1 (Wind) (CSA)			Trial	CASE-1 (Wind) (FPA)		
	o_best	f_best	g		o_best	f_best	g
Avg.	72754.3698	72766.941	12.572	Avg.	84750.3	85123.73	373.36
STD	33.7076975	50.138568	28.072	STD	4678.13	5251.145	2973.5
Worst	72850.4934	72855.311	4.8184	Worst	86557.1	98103.23	11546.0
Best	72719.3219	72719.400	0.0781	Best	86522.5	86522.50	0.0010
Median	72740.113	72752.985	1.7250	Median	86521.1	86522.49	0.4335

Table 2 Objective function and fitness function values for case-2

Trial	CASE-2 (Wind-Solar) (CSA)			Trial	CASE-2 (Wind-Solar) (FPA)		
	o_best	f_best	g		o_best	f_best	g
Avg.	72150.062	72158.481	8.4190	Avg.	72558.4	72558.91	0.4954
STD	33.737285	36.345041	11.236	STD	2646.27	2646.189	0.9279
Worst	72236.849	72248.182	11.333	Worst	72283.4	72287.58	4.0961
Best	72150.646	72150.72	0.074	Best	72254.4	72254.47	0.0071
Median	72144.22	72150.72	4.759	Median	72086.9	72086.91	0.1035

Table 3 Objective function and fitness function values for case-3

Trial	CASE-3 (Wind-Solar-Hydro) (CSA)			Trial	CASE-3 (Wind-Solar-Hydro) (FPA)		
	o_best	f_best	g		o_best	f_best	g
Avg.	60357.192	60364.46	7.2639	Avg.	60323	60366.29	43.263
STD	23.265916	27.36594	9.4058	STD	17.52	236.5547	233.59
Worst	60396.071	60403.15	7.08	Worst	60335	61637.35	1301.7
Best	60321.768	60321.82	0.051	Best	60309	60309.81	0.012176
Median	60354.025	60360.03	3.101	Median	60320	60321.00	0.249194

For the case 3 of this paper, it is considered that at bus 2 there is a wind generator and that at buses 6 and 9 there are two generators, a solar generator and a small-hydro generator respectively. It is well known that the river flow follows a Gumbel distribution [13] (Tables 1, 2, 3 and 4).

6 Conclusion

The Optimal Power Flow problem incorporating wind, solar and small-hydro energy system has been discussed and comparison between cuckoo search and flower pollination algorithm is done. The result shows that cuckoo search give better result compare to flower pollination algorithm. Also wind, solar and small-hydroelectric generator cost modeling is done. This comprises of costs associated with underestimation and overestimation of wind power by considering the uncertainty in wind speeds, solar radiation and flow of water respectively. The objective function has been presented with all physical and operation constrains and solved using cuckoo

Table 4 Optimized control variable values for all three cases

Variables	Case 1 (CSA)	Case 1 (FPA)	Case 2 (CSA)	Case 2 (FPA)	Case 3 (CSA)	Case 3 (FPA)	Lb	Ub
Pg2 (MW)	134.26	135.08	131.96	131.605	135.44	139.719	0	150
Pg3 (MW)	43.49	42	42	42.47	42.15	44.965	42	140
Pg6 (MW)	150	120.73	37.3	37.287	48.23	47.961	0	150
Pg8 (MW)	454.96	399.27	430.76	435.756	484.94	472.378	165	550
Pg9 (MW)	0	136.42	148.78	146.772	50.88	51.3966	0	150
Pg12 (MW)	354.43	314.72	347.24	340.704	364.99	369.406	123	410
V1 (p.u.)	1.049	1.0000	1.049	1.01308	1.05	1.01017	0.95	1.05
V2 (p.u.)	1.048	1.0011	1.05	1.01261	1.045	1.01047	0.95	1.05
V3 (p.u.)	1.036	0.9990	1.045	1.00315	1.035	1.00083	0.95	1.05
V6 (p.u.)	1.044	1.025	1.05	1.01103	1.033	1.01245	0.95	1.05
V8 (p.u.)	1.046	1.0312	1.049	1.02197	1.049	1.03135	0.95	1.05
V9 (p.u.)	1.021	0.9986	1.028	0.9969	1.021	0.99583	0.95	1.05
V12 (p.u.)	1.044	0.9895	1.023	0.99517	1.03	0.99449	0.95	1.05
OLTC_4-18	−9	−9	−9	−10	10	8	−10	10
OLTC_4-18	1	−4	6	−4	−7	−1	−10	10
OLTC_21-20	10	2	1	−1	5	3	−10	10
OLTC_24-26	10	1	3	2	2	1	−10	10
OLTC_7-29	−4	−4	0	−4	0	−4	−10	10
OLTC_34-32	−6	−1	2	−6	−9	−5	−10	10
OLTC_11-41	−10	−10	−4	−6	−9	−10	−10	10
OLTC_15-45	−7	−9	0	−7	−6	−9	−10	10
OLTC_14-46	−4	−9	−6	−8	−5	−9	−10	10
OLTC_10-51	−1	−8	−5	−7	−3	−8	−10	10
OLTC_13-49	−10	−10	−8	−10	−10	−10	−10	10
OLTC_11-43	−1	−7	−3	−9	−3	−8	−10	10
OLTC_40-56	−2	2	−10	2	−6	0	−10	10
OLTC_39-57	−1	−3	2	1	−4	−4	−10	10
OLTC_9-55	1	−6	3	−5	−2	−5	−10	10
SH_18	0	1	1	1	0	1	0	1
SH_25	0	1	1	1	1	1	0	1
SH_53	1	1	1	1	1	1	0	1

search algorithm. The result shows that when three wind energy sources are used operating cost is \$72719.3219 per hour. Similarly when combination of two wind energy sources and one solar energy source is used, operating cost is \$72150.646 per hour. When the combination of one wind energy source, one solar energy source and one small-hydro energy source is used, the operating cost is \$60321.768 per hour. Therefore by using different combination of renewable energy sources, operating cost is reduced.

References

1. J. Meng, G. Li, Y. Du, Economic dispatch for power systems with wind and solar energy integration considering reserve risk, in *IEEE PES Asia-Pacific Power and Energy Engineering Conference* (IEEE, 2013)
2. S. Gope, *Dynamic Optimal Power Flow with the Presence of Wind Farm* (Lambert Academic Publishing, 2012)
3. J. Hetzer, D.C. Yu, K. Bhattarai, An economic dispatch model incorporating wind power. IEEE Trans. Energy Convers. **23**(2), 603–611 (2008)
4. I.G. Damousis et al., A fuzzy model for wind speed prediction and power generation in wind parks using spatial correlation. IEEE Trans. Energy Convers. 352–361 (2004)
5. A.B. Chaib, H.R.E.H. Bouchekara, R. Mehasni, M.A. Abido, Optimal power flow with emission and non-smooth cost functions using backtracking search optimization algorithm. Int. J. Electr. Power Energy Syst. **64**, 77 (2016)
6. S.Y. Lim, M. Montakhab, H. Nouri, Economic dispatch of power system using particle swarm optimization with constriction factor. Int. J. Innov. Energy Syst. Power 29–34 (2009)
7. Y. Liu, L. Ye, I. Benoit, X. Liu, Y. Cheng, G. Morel, C. Fu, Economic performance evaluation method for hydroelectric generating units. Energy Convers. Manag. 797–808 (2003)
8. K. Kauakana, Optimal scheduling for distributed hybrid system with pumped hydro storage. Energy Convers. Manag. 253–260 (2016)
9. S. Brini, H.H. Abdallah, A. Ouali, Economic dispatch for power system included wind and solar thermal energy. Leonardo J. Sci. 204–220 (2009)
10. M.R. Patel, *Wind and Solar Power Systems* (CRC Press, Boca Raton, FL, 1999)
11. S.S. Reddy, P.R. Bijwe, A.R. Abhyankar, Real-time economic dispatch considering renewable power generation variability and uncertainty over scheduling period. IEEE Syst. J. 1440–1451 (2015)
12. T.P. Chang, Performance comparison of six numerical methods in estimating Weibull parameters for wind energy application. Appl. Energy **88**, 272–282 (2011)
13. N. Mujere, Flood frequency analysis using the Gumbel distribution. Int. J. Comput. Sci. Eng. 2774–2778 (2011)
14. Z.-L. Gaing, Particle swarm optimization to solving the economic dispatch considering the generator constraints. IEEE Trans. Power Syst. 1187–1195 (2003)
15. S. Rivera, A. Romero, J. Rueda, K.T. Lee, I. Erlich, *Evaluating the Performance of Modern Heuristic Optimizers on Smart Grid Operations Problem* (2017)
16. X.-S. Yang, Flower pollination algorithm for global optimization, in *Unconventional Computation and Natural Computation 2012*. Lecture Notes in Computer Science, vol. 7445 (2012), pp. 240–249
17. X.S. Yang, S. Deb, Cuckoo search via levy flights, in *Proceedings of World Congress on Nature and Biologically Inspired Computing* (2009), pp. 210–214

Wind Power Density Estimation Using Rayleigh Probability Distribution Function

K. S. R. Murthy and O. P. Rahi

Abstract The main aim of this paper is technical assessment of wind power for the complex hilly region of Hamirpur located in Himachal Pradesh, India. The wind data was recorded for 60 min interval for the year 2013 at this specific site at the measured height of 18.5 from above ground level. This time series wind data has been adopted to the Rayleigh distribution function for estimating the wind power density on monthly and annual basis. A Rayleigh model alone is applied for the first time to this specific site, which is the novelty and original contribution of this paper. Further, the monthly and annual disparities of the different wind resource characteristics namely, average wind speed, standard deviation, the frequency distribution, scale parameter (c) have been analyzed on the same time frames. Also, Rayleigh probability and cumulative distribution function have been computed by assuming shape parameter (k) is equal to 2, remains constant throughout the computational analysis. The monthly and year wise wind variability have been conveyed through the probability and cumulative distribution plots and it is derived from the measured data at this investigated site. Finally, the estimated annual wind power density showing 8.86 W/m^2 corresponding to the scale parameter 1.31 m/s. Hence, the outcome of this research work has been recognized that available wind resource is most suitable for micro and small wind power applications like standalone isolated systems, rural electrification, battery charging, home electricity appliances, irrigation of water by pumping through small wind mills.

Keywords Probability · Density function · Rayleigh distribution
Wind power density

K. S. R. Murthy (✉) · O. P. Rahi
National Institute of Technology, Hamirpur, Hamirpur 177005, Himachal Pradesh, India
e-mail: harikella96@gmail.com1

© Springer Nature Singapore Pte Ltd. 2019
H. Malik et al. (eds.), *Applications of Artificial Intelligence Techniques in Engineering*, Advances in Intelligent Systems and Computing 698,
https://doi.org/10.1007/978-981-13-1819-1_26

1   Introduction

In present global scenario, energy plays a key role in our regular day-to-day activities, as well as it is one of the major inputs for economic and social development of a country. The degree of development and civilization of a nation is measured by certain amount of utilization of energy by its inhabitants. Therefore, the energy demand grows gradually because population increase, rise of industrialization and urbanization, however the developing countries have a strong energy demand and resources with respect to their thriving economies. In India, the electric power generation is a key driving force for the economic growth. Although, this type of power generation causes an environmental impact due to pollution, greenhouse gases emissions, climate change, global warming, and acid rains [1–3]. As a result, the emerging green energy technologies are being accounted, efficient utilization of these technologies to reduce environmental risks, generate minimal secondary wastes [4].

In global scenario, the wind energy has been earliest energy harnessing technologies among the other green technologies that is being used for centuries. In recent times, the wind energy is commercially and operationally paramount viable source for electric power production. Therefore, wind resources on earth planet cannot be exhausted, as ever civilization exists, though oil wells might dry and coal will run out in future [5]. The wind power development in India began since 1990s and has increased tremendously in the earlier few decades. Also, India gains fourth rank in throughout world, next to China, U.S., and Germany in electricity generation through wind. National Institute of Wind Energy (NIWE) has an autonomous institution mandated by Ministry of New and Renewable Energy (MNRE), Government of India for booming a survey to evaluate the wind potential in the country, has circulated the wind atlas of India [6, 7]. This atlas provides potential estimates in the country at different mast heights of 50 m, 80 m and 100 m to the tune of 49 GW, 102 GW, and 302 GW, respectively [8].

The wind resource assessment program assists in describing and accepting the dynamics of wind at an investigated region. Further, it has been required that at least 10 years of wind data measurements to know the accurate prediction of wind resource of chosen area; nevertheless, smaller duration of wind data would still give a rough estimation of the potential. Therefore, minimum of one year duration measurements is enough in a special category of remote places, where the appropriate data measurements are not available for such longer duration [9–11].

Ramachandran and Shruti [12] have carried out the feasibility of wind power projects at Himachal Pradesh. Also identified wind rich locations through mapping for different altitude locations all over the state. However in another study, Krishnadas and Ramchandran [13] have investigated potential of wind energy by using synthesized long term surface wind data available for very few areas under consideration. Aggarwal and Chandel [14] highlighted the essential requirement of wind energy power utilization for the mountainous region of Himachal Pradesh to meet the electricity demand in remote areas. However, no initiative was taken excluding metrological masts for wind data measurements at few places. The preliminary analysis for some regions in this state at different level of wind sensors heights started from of 10 m height and beyond [15]. This program was done by NIWE, which results a low wind potential in this hilly state region. Though, these sites were arbitrarily identified without any detailed survey and missing numerous places with huge wind potential at higher altitudes.

The rest of this paper is organized as follows: Sect. 2 provides the problem formulation of the present research work. A brief review of the proposed methodology includes physical description of the examined site and computational analysis is described in Sect. 3. A brief description of study area and data analysis is presented in Sect. 3. Further, the resulting analysis of the Wind Power Density (WPD) estimation is presented in Sect. 4. Finally, the outcomes of this research work are summarized in Sect. 5.

2 Problem Formulation

The mountainous and hilly regions are also the major potential areas for electricity generation from the wind across the world, although the wind potential in complex hilly terrains in India has not been harnessed completely yet. Also, electricity generation through wind machines has not been employed for decentralized applications on large scale manner due to lack of small and micro wind resource accessibility. The wind resource assessment studies conducted by NIWE for Himachal Pradesh has shown potential that WPD is not even more than 200 W/m^2 on the basis of annual mean wind speeds. A variety of locations such as hill tops and mountain ridges should need to be investigated for the identification prominent windy areas in this region. A detailed campaign for mapping of wind resource is required to be carry out in the Himalayan region. However, the average mean wind speeds will not give the true wind potential of the any selected region.

This paper aims the preliminary investigation of wind resource potential at National Institute of Technology Hamirpur (NITH) campus, Himachal Pradesh located in India. This process is to carry out using the hourly data measurements during year 2013. The present site is located between the geographical coordinates of latitudes approximately 76°18′ E and 76°44′ E and longitudes approximately 31°25′ N and 31°52′ N having an altitude of 785 m from above mean sea level [3, 11].

In this section, the physical description of the examined site having a monitoring facility of wind data has been presented. The present proposed methodology involves the binning of hourly measured data in format of frequency distribution spanning over a year 2013 at NITH campus. From the collection of data, frequency occurrence of wind speed distribution, Rayleigh probability density functions on the basis of monthly and annual have been derived. Also, the computation of WPD using Rayleigh model has been thoroughly described in this paper.

3   Proposed Methodology

In this section, the physical description of the examined site having a monitoring facility of wind data has been presented. The present proposed methodology involves the
binning of hourly measured data in format of frequency distribution spanning over
a year 2013 at NIT Hamirpur campus. From the collection of data, frequency occurrence of wind speed distribution, Rayleigh Distribution Functions (RDF) and cumulative distribution function on the basis of monthly and annual have been derived.
Also, the computation of WPD using Rayleigh model has been thoroughly described
in this work. A RDF model alone is employed for the first time to this investigated
site, which is the original contribution of this paper.

3.1   Study Area

This paper aims to scrutinize the wind potential for the site Center for Energy and
Environmental Engineering (CEEE) rooftop of building and having a monitoring
facility of wind data which is located at the NITH campus situated in the mountainous state of Himachal Pradesh, India. The campus is geographically located with a
longitude and latitude of nearly 76.52° E and 32.46° N having an altitude of 875 m at
sea level. The monitoring station has been installed on June 2011. The hourly wind
data observations spanning over a year 2013 has been thoroughly used and analyzed
statistically for the exploitation true wind power potential of examined site under
consideration.

3.2   Frequency Distribution, Rayleigh Probability Distribution
Functions, and Wind Power Density Estimation

The observation of whole hourly wind speeds has been transformed into n number
of classes (bines) with an equal bin width (span interval) 0.5 m/s for a year 2013.
It is an easiest way of formatting to represent a frequency distribution approach
for the achievement of statistical results as revealed in Table 1. The probability of
happening each individual wind speed observation in a given bin interval (class) has
been computed as given below [16]:

$$f(v_i) = \frac{f_i}{\sum_{i=1}^{n} f_i} = \frac{f_i}{N}(i = 1, 2, \ldots, n) \tag{1}$$

where,

$f(v_i)$ The probability of each instantaneous wind speed ith of interval,
f_i Wind speed frequency of each ith interval, and

Table 1 All the wind characteristic parameters derived from the hourly wind speed time series data in frequency distribution format for the year 2013 at 18.5 m height a.g.l

i	v (m/s)	v_i	f_i	$f(v_i)$	$F(v_i)$	$f_R(v_i)$	$F_R(v_i)$
1	0.0–0.5	0.36	141	0.02	0.02	0.15	0.01
2	0.5–1.0	0.79	915	0.10	0.12	0.31	0.03
3	1.0–1.5	1.26	1788	0.20	0.32	0.46	0.08
4	1.5–2.0	1.75	2046	0.23	0.56	0.59	0.15
5	2.0–2.5	2.23	1802	0.21	0.76	0.68	0.23
6	2.5–3.0	2.72	1108	0.13	0.89	0.73	0.32
7	3.0–3.5	3.21	508	0.06	0.95	0.75	0.42
8	3.5–4.0	3.72	226	0.03	0.97	0.73	0.52
9	4.0–4.5	4.21	105	0.01	0.99	0.68	0.61
10	4.5–5.0	4.73	47	0.01	0.99	0.62	0.69
11	5.0–5.5	5.23	35	0.00	1.00	0.54	0.76
12	5.5–6.0	5.75	19	0.00	1.00	0.46	0.83
13	6.0–6.5	6.27	10	0.00	1.00	0.38	0.87
14	6.5–7.0	6.86	5	0.00	1.00	0.29	0.92
15	7.0–7.5	7.32	3	0.00	1.00	0.24	0.94
16	7.5–8.0	7.67	1	0.00	1.00	0.20	0.96
17	8.0–8.5	8.23	1	0.00	1.00	0.14	0.97

N Total number of recorded wind of data points, and n is number of bin intervals over a considered time period (month or year)

The month or yearly average wind speeds ($\bar{v}$) and standard deviation (σ) of the wind speeds are calculated as follows:

$$\bar{v} = \frac{\sum_{i=1}^{n} f_i v_i}{\sum_{i=1}^{n} f_i} = \frac{1}{N} \sum_{i=1}^{n} f_i v_i \tag{2}$$

$$\sigma = \left[\frac{1}{N-1} \sum_{i=1}^{n} f_i (v_i - \bar{v})^2 \right]^{1/2} \tag{3}$$

The cumulative frequency or probability of each bin interval is as given below:

$$F(v_i) = \sum_{i=1}^{j} f(v_i) \tag{4}$$

where, $j \leq i$ and for $i = 1, 2, 3, \ldots, n$.

$$F(v_n) = \sum_{i=1}^{j} f(v_i) = 1 \tag{5}$$

Now interchanging $f(v_i)$ in (1) and $F(v_i)$ in (4) by continuous domain two parameter Weibull density functions to get (6) and (7) as follows [17]:

$$f_W(v) = \left(\frac{k}{c}\right)\left(\frac{v}{c}\right)^{k-1} \exp\left[-\left(\frac{v}{c}\right)^k\right] \tag{6}$$

$$F_W(v) = 1 - \exp\left[-\left(\frac{v}{c}\right)^k\right] \tag{7}$$

where,

$f_W(v)$ Weibull probability distribution function,
$F_W(v)$ Weibull cumulative distribution function,
k Shape parameter (has no units),
c Scale parameter (m/s) and
v Known wind speed (m/s)

Also, the unknown parameters k and c describe how peaked the distribution is and tell the wildness of the Weibull plot. In the literature many researches have been diagnosed so many statistical models for computing these unknown Weibull parameters [18–20]. Whereas in this paper, a Rayleigh distribution model has been used throughout the analysis to determine the site specific wind characteristics and existing true wind power potential.

The fundamental equation for computing wind power $[P(v), watts]$ per given unit area in terms of air density (ρ, assumed as 1.225 kg/m3 at mean sea level) and wind speed as given follows:

$$P(v) = \frac{1}{2}\rho v^3 \tag{8}$$

The mean Rayleigh WPD $[P_R, W/m^2]$, Rayleigh probability density distribution function $[f_R(v)]$, Rayleigh cumulative distribution function $[F_R(v)]$ are given in the Eqs. (9)–(11) below [21]:

$$f_R(v) = \frac{\pi v}{2\bar{v}^2} \exp\left[-\left(\frac{\pi}{4}\right)\left(\frac{v}{\bar{v}}\right)^2\right] \tag{9}$$

$$F_R(v) = 1 - \exp\left[-\left(\frac{\pi}{4}\right)\left(\frac{v}{\bar{v}}\right)^2\right] \tag{10}$$

$$P_R = \frac{3}{\pi}\rho v_m^3 \tag{11}$$

Finally, the Rayleigh model is a special case and it is reduced from Weibull model being shape parameter (k) equal to 2. Further, the scale parameter (c) for the Rayleigh model is becomes in the form of Eq. (12) as given as follows:

$$v_m = c\sqrt{\frac{\pi}{4}} \tag{12}$$

4   Results and Discussions

The technical assessment of wind potential over the monitoring site NITH campus at 18.5 m mast height a. g. l. The wind characteristics namely frequency distribution, Rayleigh probability density distributions and WPD have been statistically computed by adopting hourly measured data for describing the wind variability over the examined site.

As observed from Table 1, the frequency distribution analysis shows that, the occurrence of observed wind speeds for 2,046 h at the 4th bin interval, i.e., 1.5–2.0 m/s has a maximum probability of 0.23 having a cumulative frequency of 56%. While, prominent bin intervals from 5th to 9th have been considered mostly, due to the cut in wind speeds for real wind machines starts from minimum of 2 m/s and below these intervals (1st to 3rd bins) power extraction will be zero. In fact that, the observations were lesser in frequency count wise but have larger in magnitude of wind speeds beyond 5th bin interval. Also, it is noticed that, bin interval 14th to 17th also neglected because even lesser the frequency count with larger magnitude, as shown in column 1–7 in a given Table 1.

As reported in Table 2, the maximum average wind velocity 2.46 m/s has been noticed in May having σ value of 1.06 and minimum of 1.64 m/s in the month of August with σ of 0.73 having an average wind velocity 1.96 m/s and value of σ is 0.88. The σ describes how degree of data measurement is near to its average value. The standard deviation close to zero gives more accurate outcomes. Rayleigh distribution is single parameter distribution and depends only on the scale parameter (c). The c parameter has been observed maximum of 1.64 m/s in May and 1.09 m/s in August month. Whereas, shape parameter (k) has been constant for all the time and is equal to 2 for the Rayleigh distribution. Consequently, the annual Rayleigh scale parameter has been observed 1.31 m/s.

The monthly and annual probability density as well as cumulative distribution plots have been computed with the help of Rayleigh parameter calculated by the Eqs. (9) and (10). In specific site, the maximum probability density function has been noticed in May at corresponding scale parameter as 1.64 m/s. The monthly and annual Rayleigh probability distributions have been depicted in Fig. 1. As well as cumulative distributions have been represented in Fig. 2. Figure 3 reveals that, mean WPD's have been computed from Rayleigh model using (11) with corresponding to their mean wind speeds on the monthly and annual time frames.

S. No.	Period	v (m/s)	σ	k	c (m/s)
1	Jan	1.86	0.80	2	1.24
2	Feb	2.12	0.92		1.42
3	Mar	2.30	0.94		1.53
4	Apr	2.12	0.94		1.41
5	May	2.46	1.06		1.64
6	Jun	2.11	0.99		1.41
7	Jul	1.74	0.81		1.16
8	Aug	1.64	0.73		1.09
9	Sep	1.72	0.76		1.15
10	Oct	1.77	0.70		1.18
11	Nov	1.87	1.35		1.25
12	Dec	1.88	0.89		1.26
13	2013	1.96	0.88		1.31

Table 2 Monthly and Annual variation of average wind velocity, standard deviation, Weibull parameters for 2013 at said site

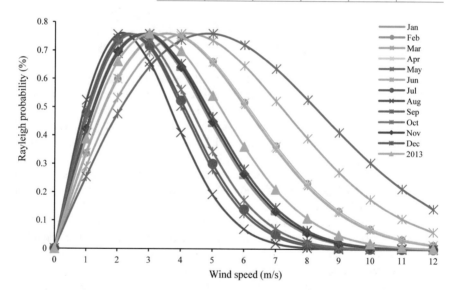

Fig. 1 Rayleigh distribution plots showing wind variability on monthly and yearly basis for 2013 over the scrutinized site

On monthly basis, the maximum WPD of 17.44 W/m^2 have been observed in the month of May and observed minimum in August month with 5.15 W/m^2. The annual wind power densities 8.86 W/m^2 for the respective Rayleigh model corresponding to the mean wind speed and c parameter of 1.96 m/s and 1.31 m/s, respectively.

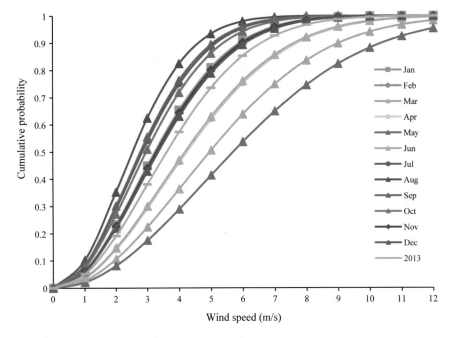

Fig. 2 Rayleigh cumulative distribution plots showing wind variability on monthly and yearly basis for 2013 over the scrutinized site

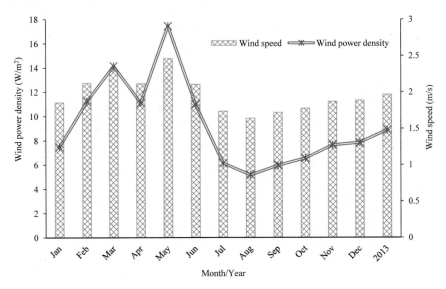

Fig. 3 Monthly and annual wind power densities determined from the Rayleigh model with respect their mean wind speeds for the year 2013

5 Conclusions

In this paper, the computation of technical potential of wind power for the CEEE rooftop building located at NITH campus in the western hilly state region Himachal Pradesh in India. A 60 min. step interval wind data has monitored at mast level of 18.5 m from above the earth's surface and employed this data to Rayleigh distribution model for achieve this objective. Also, the probability density distributions have been derived from frequency distribution analysis on monthly and annual time frames. As a result, the computed maximum wind potential of 17.44 W/m^2 has been observed in May and minimum of 5.15 W/m^2 in August, annual WPD is 8.86 W/m^2. Therefore, the specific examined site is not opted for grid inter-connected applications while it is suitable small and micro-scale wind power generation.

Acknowledgements The authors would like to acknowledge the assistance for provided the wind data facility by the CEEE at NITH located in Himachal Pradesh, India.

References

1. H.G.M. Joselin, S. Iniyanb, E. Sreevalsanc, S. Rajapandiand, A review of wind energy technologies. Renew. Sustain. Energy Rev. **11**, 1117–1145 (2007). https://doi.org/10.1016/j.rser.2005.08.004
2. M. Mohammadia, Y. Noorollahia, B. Mohammadi-ivatloob, H. Yousefi, Energy hub: from a model to a concept—a review. Renew. Sustain. Energy Rev. **80**, 1512–1527 (2017). https://doi.org/10.1016/j.rser.2017.07.030
3. K.S.R. Murthy, O.P. Rahi, A comprehensive review of wind resource assessment. Renew. Energy Sustain. Rev. **72**, 1320–1342 (2016). https://doi.org/10.1016/j.rser.2016.10.038
4. S.C. Bhattacharyya, An overview of problems and prospects for the Indian power sector. Energy **19**, 795–803 (1994). https://doi.org/10.1016/0360-5442(94)90018-3
5. S.S. Chandel, K.S.R. Murthy, P. Ramasamy, Wind resource assessment for decentralized power generation: case study of a complex hilly terrain in western Himalayan region. Sustain. Energy Technol. Assess. **8**, 18–33 (2014). https://doi.org/10.1016/j.seta.2014.06.005
6. World Wind Energy Association (WWEA), http://www.inwea.org/aboutwindenergy.html. Accessed 06 July 2013
7. The ministry of New and Renewable Energy Sources (MNRE), http://mnre.gov.in/file-manager/annual-report/2016-2017/EN/index.htm. Accessed 21 Jan 2018
8. The National Institute of Wind Energy (NIWE), http://niwe.res.in/department_wra.php. Accessed 07 Jan 2018
9. S.S. Chandel, P. Ramasamy, K.S.R. Murthy, Wind power potential assessment of 12 locations in western Himalayan region of India. Renew. Sustain. Energy Rev. **39**, 530–545 (2014). https://doi.org/10.1016/j.rser.2014.07.050
10. K.S.R. Murthy, O.P. Rahi, Preliminary assessment of wind power potential over the coastal region of Bheemunipatnam in Northern Andhra Pradesh. India. Renew. Energy **99**, 1137–1145 (2016). https://doi.org/10.1016/j.renene.2016.08.017
11. K. Vladislovas, M. Mantas, G. Giedrius, M. Antanas, Statistical analysis of wind characteristics based on Weibull methods for estimation of power generation in Lithuania. Renew. Energy **113**, 190–201 (2017). https://doi.org/10.1016/j.renene.2017.05.071
12. T.V. Ramachandra, B.V. Shruthi, Wind energy potential in Karnataka India. Wind Eng. **27**, 549–553 (2003)

13. G. Krishnadas, T.V. Ramachandra, Scope for renewable energy in Himachal Pradesh, India—a study of solar and wind resource potential. Biodivers. Clim. Change, 1–10 (2010)
14. R.K. Aggarwal, S.S. Chandel, Emerging energy scenario in Western Himalayan state of Himachal Pradesh. Energy Policy **38**, 2545–2551 (2010). https://doi.org/10.1016/j.enpol.20 10.01.002
15. N.K. Swami Naidu, B. Singh, Grid-interfaced DFIG-based variable speed wind energy conversion system with power smoothening. IEEE Trans. Sustain. Energy **8**(1), 51–58 (2017). https://doi.org/10.1109/TSTE.2016.2582520
16. K.S.R. Murthy, O.P. Rahi, Estimation of Weibull parameters using graphical method for wnd energy applications, in *18th IEEE National Power System Conference (NPSC)*, pp. 1–6, IEEE, IIT Guwahati, India (2014). https://doi.org/10.1109/npsc.2014.7103858
17. K.S.R. Murthy, O.P. Rahi, Estimation of weibull parameters using maximum likelihood method for wind power applications. Indian Wind Turbine Manuf. Assoc. Indian Wind Power **3**(3):31–36 (2017) (Vide No.: TNENG/2015/60605)
18. F. Andres, V. Daniel, Four-parameter models for wind farm power curves and power probability density functions. IEEE Trans. Sustain. Energy **8**(4), 1783–1784 (2017). https://doi.org/10.11 09/TSTE.2017.2698199
19. Kumar C. Prem, A. Siraj, W. Vilas, Comparative analysis of Weibull parameters for wind data measured from met-mast and remote sensing techniques. Renew. Energy **115**, 1153–1165 (2018). https://doi.org/10.1016/j.renene.2017.08.014
20. J. Haiyan, W. Jianzhou, W. Jie, G. Wei, Comparison of numerical methods and metaheuristic optimization algorithms for estimating parameters for wind energy potential assessmentin low wind regions. Renew. Sustain. Energy Rev. **69**, 1199–1217 (2017). https://doi.org/10.1016/j.r ser.2016.11.241
21. S.G. Jamdade, P.G. Jamdade, Analysis of wind speed data for four locations in Ireland based on Weibull distribution's linear regression model. Int. J. Renew. Energy Res. **2**, 451–455 (2012)

Fractional Order Control and Simulation of Wind-Biomass Isolated Hybrid Power System Using Particle Swarm Optimization

Tarkeshwar Mahto, Hasmat Malik and V. Mukherjee

Abstract In this work, a fractional order (FO) proportional–integral–derivative (PID) (FO-PID) controller is considered for load-frequency control (LFC) of the isolated hybrid power system, comprising of a biomass-based diesel engine generator and a wind turbine generator. The FO-PID controllers are PID controller only, and the difference lies in the order of the integral and derivative part of the controllers. In FO controllers, the order of the integral and derivative part are fractional in nature. In this paper, particle swarm optimization (PSO) algorithm has been engaged to carry out the above mentioned LFC for the considered wind-biomass isolated hybrid power system. A comparison of FO control strategy with conventional based controller techniques is made. The FO-PID controller outperforms the conventional PID controllers. And, robustness analysis is also done for the FO-PID controller.

Keywords Fractional order · Integral order · Isolated hybrid power system Proportional–integral–derivative · Particle swarm optimization

1 Introduction

The renewable energy sources (RES) are inexhaustible as per the human standards and have the prospective of receding the requirement of fossil fuels for electricity requirement with a favorable auxiliary for reducing the greenhouse gases emission levels. The catalog of RES involves solar, wind, tidal, biomass, geothermal, and

T. Mahto
Electrical and Electronics Engineering Department, BIT Mesra, Ranchi, Jharkhand 835215, India
e-mail: tara.mahto@gmail.com

H. Malik (✉)
Electrical Engineering Department, IIT Delhi, Hauz Khas, New Delhi 110016, India
e-mail: hmalik.iitd@gmail.com

V. Mukherjee
Electrical Engineering Department, IIT (ISM) Dhanbad, Jharkhand 826004, India
e-mail: vivek_agamani@yahoo.com

© Springer Nature Singapore Pte Ltd. 2019
H. Malik et al. (eds.), *Applications of Artificial Intelligence Techniques in Engineering*, Advances in Intelligent Systems and Computing 698,
https://doi.org/10.1007/978-981-13-1819-1_27

biofuel energy as different sources of energy [1]. Among these RESs, wind and solar energy has positional in most part of the world and are intensively used, that is why solar and wind energy has more developed technical potentials. But, from the viewpoint of technology and efficiency, wind energy has the superiority over the other RESs [2].

RESs have a natural property of intermittence, so, wind and solar based stand-alone has a necessity of energy storage devices [3] or a conventional energy source (like diesel engine generator (DEG)), which is increasing the cost without increasing the capacity of the isolated hybrid power system (IHPS), as energy storage devices or DEG are backups only [4]. On the other side, it is feasible to practice a hybrid system consisting of two or more RESs and are economical where extension of grid supply is limited [5].

In different parts of world, for more than thousands of years, wind energy has been used for crops grinding and water pumping. In 1891, the first wind turbine based generator was developed by Cour, for electrical power generation, thereafter application of generators based on wind energy had increased day by day [6]. Because of technological enhancement and capacity increment with time, wind energy is preferred in the modern world for power generation.

In the rural areas adequate amount if agricultural residues and biomass are produced, this may fulfil their electricity demands with the application of biomass gasifier-based power plants. Earliest literature recitation for the theory of utilizing a pressurized gasifier along with a gas turbine engine is discussed in [7]. The solid biomass is converted to more useable gases form by processing it in Biomass gasifier [8]. It is feasible to run a DEG with small modifications in the air intake or a 100% producer gas engine with the gases produced by the biomass gasifier.

In this paper, a new HPS involving a biomass gasifier as a fuel source for the DEG alongside a wind turbine generator (WTG) is undertaken due to the sufficient availability of biomass and wind resources in the rural areas and has been termed as IHPS model. The key emphasis for the present manuscript is the controlling of deviation in power and frequency for the undertaken IHPS due to variation in load demand with time by tuning the variables for the controllers and other optimizable variables of the undertaken IHPS.

2 System Model

The IHPS under consideration consists of a pitch controlled WTG rated at 150 kW coupled in parallel with a biogas based DEG (B-DEG) set consisting of a 150 kW through an ac bus-bar. Figure 1 demonstrates the block diagram based on transfer function modeling of the considered IHPS. The IHPS includes two controllers, (i.e., controller 1 on B-DEG for electronic speed governing and controller 2 on WTG for blade pitch controlling). The modeling details of B-DEG and WTG are enlightened in [9, 10]. The equation for flow of power of the studied IHPS may be articulated as in (1).

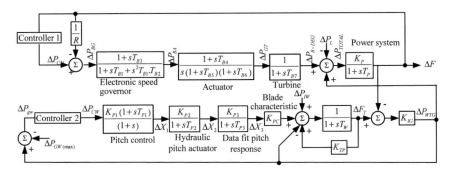

Fig. 1 Transfer function model of B-DEG and WTG-based IHPS

$$\Delta P_{TOTAL} = \Delta P_{B-DEG} + \Delta P_{WTG} - \Delta P_{L} \tag{1}$$

where, ΔP_{TOTAL} is total power output change, ΔP_{B-DEG} is B-DEG's output power deviation, ΔP_{WTG} is WTG's output power deviation, and ΔP_{L} is input load demand deviation.

3 Controller Technique

Integer order (IO) controllers are used for controlling purpose for industrial applications and IHPSs [11]. But, with the progress of fractional order (FO) controller, for the improved system performance nowadays, FO controllers are in effect [12]. The most general design of an FO controller is the $PI^{\lambda}D^{\mu}$ (i.e., FO-PID) controller [13]. FO controller offers an additional degree of liberty for designing controller gains (like, K_P, K_I, K_D) unlike IO controller along with designing orders of integral and derivative. The orders of integral and derivative are not inevitably integer always but may be any real numbers. Figure 2 shows the FO-PID controller may be streamlined to the predictable IO PID controller and scales it from point to plane. The scaling of PID increases the scheming elasticity of the controllers. The FO controller has the following transfer function form [14]:

$$G_c(s) = K_P + \frac{K_I}{s^{\lambda}} + K_D s^{\mu} \tag{2}$$

Figure 3 is a block diagram configuration of FO-PID. It is obvious that by choosing $\lambda = 1$ and $\mu = 1$, IO-PID controller is obtained. Taking $\lambda = 1$, $\mu = 0$, and $\lambda = 0$, $\mu = 1$, respectively, matches to the IO- PI & PD controllers. As all IO controllers are special cases of the $PI^{\lambda}D^{\mu}$ (i.e., FO-PID) controller.

Fig. 2 Fractional PID
controller plane

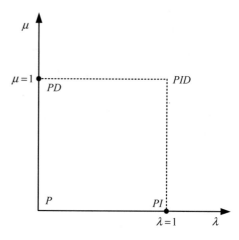

Fig. 3 Schematic diagram
of FO-PID controller

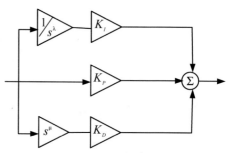

4 Performances Indices

The key emphasis of this paper is on neutralization of the power and frequency
deviation to null point as swiftly as possible after any variation in demand load.
In order to realize this, the first step is the identification of an objective function
for designing controller. In the performance measures, usually considered while
controllers designing are

$$\text{Integral absolute error (IAE)} = \int_0^\infty |\Delta F| dt \tag{3}$$

$$\text{Integral square error (ISE)} = \int_0^\infty |\Delta F|^2 dt \tag{4}$$

$$\text{Integral time absolute error (ITAE)} = \int_0^\infty t |\Delta F| dt \tag{5}$$

$$\text{Integral time square error (ITSE)} = \int_0^\infty t |\Delta F|^2 dt \qquad (6)$$

In this work, minimization of ISE has been considered the objective function for optimizing the gains for controllers, of the IHPS.

5 Optimization Methodology

In the year 1995, particle swarm optimization (PSO) was introduced by Kennedy and Eberhart as a new heuristic optimization method [15, 16]. PSO is based on swarms like bird flocking and fish schooling. As per the results obtained from research, birds do not find food individually but by flocking. So, PSO has a function for fitness evaluation that attends a random position and is assigned with a corresponding fitness value. Position having the highest fitness value for the complete iteration is called the global best (G_{best}) and there is personal best (P_{best}) for each particle having its highest fitness value. In PSO algorithm, the population for particles involves entire search space, and also remembering the best solution encountered yet. After completion of every iteration, the velocity vector of each particle is adjusted, based on its movement and impact of its best solution and the best solution of its neighbors, and then calculates a new point to be examined.

PSO algorithm may be best designated as follows:

(1) The vectors for position and velocity are initialized arbitrarily for each particle with equal dimension as of the problem.
(2) Measure, compare, and then store the fitness of each particle (P_{best}) with the best fitness (G_{best}) value.
(3) Update velocity and position vectors according to (1) and (2) for each particle.
(4) Repeat Steps 2–3 until a termination criterion is satisfied.

The block diagram illustration of the PSO algorithm is demonstrated in Fig. 4.

6 Simulation Results and Discussions

The IO and FO-PID controlled IHPS models dynamic performance of have been compacted under the changing load scenario. For the simulation work, two configurations of IHPS has been considered, "IO-PID" (i.e., IHPS being controlled by IO PID controllers) and "FO-PID" (i.e., IHPS being controlled by FO-PID controllers). The comparison has been made on the basis of the simulation results and also on the basis of performance indices (includes IAE, ISE, ITAE, and ITSE) obtained using PSO as an optimizing algorithm for the controllers. Two analyses have been considered for comparative study of the two configurations. These are:

Fig. 4 Flow diagram
illustrating the particle
swarm optimization
algorithm

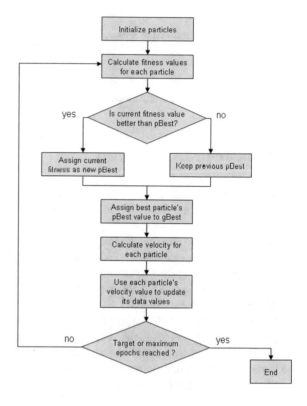

1. Performance analysis

   • Case 1: 1% step decrease in load demand and
   • Case 2: random load demand

2. Robustness analysis

6.1 Performance Analysis

Performance analysis of "IO-PID" and "FO-PID" configurations of IHPS has been directed for confirming the compatibility of the undertaken configurations under diverse input load demand circumstances for confirming the superiority of one configuration over the others. Different load patterns used in this study are depicted in Figs. 5 and 6.

Case 1: 1% *step decrease in load demand*: The "IO-PID" and "FO-PID" configurations of IHPS have been exposed to a step decrease of 1% load demand at t = 1 s., Fig. 5 shows the step decrease in load demand. Deviation profile of power and frequency formed by the two IHPS based undertaken configurations has been

Fig. 5 1% step decrease in load demand as a perturbation used for input in p.u

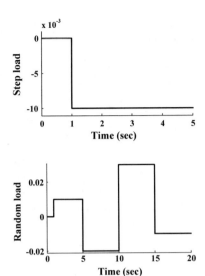

Fig. 6 Random load demand input perturbation used as input in p.u

Fig. 7 Relative dynamic responses for ΔF (Hz) of IHPS for 1% decrease in load demand

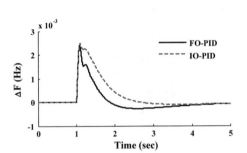

Fig. 8 Relative dynamic responses for ΔP (p.u.) of IHPS for 1% decrease in load demand

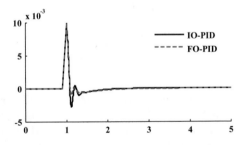

represented in Figs. 7 and 8. It is very evident from Figs. 7 and 8 that the FO-PID controller configuration (i.e., "FO-PID") minimizes the power and frequency deviation more effectively than that of IO-PID controlled IHPS (i.e., "IO-PID" configuration). In Table 1, the controller gain optimized values of IO- and FO-PID controller has been presented and Table 2 offers the performance indices values for supporting the obtained pictorial results.

Table 1 PSO-based optimized IO and FO-PID gain parameters

Gain parameters	Considered case			
	Case 1		Case 2	
	IO-PID	FO-PID	IO-PID	FO-PID
$K_{P\_B\text{-}DEG}$	0.3244	2.7970	1.0000	0.9708
$K_{I\_B\text{-}DEG}$	0.8983	4.5771	0.0010	0.9711
$\lambda_{B\text{-}DEG}$	1.0	0.8024	1.0	0.0009
$K_{D\_B\text{-}DEG}$	0.5536	1.0933	0.2893	0.7260
$\mu_{B\text{-}DEG}$	1.0	0.0231	1.0	0.7841
K_{P_WTG}	0.4477	7.0163	0.1377	0.3868
K_{I_WTG}	0.5194	0.2857	1.0000	0.9999
λ_{WTG}	1.0	0.1417	1.0	0.7640
K_{D_WTG}	0.5149	8.1507	0.0403	0.2483
μ_{WTG}	1.0	0.3794	1.0	0.2529

Table 2 Different performance indices values for different configurations of IHPS under different loading conditions

Case	Type of disturbance	Model	Performance indices			
			IAE ($\times 10^{-3}$)	ISE ($\times 10^{-5}$)	ITAE ($\times 10^{-2}$)	ITSE ($\times 10^{-4}$)
1	1% step decrease in load demand	IO-PID	1.989	0.2635	0.3804	0.03806
		FO-PID	1.349	0.1273	0.274	0.01724
2	Random load demand	IO-PID	23.01	13.54	23.74	14.76
		FO-PID	24.66	13.76	27.02	15.03

Case 2: *Random demand load*: The profile for power and frequency deviation of the undertaken configurations of IHPS under continuous varying random load demand conditions is presented in Figs. 9 and 10. Figure 6 presents the random changing demand load applied during simulation work. Random load demand change its requirement at $t = 1, 5, 10$ and 15 s, respectively, in the time axis. Figures 9, 10 and Table 2 jointly assist in noting that FO-controlled IHPS model (i.e., "FO-PID") is better in minimizing the power and frequency deviation along with diminishing the mismatch in demand and generation than that of IO-PID controlled configuration ("IO-PID").

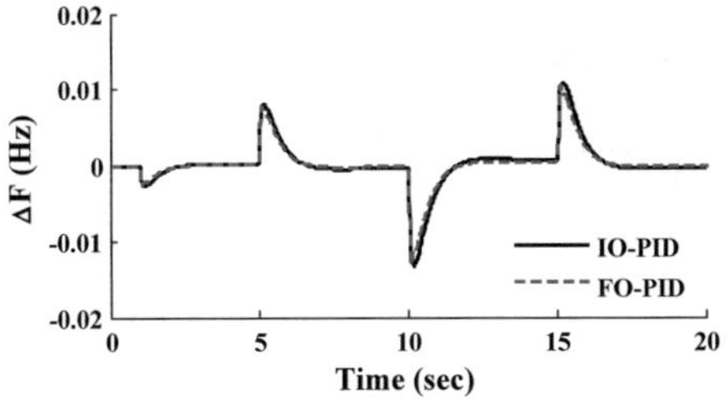

Fig. 9 Relative dynamic responses for ΔF (Hz) of IHPS for random load change

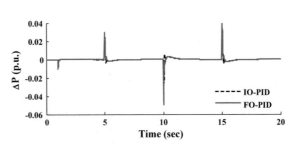

Fig. 10 Relative dynamic responses for ΔP (p.u.) of IHPS for random load change

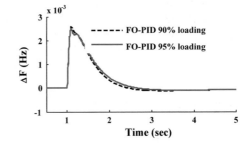

Fig. 11 Relative dynamic response for ΔF (Hz) of different loading based configurations for "FO-PID" of IHPS

6.2 Robustness Analysis

For this analysis, the FO-controlled IHPS is loaded at 95% of its rated capacity and all controllers linked to WTG and B-DEG are tweaked for 90% rated load. The deviation of power and frequency for FO-controlled system are considered and portrayed in Figs. 11 and 12 for the two different loading conditions, for the justification of PSO algorithms and FO-PID controller robustness. From Figs. 11 and 12, it might be noted that the two considered are having identical power and frequency response behavior. And, thus, provides indications to the PSO algorithm's robustness on regulating the FO controller parameters and system tunable parameters in order to regulator the power and frequency deviation in IHPS.

Fig. 12 Relative dynamic response for ΔP (p.u.) of different loading based configurations for "FO-PID" of IHPS

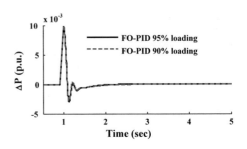

7 Conclusion

In the presented paper, an IHPS consisting of B-DEG and WTG has been considered. The considered IHPS has been subjected to different loading conditions and deviation in power and frequency has been studied. In order to control the deviation in power and frequency, IO and FO-PID controllers have been considered. And, for optimal tuning of the controller parameters, PSO has been employed. For the simulation study, step and random varying load has been considered. From the simulation results, it has been observed that the IO-controlled IHPS performance is much inferior then that of FO-controlled. And, finally, the robustness of the IO controller has been verified with the PSO algorithm for the B-DEG and WTG-based IHPS.

References

1. O. Ellabban, H. Abu-Rub, F. Blaabjerg, Renewable energy resources: current status, future prospects and their enabling technology. Renew. Sustain. Energy Rev. **39**, 748–764 (2014)
2. S. Espey, Renewables portfolio standard: a means for trade with electricity from renewable energy source. Energy Policy **29**, 557–566 (2014)
3. T. Mahto, T. Mukherjee, Energy storage systems for mitigating the variability of isolated hybrid power system. Renew. Sustain. Energy Rev. **51**, 1564–1577 (2015)
4. G. Tina, S. Gagliano, S. Raiti, Hybrid solar/wind power system probabilistic modeling for long-term performance assessment. Sol. Energy **80**, 578–588 (2006)
5. M. Deshmukh, S. Deshmukh, Modeling of hybrid renewable energy systems. Renew. Sustain. Energy Rev. **12**, 235–249 (2008)
6. Renewable energy science education manual, http://www.stemfnity.com/image/data/pdfHoriz onRenewableEnergyCurriculumManualsPreview.pdf
7. W. Gumz, *Gas producers and blast furnaces* (Wiley, New York, 1950)
8. C.M. Kinoshita, S.Q. Turn, R.P. Overend, R.L. Bain, Power generation potential of biomass gasification systems. J. Energy Eng. **128**, 88–89 (1997)
9. P. Balamurugan, S. Ashok, T.L. Jose, An optimal hybrid wind-biomass gasifier system for rural areas. Energy Sources Part A **33**, 823–832 (2011)
10. F. Jurado, J.R. Saenz, An adaptive control scheme for biomass-based diesel-wind system. Renew. Energy **28**, 45–57 (2003)
11. T. Mahto, V. Mukherjee, Evolutionary optimization technique for comparative analysis of different classical controllers for an isolated wind-diesel hybrid power system. Swarm Evolut. Computa. **26**, 120–136 (2016)

12. T. Mahto, V. Mukherjee, Fractional order fuzzy PID controller for wind energy based hybrid power system using quasi-oppositional harmony search algorithm. IET Gener. Transm. Distrib. **11**(13), 3299–3309 (2017)
13. I. Podlubny, Fractional-order systems and $PI^\lambda D^\mu$ controllers. IEEE Tran. Automa. Control **44**(1), 208–214 (1999)
14. A. Biswas, S. Das, A. Abraham, S. Dasgupta, Design of fractional-order $PI^\lambda D^\mu$ controllers with an improved differential evolution. Eng. Appl. Artif. Intell. **22**(2), 343–350 (2009)
15. J. Kennedy, R. Eberhart, Particle swarm optimization. Proc. IEEE Int. Conf. Neural Netw. **4**, 1942–1948 (1995)
16. R. Eberhart, J. Kennedy, A new optimizer using particle swarm theory, in *Sixth International Symposium on Micro Machine and Human Science* (1995), pp. 39–43

Renewable Energy Management in Multi-microgrid Under Deregulated Environment of Power Sector

Om Prakash Yadav⊙, Jasmine Kaur, Naveen Kumar Sharma
and Yog Raj Sood

Abstract Microgrid (MG) is the combination of different distributed generation units and local loads. It is a small self-sustaining power network which serves its local load. Generally it can be operated in grid connected mode or grid isolated mode. The objective of this paper is to maximize the social welfare and minimize operating cost. Social welfare/benefit with the operator may be used for giving subsidies to renewable based plant, farmers, society welfare, etc. This paper proposes an approach to maximize the social welfare of each microgrid through discriminated price auction mechanism. MATLAB Interior Point Solver (MIPS) method has been used for computing the corresponding allocations and price of each unit in the multi-microgrid system. Also, a comparison of social benefit and the operating cost of each microgrid considering both presence and absence of renewable energy sources is carried out. For analysis of this approach 49 bus system is used, which is divided as microgrid A (14 bus system), microgrid B (15 bus system), microgrid C (14 bus system) and rest of the buses are used in the main grid architecture and its dispatchable load buses. The results of this proposed method show that social welfare is maximized while optimally managing renewable-based distributed units with their conventional fuel based distributed generation units.

Keywords Microgrids (MG) · Social welfare · Renewable energy sources (RES)
Discriminated auction mechanism · Matlab interior point solver (MIPS)

O. P. Yadav (✉)
Chandigarh University, Gharaun, India
e-mail: prakash.op.39@gmail.com

J. Kaur · Y. R. Sood
NIT Hamirpur, Hamirpur, Himachal Pradesh, India
e-mail: jasminekaur.nith@gmail.com

Y. R. Sood
e-mail: yrsood.nith@gmail.com

N. K. Sharma
I.K.G Punjab Technical University, Batala, India
e-mail: naveen31.sharma@gmail.com

© Springer Nature Singapore Pte Ltd. 2019
H. Malik et al. (eds.), *Applications of Artificial Intelligence Techniques
in Engineering*, Advances in Intelligent Systems and Computing 698,
https://doi.org/10.1007/978-981-13-1819-1_28

289

1 Introduction

Microgrids (MG) are small self-sustained power network located on distribution side that make power available, to a number of residential and commercial loads. MGs may include distributed generation (DG) units and storage devices. Generally mode of operation of microgrid is a grid connected mode, but it can also be operated in island mode. In grid connected mode, a microgrid can export or import power from the main grid while in island mode, microgrid manages its demand independently. Deployment of microgrids will enhance the performance of the distribution grid in terms of efficiency, reduction of emissions, congestion management and stability. Renewable based microgrid's DG units are environmental friendly over conventional energy based generation units [1, 2]. Microgrid may also be used to provide ancillary services to the main grid. Deregulated market enables DG owners to participate in the market either through contract or bidding [3, 4]. Neuro-fuzzy approach based forecasting of price, Photovoltaic and wind power generation data were used for developing centralized model for optimal bidding strategy for the microgrids [5].

In recent years, the world is facing a number of challenges such as hike in energy demand, greenhouse gas emissions, reducing fossil fuel reserves and environmental concerns. All of these factors have led to paying more and more attention to developing power generating units based on renewable energy sources [6, 7]. Globally renewable power generation capacity has seen an increase of 93% over 2014, i.e., 154 Gigawatt (GW) which includes wind power (66 GW), solar photovoltaic power (47 GW) and hydro power (33 GW) [8]. Renewable energy plays a major role in decarbonizing the world's energy system [9].

In deregulation process system, efficiency and service quality is continuously improving and is also developing a competitive market [10]. Generally deregulated environment consists of pool and bilateral or multilateral contracts. Deregulated power market performance is measured by its social welfare which is the difference of society's willingness to pay for its demand and energy cost. Demand supply equilibrium describes the market clearing price (MCP) and market clearing volume (MCV). Graphically equilibrium is a point where incremental aggregate cost curve and incremental aggregate utility curve cross. Different types of deregulated model which are operating in different countries are pool model, pool and bilateral or multilateral contracts model [11].

In this paper, a model is proposed for maximizing the social welfare of each microgrid through discriminated price auction. The proposed model is illustrated by using a case study of three microgrids connected to main grid having dispatchable loads. Each microgrid has internally fixed load, dispatchable generating units, wind and photovoltaic based generating units. The rest of this paper is organized as follows Sect. 2 is about the basic problem formulation with detailed mathematical model and objective function along with related constraints, Sect. 3 caters to proposed model and procedure for day ahead market. Simulation analysis and results are compared in Sect. 4. Section 5 describes the discussion and conclusion.

2 Problem Formulation

2.1 Objective Function

Energy scheduling problem for maximization of social welfare of multi microgrid, in grid connected mode with various constraints can be solved by microgrid operator. Maximization of social benefit (i.e., F_1) on hourly basis of entire power network is shown in Eq. 1 and Eq. 2 shows hourly basis social welfare (F_j) of jth microgrid.

$$F_1 = \max \left\{ \sum_{i=1}^{I} A_i(Pd_i) - \sum_{j=1}^{J} B_j(Pmc_j) - \sum_{k=1}^{K} E_k(Pg_k) \right\} \qquad (1)$$

$$F_j = B_j(Pmc_j) + \sum_{n=1}^{N} (\lambda_{jn} * Pint_{jn}) - \sum_{m=1}^{M} C_{jm}(Pg_{jm}), \qquad (2)$$

where

λ_{jn} Locational marginal pricing of jth microgrid, nth bus
$Pint_{jn}$ Internal load connected of jth microgrid, nth bus
Pg_{jm} Real power of jth microgrid, mth generator
Qg_{jm} Reactive power of jth microgrid, mth generator

where $A_i(Pd_i)$ is the bid function of ith dispatchable load which shows willingness to pay for the purchase of real power i.e. Pd_i, $B_j(Pmc_j)$ is the offer function of jth microgrids for import/export of real power with main grid. While positive and negative sign of Pm_j indicate that power is exported and imported to main grid respectively. $E_k(Pg_k)$ is the bid function of kth Gencos or Independent Power Producer (IPP). Operating cost function of distributed generation of jth microgrid mth unit is represented by $C_{jm}(Pg_{jm})$.

2.2 Power Balance Constraints for jth Microgrid

$$\sum_{m=1}^{M} (Pg_{jm}) \geq Pint_j \qquad (3)$$

Equation 3 states that sum of all power of distributed generating unit, solar generating unit and wind power unit must be greater than or equal to day ahead forecasted internal load demand.

2.3 Dispatchable Unit Constraints

$$Pg_{jm}^{min} \leq Pg_{jm} \leq Pg_{jm}^{max} \tag{4}$$

$$Qg_{jm}^{min} \leq Qg_{jm} \leq Qg_{jm}^{max} \tag{5}$$

Equations (4–5) keeps the dispatchable generating unit within its boundary limit while transmission line constraint and power flow constraints equations are shown in Eqs. (6–7).

2.4 Transmission Line Constraints

$$MVA_{nq} \leq MVA_{nq}^{max} \tag{6}$$

2.5 Power Flow Constraints

$$g(V, \theta) = 0, \tag{7}$$

where

$$g(V, \theta) = \begin{cases} P_n(V, \theta) - P_{netn} & For\ each\ PQ\ bus\ n \\ Q_n(V, \theta) - Q_{netn} & For\ each\ PQ\ bus\ n \\ P_r(V, \theta) - P_{netr} & For\ each\ PV\ bus\ r \neq ref\ bus \end{cases}$$

where P_n and Q_n are calculated real and reactive power for PQ bus n respectively. Specified real and reactive power for PQ bus n is denoted by P_{netn} and Q_{netn}, P_{netr} and P_r are specified and calculated real power for PV bus r respectively. Voltage magnitude and phase angle of different buses is shown by V and θ respectively. Power limit on transmission line is described in Eq. (6) where MVA_{pq}^{max} is the maximum rating of power flow in line connecting bus n and q.

3 Proposed Approach and Solution Procedure

Figure 1 presents proposed model's schematic diagram for optimal utilization of Renewable Energy Sources and maximization of social welfare in multi microgrid in a day ahead market. The proposed model has been programmed on MATLAB's Software and MATLAB Interior Point solver (MIPS) is used to solve this model. MIPS works on Dual-primal interior point algorithm [12, 13].

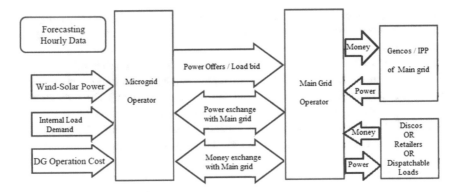

Fig. 1 Proposed approach's schematic model

Generating Company (Gencos) could be operator or owner or both, they offer power in the competitive market or sell through contracts. Distribution Company (Discos) buys wholesale electricity with Gencos either through the contracts or pool market and supplies it to the customers. Independent System operator (ISO) is the supreme entity associated with the responsibility of maintaining reliability and security of entire power network by procuring emergency reserves, reactive power, and various other services. ISO does not participate in market but it is concerned with the technical issues of the market. Power exchange (PX) conducts an energy auction for the day ahead and spot market, it is concerned with the economic aspects of the market. It conducts double sided auction where buyers as well as sellers can place their bids. The buyers and sellers are allowed to submit a portfolio bids. Microgrid Operator (MO) performs an optimal power dispatch and unit commitment for minimizing various environmental, economical and technical objectives. Detailed algorithm of proposed approach has been given below.

Step 1: MO of each MG collects hourly forecasted data of wind, solar power, internal load demand and DG operation cost.

Step 2: MO runs an optimization algorithm, satisfying its internal load demand every hour of the day and calculating its surplus/deficit power on that hour.

Step 3: Each MO submits the quantity of surplus/deficit power to the PX of main grid for open market participation along with their offers/bid price of each hour.

Step 4: PX of the main grid receives offers from other Gencos, Independent Power Producer, microgrid and bids from Discos, retailers, dispatchable loads for 24 h.

Step 5: PX runs market based on optimization algorithm and finds market clearing Price (MCP) which is based on offers, bids and available power. Also determines successful and unsuccessful market participants.

Step 6: PX allocates the price and quantity for successful participants and computes its social welfare of the main grid based on a pay-as-bid pricing during that hour.

Step 7: Successful MO who exports power to the main grid, schedules their generating unit to fulfill the total load demand. MO who imports power from main grid fulfills its internal load demand. Unsuccessful MO schedules their generating units for supplying power to its internal loads only.

Step 8: MO of each MG compute its operation cost, social welfare. Internal load price collection is based on locational marginal pricing.

4 Simulation Analysis

Simulation analysis has been carried out on 49 bus system. The entire power network architecture is classified as group of five dispatchable loads, three microgrids connected to the main grid. Each microgrid having number of internal loads connected to different buses as shown in Fig. 2 [14].

Microgrid A is 14 bus system, 15 bus and 14 bus system are microgrid B and microgrid C respectively. Only single Gencos is considered in main grid. All links between the buses in given architecture either between microgrids, microgrids to main grid having 0.003 pu resistance and 0.01 pu reactance and 500 KVA maximum power

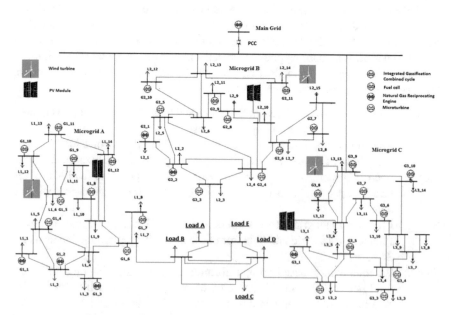

Fig. 2 Proposed Network Architecture

Table 1 Parameters of non-renewable based generating units

Type	NRE			MT			FC			IGCC		
Units	1	2	3	4	5	6	7	8	9	10	11	12
pmax (MW)	0.41	0.41	0.27	0.27	0.14	0.14	0.09	0.09	0.065	0.065	0.045	0.045
pmin (MW)	0.01	0.01	0.05	0.05	0.025	0.25	0.2	0.2	0.015	0.015	0.0.01	0.01
O/M cost ($/MWh)	55.73			43.32			55.04			32.50		

transfer limit through the lines [15]. These distributed generating units installed in microgrid are microturbine (MT), fuel cell (FC), Natural Reciprocating Engine (NRE), Integrated Gasification combined cycle (IGCC), Wind plant and Solar plant.

Microgrid A, Microgrid B, Microgrid C has three NRE, two NRE and one NRE respectively. Apart from this each MG consist three MT, three FC and three IGCC, one wind and one solar plant. Maximum and minimum power production and operation and maintenance cost coefficient of each non-renewable based units is shown in Table 1 [16].

Normalized data of photovoltaic power, wind power production and total internal load demand of each microgrid is shown in Table 2 [17]. Normalized power production is the ratio of power output to power rating of renewable energy sources. Forecasted solar irradiation temperature and wind velocity were taken from meteorological department [17, 18].

Detailed Distribution of total internal load demand of microgrids to each buses is shown in Table 3 [15]. MO of each MG in day ahead market submit, their offers/bid of surplus/deficit power for open market to PX of main grid before 24 h of actual market. Offers and bids of MG, dispatchable loads may be linear, quadratic, piecewise linear, etc. In this simulation we considered linear offers/bids [19] which is submitted to the PX of Pool market/ISO. Bidding prices and Bidding quantity of dispatchable loads are shown in Fig. 3 and Fig. 4 respectively. Microgrids offers are shown in Fig. 5.

Double auction pool model is applied in this simulation [20]. MATLAB Interior Point Solver (MIPS) is used for running this market algorithm for finding out MCP, list of successful buyer and seller, allocation of power and price for each successful player in particular hour. On the basis of market outcome generating units of each microgrid are scheduled by microgrid operator.

5   Result and Discussion

Generating units based on RES have negligible operating and maintenance cost, capital cost is not considered because that is for planning purposes, it is inexhaustible and its optimal operation in microgrids increases the bidding quantity (BQ) for open market are as shown from Figs. 6, 7 to Fig. 8. Possibilities of success in the market

Table 2 Normalized hourly forecasted renewable and forecasted internal load demand

Hour	Wind power	Solar power	Internal load demand of MG (MW)			Hour	Wind power	Solar power	Internal load demand of MG (MW)		
			A	B	C				A	B	C
1	0.0860	0	1	0.85	0.55	13	0.0995	0.8300	1.85	1.45	0.91
2	0.1465	0	1.03	0.86	0.57	14	0.0426	0.8468	1.8	1.43	0.91
3	0.2449	0	1.05	0.87	0.58	15	0.0062	0.6779	1.72	1.42	0.9
4	0.1829	0	1.07	0.88	0.6	16	0.0053	0.5933	1.7	1.41	0.89
5	0.3780	0.0935	1.09	0.89	0.62	17	0	0.2546	1.65	1.4	0.87
6	0.6424	0.2123	1.15	0.9	0.65	18	0.0040	0.2122	1.63	1.35	0.86
7	0.9835	0.2547	1.3	1.02	0.73	19	0	0.1528	1.55	1.33	0.85
8	0.6282	0.4666	1.4	1.15	0.8	20	0	0.0680	1.45	1.3	0.85
9	0.4450	0.6358	1.64	1.3	0.87	21	0.0062	0.0425	1.35	1.27	0.76
10	0.4726	0.6780	1.7	1.4	0.9	22	0.0100	0	1.2	1.2	0.63
11	0.2674	0.8131	1.87	1.48	0.99	23	0.0100	0	1.15	1.05	0.6
12	0.1927	0.8469	1.87	1.45	0.94	24	0.0344	0	1.05	0.91	0.57

Table 3 Distribution of internal load

Load (%)	Bus no. of MG A	Bus no. of MG B	Bus no. of MG C
7.5	1–6	1–6	1–6
10	7	–	7
5	8–10	7–11	8–10
7.5	11–14	12–15	11–14

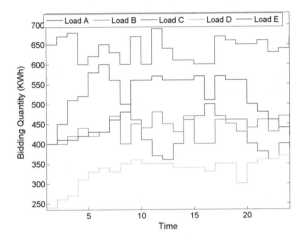

Fig. 3 Bidding quantities of dispatchable load

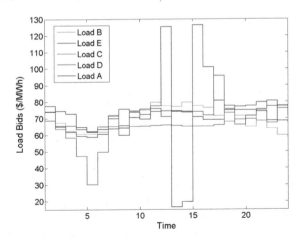

Fig. 4 Bidding Prices of dispatchable load

for MO is higher when its offer bids are low and load bids are high. Similarly when bids of dispatchable loads, discos and retailers are high then their possibilities to get success in market is also high.

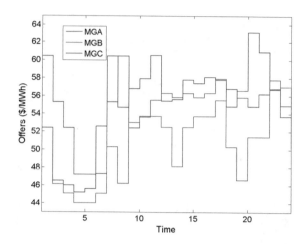

Fig. 5 Offers of microgrids

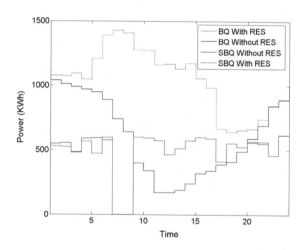

Fig. 6 Bidding quantities and successful bidding quantities of microgrid A

During 24 h in the given graphs we can see that number of market player success is different in different hours and their bidding quantity (BQ) and Successful bidding quantity (SBQ) is varying on hour to hour simulation. During the analysis of result we can find out that Operation and maintenance cost due to optimal operation of RES is reduced by 388 $/day, 130.3 $/day and 43.4 $/day in microgrid A, microgrid B, microgrid C respectively which shown in Fig. 9 and Fig. 10.

Figure 10 and Fig. 11 shows that social welfare of the entire network and microgrid A, microgrid B, microgrid C is 726.26 $/day, 714.8 $/day, 505.5 $/day and 412.26 $/day respectively. From the Table 4 we able to depict that social welfare of main grid as well as microgrids had increased with used of RES based units in their microgrids.

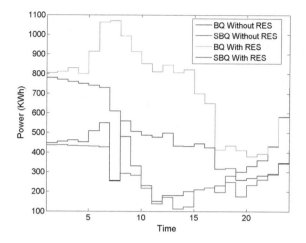

Fig. 7 Bidding quantities and successful bidding quantities of microgrid B

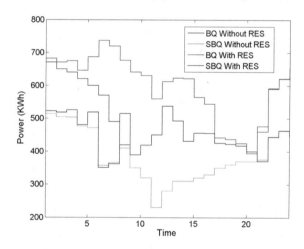

Fig. 8 Bidding quantities and successful bidding quantities of microgrid C

Social welfare of MG C had been decreased with usage of RES because during that interval dispatchable load demand was fulfilled by MG A and MG B because they have higher potential of RES based unit compare to MG C. When MG A and MG B had came in market with their RES based unit, most of the interval they became successful market player. Due to optimal operation of RES based generating units of MG their operation and maintenance cost was also reduced which is shown in Table 4. The results shows that operation of grid connected mode of microgrids and its participation in open market with renewable based generating units increases its revenue collection and social welfare (Fig. 12).

Fig. 9 Operating cost of microgrids without Renewable Energy based generation

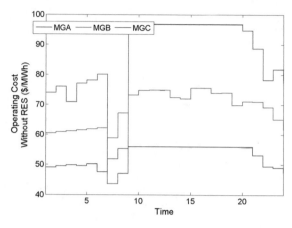

Fig. 10 Operating cost of microgrids with Renewable Energy based generation

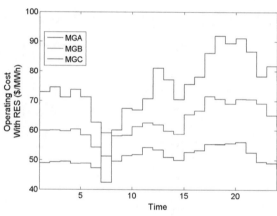

Fig. 11 Social welfare considering only non-renewable generating units

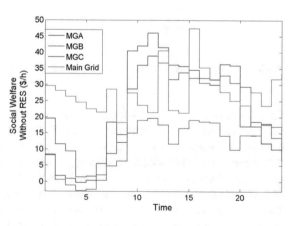

Table 4 Social welfare, Operating cost and Energy offered by Main Grid and MG

		Only non RES generating units	RES and non RES generating units
Main grid	Social welfare ($/day)	661.9	726.26
Microgrid A	Social welfare ($/day)	503.15	714.8
	Operating cost ($/day)	2030.8	1797.3
	Energy offered (MW/day)	14.69	25.03
Microgrid B	Social welfare ($/day)	265.5	505.5
	Operating cost ($/day)	1642.35	1512.05
	Energy offered (MW/day)	10.55	17.44
Microgrid C	Social welfare ($/day)	536.0	412.2
	Operating cost ($/day)	1273.88	1230.4
	Energy offered (MW/day)	10.88	14.32

Fig. 12 Social welfare considering renewable and non-renewable generating units

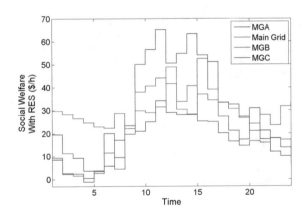

6 Conclusion

This paper presents an approach for optimal management of Renewable Energy Sources in microgrids. Network consist of sets of dispatchable loads, microgrids and main grid. Model is implemented using Matlab Interior Point Solver (MIPS) in Matpower, which is used for running the market and optimal power flow. Outcome of the market gives MCP, quantity and price allocation of successful player of the market, maintenance and operation cost and social welfare of each microgrid and the whole power network. Resulting outcome of this approach is supporting the maximization of social welfare with optimal utilization of RES in multi microgrid in a grid connected mode of operation. The model discussed in this paper is robust, scalable and easily configurable. It may effectively be used for further development and analysis in microgrids under open market participation.

References

1. N. Hatziargyriou, H. Asano, R. Iravani, and C. Marnay, Microgrids: an overview of ongoing research, development, and demonstration projects. IEEE Power Energy Mag. **5**(4), 78–94 (2007)
2. N. Nikmehr, S. Najafi-Ravadanegh, Optimal operation of distributed generations in microgrids under uncertainties in load and renewable power generation using heuristic algorithm. IET Renew. Power Gener.
3. E. Tedeschi, P. Tenti, P. Mattavelli, Synergistic control and cooperative operation of distributed harmonic and reactive compensators, in *Power Electronics Specialists Conference*, Rhodes, Greece, June 2008
4. E. Dall Anese, H. Zhu, G.B. Giannakis, Distributed optimal power flow for smart microgrids. IEEE Trans. Smart Grid **4**(3) (2013)
5. H. Shayeghi, B. Sobhani, Integrated offering strategy for profit enhancement of distributed resources and demand response in microgrids considering system uncertainties. Energy Convers. Manage. **87**, 765–777 (2014)
6. H.P. Khomami, M.H. Javidi, Energy management of smart microgrid in presence of renewable energy sources based on real-time pricing, in *IEEE Conference* (2014)
7. I. Miranda, H. Leite, N. Silva, Coordination of multifunctional distributed energy storage systems in distribution networks. IET Gener. Transm. Distrib. (2015). ISSN 1751-8687
8. R. Ferroukhi, J. Sawin, F. Sverisson, *Rethinking Energy 2017*. ISBN 978-92-95111-06-6 (Pdf). http://www.irena.org
9. World energy outlook special report 2016. International Energy Agency. http://www.iea.org/t &c
10. Y.R. Sood, N.P. Padhy, H.O. Gupta, Wheeling of power under deregulated environment of power system—a bibliographical survey. IEEE Trans. Power Syst. **17**(3), 870–878 (2002)
11. L.L. Loi, *Power System Restructuring and Deregulation* (Wiley, New York, 2001)
12. H. Wang, C.E. Murillo-Sanchez, R.D. Zimmerman, R.J. Thomas, On computational issues of market-based optimal power flow. IEEE Trans. Power Syst. **22**(3), 1185
13. H. Wang, On the computation and application of multi-period security-constrained optimal power flow for real-time electricity market operations. Ph.D. thesis, Electrical and Computer Engineering, Cornell University, May 2007. A, A.4, G.12
14. T. Logenthiran, D. Srinivasan, A.M. Khambadkone, Multi-agent system for energy scheduling of integrated microgrids in a distributed system. Electric Power Syst. Res. **81**, 138–148 (2011)
15. T. Logenthiran, D. Srinivasan, Multi-agent system for the operation of an integrated microgrid. J. Renew. Sustain. Energy **4**, 013116 (2012)
16. Lazard, Lazard's Levelized cost of energy analysis-version 10.0, Dec 2016
17. T. Logenthiran, D. Srinivasan, Short term generation scheduling of a microgrid. IEEE-TENCON (2009). 978-1-4244-4547-9/09
18. Meterological Department. http://courses.nus.edu.sg/course/geomr/front/fresearch/metstation
19. Energy Market Company of Singapore. https://www.emcsg.com/marketdata/priceinformatio n#priceDataView. Accessed 21 Jan 2017
20. G. Li, J. Shi, X. Qu, Modeling methods for GenCo bidding strategy optimization in the liberalized electricity spot market: a state-of-the-art review. Energy (2011)

Prediction of Energy Availability from Solar and Wind for Demand Side Management

Mayank Singh, R. C. Jha and M. A. Hasan

Abstract Demand side management is emerging as a potential solution for optimal power generation in a renewable fed power structure. In the optimal state, maximum electrical energy from renewable sources is utilized and rest of the load is supplied by conventional sources. This results in higher generation of clean and pollution free electrical energy. Under varying load condition, when peak load condition appears, a voluntary support from demand side is always helpful in ensuring optimal power generation. This is called demand side management. For demand side management, prediction of energy availability from all sources and load forecasting in a certain interval of time is required. Normally, energy availability from conventional sources is known in advance. However, energy from renewable sources is highly unpredictable in nature. This paper presents a predictive model for prediction of electrical energy from solar and wind sources at a particular location. Information obtained from this model can be used for demand side management.

Keywords Solar energy · Prediction · CDF · PDF · Demand side management

1 Introduction

Renewable energy sources promise a clean, environment-friendly and long-lasting energy availability. The global trend has already been set towards higher and higher penetration of renewable energy sources [1, 2]. Under higher penetration of new and clean energy sources, the area of power system operation and control has come across a new set of challenge [3–6]. The unpredictable nature of most of the renewable energy sources makes it difficult to ensure the energy availability in a certain time

M. Singh (✉) · R. C. Jha · M. A. Hasan
Department of Electrical and Electronics, Birla Institute of Technology,
Patna, Mesra, Patna Campus, Patna 800014, Bihar, India
e-mail: mayank2626@gmail.com

M. A. Hasan
e-mail: hasan.asif6@gmail.com

© Springer Nature Singapore Pte Ltd. 2019
H. Malik et al. (eds.), *Applications of Artificial Intelligence Techniques in Engineering*, Advances in Intelligent Systems and Computing 698,
https://doi.org/10.1007/978-981-13-1819-1_29

interval [7]. For efficient and reliable operation, the load dispatch centers require data for available electrical power in next dispatch time interval [8]. Therefore, a suitable mathematical model is required to predict up to satisfactory level of accuracy and the probability of electrical energy availability from all sources.

After obtaining the data of available electrical power from all participating sources, the load dispatch center allocates the power generation from different sources. The predicted loads are fed by the available energy sources. However, this scheme does not guarantee the optimal power generation and utilization of renewable energy sources. Under the occurrence of peak load condition, the share of power generation from conventional sources increases. One of the potential solutions to increase the share of renewable energy sources under peak load condition is demand side management [9, 10].

Demand side management is the scheme in which consumers are asked to shut off some of their loads when peak load condition occurs. This decreases the power generation required to satisfy all the loads and hence reduces the share of conventional power generation. The load switching off operation under peak load condition may be voluntary from consumer end, or it can be managed by utility itself. An incentive in the form of tariff in their billing has to be provided to the consumers who participate in demand side management. The optimal cost of generation can only be achieved if the power availability from all sources is known up to a good amount of certainty. For this purpose, the prediction of energy availability from renewable energy sources is required.

A number of statistical distribution functions have been discussed in literature for prediction of electrical energy from renewable energy sources [11, 12]. These functions include gamma distribution, Generalized extreme value distribution, Weibull, logistic, normal and logarithmic distribution functions [13, 14]. In this paper, extreme value distribution function has been used for energy prediction. Wind and solar are one of the most exploited and easily available source of renewable energy around the globe. The energy availability from wind governs by the local wind speed whereas the solar radiation fall at any location decides the energy availability from solar. This paper presents a predictive model based energy prediction from both solar and wind sources simultaneously. Energy availability from a hybrid energy source which includes solar and wind has been obtained using predictive model.

The paper is prepared as follows. Section 2 presents the mathematical model of extreme value distribution function used as predictive model in this paper. Section 3 presents the simulation based energy prediction from solar and wind energy sources at a specific location. Aspects of demand side management based on available data are presented in Sect. 4. The discussion of results and conclusion follows next.

2 Modeling of Extreme Value Distribution

In a power network where significant penetration of renewable energy sources is present, the advance information regarding amount of power availability from all sources is surely helpful for control and operation. Since most of the renewable energy sources are of probabilistic in nature and the power availability from these sources broadly depends upon the weather conditions, therefore a probability distribution study will be very helpful. Different renewable energy sources have different weather parameters upon which the power availability depends. Energy from solar power plant depends upon the solar radiation on a particular day at a particular time. Similarly, the power from wind farm depends upon the wind speed at a particular location at the concerned time. Therefore, the probability of any specific power output from these sources depends upon the particular weather condition.

The probability distribution of power availability from solar and wind can be prepared with the help of any of the distribution model well discussed in literature. Both solar radiation and wind speed follow a particular pattern where solar radiation or wind speed increases from morning to half day and then starts decreasing as the night reaches. This suggests that a distribution model based on maximum and minimum value of random variable will be a suitable model for probability distribution study of solar and wind sources. Extreme value distribution is one such model.

Extreme value distribution model for probability distribution is a limiting distribution scheme. This scheme searches the limiting distribution of either maximum or minimum of a large number of unbounded random variables. Using extreme value distribution, the probability distribution function (PDF) and cumulative distribution function (CDF) for a given set of data can be obtained as given below

$$f(x) = \frac{1}{\beta} e^{-(\frac{x-\mu}{\beta})} e^{-e^{-(\frac{x-\mu}{\beta})}} \tag{1}$$

$$F(x) = e^{-e^{-(\frac{x-\mu}{\beta})}} \tag{2}$$

where ß and µ are constants given as

$$\beta = \frac{\sqrt{6} \cdot \text{var}(x)}{\pi} \tag{3}$$

$$\mu = E(x) - 0.5772\,\beta \tag{4}$$

where var(x) and E(x) are the variance and mean of the random variable data.

3 Probability Distribution of Solar and Wind

3.1 Solar Power Plant

Solar power plants generate the electrical energy from the solar radiation falling on solar panels. The solar panels are made of solar cells that generate or liberate charge carriers when solar radiation falls on it. The current–voltage characteristics of solar cell are given as follows,

$$I_{pv} = I_{ph} - I_0 \left[\exp\left(\frac{qV_{pv} + I_{pv}R_s}{AkT} \right) - 1 \right] \tag{5}$$

where I_{pv} and V_{pv} are pv cell current and terminal voltages, respectively. I_{ph} is light generated current and I_0 is pv cell saturation current. R_s is series resistance of electrical equivalent circuit of pv cell. A, k, and T are diode ideality factor, Boltzmann constant, and cell temperature respectively.

If the solar panel area covered in a solar power plant is A, and efficiency of solar panels in η, then power output from the plant at a given solar radiation fall G (W = m^2), is given as follows:

$$P_{pv} = A \cdot \eta \cdot G \tag{6}$$

Using the extreme value distribution model, the PDF and CDF for solar power plant with power output as random variable can be written as follows:

$$f(p(pv)) = \frac{1}{\beta} \exp\left(-\frac{P_{pv} - \mu}{\beta} \right) \exp\left(-\exp\left(-\frac{P_{pv} - \mu}{\beta} \right) \right) \tag{7}$$

$$F(p(pv)) = \exp\left(-\exp\left(-\frac{P_{pv} - \mu}{\beta} \right) \right) \tag{8}$$

where P(pv) is the power output of the solar power plant. Since the power output is directly proportional to the solar irradiance falling on panels, therefore the nature of CDF and PDF will remain the same for solar irradiance as random variable.

Figure 1 presents the hourly variation of solar irradiance for 29 September 2017 at Patna city situated at 25.59° latitude and 85.13 longitude location. As presented in plot, the solar irradiance increases as the day progresses and falls when time moves towards night. Based on irradiance data, the CDF and PDF characteristics were obtained for a day operation. Figure 2 presents the PDF plotted against the solar irradiance. The CDF has been plotted in Fig. 3. The repeating nature of graph is due to gradual rise and fall of solar radiation throughout the day.

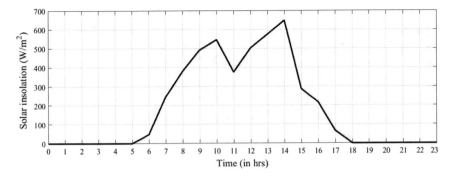

Fig. 1 Hourly variation of solar irradiance

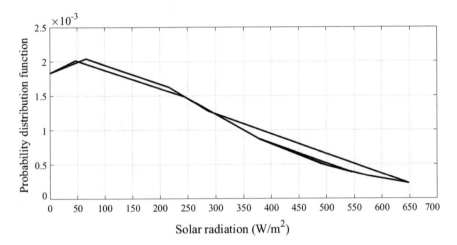

Fig. 2 Probability distribution function of solar power availability

3.2 Wind Power Plant

In wind energy conversion system, electrical energy is produced from kinetic energy of the wind. The wind blowing at some speed rotates the wind turbine and an induction generator coupled with wind turbine generates the electrical voltage. The mechanical rotation involved with wind turbine restricts the operation of wind energy conversion system to a certain speed range. The electrical energy can only be produced for a range of wind speed called cut-in speed V_{ci} and cut out speed V_{co}. The system generates a constant rated power Pr for wind speed between rated speed and cut out speed. The power generated for a wind speed between cut-in speed to rated wind speed is a function of instantaneous wind speed and is given as below,

$$P_w = P_r \frac{V_w - V_{ci}}{V_r - V_{ci}} \tag{9}$$

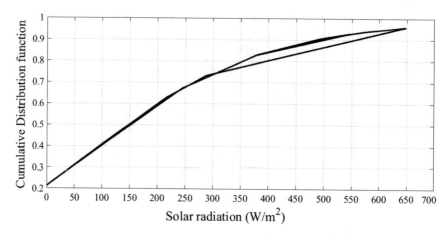

Fig. 3 Cumulative distribution function of the solar power availability

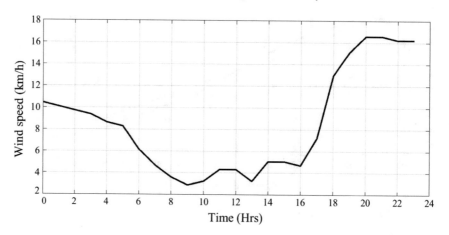

Fig. 4 Hourly variation of wind speed

where P_w is instantaneous generated power and V_w is instantaneous wind speed. Since the power output of wind energy conversion unit depends upon the wind speed, therefore the distribution function for a wind power plant is developed using wind speed as random variable. The practical data for wind speed at the same location as taken in case of solar irradiance (i.e., Patna city) has been plotted in Fig. 4. The wind speed variation on an hourly basis has been taken for probability distribution. The power output corresponding to instantaneous wind speed has been plotted in Fig. 5. Following the stochastic nature of the wind, the Rayleigh distribution, which is a particular form of Gumbel distribution has been suggested in literature to develop the distribution function for wind power plant. Taking wind speed as random variable, the probability distribution function and cumulative distribution function of wind power plant is given as follows:

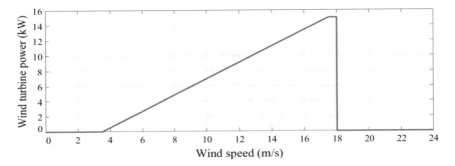

Fig. 5 Power output of wind power plant

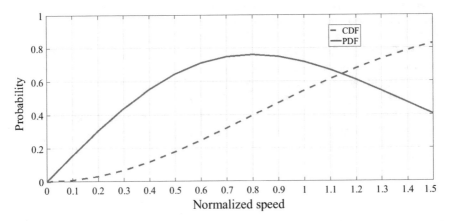

Fig. 6 Probability distribution of wind energy generation

$$f(V_w) = \frac{\pi}{2} \frac{V_w}{V_m^2} \exp\left(-\frac{\pi}{4}\left(\frac{V_w}{V_m^2}\right)^2\right) \qquad (10)$$

$$F(V_w) = 1 - \exp\left(-\frac{\pi}{4}\left(\frac{V_w}{V_m^2}\right)^2\right) \qquad (11)$$

Wind energy conversion system with data given in Table 1 has been taken in this paper for preparation of distribution function. Figure 6 presents the PDF and CDF for the wind power plant.

The power availability from solar and wind energy based power plants have been predicted through distribution function analysis. Since the energy availability from individual plants is predicted on an hourly basis, so an hourly basis demand side management can be planned. The demand response will depend upon the hourly load demand and hourly generation.

4 Conclusion

A mathematical model has been developed in this paper for obtaining probability distribution functions of solar- and wind-based power plants. The distribution function is prepared based on geographical data and the random variables are solar incidence and wind speed for the place under study. The results present the probability of a certain magnitude of power output from solar- and wind-based power plant on an hourly basis.

The power generation scheduling in an operating power network depends upon the power availability from all sources and the load demand for a schedule period. If the power availability from all renewable sources connected to power network can be predicted, then the optimum power scheduling can be done. In case of occurrence of peak load demand, the knowledge of power availability from all sources with certain probability will surely help in preparing a suitable demand side management plan.

The distribution function based prediction of power availability from renewable sources will help in the implementation of a demand side management scheme and maximum utilization of nonpolluting sources. It also helps in designing a suitable incentive based tariff scheme through which the consumers can be promoted to participate in demand side management. The probability functions will help in minimizing the cost and maximizing the usage of nonconventional energy sources.

References

1. X. Jin, Z. Wen, L. Weidong, L. Yan, T. Xinshou, L. Chao, W. Linjun, Study on the driving force and challenges of developing power grid with high penetration of renewable energy, in *2017 IEEE Transportation Electrification Conference and Expo, Asia-Pacific (ITEC Asia-Pacific)*, Aug 2017, pp. 1–5
2. C. Liu, A. Botterud, Z. Zhou, P. Du, Fuzzy energy and reserve co optimization with high penetration of renewable energy. IEEE Trans. Sustain. Energy **8**(2), 782–791 (2017)
3. P. Samadi, V.W.S. Wong, R. Schober, Load scheduling and power trading in systems with high penetration of renewable energy resources. IEEE Trans. Smart Grid **7**(4), 1802–1812 (2016)
4. N. Navid, G. Rosenwald, Market solutions for managing ramp flexibility with high penetration of renewable resource. IEEE Trans. Sustain. Energy **3**(4), 784–790 (2012)
5. A. Borghetti, M. Bosetti, S. Grillo, S. Massucco, C.A. Nucci, M. Paolone, F. Silvestro, Short-term scheduling and control of active distribution systems with high penetration of renewable resources. IEEE Syst. J. **4**(3), 313–322 (2010)
6. M.A. Hasan, S. Sourabh, S.K. Parida, Impact of a microgrid on utility grid under symmetrical and unsymmetrical fault conditions, in *2016 IEEE 7th Power India International Conference (PIICON)*, Nov 2016, pp. 1–6
7. M.A. Hasan, S.K. Parida, Temperature dependency of partial shading effect and corresponding electrical characterization of pv panel, in *2015 IEEE Power Energy Society General Meeting*, July 2015, pp. 1–3
8. S.S. Mousavi-Seyedi, F. Aminifar, S. Afsharnia, Parameter estimation of multiterminal transmission lines using joint pmu and scada data. IEEE Trans. Power Deliv. **30**(3), 1077–1085 (2015)
9. P. Palensky, D. Dietrich, Demand side management: demand response, intelligent energy systems, and smart loads. IEEE Trans. Industr. Inf. **7**(3), 381–388 (2011)

10. H. Zhao, Z. Tang, The review of demand side management and load forecasting in smart grid, in *2016 12th World Congress on Intelligent Control and Automation (WCICA)*, June 2016, pp. 625–629
11. C.D. Bussy-Virat, A.J. Ridley, Predictions of the solar wind speed by the probability distribution function model. Space Weather **12**(6), 337–353 (2014)
12. A.A. Teyabeen, Statistical analysis of wind speed data, in *IREC2015 The Sixth International Renewable Energy Congress*, March 2015, pp. 1–6
13. J. Salo, H.M. El-Sallabi, P. Vainikainen, The distribution of the product of independent rayleigh random variables. IEEE Trans. Antennas Propag. **54**(2), 639–643 (2006)
14. S. Jiang, D. Kececioglu, Graphical representation of two mixed-weibull distributions. IEEE Trans. Reliab. **41**(2), 241–247 (1992)

Fuzzy Logic Controller for DC Micro-grid Systems with Energy Supervision System

A. Kumaraswamy, Eera Thirupathi and Balakrishna Kothapalli

Abstract This article presents the implementation of Fuzzy logic controller for DC micro-grid system with energy supervision system. The modeling of the energy supervision system with Fuzzy logic controller is developed in the MATLAB and analyzed. The battery lifetime has been improved using Fuzzy logic controller. The DC micro-grid system has been tested with Fuzzy logic controller and results are presented.

Keywords Energy supervision system (ESS) · Fuzzy logic control
DC micro-grid · State of charging (SOC)

1 Introduction

Huge amount of storage is a necessary part of the future of energy. Electricity tremendous storage is required to give industry the lack of restrictions of instantly available nonconventional electricity. The advertise drivers for energy storage are set in legislation [1]. Almost all countries look forward to utilizing free energy >20% in future. Increasing micro-grids [2] especially the current development in solar photovoltaic on top of on-shore and off-shore wind energy causes problems for grid infrastructure and present generators. The proposed Fuzzy logic controller for DC micro-grid with ESS is shown in Fig. 1.

The green energy development is the latest trend, an arrangement of energy storage and dispersed power system, to decrease power losses from transmission lines to must

A. Kumaraswamy
KU College of Engineering & Technology, Warangal, India
e-mail: aksp301@gmail.com

E. Thirupathi (✉) · B. Kothapalli
SR Engineering College, Warangal, India
e-mail: thiru0805@gmail.com

B. Kothapalli
e-mail: kotapallibalakrishna@gmail.com

© Springer Nature Singapore Pte Ltd. 2019
H. Malik et al. (eds.), *Applications of Artificial Intelligence Techniques in Engineering*, Advances in Intelligent Systems and Computing 698,
https://doi.org/10.1007/978-981-13-1819-1_30

Fig. 1 Arrangement of the proposed grid system with ESS

Fig. 2 Solar panel
equivalent circuit

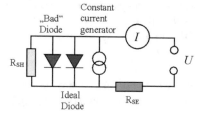

be formed DC micro-grid. Use of dispersed power system which can reduce loss of
almost 70%, space saving of 33%, investment reduced by 15%, and dependability
increased by 200% [1]. In addition, we developed 24 V (low voltage) application
such as ceiling grid for lighting. Many investigate groups have been extensive 380 V
(high voltage) distribution especially for home appliances.

2 Modeling of Energy Storage System

The following describes the design of each individual system.
    The circuit model of photovoltaic plate is shown in Fig. 2.
    The expression for current is

$$I_{pv} = n_p\, I_{ph} - n_p\, I_{rs}\left[\exp\left(\frac{q}{KTA}\frac{V_{pv}}{n_s}\right) - 1\right] \tag{1}$$

where

V_{pv} is the photovoltaic panel output voltage,
I_{pv} is the photovoltaic panel output current,
ns   is the total number of photovoltaic panels in shunt,
K   is the Boltzmann constant value 1.38×10^{-23} J/K,

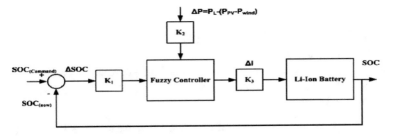

Fig. 3 Closed-loop control of Fuzzy logic controller with SOC

q   is the charge, value of 1.6×10^{-19} Coulombs,
T   is the photovoltaic panel surface temperature, and
I_{rs} is the reverse dispersal current.

In Eq. (1), if temperature varies current characteristics varies, by temperature, as expressed below:

$$I_{rs} = I_{rr}\left[\frac{T}{T_r}\right]^3 \exp\left(\frac{qE_g}{kA}\left(\frac{1}{T_r} - \frac{1}{T}\right)\right) \qquad (2)$$

where

T_r is the photovoltaic panel reference temperature,
I_{rr} is the photovoltaic panel reverse recurring current at temperature (K), and
E_g is the semiconductor material band gap.

$$I_{ph} = [I_{scr} + \alpha(T - T_r)]\frac{S}{100} \qquad (3)$$

where

I_{scr} is the short-circuit current
T_r is the reference temperature
A   is the photovoltaic panel coefficient, and
S   is the intensity of light in (kW/m2).

In this modeling, NUS0E3E photovoltaic modules are used, and rating of single module is 180 W; in my model, two parallel photovoltaic arrays are used, and total power generation is 5 kW. In each array built by 14 series photovoltaic panels, the result of this modeling, that is, output voltage versus output power of photovoltaic unit, is as shown in Fig. 3. In this, we used 1 kW/m^2 constant light intensity and temperature with Vpv varying for output results verification.

The wind generator developed power is

$$P_w = 0.5\rho A\ V^3 C_p(\lambda, \vartheta) \qquad (4)$$

where

P_w is the turbine-generated power in watts,
P is the gas density unit (kg/m),
A is the area of cross section of wind turbine blade in square meter,
V is the velocity of wind in (m/s), and
Cp is the coefficient of energy change.

The gas density and coefficient of energy change are expressed in Eqs. (5) and (6), respectively,

$$\rho = \left(\frac{353.05}{T}\right) e^{-0.034\left(\frac{Z}{T}\right)} \tag{5}$$

$$C_p(\lambda, \vartheta) = \left(\frac{116}{\lambda_i} - 0.4\, x\vartheta - 5\right) \cdot 0.5 e^{\frac{-16.5}{\lambda_i}} \tag{6}$$

where

Z is the wind turbine height,
T is the temperature of earth atmosphere,
λ_i is the first fraction at speed, and
ϑ is the tip angle.

Equation (7) is the fractional speed ratio λ_i in Eq. (6) and first tilt speed ratio λ is

$$\lambda_i = \frac{1}{1/(\lambda + 0.089\vartheta) - 0.035/(\vartheta^3 + 1)} \tag{7}$$

$$\lambda = r\frac{\omega}{V} \tag{8}$$

In this design, used wind turbine is Awv 1500 of better accuracy.

3 Energy Supervision System with Fuzzy Logic Controller

The block diagram of DC micro-grid system has five important elements. To implement an exact controller of present micro-grid, we used dynamic model of the green sources (such as photovoltaic, wind power, etc.), DC/DC converters (such that the phase-shifted full-bridge converter and buck, buck-boost converters), and bidirectional converter, such that the advanced microsystems and the full-bridge inverters are fundamental.

4 Fuzzy Logic Control

The Fuzzy logic control (FLC) exhibits enormous stability for solving various problems with an uncertainty. It is a two-layer network. The core steps in Fuzzy logic are fuzzification and de-fuzzification. Fuzzification operations map input values into Fuzzy membership functions, and the de-fuzzification operations are used to map Fuzzy output membership functions into a "crisp" output value. The important stages are as follows.

Steps involved in the Fuzzy logic control are given below:

Step 01: All input variables are fuzzified into Fuzzy membership functions.
Step 02: All rules are executed to find the Fuzzy output function.
Step 03: Fuzzy output functions are de-fuzzified to get "crisp" output values.

The Fuzzy logic controller depends upon the Fuzzy logic information in the creation of how to design a Fuzzy logic-based controller operation. A buffer region involves the conventional 0 and 1, by earnings of logic portion of none one and none zero feasible. The appearance of theoretical facts and knowledge is allowing a wider and supplier independence in the Fuzzy logic supposition. The activist controller employs a set of qualitative orders denned by comparable images deferred from Fuzzy logic-based controller.

The DC micro-grid supply system is obtainable using Fuzzy logic-based controller, as shown in Fig. 4. To be obtaining the necessities of state of charge level, Fuzzy logic-based controller is deliberated to be in charge manner or discharge manner for current micro-grid arrangement. The input Fuzzy logic control variables are ΔP and ΔSOC, where ΔI is the output variable. These are expressed as follows:

$$\Delta SOC = SOC_{command} - SOC_{now} \tag{9}$$

$$\Delta P = P_L - \left(P_{wind} + P_{pv}\right) \tag{10}$$

where ΔP is the difference in power between essential and total.

Fuzzy Logic Control Membership Rules

ΔI		ΔP				
		NB	NS	ZO	PS	PB
ΔSOC	NB	PB	PB	PB	PB	PB
	NS	PB	PB	PS	PS	PB
	ZO	ZO	ZO	ZO	PS	PB
	PS	NS	NS	NS	NS	PB
	PB	NB	NB	NB	NB	PB

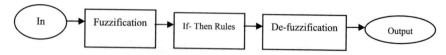

Fig. 4 Basic structure of Fuzzy logic system

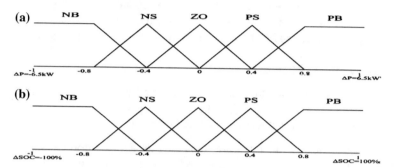

Fig. 5 Membership functions (input variables)

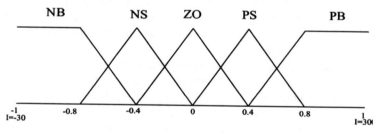

Fig. 6 Membership functions (output variable)

The membership function of Fuzzy logic control has five values:

1. Negative big (NB),
2. Negative small (NS),
3. Zero (ZO),
4. Positive small (PS), and
5. Positive big (PB).

Scale factor K_1 and K_2 are shown in the above figure. We can decide membership rank and alternative it into the Fuzzy logic control orders to gain the output current discrepancy ΔI of the lithium–ion battery. If load not receives sufficient power from green energy sources, then the value ΔP is negative. Thus, the battery has got to activate in charging mode (Figs. 5 and 6).

The supervise orders of the above-discussed prioritizing buying external power generated by the nonconventional energy sources are activated to modern Fuzzy control plan of DC micro-grid growth for buying electricity and mounting the living of lithium–ion battery.

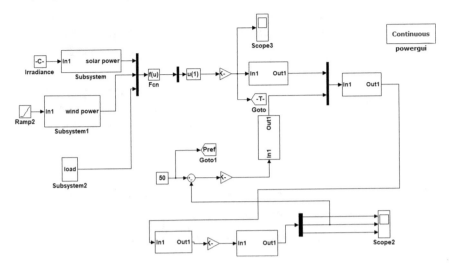

Fig. 7 Simulink model for DC micro-grid

The model, ΔI, is the output changeable for positive big (the amount of current discharging is big) after the ΔP is the input changeable for negative big (the quantity of electricity to buy is big) and ΔSOC is input changeable for negative small (better than the state of the charge demand and the membership amount is little). Conversely, ΔI is the output changeable for the negative small (the amount of current charging is little), while the ΔP is input changeable for negative big (the quantity of electricity to buy is big) and input changeable for the difference in state of charge is positive small (State of the charge demand is small as well as the membership amount is little). The negative small is output changeable for a replacement of negative big, while the oral structure is operated in the top of situations for the reason that buying electricity is the top major anxiety in this container. Orders are arranged to maintain battery state of charge build 50%. Furthermore, in the Fuzzy logic controller orders, the battery is anxious to release as the control plan of the load was more than developed power for the purpose of the power demand using the green energy sources.

5　Simulation Results

The proposed controller with DC micro-grid is implemented in MATLAB Simulink and execute on personal computer. The Simulink model is shown in Figs. 7 and 8.

This instance continuous from the accurateness of in structure with Fuzzy logic-based controller that can continue of the battery state of charge is small or big at an exact stage, while shown in Fig. 9, as well as the battery maintained 50% with an opportunity of 10%, when shown in Fig. 10, though maintained 50% with an opportunity worth of 90% while using the Fuzzy logic controller.

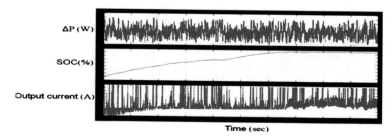

Fig. 8 Simulation marks with first battery state of charge at 10%. Consists of a 5 kW solar part, A 1.5 kW wind turbine part, 1.5 kW lithium–ion battery part, and 6.5 kW load

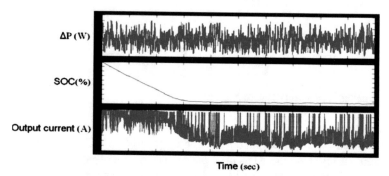

Fig. 9 Simulation result with first battery SOC at 90%

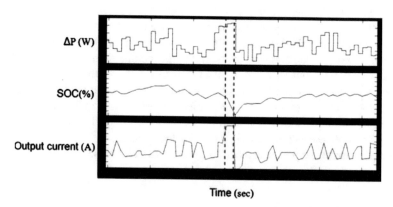

Fig. 10 Simulation results when the inverter ratings more the nominal values

The battery state of the charge control plan of the instructions on the way to buy electricity is the main concern to continuity and the battery going to down while the difference in power improved than 5 kW to stay the two-directional inverter subsystem structure was maintained the power balancing for this structure [1]. The work of inverter is more than the power rating when battery is not the maintainer for protection that is achieved.

6 Conclusion

In this research paper, modeling, design, and analysis of Fuzzy logic-based controller to attain enlargement of DC micro-grid with energy supervision system have been presented. From the output simulation results, the DC micro-grid system with ESS meets the equilibrium point, and state of charge of battery maintain in reasonable valve for life cycle of battery increases using Fuzzy logic control rules for DC micro-grid. The generated rules have been built-in intelligent DC micro-grid with ESS system and it can behave data communication and control in power position of subsystem through a closed-loop network.

References

1. H. Dongxiang, L. Sheen, C. Yoking, W. Fu, R. Gouging, DC micro-grid simulation test platform, in *Proceedings of the 9th Taiwan Power Electron. Conference* (2010), pp. 1361–1366
2. S. Morozumi, Micro-grid demonstration projects in Japan, in Proceedings of the IEEE Power Converse Conference, Apr 2007, pp. 635–642
3. Specifications for 400 V DC Power Supplies and Facility Equipment, D. Symanski, Sr. Program Manager, Electric Power Research Institute,Keiichi Hirose, NTT Facilities, and Brian Forte berry, Program Manager, Electric Power Research Institute, Green Building Power Forum, Jan 2010
4. Development of a DC Power Inlet Connector for 400 V DC IT Equipment, B. Davies, Director of Engineering, Anderson Power Products, Inc. GreenBuilding Power Forum, Jan. 2010
5. Development of Socket-outlet Bar and Power Plug for 400 V Direct Current Feeding System, T. Yuba, R&D Manager, Fujitsu Components Ltd. Green Building Power Forum, Jan 2010
6. Power Inlet Connector for 380 V DC Data Centre, B. Davies, Anderson Power Products, Green Building Power Forum, Dec 2010.
7. Intel Lab's New Mexico Energy System Research Centre 380 V DC Micro grid Tested, G. Alee, Intel Labs, Green Building Power Forum, Dec 2010
8. International Standardization of DC Power, K. Hirose, NTT Facilities,Green Building Power Forum, Dec 2010

A Semi-supervised Clustering for Incomplete Data

Sonia Goel and Meena Tushir

Abstract In the proposed work, our research focus is on semi-supervised clustering, which uses a small amount of supervised data in the form of class labels or pairwise constraints on some examples to aid unsupervised clustering. In addition, we have also included incomplete data with missing features. In the proposed clustering algorithm, we have embedded the optimal completion strategy (OCS) in semi-supervised clustering algorithm which imputes the missing values by the maximum likelihood estimate in an iterative optimization procedure. A series of detailed experimental analysis on real data sets is done to illustrate the main ideas discussed in the study work.

Keywords Fuzzy c-means clustering · Semi-supervised clustering
Missing features · Optimal control strategy

1 Introduction

Clustering is considered as an unsupervised learning algorithm, which uses a particular similarity metric to identify a group of similar data from a given data and assign them as clusters. Many research domains such as data mining and image processing sometimes contain large amount of unsupervised data but very limited labeled or supervised data, which can be difficult and costly to gather. Hence, semi-supervised clustering, which takes into account a combination of both types of data (supervised or unsupervised), has generated a lot of interest recently [1–3]. Another common problem with data collection is missing values of certain features due to measurement errors or loss of values after data acquisition. In the literature, several techniques were proposed for managing missing or incomplete data during clustering analysis.

S. Goel (✉) · M. Tushir
MSIT, C4-Janakpuri, New Delhi, India
e-mail: soniagarg23@yahoo.com

M. Tushir
e-mail: meenatushir@yahoo.com

© Springer Nature Singapore Pte Ltd. 2019
H. Malik et al. (eds.), *Applications of Artificial Intelligence Techniques in Engineering*, Advances in Intelligent Systems and Computing 698,
https://doi.org/10.1007/978-981-13-1819-1_31

The most prominent work is by Hathway and Bezdek [4] where they proposed different clustering strategies to continue the FCM clustering of incomplete data. Timm et al. [5, 6] developed the nearest cluster strategy (NCS) whereby he incorporated the concepts of Gath-Geva algorithm into fuzzy clustering. Li et al. [7] introduced variants of K-means clustering algorithm for clustering incomplete data.

In this work, we offer augmentation of semi-supervised clustering by including incomplete data with missing features. So far no work has been reported in the literature with semi-supervised clustering on incomplete data. The outline of our research paper is as follows: A review of various semi-supervised clustering algorithms is presented in Sect. 2 followed by discussion on various FCM based clustering strategies for incomplete data with missing values in Sect. 3. Details of our proposed semi-supervised clustering with incomplete data is given in Sect. 4 followed by detailed experimental analysis in Sect. 5. Conclusion along with outline for further research scope is given in Sect. 6.

2 Semi-supervised Clustering Algorithms

This section presents a survey of several semi-supervised clustering algorithms. Let the sample set be $Xs = \{x_1, x_2, \ldots, x_{Ns}\} \subset R^{Ns \times ds}$ where Ns denotes the number of samples with ds dimensions of the feature space. The sample data set Xs is divided into p classes, Ns samples associated to p classes by a membership values given in $U = [u_{ck}]_{p \times Ns}$ where $u_{ck}(1 \le c \le p, 1 \le k \le Ns)$ is the membership degree of the kth sample x_k associated to the cth class, and u_{ck} should satisfy the following two constraints:

$$\sum_{c=1}^{p} u_{ck} = 1, \quad 1 \le k \le Ns$$

$$0 \le u_{ck} \le 1, \quad 1 \le c \le p, \quad 1 \le k \le Ns$$

The FCM clustering attempts to produce fuzzy clusters of Xs via an iterative scheme by solving

$$\min Z_m(Xs, U, v) = \sum_{c=1}^{p} \sum_{k=1}^{Ns} u_{ck}^m \|x_k - v_c\|^2 \tag{1}$$

where $m > 1$ is the fuzzification parameter and the vector v_c is the center of the cth cluster.

In literature, many modifications of this objective function are proposed to formulate semi-supervised FCM algorithms. A semi-supervised clustering algorithm first introduced by Pedrycz [1, 8] minimizes the following energy function:

$$Z_m(U, v : Xs, F) = \sum_{c=1}^{p} \sum_{k=1}^{Ns} u_{ck}^m \|x_k - v_c\|^2 + \sum_{c=1}^{p} \sum_{k=1}^{Ns} (u_{ck} - f_{ck})^m \|x_k - v_c\|^2 \quad (2)$$

Pedrycz and Waletzky [2, 3] further modified this algorithm by introducing a binary vector g ($g_k = 1$ if x_k is labeled else $g_k = 0$) to distinguish between a supervised and unsupervised sample. The improved objective function is as follows:

$$Z_m(U, v : Xs, F) = \sum_{c=1}^{p} \sum_{k=1}^{Ns} u_{ck}^m \|x_k - v_c\|^2 + \beta \sum_{c=1}^{p} \sum_{k=1}^{Ns} (u_{ck} - f_{ck} g_k)^m \|x_k - v_c\|^2$$

$$(3)$$

This objective function can be remodeled by considering the effect of partial supervision is given in [3]

$$Z_m(U, v : Xs, F) = \sum_{c=1}^{p} \sum_{k=1}^{Ns} u_{ck}^m \|x_k - v_c\|^2 + \beta \sum_{c=1}^{p} \sum_{k=1}^{Ns} (u_{ck} - f_{ck})^m g_k \|x_k - v_c\|^2$$

$$(4)$$

It is assumed that the number of classes which are known in advance is the same as that of number of clusters and the same is shown in matrix F. Stutz and Runkler [9] inserted the scale coefficient β into the credibility of the labeled features in the data. Its updated objective function is as follows:

$$Z_m(U, v : Xs, F) = (1 - \beta) \sum_{c=1}^{p} \sum_{k=1}^{Ns} u_{ck}^m \|x_k - v_c\|^2 + \beta \sum_{c=1}^{p} \sum_{k=1}^{Ns} (u_{ck} - f_{ck})^m \|x_k - v_c\|^2$$

$$(5)$$

Pedrycz and Vukovich [10] also proposed a modified objective function, as given in the following equation:

$$Z_m(U, v : Xs, F) = \sum_{c=1}^{p} \sum_{k=1}^{Ns} u_{ck}^m \|x_k - v_c\|^2 + \beta \sum_{c=1}^{p} \sum_{k=1}^{Ns} u_{co(k)k}^m \left(f_{co(k)} - y_k\right)^2 \|x_k - v_{co(c)}\|^2 \quad (6)$$

3 FCM Algorithms for Incomplete Data

It is not possible to apply FCM clustering directly to the data with missing features because it is necessary to have all feature values of each data sample for calculation of the cluster centers and the distance metrics. Hathway and Bezdek [4] proposed several strategies using fuzzy clustering algorithms to deal with incomplete data. Whole data strategy fuzzy c-means (WDSFCM) clustering is based on omitting either the incomplete data samples or the missing feature value of the data. The cluster centers are then calculated with conventional FCM. When the incomplete data samples are small in proportion, WDSFCM algorithm can be adopted for clustering of incomplete data. In [4], the authors proposed that WDS should be used only when the availability of complete data items in the data set is at least 75%.

When the proportion of incomplete data is sufficiently large, whole data strategy cannot be justified. Dixon [11] proposed the partial distance function for analysis of large incomplete data. The sum of the squared Euclidean distances between all non-missing features is calculated with the help of partial distance function and it is then scaled by the reciprocal of the proportion of values used in the calculation. In [9], FCM is restructured by incorporating the *partial distance strategy (PDS)*, resulting in new algorithm as PDSFCM. PDSFCM is adopted with two modifications. These are (1) distances are calculated using Eq. (7). (2) Cluster centers are updated using Eq. (8). Partial distances are calculated using the following equations:

$$D_{par}(x_k, v_c) = \frac{ds}{\sum_{d=1}^{ds} c_{kd}} \sum_{d=1}^{ds} (x_{kd} - v_{cd})^2 c_{kd} \tag{7}$$

where

$$c_{kd} = \begin{cases} 1, \text{ if } x_{kd} \text{ is available } & \text{for } 1 \le d \le ds, 1 \le k \le Ns \\ 0, & else \end{cases} \tag{7a}$$

The cluster center is calculated as

$$v_{cd} = \frac{\sum_{k=1}^{Ns} (u_{ck})^m c_{kd} x_{kd}}{\sum_{k=1}^{Ns} (u_{ck})^m c_{kd}} \text{ for } 1 \le c \le p, 1 \le d \le ds \tag{8}$$

The third algorithm namely optimal completion strategy is based on minimization of objective function by estimating the missing values as the additional variables [4]. The partial derivative of the objective function with respect to the missing values is taken to estimate the missing values and setting them to zero. The updated value of x_{kd} is as follows:

$$x_{kd} = \frac{\sum_{c=1}^{p} (u_{ck})^m v_{cd}}{\sum_{c=1}^{p} (u_{ck})^m} \text{ for } 1 \le k \le Ns \text{ and } 1 \le d \le ds \tag{9}$$

This algorithm is based on the initialization of cluster centers. Additionally, random values are assigned to missing features. In the first iteration step, the membership values of data points and the cluster centers are calculated in the same way as in the conventional fuzzy c-means clustering. The cluster prototypes with incomplete data item of higher membership degree are more significant when estimation of missing values of the data item is done. The FCM algorithm modified in this way is termed as OCS_FCM.

OCS based FCM clustering is further modified by taking into account the concept of the smallest partial distance between cluster centers and data point. This new algorithm is referred to as the nearest prototype strategy (NPS) and FCM algorithm described for this strategy is termed as NPSFCM. The missing values of an incomplete dataset are calculated as follows:

$$x_{kd} = v_{cd} \text{ with } D_{par}(x_k, vc) = \min\{D_{par}(x_k, v_1), D_{par}(x_k, v_2), \ldots, D_{par}(x_k, v_p)\} \tag{10}$$

for $1 \leq k \leq n$ and $1 \leq d \leq ds$

4 Proposed Semi-supervised Clustering for Incomplete Data

Here, we developed a semi-supervised clustering algorithm that explicitly accounts for incomplete data when some of the data is supervised (labeled). Our proposed algorithm extends semi-supervised clustering proposed by Pedrycz [1] to obtain a semi-supervised algorithm that handles incomplete data as well.

Let $Xs = \{x_1, x_2, \ldots, x_{Ns}\}$ be ds-dimensional data vector with Ns samples
We divide the data into two categories, supervised data and unsupervised data Such that $Xs = Xs^L \cup Xs^u$

Xs^L supervised data
Xs^u unsupervised data

Further
$Xs_W = \{x_k \in Xs | x_k \text{ is a complete datum}\}$
$Xs_M = \{x_{kd} = * \text{ for } 1 \leq k \leq Ns, 1 \leq d \leq ds | x_{kd}\}$ is a missing feature from Xs
We assign initial values of missing data randomly and optimize the function

$$Z_m(U, V) = \sum_{c=1}^{p} \sum_{k=1}^{Ns} u_{ck}^m \|x_k - v_c\|^2 + \beta \sum_{c=1}^{p} \sum_{k=1}^{Ns} (u_{ck} - f_{ck} \, g_k)^m \|x_k - v_c\|^2 \tag{11}$$

Missing features of the data may be from labeled data or unlabeled data. The membership grades of supervised pattern are arranged in $F = [f_{ck}]$ where $1 \leq c \leq p, 1 \leq k \leq Ns$

Here $g_k = 1$, if pattern x_k is labeled otherwise assigned a value 0 if unlabeled. The scaling factor β is introduced to maintain the balance between supervised and unsupervised components of the objective function. We embedded optimal completion strategy into semi-supervised clustering algorithm. The membership value of labeled pattern is always assigned by domain expert.

Using the standard technique of Lagrange multipliers, we have the following equation for objective function,

$$Z_m(U, V, \lambda) = \sum_{c=1}^{P} \sum_{k=1}^{Ns} u_{ck}^m \|x_k - v_c\|^2 + \beta \sum_{c=1}^{P} \sum_{k=1}^{Ns} (u_{ck} - f_{ck} g_k)^m \|x_k - v_c\|^2 - \lambda \left(\sum_{c=1}^{P} u_{ck} - 1 \right)$$

(12)

Partially differentiating $Z_m(U, V)$ with respect to membership matrix U and the prototype v_c, we have the following equations for membership matrix U and v_c:

$$u_{ck} = \frac{1}{\beta + 1} \left\{ \frac{1 + \beta(1 - g_k \sum_{c=1}^{P} f_{ck})}{\sum_{c=1}^{P} \frac{ds_{ck}^2}{ds_{ck}^2}} + \beta f_{ck} g_k \right\}$$

(13)

$$v_c = \frac{\sum_{k=1}^{Ns} u_{ck}^m x_k + \beta \sum_{d=1}^{Ns} (u_{ck} - f_{ck} g_k)^m x_k}{\sum_{k=1}^{Ns} u_{ck}^m + \beta \sum_{k=1}^{Ns} (u_{ck} - f_{ck} g_k)^m}$$

(14)

Calculate Xs_M for $x_{kd} \in Xs_M$ using

$$x_{kd} = \frac{\sum_{c=1}^{P} \left(\left(u_{ck} - f_{ck} g_k \right)^m + u_{ck}^m \right) v_c}{\sum_{c=1}^{P} \left(u_{ck} - f_{ck} g_k \right)^m + u_{ck}^m}$$

(15)

5   Numerical Examples

In this section, we compared our proposed clustering with standard FCM, standard semi-supervised FCM (SSFCM) clustering and OCS version of FCM (OCM_FCM) using real data sets. In all these experiments, missing features were artificially created with a constraint that at least one feature is present in each sample of incomplete data.

5.1   Real-Life Data Set Description

We applied these techniques to benchmark problems consisting of 02 famous datasets from repository [12] that can be easily found in the literature. Table 1 describes the characteristics of these data sets and the results obtained from various clustering algorithms are presented in Table 2 for further analysis and comparison of their performance. Iris data set consists of total 150 samples of 3 classes with each class consisting of 50 samples. Each class represents a variety of iris plant and can be treated as a cluster. Each sample has 4 features. The seed data set consists of kernels representing three different varieties of wheat, each having 70 elements. Seven features of wheat kernels were measured.

Table 1 Summary of characteristics of the IRIS and SEED data sets

Data set	No of objects	Feature	Classes
Iris	150	4	3
Seed	210	7	3

5.2 Results and Discussion

We now report results involving various real-life data sets. Data is first divided into two parts: supervised data and unsupervised data. Initially, complete data Xs is chosen and randomly a small percentage of data is marked as supervised. Xs is further transformed by randomly removing certain feature values of its components X_{kd}, thus resulting in incomplete data. Throughout this paper, we have labeled 10% data, and considered three cases of 10% incomplete (missing) data:

(i) All the missing features are from labeled data.
(ii) 50% of the missing data is from labeled data and 50% of missing data is from unlabeled data
(iii) All the missing data is unlabeled data.

Table 2 gives comparative analysis of our proposed algorithm with some of the existing clustering methods on the basis of mean results of 20 trials using partially labeled incomplete data. The third column gives the average number of misclassifications and the fourth column corresponds to accuracy measure to assess the quality of clustering. Last column gives the mean number of iterations required over 20 trials. Results in Table 2 show that SSFCM gives better results than standard FCM. It is evident that increase in number of labeled data will improve the performance of SSFCM. Next, we have applied OCS_FCM on incomplete data and as expected, the performance deteriorates with missing feature values. However, when we include supervised or labeled data and apply our proposed algorithm, the performance is better as compared to OCS_FCM. Accuracy of our proposed method needs further exploration by varying the degree of incomplete data of supervised and unsupervised data.

6 Conclusion

This paper considers the clustering problem for partially labeled data with some of the features missing. The missing features can be from labeled data or from unlabeled data or from both. Numerical experiments on artificially generated real-world data set demonstrate that the proposed clustering shows better results than the existing OCS based FCM algorithm. To further study the proposed algorithm in the future,

Table 2 Comparative analysis of SSFCM_OCS algorithm with other clustering methods in terms of misclassifications, accuracy, no of iterations and performance index

Iris data set (10% Labeled data and all missing data is labeled data)

Clustering method	Parameters	Mean number of misclassifications	% accuracy	Mean number of iterations to termination
FCM	$m=2$	16.8	88.8	16.4
SSFCM	$m=2$	13	91.3	5.2
OCS_FCM	$m=2$	21	86.0	21
Proposed	$m=2, \beta=0.5$	19.5	87	6.6

Iris data set (10% Labeled data, 50% of the missing data is labeled and 50% of missing data is from unlabeled data)

FCM	$m=2$	16.8	88.8	16.4
SSFCM	$m=2$	13	91.3	5.2
OCS_FCM	$m=2$	21	86.0	21
Proposed	$m=2, \beta=0.5$	15.4	89.7	5.2

Iris data set (10% Labeled data, All missing data is from unlabeled data)

FCM	$m=2$	16.8	88.8	16.4
SSFCM	$m=2$	13	91.3	5.2
OCS_FCM	$m=2$	21	86.0	21
Proposed clustering method	$m=2, \beta=0.5$	19.4	87	6.5

Seed data set (10% Labeled data and all missing data is labeled data)

	Parameters	Mean number of misclassifications	Accuracy	Mean number of iterations to termination
FCM	$m=2$	22	89.5	17
SSFCM	$m=2$	16	92.4	7
OCS_FCM	$m=2$	53	74.8	25
Proposed	$m=2, \beta=0.5$	43	79.5	5

Seed Data set (10% Labeled data, 50% of the missing data is labeled and 50% of missing data is from unlabeled data)

FCM	$m=2$	22	89.5	17
SSFCM	$m=2$	16	92.4	7
OCS_FCM	$m=2$	53	74.8	25
Proposed	$m=2, \beta=0.5$	40	81	5

Seed Data set (10% Labeled data, All missing data is from unlabeled data)

FCM	$m=2$	22	89.5	17
SSFCM	$m=2$	16	92.4	7
OCS_FCM	$m=2$	53	74.8	25
Proposed	$m=2, \beta=0.5$	42	80	7

we will incorporate other approaches regarding clustering of incomplete data into semi-supervised clustering. Also, the effect of varying degree of missing features in the data can further be analyzed.

References

1. W. Pedrycz, Algorithms of fuzzy clustering with partial supervision. Pattern Recognit. Lett. **3**(1), 13–20 (1985)
2. W. Pedrycz, J. Waletzky, Fuzzy clustering with partial supervision. IEEE Trans. Syst. Man Cybern. Part B (Cybernetics) **27**(5), 787–795 (1997)
3. W. Pedrycz, J. Waletzky, Neural-network front ends in unsupervised learning. IEEE Trans. Neural Netw. **8**(2), 390–401 (1997)
4. R.J. Hathaway, J.C. Bezdek, Fuzzy c-means clustering of incomplete data. IEEE Trans. Syst. Man Cybern. Part B (Cybernetics) **31**(5), 735–744 (2001)
5. H. Timm, C. Döring, R. Kruse, Differentiated treatment of missing values in fuzzy clustering, *International Fuzzy Systems Association World Congress* (Springer, Berlin, Heidelberg, 2003), pp. 354–361
6. H. Timm, C. Döring, R. Kruse, Different approaches to fuzzy clustering of incomplete datasets. Int. J. Approx. Reason. **35**(3), 239–249 (2004)
7. J. Li, S. Song, Y. Zhang, Z. Zhou, Robust k-median and k-means clustering algorithms for incomplete data. *Mathematical Problems in Engineering* (2016)
8. W. Pedrycz, *Knowledge-Based Clustering: from Data to Information Granules* (Wiley, 2005)
9. C. Stutz, T.A. Runkler, Classification and prediction of road traffic using application-specific fuzzy clustering. IEEE Trans. Fuzzy Syst. **10**(3), 297–308 (2002)
10. W. Pedrycz, G. Vukovich, Fuzzy clustering with supervision. Pattern Recogn. **37**(7), 1339–1349 (2004)
11. J.K. Dixon, Pattern recognition with partly missing data. IEEE Trans. Syst. Man Cybern. **9**(10), 617–621 (1979)
12. C.L. Blake, *UCI Repository of Machine Learning Databases* (University of California, Irvine, 1998). http://www.ics.uci.edu/mlearn/MLRepository.html

A Vision-Based System for Traffic Light Detection

Altaf Alam and Zainul Abdin Jaffery

Abstract Traffic detection and interpretation of its correct state is one of the most important information for developing an autonomous vehicle navigation system. Traffic light detection helps the autonomous vehicle to navigate safely in outdoor environment. In this paper, a vision-based algorithm is developed for traffic light detection and recognition. A monocular camera is used for capturing the surrounding outdoor environment. Intensity features are extracted from the templates of the traffic light and, the detection system is trained with these features. Similar features are searched in a predefine region of interest in the acquired image. Highly match candidates are considered as suitable traffic light candidate. The proposed algorithm is implemented in Labview NI-VISION system. Labview acquires data very effectively and effectively performs real-time processing. Using NI-VISION platform based system can go from design to test with minimum system changes. Adaptation of changing requirement is easy and less time consuming for Labview-based system. The algorithm is tested on different light condition images to check the reliability of system. Results show that developed algorithm is highly effective in real-time application for autonomous vehicle.

Keywords Traffic light signal detection · Template matching · Safe navigation
Autonomous vehicle · NI-Vision

1 Introduction

Many of researchers have been focusing on to develop the autonomous vehicle which can navigate safely in outdoor environment. Traffic light signals play very important role in traffic control system for urban environment. Traffic light detection and correct

A. Alam (✉) · Z. A. Jaffery
Department of Electrical Engineering, Jamia Millia Islamia, New Delhi, India
e-mail: Alam.altaf07@gmail.com

Z. A. Jaffery
e-mail: Zjaffery@jmi.ac.in

© Springer Nature Singapore Pte Ltd. 2019
H. Malik et al. (eds.), *Applications of Artificial Intelligence Techniques in Engineering*, Advances in Intelligent Systems and Computing 698,
https://doi.org/10.1007/978-981-13-1819-1_32

state interpretation help autonomous vehicle to follow their desired path. Many of researcher and scientist are working on vision-based autonomous vehicle detection system [1–14]. Mostly traffic light detection system has two stages of working, suitable candidate detection and recognizing their states. Suitable candidate region extraction method proposed by the system [1–7]. All proposed model used different types of color space for features extraction. Tae-Hyun et al. [1, 2, 6, 7] presented Hue saturation and intensity (HSI) color space based model while [3–5] proposed RGB based color space model to extract the region of interest. Threshold value would select manually in most of the color based segmentation methods hence, colo-based model has less flexibility [8]. A two-dimensional Gaussian model [9] detects the traffic light candidate and extracts the features. These epitome features utilized in recognition system for training purpose. Feature extraction and then trained the system based model take more time hence, these systems are less suitable in real-time processing. Hough transform, and radial transform based feature extraction proposed in [10, 11] for traffic light detection. Fixed camera and red light runner based intelligent transportation system in [12, 13] also proposed for detecting the traffic signal. Shape-based features are also common for detecting the traffic signal. Red light runner and shape features based models detect false candidate when similar shape and vehicle's rear light present in the acquired image. Most of the proposed methods do not perform appropriate detection when similar objects presents in acquired image and also do not work properly in real-time processing.

This research work proposed a template matching based traffic light signal detection and recognition. All the experimental work performed on LABVIEW system toolbox to achieve real-time processing. Data acquisition ability and real-time processing ability are very good for NI-LABVIEW. Traffic light candidate searched over a predefined high probable region besides of searching over whole pixel into acquiring image. Searching over a predefined region reduces the computational cost hence propose system performance is satisfactory in real-time processing. In addition, to interpret the traffic signal status case structure is evolved for every status.

Rest of the paper is arranged as follows. Section 2 described the architecture of the proposed system. Simulation and experimental results are discussed in Sect. 3 and finally, Sect. 4 concluded the work.

2   System Architecture

Proposed system layout is demonstrated by Fig. 1. It is quite clear from the figure that the detection process required three stages processing to complete the task. (1) Collects the templates of traffic light signals and extract the key point features. (2) Matches these extracted features with acquired image and find the similarity between image captured and template image. (3) In this stage, according to the matches result, a case structure is developed which interprets the traffic light signal states. A brief description of steps involves in light detection system is given below.

Traffic Light Signal and it's Status Detection

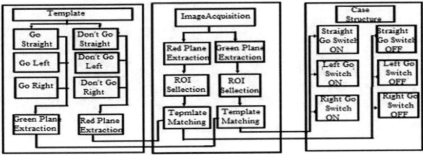

Fig. 1 Proposed system layout

Fig. 2 a–c Status of motion control signal

(a) Don't go and go straight

(b) Don't turn and turn left   (c) Don't turn and turn right

2.1 Templates

Templates are images, which have traffic signal texture used for extracting the features. Some traffic signal and their states related templates are pointed out in Fig. 2 and Eq. (1) shows the template image representation.

$$T(x, y) = g(x, y) \tag{1}$$

where T(x, y) = Template image

2.2 Image Acquisition

A digital camera, mounted on the top of the vehicle, is acting as scene information acquiring device. The acquired images are in RGB color form. Let the input image be F(x, y) which contain red, green, and blue channel as shown in Eq. (2).

$$F(x, y) = \begin{bmatrix} f_R(x, y) \\ f_G(x, y) \\ f_B(x, y) \end{bmatrix} \tag{2}$$

(a)   **(b)**

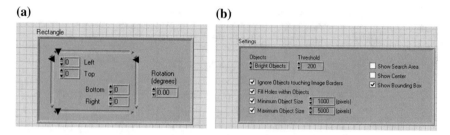

Fig. 3 **a, b** Region of interest and traffic signal parameter selection block

2.3 Intensity Image Plane Extraction

The acquired image is RGB image plane while features extracted from the template are intensity features hence; captured image is converted into intensity image. RGB image consists three individual intensity plane of red, green, and blue. Red and green plane intensity information is extracted for further processing. Equation (3) shows the intensity plane.

$$I(x, y) = f_{(R,G,B)}(x, y) \tag{3}$$

where I(x, y) is intensity plane and $f_{(R,G,B)}$ (x, y) is acquired red, green, and blue plane.

2.4 ROI Selection

Matching search in whole pixel is a time-consuming process; therefore, a high probable region is predefined for matching search. Mostly traffic light mounted at some height of road intersection. Hence, high probable region for traffic light occurrence in the image is upper part of the image. This part of image is the predefined region for matching search. Figure 3a, b shows Labview front panel window where ROI and another specification of traffic light can set.

2.5 Template Matching

In template matching, template slides over predefined region in acquire image and check the similarity. Intensity profile of entity seeks to match the image. Traffic light signal is differentiated from the other objects in acquired image. Besides the measurement of match, this work used an alternate measure of mismatch. The template

Table 1 The states of a traffic light

Movement decision/light status	Red	Green
Go straight	OFF	ON
Don't go straight	ON	OFF
Turn left	OFF	ON
Don't turn left	ON	OFF
Turn right	OFF	ON
Don't turn right	ON	OFF

plane g(x, y) slides over predefined region on f(x, y), and measures the mismatch. Equation (5) indicates the mismatching calculation, M.

$$M = \sum \sum (f(x, y) - g(x, y))^2 \tag{5}$$

where M is matching metric. Whenever f and g will be identical then M will be small; otherwise, it will be large.

2.6 Case Structure

Six type of traffic light status identified by this work such as, straight go or don't go, left turn or not turn and right turn or not turn. Digital switch used to display the status of light in Labview front panel. Table 1 shows the case structure of used status of traffic light and Fig. 4 Show the case structure display in Labview panel.

3 Simulation and Experimental Results

Figure 5 shows the simulation diagram of traffic light detection and recognition. Image grabber extracts the image frame from the acquired image while traffic light signal templates are read from the computer memory. Pattern learning block is connected with the template which extracts the information of templates, and pattern search block is connected with the acquired image which extracts the information from input images. Object size and pre-allocation region for matching search is defined to reduce the false detection. Traffic light signal has calibrated size hence too large and too small candidate avoided by the system.

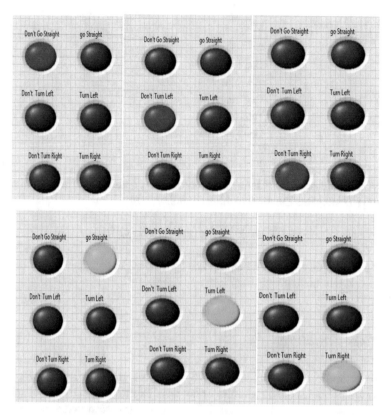

Fig. 4 Case structure of signal status

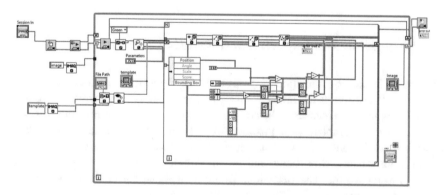

Fig. 5 Simulation block of proposed system

3.1   Performance Analysis

Figure 6 shows the different types of templates used by the system; while Fig. 7 shows the detected traffic signal in daytime and Fig. 8 shows detected result of night time. More than 100 images of different light conditions are taken to check the reliability and accuracy of system. System performance is analyzed over detection rate and accurate result.

Detection rate: Detection rate is a rate of change. It is basically ratio between true positive detection and total positive detection. Mathematically, it can be defined as

$$DR = \frac{TP}{TP + FP} \tag{6}$$

Detection accuracy: The accuracy of any system defines the degree of closeness of a measuring value and true value. Mathematically it can define as

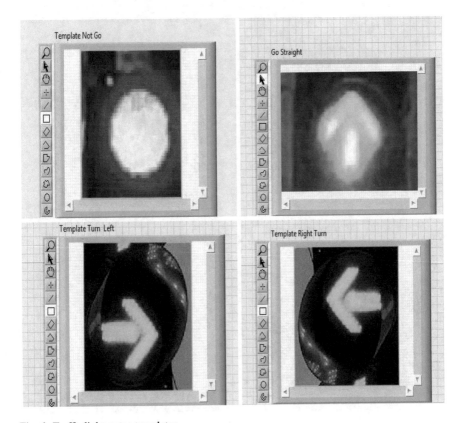

Fig. 6 Traffic light status templates

$$DA = \frac{TP}{TP + FN} \tag{7}$$

Fig. 7 Detected traffic light signals in daytime

where TP, FP, and TN is true positive, false positive, and false negative. DA is detection accuracy and DR is the detection rate.

Table 2 shows the performance analysis results. Detection rate and detection accuracy is 96% and 97% during daytime; while, it is 94% and 96% during night time. Result shows that accuracy and detection rate is better at daytime than night time. Vehicle light and other light cause fault detection during night time. Total

Fig. 8 Detected traffic light signals in night time

Table 2 Performance of traffic light detection system

Light condition	Performance measuring parameters	
	Detection rate (DR) (%)	Detection accuracy (DA) (%)
Daytime	96	97.00
Night time	94	96.00

Table 3 Processing time

Process	Processing time of existing method	Processing time of proposed method
Image acquisition	2.00	1.8
Image correction	7.84	5.5
Candidate detection	27.39	25.5
Stereo matching	1.39	NA
Tracking	3.37	3.00
Image O/P	13.52	10.2
Total	55.51	46.00

processing time of the proposed algorithm for one frame is shown in Table 3 which is about to 46.00 ms, while the processing time of the existing method [14] was 55.51 ms. Results show that the proposed method yielded less processing time and more accuracy than the existing method.

4 Conclusion

This paper proposed a real-time traffic light signal detection and recognition system. System implemented on the Labview software to achieve the real-time processing. Matching is searched over a predefined highly suitable candidate region. Therefore computational time reduced. Over and above system performance is good in real-time detection. Whenever vision-based algorithm developed usually they faced fault detection problem due to Illumination variation. So in this work, image of different light condition used to check the system reliability. A case structure is prepared for every status of detected traffic signal which interprets the signal status. Result shows that overall performance of the proposed system is good.

References

1. H. Tae-Hyun et al., Detection of traffic lights for vision-based car navigation system. Lect. Notes Comput. Sci. 682–691 (2006)
2. K.H. Lu et al., Traffic light recognition. J. China Inst. Eng. 1069–1075 (2008)
3. M. Omachi, S. Omachi, Traffic light detection with color and edge information, in *Proceeding of the IEEE International Conference, Computer Science Information Technology*, 2009, pp. 284–287
4. M. Omachi, S. Omachi, Detection of traffic light using structural information, in *Proceeding of the IEEE International Conference Signal Processing*, 2010, pp. 809–812
5. J. Gong, Y. Jiang, G. Xiong, C. Guan, G. Tao, H. Chen, The recognition and tracking of traffic lights based on color segmentation and cam shift for intelligent vehicles, in *Proceedings of the IEEE Intelligent Vehicle Symposium*, 2010, pp. 431–435
6. C. Wang, T. Jin, M. Yang, B. Wang, Robust and real-time traffic lights recognition in complex urban environments. Int. J. Comput. Intell. Syst. 1383–1390 (2011)
7. M. Diaz-Cabrera et al., Suspended traffic lights detection and distance estimation using color features, in *Proceedings of the IEEE Conference Intelligent Transport System*, 2012, pp. 1315–1320
8. Y. Shen, U. Ozguner, K. Redmill, J. Liu, A robust video based traffic light detection algorithm for intelligent vehicles, in *Proceedings of the IEEE Intelligent Vehicle Symposium*, 2009, pp. 521–526
9. Z. shi et al., Real time traffic light detection with adaptive background suppression filter, in *IEEE Transaction on Intelligent Transportation System*, 2015, pp. 690–700
10. G. Siogkas, E. Skodras, E. Dermatas, Traffic lights detection in adverse conditions using color, symmetry and spatio temporal information, in *Proceeding of the International Conference Computational Vision Theory*, 2012, pp. 620–627
11. Y.-C. Chung, J.-M. Wang, S.-W. Chen, A vision-based traffic light system at intersections. J. Taiwan Norm. Univ. Math. Sci. Technol. 67–86 (2002)
12. J. Levinson et al., Traffic light mapping, localization and state detection for autonomous vehicle, in *IEEE Conference on Robotics and Automation*, 2011, pp. 5784–5791
13. N.Y.H.S. Lai, A video-based system methodology for detecting red light runners, in *Proceedings of the IAPR Workshop on Machine Vision Applications*, 1998, pp. 23–26
14. H. Moizumi et al., Traffic light detection considering color saturation using in-vehicle stereo camera, J. Inf. Process. 349–357 (2016)

Analysis of the Performance of Learners for Change Prediction Using Imbalanced Data

Ankita Bansal, Kanika Modi and Roopal Jain

Abstract Software change prediction is important to economically schedule alloca-tion of resources during various phases of software maintenance and testing. Further-more, exact characterization of progress inclined and non-change inclined classes is significant in beginning times of programming advancement life cycle since that helps with creating financially savvy quality programming for real-time use. A good prediction model should predict both the change and non-change prone classes with high accuracy. However, most practical datasets have underrepresented information and serious class appropriation skews. Due to imbalanced data, the minority classes are not predicted accurately causing poor planning of resources. Popular operating systems like Android get updated very fast. In the current scenario, it is essential to recognize change prone and non-change prone classes with precision in newer versions of such software that are updated very frequently. In this paper, we give a complete survey of various machine learning models to predict change prone classes algorithms using sampling technologies like resampling and spreadsubsampling on six open source datasets having imbalanced data. The experimental result of the study advocates that resampling technique consistently and significantly improves the performance of all the models.

Keywords Software change prediction · Sampling · Change prone classes
Imbalanced learning · Object-oriented metrics · K-fold cross validation

1 Introduction

Changes in software have many causes. These include the addition of new features, increasing the scalability of the software, modification in the design, or technology being used or changes done for maintenance. Whenever revisions in the code are made, it is essential to check for bugs and integrate all the modules as a whole.

A. Bansal (✉) · K. Modi · R. Jain
Division of IT, NSIT, New Delhi, Delhi, India
e-mail: ankita.bansal06@gmail.com

© Springer Nature Singapore Pte Ltd. 2019
H. Malik et al. (eds.), *Applications of Artificial Intelligence Techniques
in Engineering*, Advances in Intelligent Systems and Computing 698,
https://doi.org/10.1007/978-981-13-1819-1_33

Operating systems and kernels are updated regularly. Every new update comes with enhanced core functionality with improvised features. Android is a well-known operating system that rolls out a new version almost every year with bug improvements and the addition of new features in already existing applications. If during the designing phase, these change prone classes are somehow recognized, it will simplify the testing and maintenance of entire software. With the help of internal attributes or metrics, prediction models can be constructed with the help of which change prone classes can be identified. An efficient prediction model is one which predicts both the change and non-change prone classes with high accuracy. In a practical scenario, the number of modules undergoing changes can be either very high or very less leading to an imbalanced data. Because of imbalanced nature of the data, it is observed that the minority classes are not predicted accurately. In other words, machine learning (ML) and statistical models trained on these imbalanced datasets yield poor efficiency, they are alone not enough. Such datasets require more sophisticated method to correctly identify change prone and non-prone classes with a high accuracy. In this paper, various techniques of sampling are employed to obtain a balanced dataset from imbalanced data. Afterwards, several ML techniques like Bagging, KStar, Random Tree, and Random Forest, etc. are applied and the performances are compared with the performance of the models on imbalanced data.

The following research questions (RQ) has been studied in this paper:

RQ1: After employing various sampling techniques to imbalanced data, do performance measures improve?
RQ2: Which sampling approach gives optimum results?

For evaluating above RQs, we have used two open source datasets (Android and JNode). Primarily, predictions are done on the original datasets which contain vast variations in the instances of change prone and non-change prone classes. The imbalanced learning problem (ILP) is then addressed by using two data sampling methods, namely, resampling and SpreadSubsample. The performances of all the ML techniques are evaluated and compared using area under the receiver operating characteristics (ROC) curve (AUC). The results of the study showed that among all classifiers, KStar and Tree Classifiers have outperformed. Henceforth, we propose the use of aforementioned techniques to predict change prone classes in related projects.

The organization of paper is as follows. Section 2 summarizes the related work. Section 3 gives the background of the research which focuses on the class imbalance problem and its solution. Section 4 depicts the dataset utilized as a part of this investigation. Various data analysis methods and performance measures are also discussed in this section. The outcomes of the study are summarized in Sect. 5. Finally, the work is finished up in Sect. 6.

2 Related Work

This section briefly states studies related to Imbalanced Learning Problem (ILP). ILP is well-acknowledged where substantial disparity is found between instances of individual classes. This paper is primarily centered on binary class imbalance problem, and it is observed that count of non-change prone classes takes up almost 80% of total data. This imbalance efficiently needs to be taken care for accurate results. Consequently, numerous studies have been done to confer aspects related to imbalanced data. These include studies on developing efficient models on imbalanced datasets by Weiss [1], He and Garcia [2], Zhang and Li [3] and Lopez et al. [4]. These researches have suggested use of sampling, cost sensitive, kernel-based and ML techniques. Bekkar [5], Jeni [6], Weng and Poon [7] suggested the use of appropriate evaluation metrics. Methods like Mahalanobis–Taguchi method [8, 9] are widely used for evaluating imbalanced data in the field of quality engineering.

ILP is found in numerous real-world applications as well. These include fraud text mining [10], sentiment analysis [11], bioinformatics [12], fraud detection [13], and several others. In the field of software engineering, change prediction is an active area of research. Some of the studies in this area have explored various sampling methods to obtain higher efficiency predictive models. These include studies by Liu [14], Kamai et al. [15], Shatnawi [16], Wang and Yao [17], Seiffert et al. [18].

Predicting change in software is essential for both maintenance of code and for future versions of the software. This field have software change prediction have been studied by Gemma et al. [19], Xin et al. [20], Elvira et al. [21], Lov [22], Di et al. [23] and Briand et al. [24].

3 Research Background

In this segment, the issue of imbalanced classes is discussed. A brief outline of the problem addressed and how we have approached has been shown in Fig. 1. Figure 1 shows the concept of imbalanced data and how it can be made balanced using various sampling techniques. We will talk about sampling techniques to overcome the problem of imbalanced classes in this section. After this, we also discuss the independent variables used in this study.

We have considered datasets which have a large variation in the instances of classes that are prone to change and non-change prone classes. Several ML models are trained and tested on the datasets to yield results. To make equal instances of both the classes, sampling techniques are applied. Same ML models are applied on the same dataset. It is found that the accuracy of the models has increased.

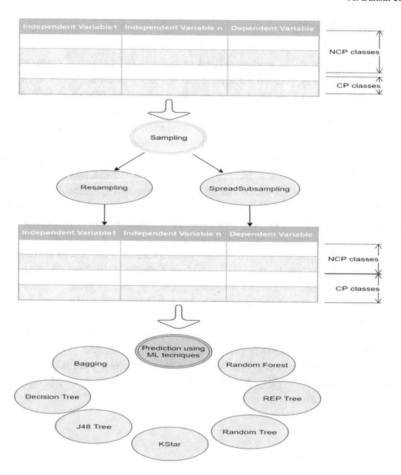

Fig. 1 Basic methodology followed

3.1 Class Imbalance Problem

Class imbalance is a problem in ML where the total number of a one type class data is very less or a lot more than the total number of another type class data. This paper focuses on the binary class imbalance problem. A binary class imbalance problem is where only two types of classes are present—change prone and non-change prone. The problem arises when the occurrences of one class far surpass the other. Because our training data is imbalanced, the accuracy of identifying change prone classes decreases substantially. So, basically while learning from imbalanced data, i.e., when instances of some classes (minority) are more compared to others (majority), we need to apply a technique that will give us higher accuracy for the minority class without impairing the majority class's accuracy. To deal with this problem, we have used sampling to make our dataset balanced. For instance, a basic resampling procedure copies positive instances in the training set to adjust the two classes.

3.2   Use of Sampling to Balance Dataset

We can modify the dataset that we use to build the predictive model by making the data more balanced to improve the performance of our model. This modification is known as sampling the dataset and there are two main techniques used:

1. Oversampling or sampling with replacement—Adds copies of samples from the underrepresented class to over-represented class
2. Under-sampling—Deletes instances from the overrepresented class

These methods adjust the class distribution of a dataset (i.e., the ratio between the different classes represented). Both the techniques involve setting a bias to select more instances from one class than from another based on the distribution skew.

Here is an outline of two useful approaches:

3.2.1   Resampling

Resampling is a method used to reconstruct the balanced sample datasets, including training sets and validation sets. In this method, we further divide the training set into its training and testing subsets and calculate the accuracy of each subset and continue this process for many subsets repeatedly. Finally, we choose those coefficients (model) which gives us maximum accuracy among these subsets and assume that this model will give maximum accuracy to final testing set. Unique sampling distribution is generated from actual data in resampling. There is no specific sample size requirement but better results are generated when the sample size is large. It improves the accuracy of the model.

3.2.2   SpreadSubSampling

SpreadSubSampling is a method that produces a random subsample from a given spread (maximum difference between the most common class and the most uncommon class) between class frequencies. It supports both types of classes—binary and normal but we are using binary classes. There is one precondition that the dataset must fit in memory completely. We can set various parameters like adjustWeights (used to adjust weight of an instance to maintain the total weight), distributionSpread (to set the maximum spread) etc.

3.3   Variables Used

Independent Variables

Metrics or independent variables are those parameters that provide information of
the lower level of granularity. There are many types of metrics available such as CK,
Lorenz and Kidd, MOOD, Li and Henry, QMOOD, etc. Among these, CK metric
suite [25] is the most popular. Thus, we have used CK metrics along with few other
size metrics as independent variables for our study. They are defined as below:

- *Depth of Inheritance Tree (DIT)*—Maximum count of edges between a given class
  and a parent class.
- *Response for a Class (RFC)*—Aggregate of local methods of a class and all the
  external methods directly called by any local method.
- *Coupling between Objects (CBO)*—Aggregate of external classes whose attributes
  and methods are used from the measured class.
- *Lack of Cohesion (LCOM)*—Average percentage of the methods in a class using
  that data field minus 100%.
- *Number of Private methods (NPROM)*—Total count of local protected methods in
  a class.
- *Number of Public Methods (NPM)*—Total count of public methods in a class.
- *Number of Private methods (NPRM)*—Total count of local private methods in a
  class
- *Lines of Code (LOC)*—Total count of lines of code in a class.
- *Number of Children (NOC)*—Total count of immediate subclasses of a class.
- *Weighted Methods Per Class (WMC)*—Total sum of complexities of all methods
  in a class.

Dependent Variables

This study uses a binary dependent variable that is used to represent the change
proneness of a class. This variable can assume two types of values—"Yes" or "No"
based on the change prone nature of a class. A class is said to be change prone if
it has changed in terms of LOC added, deleted, or modified in the recent version of
software as compared to the previous version of the same software.

4   Research Methodology

In this section, we discuss the datasets used in the study along with the data collection
process. Various analysis methods used to construct the models are also discussed.
The performance measure used to evaluate the performance of the models is also
explained.

4.1 Subject Selection

We have used two application packages of open source software developed in Java, namely Android and JNode. Android is a wide-known operating system created for touchscreen devices by Google.JNode (Java New Operating System Design Effort) is also an easy to use operating system, designed for personal use on modern devices. We have analyzed the last three latest versions of Android, i.e., Android 4.4 (KitKat), 5.0 (Lollipop) and 6.0 (Marshmallow). Furthermore, three versions of JNode are also analyzed, i.e., JNode 0.2.6, 0.2.7 and 0.2.8.

Data is collected between the successive versions of the software using CRG tool (Change Report Generator) developed by one of the authors [26]. We have considered change logs in GIT repository between two successive versions of the same software which are collected using CRG tool. Using this tool, we collect data between two versions of software between every pair of same classes.

Following steps are taken to calculate object-oriented metrics and change data between two consecutive versions of software:

- Java files are extracted from both the versions and CRG checks that they have "java" extension and common Java files between the two versions are found by sorting file names lexicographically.
- Common Java files are preprocessed by removing the comments from the code and code for common classes are extracted in separate files for each version and saved as Package name + class name.
- Common classes sorted by their name lexicographically are compared for both the versions.
- Change (total count of inserted, deleted and modified source code lines) between these common classes is calculated using the UNIX diff command. Total changes = added lines + deleted lines + (2 * modified lines)
- Changes between the two versions are reported and stored in two columns (class name and total count of changes between the two versions).
- Metrics computation and project creation are automated using Understand command line tool und.
- Change report, metrics data is merged for each class by combining object-oriented metrics data. A Perl script, developed by us for Understand Perl API, collects and merges the metrics data for classes in change report.

The details of software listing the number of total classes and the number of change prone classes are given in Table 1. We can notice the imbalanced nature of all the datasets where the number of change prone classes is very few as compared to the number of non-change prone classes.

Table 1 Software details

Software	No. of classes	No. of change prone classes	No. of non-change prone classes
Android 4.4	7653	2119	5534
Android 5.0	8262	2602	5660
JNode 0.2.6	14,935	2132	12,803
JNode 0.2.7	15,367	1546	13,281

4.2 Data Analysis Methods Used

The seven ML techniques used in the study are Bagging (BG), Decision Table (DT), J48 Decision Tree (J48), KStar, Random Tree (RT), Reduced Error Pruning Tree (REP) and Random Forest (RF). The default parameter settings of WEKA tool (http://www.cs.waikato.ac.nz/ml/weka/) are used for carrying out experiments.

Bagging (BG) or Bootstrap aggregating, is a technique which continually improves classification by combining random versions of training set [27]. The parameters used for running the experiments include bag size of 100, random seed 1 and 10 iterations. Bagging is very efficient in dealing with the problem of overfitting. The decision is made by the classification tree trained on bootstrap specimen of training dataset.

Decision Tables (DT) classifiers consist of a hierarchical table in which at every level, entry is broken down by attributes. The default settings have the nearest neighbor, forward search, and cross validation. The classifier works iteratively and considers that every possible combination of condition values is considered. Decision Tables are mainly used for deducing logical relationship between variables.

The J48 classifier is an open source Java version of C4.5 algorithm [28]. The parameters used for running the experiments include pruning confidence as 0.25 and number of folds as 3. J48 Decision tree are capable of handling non-relevant attributes by information gain. Another advantage of using J48 classifier is to its ability of handling skewed distribution. The imbalanced data problem is concerned with skewed distribution, so decision tree classifiers are a good choice.

KStar is a lazy learning ML technique, most useful for large datasets having fewer features. This algorithm has lesser training time but slower prediction time and has more memory requirements. It provides an efficient approach to handle missing values, symbolic attributes, and real-valued attributes.

Random Trees (RT) takes input feature vectors, classifies with each tree in the forest and outputs the class label which gets majority. The classifier response is mean of responses of all the trees in case of regression.

Reduced Error Pruning (REP) Tree is another decision tree which learns very fast. Using information variance/gain, a regression/decision tree is built, which is pruned using back fitting, also known as reduced error pruning. This classifier accurately deals with missing values by splitting instances.

Random Forest (RF) is an ensemble classifier used for classification and prediction that outputs one class which is the mode (classification) or mean (regression) of classes. This property of Random Forest overcomes the problem of overfitting in the training set. The parameters used for running the experiments include bag size of 100 and 100 iterations. Random Forest has lesser variance as compared to decision trees and overcomes several shortcomings of decision trees. In majority of the cases, it gives more accuracy than decision trees.

4.3 Performance Measures Used

Given the imbalanced nature of the datasets, the selection of suitable performance measures is very important. The measures like precision, recall and accuracy have been reported as the poor indicators of performance by the researchers when the models are evaluated on imbalanced data [2, 29, 30]. Instead, the studies [2, 16, 31] and have suggested the use of area under the ROC curve (AUC) to evaluate and compare the models developed using imbalanced data. Thus, we have used AUC obtained from ROC analysis in this study.

ROC curve is a plot of true positive rate (taken on the y-axis) and false positive rate (taken on the x-axis) for different cutoff points between 0 and 1. AUC is the ability of the test to correctly identify change prone and non-change prone classes. Area of 1 represents highest (perfect) accuracy.

5 Result Analysis

For prediction of change prone and non-change prone classes, several models are constructed. The performance is evaluated before applying the sampling techniques and after the sampling techniques have been applied. The balanced datasets obtained after applying sampling techniques is shown in Table 2. We can observe from the table that same number of Change Prone (CP) and Non-Change Prone (NCP) classes is obtained after each of the sampling technique.

Table 2 Details of software after sampling

Sampling	After resampling technique		After spreadsubsampling technique	
Software	CP classes	NCP classes	CP classes	NCP classes
Android 4.4	3826	3826	2119	2119
Android 5.0	4131	4131	2602	2602
JNode 0.2.6	7467	7467	2132	2132
JNode 0.2.7	7683	7683	1546	1546

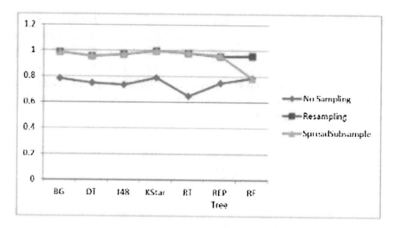

Fig. 2 Comparison of AUC of the models for Android 4.4–Android 5.0

Table 3 Validation results of models for Android 4.4–Android 5.0

ML models	No sampling	Resampling	SpreadSubsample
BG	0.782	0.986	0.986
DT	0.748	0.954	0.955
J48	0.734	0.968	0.971
KStar	0.789	0.994	0.994
RT	0.648	0.978	0.980
REP Tree	0.748	0.954	0.959
RF	0.788	0.957	0.785

Correlation-Based Feature Selection (CFS) technique is applied before developing an ensemble model to obtain a subset of significant independent variables which are major contributors in predicting the dependent variable [32]. The validation method used is 10-cross validation as the results would be highly optimistic if training and testing are done on the same datasets.

Tables 3, 4, 5 and 6 show the results of 10-cross validation of all the models before and after sampling. It can be seen from Table 3 that all the models of Android 4.4 have shown significant improvement after resampling technique is used. When spreadsubsample is used, except RF, all the models have again shown superior performance as compared to the performance of the models when no sampling is used. This can be clearly seen from the Fig. 3 where all the models without sampling have shown lower performance than the models after sampling. Best results are observed when the data is resampled and KStar technique is applied on it.

Similar observations can be inferred from Tables 4, 5, and 6. The results show that the AUC has improved after resampling and spreadsubsampling compared to unsampled data. KStar has shown the best performance among the models of Android 5.0 (Table 4), whereas RF has outperformed all other models of JNode 0.2.6 with highest

Table 4 Validation results of models for Android 5.0–Android 6.0

ML models	No sampling	Resampling	SpreadSubsample
BG	0.771	0.905	0.910
DT	0.759	0.786	0.787
J48	0.750	0.857	0.859
KStar	0.778	0.947	0.949
RT	0.629	0.631	0.880
REP Tree	0.748	0.838	0.844
RF	0.763	0.938	0.763

Table 5 Validation results of models for JNode 0.2.6–JNode 0.2.7

ML models	No sampling	Resampling	SpreadSubsample
BG	0.811	0.966	0.795
DT	0.711	0.848	0.714
J48	0.709	0.927	0.718
KStar	0.832	0.989	0.831
RT	0.652	0.935	0.674
REP Tree	0.738	0.902	0.748
RF	0.851	0.994	0.822

value of AUC (Table 5). KStar follows RF with second highest AUC. The results from Table 6 show that both RF and KStar have comparable AUCs with value of 0.997 and 0.0996 after resampling. In addition to this, the models of Tables 5 and 6 show that the models have not shown significant improvement after spreadsubsampling is used as compared to the models without sampling. This can also be graphically analyzed from Figs. 2, 3, 4 and 5.

General Observations:

We have used one meta-algorithm; one lazy learning and rest are decision trees. From the above tables, it can be concluded AUC has increased tremendously after sampling especially resampling compared to spreadsubsampling. Also, KStar, Bagging and

Table 6 Validation results of models for JNode 0.2.7–JNode 0.2.8

ML models	No sampling	Resampling	SpreadSubsample
BG	0.795	0.977	0.794
DT	0.723	0.865	0.727
J48	0.703	0.942	0.708
KStar	0.946	0.996	0.945
RT	0.841	0.944	0.822
REP Tree	0.810	0.850	0.808
RF	0.955	0.997	0.822

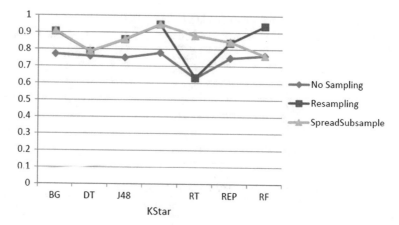

Fig. 3 Comparison of AUC of the models for Android 5.0–Android 6.0

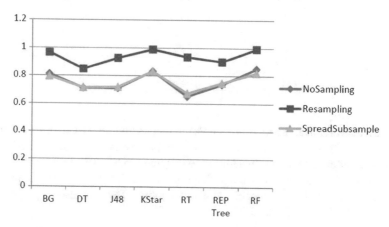

Fig. 4 Comparison of AUC for the models of JNode 0.2.6–Jnode 0.2.7

Random Forest have the best performance among all the ML models. These models should be used for research and practice purposes.

After analyzing results, we are now able to address the research questions previously mentioned. **It is found that area under ROC curve has shown vast increase after sampling. Hence, it can be inferred that sampling improves performance of ML models. Among the applied techniques, resampling performs best with tuning of parameter BiasToUniform as 1.0.**

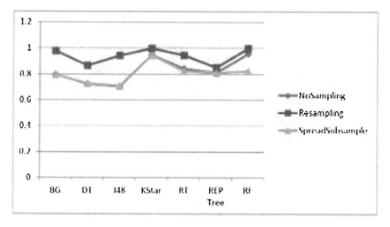

Fig. 5 Comparison of AUC for the models of JNode 0.2.7–Jnode 0.2.8

6 Conclusion

Prediction of change prone classes in early phases of life cycle of software development is very crucial because it not only saves time and money but also helps in
maintaining quality code. Apart from predicting change prone classes, this study has
also efficiently handled the issue of imbalanced datasets. The study has employed
the use of two sampling techniques, resampling, and spreadsubsampling. The change
prediction models were constructed using seven machine learning techniques. The
performances of all the models were evaluated using area under the receiver operating
characteristics curve. Subsequently, the performance was compared before sampling
and after the sampling technique was applied. The empirical validation is carried on
two open source datasets, both of which are operating systems in Java, Android and
JNode. We analyzed two versions of each operating system, Android 4.4, 5.0 and
JNode 0.2.7 and 0.2.8. Experimental results demonstrate that the use of resampling
technique consistently and significantly improves the performance of all the models
when sampling technique is not used. After applying various machine learning techniques to sampled and unsampled data, it was found that KStarand Random Forest
has the best performance among all the machine learning models. These models can
be used by the researchers and practitioners for similar type of problem statements
and for predicting change prone to change of similar nature datasets.

References

1. G.M. Weiss, Mining with rarity: a unifying framework. ACM SIGKDD Explor. Newslett. **6**(1),
   1–7 (2014)

2. H. He, Garcia EA learning from imbalanced data. IEEE Trans. Knowl. Data Eng. **21**(9), 1263–1284 (2009)
3. X. Zhang, Y. Li, An empirical study of learning from imbalanced data, in *Proceedings of the 22nd Australasian Database Conference* (2011), pp. 85–94
4. V. Lopez, A. Fernandez, S. Garcia, V. Palade, F. Herrera, An insight into classification with imbalanced data: empirical results and current trends on using data intrinsic characteristics. Inf. Sci. **250**(20), 113–141 (2013)
5. M. Bekkar, H.K. Djemaa, T.A. Alitouche, Evaluation measures for models assessment over imbalanced data sets. J. Inf. Eng. Appl. **3**(10), 27–38 (2013)
6. L. Jeni, J.F. Cohn, F. De La Torre, Facing imbalanced data—recommendations for the use of performance metrics, in *Proceedings of the Humane Association Conference on Affective Computing and Intelligent Interaction* (2013), pp. 245–251
7. C.G. Weng, J. Poon, A new evaluation measure for imbalanced datasets, in *Proceedings of the 7th Australian Data Mining Conference* (2008), pp. 27–32
8. A. Hirohisa, N. Mochiduki, H.A. Yamada, Model for detecting cost-prone classes based on Mahalanobis-Taguchi method. IEICE Trans. Inf. Syst. **89**(4), 1347–1358 (2006)
9. C.T. Su, Y.H. Hsiao, An evaluation of the robustness of MTS for imbalanced data. IEEE Trans. Knowl. Data Eng. **19**(10), 1321–1332 (2007)
10. T. Munkhdalai, O.E. Namsrai, K.H. Ryu, Self-training in significance space of support vectors for imbalanced biomedical event data. BMC Bioinf. **16**(7), 1–2 (2015)
11. R. Xu, T. Chen, Y. Xia, Q. Lu, B. Liu, X. Wang, Word embedding composition for data imbalances in sentiment and emotion classification. Cogn. Comput. **7**(2), 226–240 (2015)
12. P. Yang, P.D. Yoo, J. Fernando, B.B. Zhou, Z. Zhang, A.Y. Zomaya, Sample subset optimization techniques for imbalanced and ensemble learning problems in bioinformatics applications. IEEE Trans. Cybern. **44**(3), 445–455 (2014)
13. C. Phua, D. Alahakoon, V. Lee, Minority report in fraud detection: classification of skewed data. SIGKDD Explor. **6**(1), 50–59 (2004)
14. Y. Liu, A. An, X. Huang, Boosting prediction accuracy on imbalanced datasets with SVM ensembles, in *Advances in Knowledge Discovery and Data Mining* (2006), pp. 107–118
15. Y. Kamei, A. Monden, S. Matsumoto, T. Kakimoto, K. Matsumoto, The effects of over and under sampling on fault-prone module detection, in *Proceedings of the 1st International Symposium on Empirical Software Engineering and Measurement* (2007), pp. 196–204
16. R. Shatnawi, Improving software fault-prediction for imbalanced data, in *Proceedings of the International Conference on Innovations in Information Technology* (2012), pp. 54–59
17. S. Wang, X. Yao, Using class imbalance learning for software defect prediction. IEEE Trans. Reliab. **62**(2), 434–443 (2013)
18. C. Seiffert, T.M. Khoshgoftaar, J.V. Hulse, A. Folleco, An empirical study of the classification performance of learners on imbalanced and noisy software quality data. Inf. Sci. **259**(1), 571–595 (2014)
19. G.Catolino, F. Palomba, A.D. Lucia, F. Ferrucci, A. Zaidman, Developer-related factors in change prediction: an empirical assessment, in *Proceedings of the 25th International Conference on Program Comprehension*, Argentina (2017)
20. X. Xia, D. Lo, S. McIntosh, E. Shihab, A.E. Hassan, Cross-project build co-change prediction, in *SANER* (2015), pp. 311–320
21. E.M. Arvanitou, A. Ampatzoglou, A. Chatziogeorgiou, P. Avgeriou, *A Method for Assessing Class Change Proneness, Evaluation and Assessment in Software Engineering* (ACM, Sweden, 2017)
22. L. Kumar, Transfer learning for cross-project change-proneness prediction in object-oriented software systems: a feasibility analysis. ACM SIGSOFT Softw. Eng. Notes **42**(1), 1–11 (2017)
23. L. Briand, J. Daly, V. Porter, J. Wüst, Predicting fault-prone classes with design measures in object-oriented systems, *in Proceedings of the Ninth International Symposium Software Reliability Engineering* (ISSRE 1998)
24. D. Di Nucci, F. Palomba, G. De Rosa, G. Bavota, R. Oliveto, A. De Lucia, A developer centered bug prediction model, IEEE Trans. Softw. Eng. (2017)

25. S.R. Chidamber, C.F. Kemerer, A metrics suite for object-oriented design. IEEE Trans. Softw. Eng. **20**(6), 476–493 (1994)
26. R. Malhotra, A. Bansal, S. Jajoria, An automated tool for generating change report from open-source software, in *International Conference IEEE Advances in Computing, Communications and Informatics (ICACCI)* (2016), pp. 1576–1582
27. L. Breiman, Bagging predictors. Mach. Learn. **24**(2), 123–140 (1996)
28. J.R. Qinlan, *C4.5: Programs for Machine Learning* (Morgan Kaufmann Publishers, San Mateo, CA, 1993)
29. K. Gao, T.M. Khoshgoftaa, A. Napolitano, Combining feature subset selection and data sampling for coping with highly imbalanced software data, in *Proceedings of the 27th International Conference on Software Engineering and Knowledge Engineering*, Pittsburgh (2015)
30. T. Menzies, A. Dekhtyar, J. Distefance, J. Greenwald, Problems with precision: a response to comments on 'data mining static code attributes to learn defect predictors'. IEEE Trans. Softw. Eng. **33**(9), 637–640 (2007)
31. S. Lessmann, B. Baesans, C. Mues, S. Pietsch, Benchmarking classification models for software defect prediction: a proposed framework and novel finding. IEEE Trans. Softw. Eng. **34**(4), 485–496 (2008)
32. M.A. Hall, Correlation-based feature selection for discrete and numeric class machine learning, in *Proceedings of the Seventeenth International Conference on Machine Learning* (2008), pp. 359–366

Characterizations of Right Regular Ordered LA-Semihypergroups via Soft Hyperideals

M. Y. Abbasi, Sabahat Ali Khan and Aakif Fairooze Talee

Abstract In this paper, we introduce soft hyperideals in ordered LA-semihypergroups through new approach and investigate some useful results. Further, we define right regular ordered LA-semihypergroup and characterize right regular ordered LA-semihypergroups using soft hyperideals.

Keywords Ordered LA-semihypergroups · Soft intersection hyperideals · Right regular ordered LA-semihypergroups

1 Introduction and Preliminaries

Marty [12] introduced the notion of algebraic hyperstructures as natural generalization of classical algebraic structures. In algebraic structures, binary operation on a set A is a map from A × A to A but in an algebraic hyperstructure binary hyperoperation on a set A is a map from A × A to the power set of A excluding empty set. Hasankhani [7] defined ideals in right (left) semihypergroups and discussed some hyper versions of Green's relations. The concept of ordering hypergroups investigated by Chvalina [4] as a special class of hypergroups. Heidari and Davvaz [8] studied a semihypergroup (S, o) besides a binary relation '≤', where '≤' is a partial order relation such that satisfies the monotone condition.

The concept of LA-semigroup was given by Kazim and Naseeruddin [10]. Mushtaq and Yusuf [14] studied some properties of LAsemigroups. Khan et al. gave some characterizations of right regular LA-semigroups in terms of fuzzy ideals. Hila and Dine [9] defined LA-semihypergroups and studied several properties of

M. Y. Abbasi · S. A. Khan (✉) · A. F. Talee
Department of Mathematics, Jamia Millia Islamia, New Delhi 110025, Delhi, India
e-mail: khansabahat361@gmail.com

M. Y. Abbasi
e-mail: yahya_alig@yahoo.co.in

A. F. Talee
e-mail: fuzzyaakif786.jmi@gmail.com

© Springer Nature Singapore Pte Ltd. 2019
H. Malik et al. (eds.), *Applications of Artificial Intelligence Techniques in Engineering*, Advances in Intelligent Systems and Computing 698,
https://doi.org/10.1007/978-981-13-1819-1_34

361

hyperideals in LA-semihypergroups. Yaqoob et al. [16] gave some characterizations of LA-semihypergroups using left and right hyperideals. Yaqoob and Gulistan [17] studied the concept of ordered LA-semihypergroup and defined a binary relation on LA-semihypergroup such that to become partially ordered LA-semihypergroups.

Molodtsov [13] introduced a mathematical tool for dealing with hesitant, fuzzy, unpredictable and unsure articles known as soft set. Further, Maji et al. [11] defined many applications in soft sets. Cagman and Aktas [1] introduced soft group theory and correlate soft sets with rough sets and fuzzy sets and gave [3] a new approach to soft group called soft intersection group. Sezgin [15] studied soft set theory in LA-semigroup with the concept of soft intersection LA-semigroups and soft intersection LA-ideals.

We first recall the basic terms and definitions from the LA-semihypergroup theory and soft set theory.

Definition 1 [5, 6] Let $S(\neq \emptyset)$ be a set and let $\wp^*(S)$ be the set of all non-empty subsets of S. A hyperoperation on S is a map $o: S \times S \to \wp^*(S)$ and (S, o) be called a hypergroupoid.

Definition 2 [5, 6] A hypergroupoid (S, o) is called a semihypergroup if $\forall x, y, z \in S$, we have $(x \circ y) \circ z = x \circ (y \circ z)$, which means that

$$\bigcup_{u \in x o y} u \circ z = \bigcup_{v \in y o z} x \circ v.$$

If $x \in S$ and $A(\neq \emptyset)$, $B(\neq \emptyset)$ are subsets of S, then we denote

$$A \circ B = \bigcup_{a \in A, b \in B} a \circ b, x \circ A = \{x\} \circ A \text{ and } A \circ x = A \circ \{x\}.$$

Definition 3 [9] Let $S(\neq \emptyset)$ be set. A hypergroupoid S is called an LA-semihypergroup if for every $x, y, z \in S$, we have

$$(x \circ y) \circ z = (z \circ y) \circ x.$$

This law is called the left invertive law and it also satisfies the following law:

$$(x \circ y) \circ (z \circ w) = (x \circ z) \circ (y \circ w),$$

for all $x, y, z, w \in S$. This law is called medial law.

Definition 4 [16] Let S be an LA-semihypergroup, then an element $e \in S$ is called left identity (briefly, l.i.) (resp., pure left identity) if $\forall a \in S$, $a \in e \circ a$ (resp., $a = e \circ a$).

An LA-semihypergroup (S, o) with l.i. 'e' satisfies the following laws, $\forall x, y, z, w \in S$.

$$(x \circ y) \circ (z \circ w) = (w \circ z) \circ (y \circ x),$$

called a paramedial law, and

$$x \circ (y \circ z) = y \circ (x \circ z).$$

Definition 5 [17] An ordered LA-semihypergroup $(\mathbf{S}, \circ, \leq)$ is a poset $(\mathbf{S}, \leq)$ at the same time an LA-semihypergroup $(\mathbf{S}, \circ)$ such that: for any $a, b, x \in \mathbf{S}, a \leq b$ implies $x \circ a \leq x \circ b$ and $a \circ x \leq b \circ x$.

If A and B are non-empty subsets of $\mathbf{S}$, then we say that $A \leq B$ if for every $a \in A$ there exists $b \in B$ such that $a \leq b$.

Definition 6 [17] Let $(\mathbf{S}, \circ, \leq)$ be an ordered LA-semihypergroup. For non-empty subset H of $\mathbf{S}$, we define

$$(\mathsf{H}] := \{t \in \mathbf{S} | t \leq h \text{ for some } h \in \mathsf{H}\}.$$

For $\mathsf{H} = \{h\}$, we write $(h]$ instead of $(\{h\}]$.

Lemma 1 [17] *Let* $(\mathbf{S}, \circ, \leq)$ *be an ordered LA-semihypergroup. Then the following assertions hold:*

1. $A \subseteq (A]$ *and* $((A]] = (A], \forall A \subseteq \mathbf{S}$.
2. *If* $A \subseteq B \subseteq \mathbf{S}$, *then* $(A] \subseteq (B]$.
3. $(A] \circ (B] \subseteq (A \circ B]$ *and* $((A] \circ (B]] = (A \circ B], \forall A, B \subseteq \mathbf{S}$.
4. *For any two non-empty subsets* A, B *of* $\mathbf{S}$ *such that* $A \leq B$, *we have* $C \circ A \leq C \circ B$ *and* $A \circ C \leq B \circ C$ *for non-empty subset* C *of* $\mathbf{S}$.

Definition 7 [17] A subset $A (\neq \emptyset)$ of an ordered LA-semihypergroup $\mathbf{S}$ is called a sub-LA-semihypergroup of $\mathbf{S}$ if $(A \circ A] \subseteq A$

Definition 8 [17] A subset $|(\neq \emptyset)$ of an ordered LA-semihypergroup $\mathbf{S}$ is called a left (resp., right) hyperideal of $\mathbf{S}$ if:

1. $\mathbf{S} \circ | \subseteq |$ ($| \circ \mathbf{S} \subseteq |$, respectively) and
2. If $i \in |$ and $s \leq i$, then $s \in |$ for every $s \in \mathbf{S}$.

$|$ is called a hyperideal of $\mathbf{S}$ if it is a left and a right hyperideal.

Definition 9 [17] A sub-LA-semihypergroup B of an ordered LA-semihypergroup $(\mathbf{S}, \circ, \leq)$ is called a bi-hyperideal of $\mathbf{S}$ if:

1. $(\mathsf{B} \circ \mathbf{S}) \circ \mathsf{B} \subseteq \mathsf{B}$;
2. If $b \in \mathsf{B}$ and $s \leq b$, then $s \in \mathsf{B}$ for every $s \in \mathbf{S}$.

Lemma 2 [17] *Let* $(\mathbf{S}, \circ, \leq)$ *be an ordered LA-semihypergroup with l.i. such that* $\mathbf{S} = \mathbf{S} \circ \mathbf{S}$, *then* $(\mathbf{S} \circ s]$ *is a left hyperideal of* $\mathbf{S}$, *for all* $s \in \mathbf{S}$.

Cagman and Enginoglu [2] gave the following concept of soft sets. Throughout this paper we represent:

S: an ordered LA-semihypergroup; V: *an initial universe; E: a set of parameters*
F(S): *set of all soft sets of* S *over* V; *P*(V): *the powerset of* V.

A soft set F_A *over* V *is a set defined by* $F_A : E \rightarrow P(V)$ *such that* $F_A(x) = \emptyset$ *if*
$x \notin A$.

Here F_A *is also called an approximate function. A soft set over* V *can be repre-*
sented by the set of ordered pairs

$$F_A = \{(x, F_A(x)) : x \in E, F_A(x) \in P(V)\}.$$

Clearly, a soft set is a parameterized family of subsets of the set V.

Definition 11 Let $F_A, F_B \in F(S)$. Then, F_A is called a soft subset of F_B and denoted
by $F_A \sqsubseteq F_B$, if $F_A(x) \subseteq F_B(x)$ for all $x \in E$.

Definition 12 Let $F_A, F_B \in F(S)$. Then, the union of F_A and F_B denoted by $F_A \widetilde{\cup} F_B$,
is defined as $F_A \widetilde{\cup} F_B = F_{A\widetilde{\cup}B}$, where $F_{A\widetilde{\cup}B}(x) = F_A(x) \cup F_B(x)$ for all $x \in E$ and
the intersection of F_A and F_B denoted by $F_A \widetilde{\cap} F_B$, is defined as $F_A \widetilde{\cap} F_B = F_{A\widetilde{\cap}B}$,
where $F_{A\widetilde{\cap}B}(x) = F_A(x) \cap F_B(x)$ for all $x \in E$.

2 Soft Characteristic Function, Soft Intersection (S.I.) Product and Soft Intersection (S.I.) Hyperideals

In this section, we define soft characteristic function, soft intersection (S.I.) product,
and soft intersection (S.I.) hyperideals. Further, we study S.I. hyperideals with S.I.
product and some interesting results.

Definition 13 Let Y be a subset of an ordered LA-semihypergroup $(S, \circ, \leq)$. We
denote the soft characteristic function of Y by S_Y and is defined as:

$$S_Y(y) = \begin{cases} V, & \text{if } y \in Y \\ \phi, & \text{if } y \notin Y \end{cases}$$

In this paper, we denote an ordered LA-semihypergroup $(S, \circ, \leq)$ as a set of
parameters.

Let $(S, \circ, \leq)$ be an ordered LA-semihypergroup.

For $x \in S$, we define $S_x = \{(y, z) \in S \times S | x \leq y \circ z\}$.

Definition 14 Let F_S and G_S be two soft sets of an ordered LA-semihypergroup
$(S, \circ, \leq)$ over V. Then the soft product $F_S \overset{\wedge}{\diamond} G_S$ is a soft set of S over V, defined by

$$\left(F_S \overset{\wedge}{\diamond} G_S\right)(x) = \begin{cases} \bigcup_{(y,z) \in S_x} \{F_S(y) \cap G_S(z)\} & \text{if } S_x \neq \phi \\ \phi & \text{if } S_x = \phi \end{cases} \quad \forall x \in S$$

Theorem 1 *Let X and Y be non-empty subsets of an ordered LA-semihypergroup $(\mathbf{S}, \circ, \leq)$. Then*

(1) If $X \subseteq Y$, then $\mathsf{S}_X \sqsubseteq \mathsf{S}_Y$.
(2) $\mathsf{S}_X \tilde{\cap} \mathsf{S}_Y = \mathsf{S}_{X \cap Y}$, $\mathsf{S}_X \tilde{\cup} \mathsf{S}_Y = \mathsf{S}_{X \cup Y}$.
(3) $\mathsf{S}_X \overset{\wedge}{\diamond} \mathsf{S}_Y = \mathsf{S}_{(X \circ Y]}$.

Proof (1) and (2) are trivial.

(3) Let $s \in S$ such that $s \in (X \circ Y]$, then $\exists x \in X$ and $y \in Y$ such that $s \leq x \circ y$. Thus, $(x, y) \in S_s$. So, S_s is non-empty. Then, we have

$$(\mathsf{S}_X \overset{\wedge}{\diamond} \mathsf{S}_Y)(s) = \bigcup_{(p,q) \in S_s} \{\mathsf{S}_X(p) \cap \mathsf{S}_Y(q)\} \supseteq \mathsf{S}_X(x) \cap \mathsf{S}_Y(y) = \mathsf{V}$$

It implies $\mathsf{S}_X \overset{\wedge}{\diamond} \mathsf{S}_Y = \mathsf{V}$. Hence, $\mathsf{S}_X \overset{\wedge}{\diamond} \mathsf{S}_Y = \mathsf{S}_{(X \circ Y]}$.

Definition 15 A non-null soft set $\mathsf{F_S}$ is said to be an S.I. sub-LA-semihypergroup of an ordered LA-semihypergroup $(\mathbf{S}, \circ, \leq)$ over V if

1. $\bigcap_{\vartheta \in x \circ y} \mathsf{F_S}(\vartheta) \supseteq \mathsf{F_S}(x) \cap \mathsf{F_S}(y) \, \forall x, y \in \mathbf{S}$ and
2. For any $x, y \in \mathbf{S}$ such that $x \leq y$ implies $\mathsf{F_S}(x) \supseteq \mathsf{F_S}(y)$.

Definition 16 A non-null soft set $\mathsf{F_S}$ is said to be an S.I. left (resp., right) hyperideal of an ordered LA-semihypergroup $(\mathbf{S}, \circ, \leq)$ over V if

1. $\bigcap_{\vartheta \in x \circ y} \mathsf{F_S}(\vartheta) \supseteq \mathsf{F_S}(y)$ (resp. $\bigcap_{\vartheta \in x \circ y} \mathsf{F_S}(\vartheta) \supseteq \mathsf{F_S}(x)$) $\forall x, y \in \mathbf{S}$ and
2. For any $x, y \in \mathbf{S}$ such that $x \leq y$ implies $\mathsf{F_S}(x) \supseteq \mathsf{F_S}(y)$.

Definition 17 A non-null soft set $\mathsf{F_S}$ is said to be an S.I. hyperideal of an ordered LA-semihypergroup $(\mathbf{S}, \circ, \leq)$ over V if $\mathsf{F_S}$ is an S.I. left as well as an S.I. right hyperideal of $\mathbf{S}$ over V.

Example 1 Let $(\mathbf{S}, \circ, \leq)$ be an ordered LA-semihypergroup, where $\mathbf{S} = \{1, 2, 3, 4\}$ with a hyperoperation '$\circ$' is given by the following table:

$\circ$	1	2	3
1	$\{1, 3\}$	3	$\{2, 3\}$
2	$\{2, 3\}$	3	3
3	$\{2, 3\}$	$\{2, 3\}$	$\{2, 3\}$

Order relation is defined by: $\leq = \{(1, 1), (2, 2), (3, 3)\}$.
Let $\mathsf{V} = \{x, y, z\}$. Define a soft set $\mathsf{F_S} : \mathbf{S} \to P(\mathsf{V})$ by

$$\mathsf{F_S}(1) = \{x\}, \ \mathsf{F_S}(2) = \{x, y, z\} \text{ and } \mathsf{F_S}(3) = \{x, y, z\}.$$

Then we can verify that $\bigcap_{\vartheta \in aob} F_S(\vartheta) \supseteq F_S(b)$ and $\bigcap_{\vartheta \in aob} F_S(\vartheta) \supseteq F_S(a) \forall a,$ $b \in S$. Therefore, F_S is an S.I. hyperideal of S over V.

Definition 18 An S.I. sub-LA-semihypergroup F_S is said to be an S.I. bi-hyperideal of an ordered LA-semihypergroup $(S, \circ, \leq)$ over V if

1. $\bigcap_{\vartheta \in (xoy)oz} F_S(\vartheta) \supseteq F_S(x) \cap F_S(z) \forall x, y, z \in S$ and
2. For any $x, y \in S$ such that $x \leq y$ implies $F_S(x) \supseteq F_S(y)$.

Theorem 2 *A non-null soft set F_S is an S.I. left hyperideal of an ordered LA-semihypergroup $(S, \circ, \leq)$ over V if and only if*

$$S_S \overset{\wedge}{\diamond} F_S \sqsubseteq F_S$$

Proof Suppose F_S is an S.I. left hyperideal of an ordered LA-semihypergroup $(S, \circ, \leq)$ over V. Then $\bigcap_{\vartheta \in xoy} F_S(\vartheta) \supseteq F_S(y) \forall x, y \in S$. Now, if $S_x = \emptyset$. Then $(S_S \overset{\wedge}{\diamond} F_S)(x) = \emptyset$. In this case, $(S_S \overset{\wedge}{\diamond} F_S)(x) \subseteq F_S(x)$, therefore $S_S \overset{\wedge}{\diamond} F_S \sqsubseteq F_S$. If $S_x \neq \emptyset$. Then $\exists u, v \in S$ such that $x \leq u \circ v$. Hence, we have

$$(S_S \overset{\wedge}{\diamond} F_S)(x) = \bigcup_{(u,v) \in S_x} \left\{ S_S(u) \bigcap F_S(v) \right\} = \bigcup_{(u,v) \in S_x} \left\{ V \bigcap F_S(v) \right\}$$

$$= \bigcup_{(u,v) \in S_x} \{F_S(v)\} \subseteq \bigcup_{(u,v) \in S_x} \left\{ \bigcap_{\vartheta \in pov} F_S(\vartheta) \right\} \text{As } F_S \text{ is an S.I. left hyperideal}$$

$$\subseteq \bigcup_{(u,v) \in S_x} \left\{ \bigcap_{x \in pov} F_S(x) \right\} \subseteq \bigcup_{(u,v) \in S_x} \left\{ \bigcap_{x \in uov} F_S(x) \right\} = F_S(x).$$

Therefore, $S_S \overset{\wedge}{\diamond} F_S \sqsubseteq F_S$.

Conversely, suppose that $S_S \overset{\wedge}{\diamond} F_S \sqsubseteq F_S$. Now, we have to show that F_S is an S.I. left hyperideal of an ordered LA-semihypergroup $(S, \circ, \leq)$ over V. Then

$$\bigcap_{\vartheta \in xoy} F_S(\vartheta) \supseteq \bigcap_{\vartheta \in xoy} (S_S \overset{\wedge}{\diamond} F_S)(\vartheta) = \bigcap_{\vartheta \in xoy} \left\{ \bigcup_{(u,v) \in S_\vartheta} [S_S(u) \bigcap F_S(v)] \right\}$$

$$= \bigcap_{\vartheta \in xoy} \left\{ \bigcup_{(u,v) \in S_\vartheta} [V \bigcap F_S(v)] \right\} = \bigcap_{\vartheta \in xoy} \left\{ \bigcup_{(u,v) \in S_\vartheta} [F_S(v)] \right\}$$

$$\supseteq \bigcap_{\vartheta \in xoy} \left\{ \bigcup_{(u,v) \in S_\vartheta} [F_S(y)] \right\} \supseteq \bigcap_{\vartheta \in xoy} \left\{ \bigcup_{(x,y) \in S_\vartheta} [F_S(y)] \right\} = F_S(y).$$

It follows that F_S is an S.I. left hyperideal of S over V.

Theorem 3 *A non-null soft set F_S is an S.I. sub-LA-semihypergroup (resp., S.I. right hyperideal, S.I. bi-hyperideal) of an ordered LA-semihypergroup $(S, \circ, \leq)$ over V if and only if $F_S \overset{\wedge}{\diamond} F_S \sqsubseteq F_S$ (resp., $F_S \overset{\wedge}{\diamond} S_S \sqsubseteq F_S$, $((F_S \overset{\wedge}{\diamond} S_S) \overset{\wedge}{\diamond} F_S) \sqsubseteq F_S$).*

Proof Proof is analogous to the Theorem 2.

Theorem 4 *Every S.I. left hyperideal of an ordered LA-semihypergroup* $(S, \circ, \leq)$ *over* V *is an S.I. bi-hyperideal of* S *over* V.

Lemma 3 *Let* $X (\neq \emptyset)$ *be any subset of an ordered LA-semihypergroup* $(S, \circ, \leq)$. *Then* X *is a left (resp. right) hyperideal of* S *if and only if* S_X *is an S.I. left (resp. right) hyperideal of* S *over* V.

Proof Let X be a left hyperideal of an ordered LA-semihypergroup $(S, \circ, \leq)$. Then $(S \circ X] \subseteq X$. Now $\mathcal{S}_S \overset{\wedge}{\diamond} \mathcal{S}_X = \mathcal{S}_{(S \circ X]} \sqsubseteq \mathcal{S}_X$. This shows that S_X is an S.I. left hyperideal of S over V.

Conversely, suppose that S_X is an S.I. left hyperideal of an ordered LA-semihypergroup $(S, \circ, \leq)$ over V. Let $x \in (S \circ X]$, then $S_X(x) \supseteq (\mathcal{S}_S \overset{\wedge}{\diamond} \mathcal{S}_X)(x) = S_{(S \circ X]}(x) = V$. It implies $x \in X$. Hence $(S \circ X] \subseteq X$. Therefore X is a left hyperideal of S.

Lemma 4 *Let* $X (\neq \emptyset)$ *be any subset of an ordered LA-semihypergroup* $(S, \circ, \leq)$. *Then* X *is a bi-hyperideal of* S *if and only if* S_X *is an S.I. bi-hyperideal of* S *over* V.

Proof Proof is analogous to the Lemma 3.

Definition 19 A non-null soft set F_S is said to be an S.I. quasi hyperideal of an ordered LA-semihypergroup $(S, \circ, \leq)$ over V if

1. $(F_S \overset{\wedge}{\diamond} S_S) \widetilde{\cap} (S_S \overset{\wedge}{\diamond} F_S) \sqsubseteq F_S$ and
2. For any $x, y \in S$ such that $x \leq y$ implies $F_S(x) \supseteq F_S(y)$.

Proposition 1 *If* F_S, G_S *are S.I. right and S.I. left hyperideals of an ordered LA-semihypergroup* $(S, \circ, \leq)$ *over* V, *respectively. Then* $F_S \widetilde{\cap} G_S$ *is an S.I. quasi hyperideal of* S *over* V.

Theorem 5 *Let* $(S, \circ, \leq)$ *be an ordered LA-semihypergroup and* $F(S)$ *be the set of all soft sets of* S *over* V. *Then* $(F(S), \overset{\wedge}{\diamond})$ *is an LA-semigroup.*

Proof Let us suppose that $F_S, G_S, H_S \in F(S)$ and $x \in S$ such that $x \leq y \circ z$, $y \leq p \circ q$ and $t \leq y \circ p$, where $y, z, p, q, t \in S$. Then, we have

$$((F_S \overset{\wedge}{\diamond} G_S) \overset{\wedge}{\diamond} H_S)(x) = \bigcup_{(y,z) \in S_x} [(F_S \overset{\wedge}{\diamond} G_S)(y) \cap H_S(z)]$$

$$= \bigcup_{(y,z) \in S_x} [\bigcup (\bigcup_{(p,q) \in S_y} \{F_S(p) \cap G_S(q)\}) \cap H_S(z)] = \bigcap_{x \leq y \circ z} [(\bigcup_{y \leq p \circ q} \{F_S(p) \cap G_S(q)\}) \cap H_S(z)]$$

$$= \bigcup_{x \leq (p \circ q) \circ z} \{\{F_S(p) \cap G_S(q)\} \cap H_S(z)\} = \bigcup_{x \leq (z \circ q) \circ p} \{\{H_S(z) \cap G_S(q)\} \cap F_S(p)\}$$

$$= \bigcup_{x \leq t \circ p} \{(\bigcup_{t \leq z \circ q} \{H_S(z) \cap G_S(q)\}) \cap F_S(p)\} = \bigcup_{(t,p) \in S_x} \{(\bigcup_{(z,q) \in S_t} \{H_S(z) \cap G_S(q)\}) \cap F_S(p)\}$$

$$= \bigcup_{(t,p) \in S_x} \{(H_S \overset{\wedge}{\diamond} G_S)(t) \cap F_S(p)\} = ((H_S \overset{\wedge}{\diamond} G_S) \overset{\wedge}{\diamond} F_S)(x).$$

Therefore, $((F_S \overset{\wedge}{\diamond} G_S) \overset{\wedge}{\diamond} H_S) = ((H_S \overset{\wedge}{\diamond} G_S) \overset{\wedge}{\diamond} F_S)$.

Theorem 6 *If* $(S, \circ, \leq)$ *is an ordered LA-semihypergroup. Then medial law holds in* $F(S)$.

Proof Let us suppose that S is an ordered LA-semihypergroup and $F(S)$ be the set of all soft sets of S over V and let $F_S, G_S, H_S, K_S \in F(S)$, Then by applying invertive law, we have

$$(F_S \hat{\diamond} G_S) \hat{\diamond} (H_S \hat{\diamond} K_S) = ((H_S \hat{\diamond} K_S) \hat{\diamond} G_S) \hat{\diamond} F_S$$

$$= ((G_S \hat{\diamond} K_S) \hat{\diamond} H_S) \hat{\diamond} F_S = (F_S \hat{\diamond} H_S) \hat{\diamond} (G_S \hat{\diamond} K_S).$$

Theorem 7 *Let* $(S, \circ, \leq)$ *be an ordered LA-semihypergroup with l.i. and* $F_S, G_S, H_S \in F(S)$. *Then following holds:*

(i) $F_S \hat{\diamond} (G_S \hat{\diamond} H_S) = G_S \hat{\diamond} (F_S \hat{\diamond} H_S).$

(ii) $(F_S \hat{\diamond} G_S) \hat{\diamond} (H_S \hat{\diamond} K_S) = (K_S \hat{\diamond} H_S) \hat{\diamond} (G_S \hat{\diamond} F_S).$

Proof (i) Let us suppose that $F_S, G_S, H_S, K_S \in F(S)$ and $x \in S$ such that $x \leq y \circ z$, $z \leq p \circ q$ and $t \leq y \circ p$, where $y, z, p, q, t \in S$. Then

$$(F_S \hat{\diamond}(G_S \hat{\diamond} H_S))(x) = \bigcup_{(y,z) \in S_x} [F_S(y) \cap (G_S \hat{\diamond} H_S)(z)]$$

$$= \bigcup_{(y,z) \in S_x} [F_S(y) \cap (\bigcup_{(p,q) \in S_z} \{G_S(p) \cap H_S(q)\})] = \bigcup_{x \leq y \circ z} [F_S(y) \cap (\bigcup_{z \leq p \circ q} \{G_S(p) \cap H_S(q)\})]$$

$$= \bigcup_{x \leq y \circ (p \circ q)} \{F_S(y) \cap \{G_S(p) \bigcap H_S(q)\}\} = \bigcup_{x \leq p \circ (y \circ q)} \{\{G_S(p) \cap F_S(y)\} \cap H_S(q)\}$$

$$= \bigcup_{x \leq p \circ t} \{\{G_S(p) \cap \bigcup_{t \leq y \circ q} \{F_S(y)\} \bigcap H_S(q)\}\} = \bigcup_{(p,t) \in S_x} \{\{G_S(p) \cap \bigcup_{(y,q) \in S_t} \{F_S(y) \cap H_S(q)\}\}$$

$$= \bigcup_{(p,t) \in S_x} \{\{G_S(p) \cap (F_S \hat{\diamond} H_S)(t)\} = (G_S \hat{\diamond} (F_S \hat{\diamond} H_S))(x).$$

    Therefore, $F_S \hat{\diamond} (G_S \hat{\diamond} H_S) = G_S \hat{\diamond} (F_S \hat{\diamond} H_S).$

(ii) Let us suppose that $F_S, G_S, H_S, K_S \in F(S)$ and $x \in S$ such that $x \leq y \circ z$, $y \leq p \circ q$ and $z \leq s \circ t$, where $y, z, p, q, s, t \in S$. Then, we have

$$((F_S \hat{\diamond} G_S) \hat{\diamond} (H_S \hat{\diamond} K_S))(x) = \bigcup_{(y,z) \in S_x} [(F_S \hat{\diamond} G_S)(y) \cap (H_S \hat{\diamond} K_S)(z)]$$

$$= \bigcup_{(y,z) \in S_x} [\bigcup_{(p,q) \in S_y} \{F_S(p) \cap G_S(q)\} \bigcap \bigcup_{(s,t) \in S_z} \{H_S(s) \cap K_S(t)\}]$$

$$= \bigcup_{x \leq y \circ z \ y \leq p \circ q} [\bigcup \{F_S(p) \cap G_S(q)\} \bigcap \bigcup_{z \leq s \circ t} \{H_S(s) \cap K_S(t)\}]$$

$$= \bigcup_{x \leq (p \circ q) \circ (s \circ t)} \{F_S(p) \cap G_S(q) \cap H_S(s) \cap K_S(t)\}$$

$$= \bigcup_{x \leq (t \circ s) \circ (q \circ p)} \{K_S(t) \cap H_S(s) \cap G_S(q) \cap F_S(p)\}$$

$$= \bigcup_{x \leq (m \circ n)} \{\bigcup_{m \leq (t \circ s)} \{K_S(t) \cap H_S(s)\} \bigcap \bigcup_{n \leq (q \circ p)} \{G_S(q) \cap F_S)(p)\}\}$$

$$= \bigcup_{(m,n) \in S_x} \{\bigcup_{(t,s) \in S_m} \{K_S(t) \cap H_S(s)\} \bigcap \bigcup_{(q,p) \in S_n} \{G_S(q) \cap F_S)(p)\}\}$$

$$= \bigcup_{(m,n)\in S_x} \{(\mathcal{K}_S \,\hat{\diamond}\, \mathcal{H}_S)(m) \cap (\mathcal{G}_S \,\hat{\diamond}\, \mathcal{F}_S)(n)\}$$

$$= ((\mathcal{K}_S \,\hat{\diamond}\, \mathcal{H}_S) \,\hat{\diamond}\, (\mathcal{G}_S \,\hat{\diamond}\, \mathcal{F}_S))(x).$$

Hence, $(F_S \overset{\wedge}{\diamond} G_S) \overset{\wedge}{\diamond} (H_S \overset{\wedge}{\diamond} K_S) = (K_S \overset{\wedge}{\diamond} H_S) \overset{\wedge}{\diamond} (G_S \overset{\wedge}{\diamond} F_S)$.

Proposition 2 *In an ordered LA-semihypergroup* $(S, \circ, \leq)$ *with l.i., for every S.I. left hyperideal* F_S *of* S *over* V, $S_S \overset{\wedge}{\diamond} F_S = F_S$.

Proof Let F_S be an S.I. left hyperideal of an ordered LA-semihypergroup $(S, \circ, \leq)$ over V, then $S_S \overset{\wedge}{\diamond} F_S \sqsubseteq F_S$. Thus, we only need to prove that $F_S \sqsubseteq S_S \overset{\wedge}{\diamond} F_S$. Since S is an ordered LA-semihypergroup with l.i., then for any $x \in S$, $x \leq e \circ x$, where 'e' is the l.i. of S. So, $(e, x) \in S_x$. Thus S_x is non-empty. Now, we have

$$(S_S \overset{\wedge}{\diamond} F_S)(x) = \bigcup_{(y,z)\in S_x} \{S_S(y) \cap F_S(z)\} \supseteq \{S_S(e) \cap F_S(x)\} = F_S(x).$$

Therefore, $S_S \overset{\wedge}{\diamond} F_S = F_S$.

Proposition 3 *In an ordered LA-semihypergroup* $(S, \circ, \leq)$ *with l.i., for every S.I. right hyperideal* F_S *of* S *over* V, $F_S \overset{\wedge}{\diamond} S_S = F_S$.

Proof Let F_S be an S.I. right hyperideal of an ordered LA-semihypergroup $(S, \circ, \leq)$ over V, then $F_S \overset{\wedge}{\diamond} S_S \sqsubseteq F_S$. Thus, it is only remains to show that $F_S \sqsubseteq F_S \overset{\wedge}{\diamond} S_S$. Since S is an ordered LA-semihypergroup with l.i., thus for any $x \in S$, $x \leq e \circ x \leq (e \circ e) \circ x = (x \circ e) \circ e$, where '$e$' is the l.i. of S. Thus, $\exists\, y \in x \circ e$ such that $x \leq y \circ e$. So, $(y, e) \in S_x$. Thus S_x is non-empty. Now, we have

$$(F_S \overset{\wedge}{\diamond} S_S)(x) = \bigcup_{(p,q)\in S_x} \{F_S(p) \cap S_S(q)\} \supseteq F_S(y) \cap S_S(e) = F_S(y) \tag{1}$$

As F_S is an S.I. right hyperideal of an ordered LA-semihypergroup $(S, \circ, \leq)$ over V, we have $\bigcap_{\vartheta \in x \circ e} F_S(\vartheta) \supseteq F_S(x)$ $\forall x, e \in S$. Since $y \in x \circ e$, thus we have $F_S(y) \supseteq F_S(x)$, therefore from (1)

$$(F_S \overset{\wedge}{\diamond} S_S)(x) \supseteq F_S(y) \supseteq F_S(x).$$

It implies, $F_S \sqsubseteq F_S \overset{\wedge}{\diamond} S_S$. Hence, $F_S \overset{\wedge}{\diamond} S_S = F_S$.

Corollary 1 *In an ordered LA-semihypergroup* $(S, \circ, \leq)$ *with l.i.,* $S_S \overset{\wedge}{\diamond} S_S = S_S$.

Proposition 4 *Let* $(S, \circ, \leq)$ *be an ordered LA-semihypergroup with l.i.,* F_S *be any soft set and* K_S *an S.I. left hyperideal of* S *over* V. *Then, for any soft set* H_S *and an*

S.I. left hyperideal G_S *of* **S** *over* V, $F_S \overset{\wedge}{\diamond} G_S = H_S \overset{\wedge}{\diamond} K_S$ *implies that* $G_S \overset{\wedge}{\diamond} F_S = K_S \overset{\wedge}{\diamond} H_S$.

Proof Suppose G_S and K_S are S.I. left hyperideals of an ordered LA-semihypergroup $(\mathbf{S}, \circ, \leq)$ over V. Then by the Proposition 2, $S_S \overset{\wedge}{\diamond} G_S = G_S$ and $S_S \overset{\wedge}{\diamond} K_S = K_S$. Now, we have

$$G_S \overset{\wedge}{\diamond} F_S = (S_S \overset{\wedge}{\diamond} G_S) \overset{\wedge}{\diamond} F_S = (F_S \overset{\wedge}{\diamond} G_S) \overset{\wedge}{\diamond} S_S = (H_S \overset{\wedge}{\diamond} K_S) \overset{\wedge}{\diamond} S_S = (S_S \overset{\wedge}{\diamond} K_S) \overset{\wedge}{\diamond} H_S = K_S \overset{\wedge}{\diamond} H_S.$$

Proposition 5 *Let* $(\mathbf{S}, \circ, \leq)$ *be an ordered LA-semihypegroup with l.i.. Then, every S.I. right hyperideal of* **S** *over* V *is an S.I. hyperideal of* **S** *over* V.

Proof Let F_S be an S.I. right hyperideal of an ordered LA-semihypergroup $(\mathbf{S}, \circ, \leq)$ over V. Then $F_S \overset{\wedge}{\diamond} S_S \sqsubseteq F_S$. Thus, we have

$$S_S \overset{\wedge}{\diamond} \mathcal{F}_S = (S_S \overset{\wedge}{\diamond} S_S) \overset{\wedge}{\diamond} \mathcal{F}_S = (\mathcal{F}_S \overset{\wedge}{\diamond} S_S) \overset{\wedge}{\diamond} S_S \sqsubseteq \mathcal{F}_S \overset{\wedge}{\diamond} S_S \sqsubseteq \mathcal{F}_S.$$

So, F_S is an S.I. left hyperideal of **S** over V, hence an S.I. hyperideal of **S** over V.

Theorem 8 *If* F_S *is an S.I. right hyperideal of an ordered LA-semihypergroup* $(\mathbf{S}, \circ, \leq)$ *over* V *and* G_S *an S.I. left hyperideal of* **S** *over* V. *Then*

$$F_S \overset{\wedge}{\diamond} G_S \sqsubseteq F_S \tilde{\cap} G_S$$

Proof Proof is straight forward.

Proposition 6 *If* F_S *and* G_S *are S.I. left (resp. right, two sided) hyperideals of an ordered LA-semihypergroup* $(\mathbf{S}, \circ, \leq)$ *over* V *with l.i.. Then the S.I. product* $F_S \overset{\wedge}{\diamond} G_S$ *is an S.I. left (resp. right, two sided) hyperideal of* **S** *over* V.

Proof Let F_S and G_S be S.I. left hyperideals of an ordered LA-semihypergroup $(\mathbf{S}, \circ, \leq)$ over V. Then, $S_S \overset{\wedge}{\diamond} F_S \sqsubseteq F_S$ and $S_S \overset{\wedge}{\diamond} G_S \sqsubseteq G_S$. Now, we have

$$S_S \overset{\wedge}{\diamond} (\mathcal{F}_S \overset{\wedge}{\diamond} \mathcal{G}_S) = \mathcal{F}_S \overset{\wedge}{\diamond} (S_S \overset{\wedge}{\diamond} \mathcal{G}_S) \sqsubseteq \mathcal{F}_S \overset{\wedge}{\diamond} \mathcal{G}_S.$$

It follows that $F_S \overset{\wedge}{\diamond} G_S$ is an S.I. left hyperideal of **S** over V.

Proposition 7 *If* F_S *is an S.I. left (resp. right, two sided) hyperideal of an ordered LA-semihypergroup* $(\mathbf{S}, \circ, \leq)$ *over* V *with l.i.. Then the S.I. product* $F_S \overset{\wedge}{\diamond} F_S$ *is an S.I. hyperideal of* **S** *over* V.

Theorem 9 *Let* F_S *and* G_S *be any S.I. right hyperideal of an ordered LA-semihypergroup* $(\mathbf{S}, \circ, \leq)$ *over* V *with l.i.. Then the soft product* $F_S \overset{\wedge}{\diamond} G_S$ *and* $G_S \overset{\wedge}{\diamond} F_S$ *are S.I. bi-hyperideals of* **S** *over* V.

Theorem 10 *Let* F_S *be an S.I. left hyperideal of an ordered LA-semihypergroup* $(S, \circ, \leq)$ *over* V *with l.i.. Then* $F_S \overset{\wedge}{\diamond} F_S$ *is an S.I. bi-hyperideal of* S *over* V.

3 Characterizations of Right Regular Ordered LA-Semihypergroups

In this section, we introduce soft (1,2)-hyperideal of ordered LA-semihypergroups. Further, we define right regular ordered LA-semihypergroup and characterize right regular ordered LA-semihypergroup using various soft hyperideals.

Definition 20 An S.I. sub-LA-semihypergroup F_S of an ordered LA-semihypergroup $(S, \circ, \leq)$ over V is said to be an S.I. (1,2)-hyperideal of S over V if

1. $\bigcap_{\vartheta \in (x \circ a) \circ (y \circ z)} F_S(\vartheta) \supseteq F_S(x) \cap F_S(y) \cap F_S(z) \, \forall \, a, x, y, z \in S$.
2. For any $x, y \in S$ such that $x \leq y$ implies $F_S(x) \supseteq F_S(y)$.

Definition 21 An element 'h' of an ordered LA-semihypergroup $(S, \circ, \leq)$ is called a right regular element if there exists an element $x \in S$ such that $h \leq (h \circ h) \circ x$. If every element of S is right regular, then S is called a right regular ordered LA-semihypergroup.

Lemma 5 *If* $(S, \circ, \leq)$ *is a right regular ordered LA-semihypergroup, then for any soft set* F_S *of* S, *we have* $S_S \overset{\wedge}{\diamond} F_S \sqsupseteq F_S$.

Proof Suppose that $(S, \circ, \leq)$ is a right regular ordered LA-semihypergroup, thus for any $h \in S$, $\exists \, x \in S$ such that $h \leq (h \circ h) \circ x = (x \circ h) \circ h$. Then, $\exists \, b \in x \circ h$ such that $h \leq b \circ h$. Thus, $(b, h) \in S_h$. So, S_h is non-empty. Now, we have

$$(S_S \overset{\wedge}{\diamond} F_S)(h) = \bigcup_{(y,z) \in S_h} \{S_S(y) \cap F_S(z)\} \supseteq S_S(b) \cap F_S(h) = F_S(h).$$

Hence, $S_S \overset{\wedge}{\diamond} F_S \sqsupseteq F_S$.

Corollary 2 *If* $(S, \circ, \leq)$ *is a right regular ordered LA-semihypergroup, then for any S.I. left hyperideal* F_S *of* S *over* V, *we have* $S_S \overset{\wedge}{\diamond} F_S = F_S$.

Lemma 6 *If* $(S, \circ, \leq)$ *is a right regular ordered LA-semihypergroup, then for any S.I. left hyperideal* F_S *of* S *over* V, *we have* $F_S \overset{\wedge}{\diamond} S_S \sqsupseteq F_S$.

Proof Suppose that $(S, \circ, \leq)$ is a right regular ordered LA-semihypergroup, thus for any $h \in S$, $\exists \, x \in S$ such that $h \leq (h \circ h) \circ x = (x \circ h) \circ h$. Then, $\exists \, b \in x \circ h$ such that $h \leq b \circ h$. Thus, $(b, h) \in S_h$. So, S_h is non-empty. Now, we have

$$(F_S \overset{\wedge}{\diamond} S_S)(h) = \bigcup_{(y,z) \in S_h} \{F_S(y) \cap S_S(z)\} \supseteq F_S(b) \cap S_S(h) = F_S(b). \qquad (2)$$

As F_S is an S.I. left hyperideal of S over V, we have $\bigcap_{\vartheta \in xoh} F_S(\vartheta) \supseteq F_S(h)$. Since $b \in x \circ h$, it would imply that $F_S(b) \supseteq F_S(h)$. Therefore from (2), we have

$$(F_S \overset{\wedge}{\diamond} S_S)(h) \supseteq F_S(b) \supseteq F_S(h). \qquad (3)$$

Hence, $F_S \overset{\wedge}{\diamond} S_S \sqsupseteq F_S$.

Corollary 3 *If F_S is an S.I. hyperideal of a right regular ordered LA-semihypergroup S over V. Then, $F_S \overset{\wedge}{\diamond} S_S = F_S$ and $S_S \overset{\wedge}{\diamond} F_S = F_S$.*

Proposition 8 *Let $(S, \circ, \leq)$ be a right regular ordered LA-semihypegroup. Then, every S.I. left hyperideal of S over V is an S.I. right hyperideal of S over V and vice versa.*

Proof Suppose that F_S is an S.I. left hyperideal of a right regular ordered LA-semihypegroup $(S, \circ, \leq)$ over V. Then $F_S \overset{\wedge}{\diamond} S_S \sqsubseteq F_S$. As S is a right regular ordered LA-semihypegroup, thus for any $h \in S$, there exists $x \in S$ such that $h \leq (h \circ h) \circ x$. Then, we have

$$h \circ g \leq ((h \circ h) \circ x) \circ g = ((g \circ x) \circ (h \circ h)).$$

Now, there exists $\vartheta \in h \circ g, k \in g \circ x, l \in h \circ h$ and $m \in k \circ l$ such that $\vartheta \leq m$. Then, we have

$$F_S(\vartheta) \supseteq F_S(m) \supseteq \bigcap_{m \in (k \circ l)} F_S(m) \supseteq \bigcap_{m \in ((g \circ x) \circ (h \circ h))} F_S(m)$$

$$\supseteq \bigcap_{m \in (h \circ h)} F_S(m) \text{ as } F_S \text{ is an S.I. left hyperideal}$$

$$\supseteq F_S(h).$$

It implies $\bigcap_{\vartheta \in h \circ g} F_S(\vartheta) \supseteq F_S(h)$. Hence, F_S is an S.I. right hyperideal of S over V.

Theorem 11 *Let $(S, \circ, \leq)$ be a right regular ordered LA-semihypergroup with l.i., then the following statements are equivalent:*

(1) F_S is an S.I. (1, 2)-hyperideal of S over V;
(2) F_S is an S.I. hyperideal of S over V.

Proof (1) $\Rightarrow$ (2): Suppose that F_S is an S.I. (1, 2)-hyperideal of a right regular ordered LA-semihypergroup $(S, \circ, \leq)$ over V with l.i. and let $b, h \in S$, then $\exists x \in S$ such that $h \leq (h \circ h) \circ x$. Thus, we have

$$b \circ h \leq b \circ ((h \circ h) \circ x)) = ((h \circ h) \circ (b \circ x)) \leq ((((h \circ h) \circ x) \circ h) \circ (b \circ x))$$
$$= (((h \circ x) \circ (h \circ h)) \circ (b \circ x)) = (((h \circ h) \circ (x \circ h)) \circ (b \circ x)) = (((b \circ x) \circ (x \circ h)) \circ (h \circ h))$$
$$= (((h \circ x) \circ (x \circ b)) \circ (h \circ h)) \leq (((((h \circ h) \circ x) \circ x) \circ (x \circ b)) \circ (h \circ h))$$
$$= ((((x \circ x) \circ (h \circ h)) \circ (x \circ b)) \circ (h \circ h)) = ((((h \circ h) \circ (x \circ x)) \circ (x \circ b)) \circ (h \circ h))$$
$$= ((((x \circ b) \circ (x \circ x)) \circ (h \circ h)) \circ (h \circ h)) = ((h \circ (((x \circ b) \circ (x \circ x)) \circ h)) \circ (h \circ h)).$$

Now, there exists $\vartheta \in b \circ h$, $c \in (((x \circ b) \circ (x \circ x)) \circ h)$ and $\delta \in (h \circ c) \circ (h \circ h)$ such that $\vartheta \leq \delta$. Thus, we have

$$F_S(\vartheta) \supseteq F_S(\delta) \supseteq \bigcap_{\delta \in (((h \circ c) \circ (h \circ h))} F_S(\delta) \supseteq \bigcap_{\delta \in (((h \circ (((x \circ b) \circ (x \circ x)) \circ h)) \circ (h \circ h))} F_S(\delta)$$

$$\supseteq F_S(h) \cap F_S(h) \cap F_S(h) \text{ As } F_S \text{ is an S.I. } (1,2)-\text{hyperideal}$$

$$= F_S(h).$$

It implies $\bigcap_{\vartheta \in b \circ h} F_S(\vartheta) \supseteq F_S(h)$. Therefore, F_S is an S.I. left hyperideal of S over V. By Proposition 8, F_S is an S.I. right hyperideal of S over V. It follows that F_S is an S.I. hyperideal of S over V.

$(2) \Rightarrow (1)$: It is obvious.

Theorem 12 *Let* $(S, \circ, \leq)$ *be a right regular ordered LA-semihypergroup with l.i., then the following statements are equivalent:*

(1) F_S *is an S.I.* $(1, 2)$-*hyperideal of* S *over* V;
(2) F_S *is an S.I. quasi hyperideal of* S *over* V.

Proof $(1) \Rightarrow (2)$: It is easy to prove by Theorem 11.

$(2) \Rightarrow (1)$: Suppose that F_S is an S.I. quasi hyperideal of a right regular ordered LA-semihypergroup $(S, \circ, \leq)$ over V with l.i. 'e'. Let $h \in S$, then $\exists x \in S$ such that $h \leq (h \circ h) \circ x$. Thus, we have

$$h \leq (h \circ h) \circ x = (x \circ h) \circ h \subseteq (x \circ h) \circ (e \circ h)$$
$$= (h \circ e) \circ (h \circ x) = h \circ ((h \circ e) \circ x)$$

Now, there exists $b \in (h \circ e)$ such that $h \leq h \circ (b \circ x)$. Again, $\exists c \in b \circ x$ such that $h \leq h \circ c$. It implies $(h, c) \in S_h$. So, S_h is non-empty. Thus, we have

$$(F_S \overset{\wedge}{\diamond} S_S)(h) = \bigcup_{(y,z) \in S_h} \{F_S(y) \cap S_S(z)\} \supseteq F_S(h) \cap S_S(c) = F_S(h).$$

Also, we have

$$\mathcal{F}_S \overset{\wedge}{\diamond} \mathcal{S}_S \supseteq (\mathcal{S}_S \overset{\wedge}{\diamond} \mathcal{F}_S) \overset{\wedge}{\diamond} (\mathcal{S}_S \overset{\wedge}{\diamond} \mathcal{S}_S) = (\mathcal{S}_S \overset{\wedge}{\diamond} \mathcal{S}_S) \overset{\wedge}{\diamond} (\mathcal{F}_S \overset{\wedge}{\diamond} \mathcal{S}_S)$$

$$= \mathcal{S}_S \overset{\wedge}{\diamond} (\mathcal{F}_S \overset{\wedge}{\diamond} \mathcal{S}_S) \supseteq \mathcal{S}_S \overset{\wedge}{\diamond} \mathcal{F}_S.$$

It follows that $S_S \overset{\wedge}{\diamond} F_S \sqsubseteq (S_S \overset{\wedge}{\diamond} F_S) \tilde{\cap} (F_S \overset{\wedge}{\diamond} S_S)$. Since F_S is an S.I. quasi hyperideal of S over V, it implies $S_S \overset{\wedge}{\diamond} F_S \sqsubseteq F_S$. By Proposition 8, F_S will S.I. hyperideal of S over V. Hence, by Theorem 11, F_S will be an S.I. $(1, 2)$-hyperideal of S over V.

Theorem 13 *Let $(S, \circ, \leq)$ be a right regular ordered LA-semihypergroup with l.i., then the following statements are equivalent:*

(1) F_S is an S.I. $(1, 2)$-hyperideal of S over V;
(2) F_S is an S.I. bi-hyperideal of S over V.

Proof $(1) \Rightarrow (2)$: Let $(S, \circ, \leq)$ be a right regular ordered LA-semihypergroup with l.i. and F_S an S.I. $(1, 2)$-hyperideal of S over V. As S is right regular, thus for h, a, $b \in S$, $\exists \, h', a', b' \in S$ such that $h \leq (h \circ h) \circ h', a \leq (a \circ a) \circ a'$ and $b \leq (b \circ b) \circ b'$. Then, we have

$$(a \circ h) \circ b) \leq (a \circ h) \circ ((b \circ b) \circ b') = (b \circ b) \circ ((a \circ h) \circ b')$$
$$= (b' \circ (a \circ h)) \circ (b \circ b) = (a \circ (b' \circ h)) \circ (b \circ b).$$

Now, there exists $\vartheta \in ((a \circ h) \circ b)$, $c \in (b' \circ h)$ and $\delta \in (a \circ c) \circ (b \circ b)$ such that $\vartheta \leq \delta$. Thus, we have

$$F_S(\vartheta) \supseteq F_S(\delta) \supseteq \bigcap_{\delta \, \in \, (aoc)o(bob)} F_S(\delta) \supseteq \bigcap_{\delta \, \in \, (ao(b'oh))o(bob)} F_S(\delta)$$
$$\supseteq F_S(a) \cap F_S(b) \cap F_S(b) \text{ As } F_S \text{ is an S.I. } (1, 2)-\text{hyperideal}$$
$$= F_S(a) \cap F_S(b).$$

It implies $\bigcap_{\vartheta \, \in \, ((aoh)ob)} F_S(\vartheta) \supseteq F_S(a) \cap F_S(b)$. Hence, F_S is an S.I. bi-hyperideal of S over V.

$(2) \Rightarrow (1)$: Suppose that $(S, \circ, \leq)$ is a right regular ordered LA-semihypergroup with l.i. 'e' and F_S an S.I. bi-hyperideal of S over V. Let $x, h, y, z \in S$, then $\exists \, x' \in S$ such that $x \leq (x \circ x) \circ x'$. Then, we have

$$((x \circ h) \circ (y \circ z)) = ((z \circ y) \circ (h \circ x) \leq ((z \circ y) \circ (h \circ ((x \circ x) \circ x')))$$
$$= ((z \circ y) \circ ((x \circ x) \circ (h \circ x'))) = ((z \circ y) \circ (((h \circ x') \circ x) \circ x))$$
$$\leq ((z \circ y) \circ (((h \circ x') \circ ((x \circ x) \circ x')) \circ x)) \leq ((z \circ y) \circ (((h \circ x') \circ ((x \circ x) \circ (e \circ x'))) \circ x))$$
$$= ((z \circ y) \circ (((h \circ x') \circ ((x' \circ e) \circ (x \circ x))) \circ x)) = ((z \circ y) \circ (((h \circ x') \circ (x \circ ((x' \circ e) \circ x))) \circ x))$$
$$= ((z \circ y) \circ ((x \circ ((h \circ x') \circ ((x' \circ e) \circ x))) \circ x))$$

Now, there exists $\vartheta \in ((x \circ h) \circ (y \circ z))$, $c \in ((h \circ x') \circ ((x' \circ e) \circ x))$, $d \in ((x \circ c) \circ x)$ and $\delta \in (z \circ y) \circ d$ such that $\vartheta \leq \delta$. Thus, we have

$$F_S(\vartheta) \supseteq F_S(\delta) \supseteq \bigcap_{\delta \in ((zoy)od)} F_S(\delta) \supseteq \bigcap_{\delta \in ((zoy)o((xoc)ox))} F_S(\delta)$$

$$\supseteq \bigcap_{\delta \in (zoy)} F_S(\delta) \cap \bigcap_{\delta \in ((xoc)ox)} F_S(\delta) \supseteq F_S(z) \cap F_S(y) \cap F_S(x) \cap F_S(x)$$

$$= F_S(x) \cap F_S(y) \cap F_S(z).$$

It implies $\bigcap_{\vartheta \in ((xoh)o(yoz))} F_S(\vartheta) \supseteq F_S(x) \cap F_S(y) \cap F_S(z)$. Hence, F_S is an S.I. $(1, 2)$-hyperideal of S over V.

References

1. H. Aktas, N. Cagman, Soft sets and soft groups. Inform. Sci. **177**, 2726–2735 (2007). https://doi.org/10.1016/j.ins.2006.12.008
2. N. Cagman, S. Enginoglu, Soft set theory and uni-int decision making. Eur. J. Op. Res. **207**, 848–855 (2010). https://doi.org/10.1016/j.ejor.2010.05.004
3. N. Cagman, F. Çitak, H. Aktas, Soft int-group and its applications to group theory. Neural Comput. Appl. **21**, 151–158 (2012). https://doi.org/10.1007/s00521-011-0752-x
4. J. Chvalina, Commutative hypergroups in the sense of Marty and ordered sets, in *Proceedings of International Conference General Algebra and Ordered Sets*, Olomouc (1994), pp. 19–30
5. P. Corsini, *Prolegomena of Hypergroup Theory*, Aviani editor, 2nd edn. (1993)
6. B. Davvaz, V.L. Fotea, *Hyperring Theory and Applications* (International Academic Press, Palm Harber, Fla, USA, 2007), p. 115 (1993)
7. A. Hasankhani, Ideals in a semihypergroup and Greens relations. Ratio Math. **13**, 29–36 (1999)
8. D. Heidariand, B. Davvaz, On ordered hyperstructures. U.P.B. Sci. Bull. Ser. A. **73**(2), 85–96 (2011)
9. K. Hila, J. Dine, On hyperideals in left almost semihypergroups. ISRN Algebra **2011**, Article ID953124, 8 pp. (2011). https://doi.org/10.5402/2011/953124
10. M.A. Kazimand, M. Naseeruddin, On almost semigroups. Alig. Bull. Math. **2**, 1–7 (1972)
11. P.K. Maji, R. Biswas, A.R. Roy, Soft set theory. Comput. Math. Appl. **45**, 555–562 (2003). https://doi.org/10.1016/S0898-1221(03)00016-6
12. F. Marty, Sur une generalization de la notion de group, in *8th Congress des Mathematiciens Scandinaves* (1934), pp. 45–49
13. D. Molodtsov, Soft set theory-first results. Comput. Math. Appl. **37**, 19–31 (1999). https://doi.org/10.1016/S0898-1221(99)00056-5
14. Q. Mushtaq, S.M. Yusuf, On LA-semigroups. Alig. Bull. Math. **8**, 65–70 (1978)
15. A. Sezgin, A new approach to LA-semigroup theory via the soft sets. J. Intell. Fuzzy Syst. **26**, 2483–2495 (2014). https://doi.org/10.3233/IFS-130918
16. N. Yaqoob, P. Corsini, F. Yousafzai, On intra-regular left almost semihypergroups with pure left identity. J. Math. 10 pp. (2013). https://doi.org/10.1155/2013/510790
17. N. Yaqoob, M. Gulistan, Partially ordered left almost semihypergroups. J. Egypt. Math. Soc. **23**, 231–235 (2015). https://doi.org/10.1016/j.joems.2014.05.012
18. M. Khan, Y.B. Jun, F. Yousafzai, Fuzzy ideals in right regular LA-semigroups. Hacettepe J. Math. Stat. **44**(3), 569–586 (2015). https://doi.org/10.15672/HJMS.2015449419

Score Improvement Using Backpropagation in Biometric Recognition System

Gopal, Monika Gupta, Akshay Sahai, Shivam Verma
and Vikramaditya Agarwal

Abstract In this paper, a novel score improvement technique is proposed for person biometric authentication based on backpropagation. To obtain this, threshold values are optimized to reduce the equal error rate (EER) and to trade off false rejection rate (FRR) with false acceptance rate (FAR). To validate the proposed method, palmprint recognition has been tested on IITD database of 230 persons and PolyU database of 386 persons. Experimental results show the primacy of the proposed technique over the existing ones in the literature and achieved higher accuracy.

Keywords k-Nearest neighbor · Palmprint · Neural networks · Biometrics

1 Introduction

For the last few decades, biometrics has been a progressive field due to its cardinal contribution in the field of security, surveillance, and banking [1–3]. Biometric systems include measurement of various human physiological characteristics and traits such as fingerprint scanning, facial recognition, palmprint recognition, retinal and iris scanning. Many innovative and stout technologies have been seen in literature [4, 5]. It is noticeable from the existing literature that palmprint and faceprint technologies are highly efficient and secure [6–8]. A typical biometric system is shown in Fig. 1.

The accuracy of biometric systems can be improved using several methods, some of them are image enhancement, feature extractions, better classifier, and fusions. Lots of research is done in image enhancements, feature extractions, and fusions. In most biometric systems, classifiers finally give scores to evaluate the performance

Gopal (✉)
Bharati Vidyapeeth's College of Engineering, New Delhi, Delhi, India
e-mail: gopal.chaudhary88@gmail.com

M. Gupta · A. Sahai · S. Verma · V. Agarwal
Maharaja Agrasen Institute of Technology, New Delhi, Delhi, India
e-mail: monikagupta@mait.ac.in

© Springer Nature Singapore Pte Ltd. 2019
H. Malik et al. (eds.), *Applications of Artificial Intelligence Techniques in Engineering*, Advances in Intelligent Systems and Computing 698,
https://doi.org/10.1007/978-981-13-1819-1_35

377

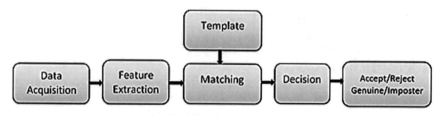

Fig. 1 A typical biometric system

of system. These scores are helpful in finding recognition rate, success rate, genuine acceptance rate (GAR), etc. Euclidian Distance, to calculate the scores of training and testing samples, is obtained from features of each biometric process. Generally, receiver operating characteristic (ROC) curves are employed to visually analyze the use of biometric systems in verification. To plot ROC, first threshold value *th* is selected. If the value of score is greater than threshold value *th* then this score is assumed to be a genuine score, that is, claimed identity is a true person. If the value of score is lesser than threshold value *th* then this score is assumed to be an imposter score, that is, claimed identity is a false person. Curve between the genuine acceptance rate (GAR) and false acceptance rate (FAR) is usually referred to as ROC; GAR = 100-FRR. FAR is the ratio of erroneously recognized person, while FRR is the ratio of genuine individuals which are mistakenly excluded.

Not a copious amount of research has been done on score improvement methods. Cohort scores have been used in the literature before, such as, for normalization [9] in which ranking is used to distinguish the imposter and genuine. In literature, refined scores are used to improve the results. But in both the methods, score improvement has not taken place.

In this paper, a novel score improvement technique is proposed for person authentication based on backpropagation. To obtain this, the threshold value is optimized to reduce the equal error rate (EER) and to trade off false rejection rate (FRR) with false acceptance rate (FAR). To validate the proposed method, palmprint recognition has been tested on IITD database of 230 persons and PolyU database of 386 persons. Experimental results showed primacy of the proposed technique over the existing ones in the literature and achieved higher accuracy.

1.1 Novelties and Contributions of the Paper

1. A novel score improvement technique is proposed for person authentication based on backpropagation.
2. Threshold value is optimized to reduce the Equal Error Rate (EER).
3. Score matrix is normalized.
4. Proposed technique is tested on IIT Delhi palmprint database of 230 persons and PolyU database of 386 persons.

The organization of the paper is as follows. Section 2 describes the method of preprocessing involved for ROI extraction from database. Section 3 presents the methodology. Section 4 demonstrated simulations and result analysis. Section 5 concludes the suggested work.

2 Preprocessing

For feature extraction, the orientation of data samples must be same to withdraw the same set of information. For any biometric system, the first requirement is the acquisition of enough databases for proper training and testing. After data acquisition, the second step is to find out the region of interest (ROI) from the data sample. Most of the general biometric system are made to have interclass variations to make the system more real time and includes those mistakes that can be added by users at the verification point, such as, in offices or banks. Hence orientation or rotational variations are made. In case where data acquisition is constrained by peg or pins, ROI can be easily extracted. If the database has rotational variation, then hand samples are normalized before ROI extraction to make system robust to such variations. All the steps that align database samples for feature extraction are referred as preprocessing. In most of the preprocessing methods, valley points between the fingers are extracted to crop the ROI. The basic steps are: binarizing to extract the boundary of hand which also helps in masking of hand on a background free image; finding the key-points; generating the coordinate system; cropping the ROI.

Generally, finding the fingertips and centroid is followed by masking of original bounded image. The ordering of fingertips is done by doing circular traversal with the centroid as the center of the hand in clockwise direction so that the first coordinates are that of small finger and last coordinates are of thumb. In this way, the coordinates of all the fingertips are obtained. A line is drawn joining the index and ring finger and rotated in such way that it becomes parallel to the horizontal axis. After this, finger valleys are calculated indicating the rate of change of the slope of the boundary. Then each palm is further straightened using the coordinates of finger valleys. Lastly, the ROI is cropped from the processed image.

3 Methodology

Feature extraction is performed on the preprocessed input. Matrices of dataset are then divided into grids, with each cell of size 5 * 5 and resultant matrices of size 30 * 30 were obtained. Mean value of each cell is calculated, and cell matrices 5 * 5 are replaced with their mean values and matrices of 30 * 30 are left. These matrices are then converted or resized into a single column matrix of size 1 * 900. There are 5 palmprints of 230 different people, therefore matrix of size 230 * 5 is created with each cell of size 1 * 900. The input is then provided to the classifier. Classifier is

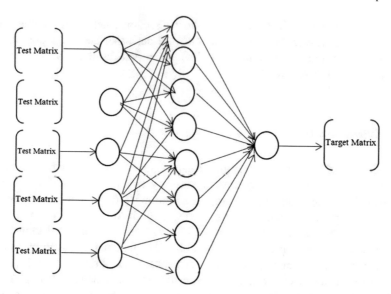

Fig. 2 Architecture of the proposed method used for backpropagation to update the threshold value

the algorithm that produces score matrix, which is used to evaluate performance of system.

K-NN is applied on the matrices, i.e., first 4 columns are now considered for training and last column is considered for testing. Value of each cell of testing column is compared to each cell of all the training columns (testing-training). The values of each set of palmprint of each person is considered separately. The minimum error value among the 4 values for each person are taken. This whole process is started again in clockwise direction, i.e., the testing and training columns are changed in clockwise direction. The minimum value, thus taken, corresponds to the genuine palmprint. What is being done above is training and testing palmprints of every single person with each other. Then these values are placed together in a matrix. We make a matrix of size 230 * 230, these minimum values are taken in the diagonal cells, i.e., the diagonal cells become the genuine cells, and the rest becomes the imposter cells. This matrix of size 230 * 230 is known as score matrix. Score matrix is the result matrix of k-NN, with ambiguous result. Score matrix is used to evaluate the performance of system. It depends on one parameter known as threshold value. Therefore, to increase the efficiency of system, score matrix must be optimized and to optimize score matrix, threshold value is optimized. Optimization of score matrix leads to reduction of equal error rate (ERR) and trade-off of false rejection rate (FRR) with false acceptance rate (FAR). The algorithm proposed to optimize the threshold value is shown in Algorithm 1. Architecture of the proposed method used for backpropagation to update the threshold value is shown in Fig. 2.

Algorithm 1 Threshold optimization using backpropagation

```
1: procedure
2:    th 0.01
3:    total test score matrix=5
4:    target score matrix=diogonal matrix
5:    s_{i=j} = 0
6:    s_{i≠j} = 1
7:    If
8:    number of((s_{i=j}<th) < max(i; j)
9:    th ← th_optimized
10:   Else
11:   Update  th
```

4 Experimental Results

To judge the performance quantitatively, decidability index (DIFactor) [10] is used.
Separation, and if required, overlap of similarity scores is given by the value of DI
factor [11]. It is defined as

$$DI = \frac{|\mu_1 + \mu_2|}{\sqrt{\frac{\sigma_1^2 + \sigma_2^2}{2}}}$$

where $\mu 1$ and $\mu 2$ denote the mean of the genuine and imposter distributions, respectively, and σ_1^2 and σ_2^2 symbolize the variances of the genuine and imposter distributions. Elevated values of *DI Factor* illustrate more separation similarity distributions.

Equal error rate (EER) is considered here to support the performance of the proposed results in verification. EER is picked as a performance measurement quantity
where poorer values show the high performance of the proposed method where EER
is calculated; FAR equals FRR.

It is clearly shown in Table 1 that EER is 1.686 for threshold *th* = 0.01 which is
improved to 0.972 and 0.881 for th = 0.015 and th = 0.023 respectively, for IITD
database and DI factor varied from 3.1684 to 4.544 and 4.888 for th = 0.015 and
th = 0.023 respectively. It is seen that EER is reduced on optimizing threshold value
of threshold *th*. Also, DI Factor is also increased.

Table 1 Performance of the proposed technique

	IITD database		PolyU database	
	EER	DI factor	EER	DI factor
Th = 0.01	1.686	3.1684	0.3464	4.0426
Th = 0.015	0.972	4.544	0.271	4.898
Th = 0.023	0.881	4.888	0.182	6.668

Table 2 Identification results of different values of the threshold for IITD database

| | Genuine acceptance rate (GAR %) | | Identification results |
	FAR = 0.1	FAR = 1	
Th = 0.1	79.82	84.18	91.1
Th = 0.015	81.96	87.09	93
Th = 0.023	93.95	96.05	96

Table 3 Identification results of different values of the threshold for PolyU database

| | Genuine acceptance rate (GAR %) | | Identification results |
	FAR = 0.1	FAR = 1	
Th = 0.1	97.6	98.5	99
Th = 0.015	97.5	99.2	100
Th = 0.023	98.2	99.4	100

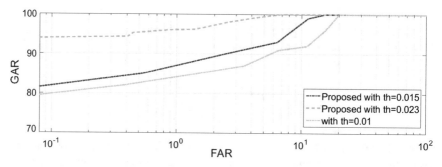

Fig. 3 Results for IITD database

From Tables 2 and 3, it can be seen that recognition rate is also dependent on threshold value. It is also seen that the convergence of the proposed method is faster.

In Figs. 3 and 4, it is clear that the proposed result covers the maximum area under the curve and reaches 100% GAR at a faster rate as compared to ROCs of other techniques.

5 Conclusion

In this paper, a novel score improvement technique is proposed for person authentication based on backpropagation. To obtain this, the threshold value is optimized to reduce the equal error rate (EER) using backpropagation. The ROC plots have shown improvement by using improved threshold value. It is clear that the optimized value of threshold can clearly distinguish the genuine score to imposter score. In future, score optimization can be done using bioinspired optimization algorithm also.

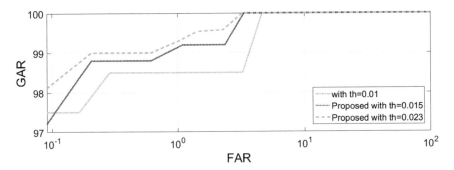

Fig. 4 Results for PolyU database

References

1. H. Kikuchi, K. Nagai, W. Ogata, M. Nishigaki, Privacy-preserving similarity evaluation and application to remote biometrics authentication. Soft Comput. **14**, 529–536 (2010)
2. K. Choi, K.-A. Toh, U. Youngjung, H. Byun, Service-oriented architecture based on biometric using random features and incremental neural networks. Soft. Comput. **16**, 1539–1553 (2012)
3. D. Bhattacharjee, D.K. Basu, M. Nasipuri, M. Kundu, Human face recognition using fuzzy multilayer perceptron. Soft Comput. **14**, 559–570 (2010)
4. S. Srivastava, S. Bhardwaj, S. Bhargava, Fusion of palm-phalanges print with palmprint and dorsal hand vein. Appl. Soft Comput. **47**, 12–20 (2016)
5. R. Purkait, External ear: an analysis of its uniqueness. Egypt. J. Forensic Sci. **6**, 99–107 (2016)
6. L. Fei, Y. Xu, W. Tang, D. Zhang, Double orientation code and non-linear matching scheme for palmprint recognition. Pattern Recognit. **49**, 89–101 (2016)
7. Y.X. LunkeFei, D. Zhang, Half-orientation extraction of palmprint features. Pattern Recogn. Lett. **69**, 35–41 (2016)
8. Y. Bi, M. Lv, Y. Wei, N. Guan, W. Yi, Multi-feature fusion for thermal face recognition. Infrared Phys. Technol. **77**, 366–374 (2016)
9. R.A. Finan, A.T. Sapeluk, R.I. Damper, Impostor cohort selection for score normalization in speaker verification. Pattern Recogn. Lett. **18**(9), 881–888 (1997)
10. J.G. Daugman, High confidence visual recognition of persons by a test of statistical independence. IEEE Trans. Pattern Anal. Mach. Intell. **15**(11), 1148–1161 (1993)
11. J. Daugman, How iris recognition works. The essential guide to image processing. 715–739 (2009)

Attribute-Based Access Control in Web Applications

Sadia Kauser, Ayesha Rahman, Asad Mohammed Khan and Tameem Ahmad

Abstract Nowadays, controlling access to resources is a challenge and the security policies are also needed to be flexible. For providing access control, many access control models can be implemented like Discretionary Access control (DAC), Mandatory Access Control (MAC) and Role-Based Access Control (RBAC) but there are certain limitations in them. Attribute-Based Access Control (ABAC) is the general model which could comprehend the benefits of these models while surpassing their demerits. In this work, we have explained how ABAC can be implemented on Web Application. For implementing ABAC, we have developed a sample web application for an organization to provide users and members access to different resources like course materials and project files. We have defined subject's attributes, object's attributes, and policies for it. In conclusion, ABAC can provide dynamic access permissions and flexibility to the access control mechanism in comparison to RBAC and is more secure than Trust-Based Access Control (TrustBAC).

1 Introduction

In today's world of computer security, access control is a challenge. Access control is a mechanism which enables us to specify certain limitations on access of information to authorized users. Access control components decide using policies that whether subject can be granted permission to access the objects or not.

S. Kauser · A. Rahman · A. M. Khan · T. Ahmad (✉)
Department of Computer Engineering, Zakir Husain College of Engineering
and Technology, A.M.U, Aligarh, Uttar Pradesh, India
e-mail: tameemahmad@gmail.com

S. Kauser
e-mail: sadiakauser69@gmail.com

A. Rahman
e-mail: ayesharahma.786@gmail.com

A. M. Khan
e-mail: masadiitr@gmail.com

© Springer Nature Singapore Pte Ltd. 2019
H. Malik et al. (eds.), *Applications of Artificial Intelligence Techniques
in Engineering*, Advances in Intelligent Systems and Computing 698,
https://doi.org/10.1007/978-981-13-1819-1_36

385

Due to the drawbacks and limitations of RBAC, researchers have proposed many extended RBAC models [1, 2]. Many Trust-Based Access Control Models [3, 4] are also proposed. In TrustBAC [5], instead of roles trust level are assigned to users based on their identities, behavior history, recommendation, etc. But this model also fails to address all issues. They also integrated role-based access control with attributes [6, 7]. Though they solve the problem up to a great extent but shortcomings are still there.

Due to this, there is a need of a general model which covers the manifested advantages of all these models while surpassing their limitations. ABAC meets all these demands. It has implemented in cloud infrastructure [8]. In this work, our purpose is to enable Attribute-Based Access Control in Web Application.

The next section gives details about DAC, MAC and RBAC models and their limitations. In Sect. 3, there is given general concepts necessary to understand new general model ABAC and its generic features and components, and Sect. 4 includes the detail about policy language required for implementation, i.e., XACML.

In Sect. 5, all the details regarding the implementation of ABAC in web application are given including attributes and policies used in our system, Sect. 6 presents the result, and Sect. 7 concludes the work.

2 Backgrounds

There are many types of access control models available that could be implemented in information security realm. Only three are widespread in security practices: DAC, MAC, and RBAC.

Discretionary Access Control (DAC)
DAC is the model in which access policies are specified by the owner of the resources. Therefore using this model, decisions are generated directly for subjects. This model allows the owner of the resources to change its permission at its own discretions. Though grants larger flexibility but suffers from many drawbacks. Unauthorized users can settle unsafe access rights to the information.

Mandatory Access Control (MAC)
MAC is a static access control method [9]. It lists security levels on both objects and subjects. MAC is based on the principle that access control is provided according to certain clearance required by users to access objects. MAC provides information security by propagating sensitivity labels on derivative objects, including copies also. In MAC, access control policies are provided by the operating system rather than user itself. But both DAC and MAC have fixed set of policies and there is no flexibility.

Role-Based Access Control (RBAC)

RBAC was developed as an authorization system [10] to be used by organizations and has become a prominent advance access control model [11].

This model was primarily designed to be used in organization system where employees were provided with access rights to the organization's resources. Employees in an organization, who perform similar tasks, require similar access rights to the resources. Therefore, some predefined "roles" are set and assigned to the designated members of the group and permissions to access were provided to these "roles". In this way, access management becomes much easier. But in a very large organization, role management become a huge task. In a large system with huge range of users, the RBAC model undergoes a problem known as "role explosion" where size of roles incremented exponentially.

3 Attribute-Based Access Control (ABAC)

ABAC is the general model which could comprehend the benefits of the models previously discussed while surpassing their demerits.

Terminologies used in ABAC [12]

- *Attributes* represent the subject, object, or environment properties in the form of name-value pair.
- A *subject* is any entity which is demanding an access request to an information asset that can be any process, or computer or human user.
- An *object* can be any resource of the computer system.
- An *operation* is the function that the subject wants to perform on an object.
- *Policy* represents rules or relationships that govern the possibility to determine whether a request should be granted or denied.
- *Environment conditions* are dynamic factors identifying the context in which access is requested.

The basic ABAC scenario is depicted in Fig. 1 in which the ABAC mechanism on receiving the subject's access request, and determines whether subject can perform the specific operation upon that object by examining the attributes of subject and object according to a specific policy set.

4 eXtensible Access Control Markup Language (XACML)

ABAC uses a processing model called XACML [13] developed by OASIS describing the procedure to evaluate requests to access resources according to the policy set rules and relationships. XACML defines both a security policy and an access control decision request/response language. As per OASIS standards, the XAMCL carries

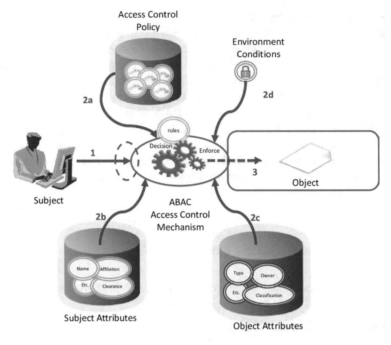

Fig. 1 Basic ABAC scenario [12]

out the function of data flow as follows (also depicted in Fig. 2). It has four layers for access policy control.

The Policy Administration Point (PAP): The role of PAP is to set up securities criteria and the selected policies are saved in the selected repository. The Policy Enforcement Point (PEP): The function of PEP is to carry out access to control. This is done by emphasizing authority decisions. Policy Information Point (PIP): The purpose of PIP is to save values of attributes or the data needed for examining policies. The Policy Decision Point (PDP): PDP checks out the appropriate policies and examines the appropriate policy and gives the authorization decision.

5 Implementation

5.1 Overview

For implementing ABAC, we have developed a sample web application for an organization to provide access to different resources like course materials and project files of the organization. In this application, subscribed users can access the courses

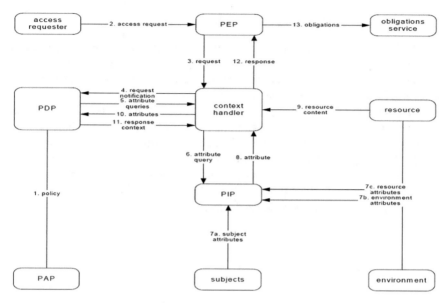

Fig. 2 XACML usage scenario [13]

according to the policies defined (which are explained later). And members can access the project files depending upon the attributes and policies.

5.2 Attributes

For implementing ABAC in our application, we have defined these attributes (Table 1).

In our system, we have taken two subjects, i.e., user and member. So we have taken "member name" and "user name" as subject's attributes to authenticate or distinguish them from other members and users respectively when they want to access the resources. Another attribute we have taken is "score" which decides the level (i.e., beginner, intermediate or expert) of user. Score of user depends on its performances in courses.

- If score ≤ 50 then user is Beginner.
- If 50 < score ≤ 100 then user is Intermediate.

Table 1 Subject's attribute

Subject	Attributes
User	User name, score
Member	Member name

- If $100 < \text{score} \leq 150$ then user is Expert (Table 2).

For users, there are three types of resources. For every course, we have defined an attribute "level" whose value depends upon its type, i.e., for Level 1 Courses (level = 1), Level 2 Courses (level = 2), Level 3 Courses (level = 3). Another attribute "name" is used to identify the course requested.

For members there are four types of resources: (1) Project code file, (2) Project test file, (3) Project executable file and (4) Project bug report. In project, we have defined six attributes, "Project name" contains the name of project which member wants to access, "File type" contains the type of the file which member want to access, i.e., code file, test file, exe file, bug report, "Head" contains name of the project head, "Tester" contains the name of the tester, "Developer" contains name of the developer, and "Debugger" contains name of the debugger.

5.3 Policy and Rules

Pseudo code of policies is given below:

```
<Policyid=1 rulecombalgo=permit override>
      <Target> subject_id=="user", resource_id=="course", action_id=="read"
      <Rule id=1 effect=Permit>
               <Condition>          subject.score<=50 and resource.level==1
      <Ruleid=2 effect Permit>
               <Condition>          50<subject.score<=100 and resource.level<=2
      <Ruleid=3effect =Permit>
               <Condition>          100<subject.score<=150and resource.level<=3
      <Rule id=4 effect=Deny>
```

To provide user access to the course, this policy will be used along with attribute values. If the user has score less than or equal to 50, then they can access only course files whose level is 1, while user having score greater than 50 but less than 100 can access both level 1 and level 2 courses. Similarly, user having score greater than 100 but less than 150 can access all three types of courses.

Table 2 Object's attribute

Object	Attributes
Course	Level, name
Project	File type, project name, head, tester, developer, debugger

```
<Policy id=2 rulecombalgo=permit override>
        <Target> subject_id=="member", resource_id=="course", action_id=="read"
        <Rule id=1 effect=Permit>
                <Target> resource_id=="codefile"
                <Condition>          membername=head||developer||debugger
        <Rule id=2 effect=Permit>
                <Target> resource_id=="testfile"
                <Condition>          membername=head||tester
        <Rule id=3 effect=Permit>
                <Target> resource_id=="bugreport"
                <Condition>          membername=head||tester||debugger
        <Rule id=4 effect=Permit>
                <Target> resource_id=="exefile"
                <Condition>          membername=head||tester||debugger||developer
        <Rule id=5 effect=Deny>
```

This policy is targeted for members, they can access code files if he/she is head/developer/debugger of the project and can access test files if he/she is head/tester. If the resource is bug report than head/tester/debugger can access it and exe file can be accessed by head/tester/debugger/developer.

6 Results

In our project, we have implemented ABAC and tried to analyze how ABAC provides better and more secure access control policies than conventional RBAC model. ABAC is the next generation access control model and is more appealing in terms of security and flexibility. In ABAC, permission to access resource will be granted if attributes of user have the same values as the attributes of resources. In RBAC, access is limited to the roles where permissions to resource are predefined or limited. But this will not be the case with ABAC, for changing permissions in it, only values of attributes need to be manipulated without changing policy sets.

RBAC has been the most widespread model and renders security and administrative advantages. However, its implementation is expensive; it is outdated, and ineffective to suit real-time access control parameters.

TrustBAC extends the conventional RBAC. It is based on trust calculations using fuzzy approach where access feedback is used for access control and trust levels are assigned to users.

In TrustBAC model, during a particular type of session instance invoked by user at an instant of time trust level is assigned to the user. Then for this session, the role associated with that trust level is assigned to him/her. Any user can play any role depending upon its trust level for that session. Also, more than one user can play a single role. Every role is specified by set of permissions in order to restrict the access of users only for particular operations on particular resources [14].

Though it is more flexible than RBAC it does not address security issues at fullest as role assignment is dependent on trust level which are calculated using fuzzy approach which makes it sometime hard to predict the changes or behavior of access model which is a very crucial issue.

Moreover, ABAC is newer, easier to implement, and suits for real-time access control parameters. Furthermore, RBAC is not able to accommodate dynamically varying attributes. Therefore, ABAC proves to be a better access control model than both RBAC and TrustBAC in term of both flexibility and security.

7 Conclusion and Future Work

This paper gives details about some access control models like DAC, MAC and RBAC and their limitations, which tells the need for new general model, i.e., ABAC. It also includes detail about policy language required for implementation, i.e., XACML. In the last part, it gives an overview of our work, i.e., implementation of ABAC in our web application, gives detail about the application and attributes needed to provide access control. This part also includes detail about policies which are used in our system. Overall it includes all detail how we have implemented ABAC in our web application.

Therefore, ABAC can be considered as next generation authorization model which can provide dynamic access permissions.

In the future, implementation of ABAC on more complex web applications can be done.

References

1. H. Li, S. Wang, X. Tian, W. Wei, C. Sun, A survey of extended role-based access control in cloud computing, in *Proceedings of the 4th International Conference on Computer Engineering and Networks. Lecture Notes in Electrical Engineering*, vol. 355 ed by W. Wong (Springer, Cham, 2015). https://doi.org/10.1007/978-3-319-11104-9_95
2. G. Kaur, E. Bharti, Securing Multimedia on Hybrid Architecture with Extended Role-Based Access Control. J. Bioinform. Intell. Control 3(3), 229–233 (2014). https://doi.org/10.1166/jbic.2014.1085
3. P.K. Behera, P.M. Khilar, A Novel Trust Based Access Control Model for Cloud Environment, in *Proceedings of the International Conference on Signal, Networks, Computing, and Systems. Lecture Notes in Electrical Engineering,* vol. 395, ed. by D. Lobiyal, D. Mohapatra, A. Nagar, M. Sahoo (Springer, New Delhi, 2017). https://doi.org/10.1007/978-81-322-3592-7_29
4. L. Zhou, V. Varadharajan, M. Hitchens, Trust enhanced cryptographic role-based access control for secure cloud data storage. IEEE Trans. Inf. Forensics Secur. 10(11), 2381–2395 (2015). https://doi.org/10.1109/TIFS.2015.2455952
5. S. Chakraborty, I. Ray, TrustBAC: integrating trust relationships into the RBAC model for access control in open systems, in *2006 SACMAT* (2006). https://doi.org/10.1145/1133058.11 33067
6. Q.M. Rajpoot, C.D. Jensen, R. Krishnan (2015) Integrating attributes into role-based access control, in *Data and Applications Security and Privacy XXIX. DBSec 2015*, vol. 9149, ed. by P. Samarati (Springer, Cham). https://doi.org/10.1007/978-3-319-20810-7_17

7. Q.M. Rajpoot, C.D. Jensen, R. Krishnan, Attributes enhanced role-based access control model, in trust, privacy and security in digital business, in *TrustBus 2015. Lecture Notes in Computer Science,* vol 9264, ed. by S. Fischer-Hübner, C. Lambrinoudakis, J. López (Springer, Cham, 2015). https://doi.org/10.1007/978-3-319-22906-5_1

8. X. Jin, Attribute-based access control models and implementation in cloud infrastructure as a service, May 2014 (The University of Texas, San Antonio), 160 .pp

9. D.E. Bell, L.J. LaPadula, Secure computer systems: mathematical foundations, vol. 1 (MITRE Corporation Bedford Massachusetts, 1973)

10. R.S. Sandhu, E.J. Coyne, H.L. Feinstein, C.E. Youman, Role-based access control models. Computer **29**(2), 38–47 (1996). https://doi.org/10.1109/2.485845

11. D.F. Ferraiolo, R. Sandhu, S. Gavrila, D.R. Kuhn, R. Chandramouli, Proposed NIST standard for role-based access control. ACM Trans. Inf. Syst. Secur. (TISSEC), **4**(3), 224–274 (2001). https://doi.org/10.1145/501978.501980

12. V.C. Hu, D. Ferraiolo, R. Kuhn, A. Schnitzer, K. Sandlin, R. Miller, K. Scarfone, Guide to attribute based access control (ABAC) definition and considerations, in NIST Special Publication 800-162, Jan 2014, NIST, Gaithersburg, Maryland (2014), 45 .pp. https://doi.org/10.602 8/nist.sp.800-162

13. eXtensible Access Control Markup Language (XACML) Version 3.0. 22 Jan 2013. OASIS Standard. http://docs.oasis-open.org/xacml/3.0/xacml-3.0-core-spec-os-en.html

14. O. Folorunso, O.A. Mustapha, A fuzzy expert system to trust-based access control in crowd-sourcing environments. Appl. Comput. Inform. **11**(2), 116–129 (2015). https://doi.org/10.101 6/j.aci.2014.07.001

An Approach to Implement Quality Tools in Inventory Management

Swadhin Kumar Nayak and D. K. Singh

Abstract The market in today's world has become lucrative and more sensitive towards time management. From the manufacturers' and suppliers' point of view, the strive is always towards increasing the productivity by the optimal utilization of resources. But on the other hand from a customer point of view, it is more oriented towards responsiveness of the suppliers. To gain the competitive advantage in the business, the focus is on cost advantage as well as value advantage. The value advantage is also aligned towards maintaining the inventory so that the right delivery at right time can be assured and also the inventory can be leveled up as soon as it reaches the predefined point. In this paper, the concept of quality tools is emphasized which includes employee involvement, continuous improvement in quality, proper handling of materials and it is related to the waste-free "Inventory management". The inventory as a waste being said earlier is highlighted here and the seven magnificent tools of quality are used to monitor and analyze the causes of wastage in the inventory. In this paper, the main focus has been put on Pareto analysis, Check sheet, Scatter diagram, and Fishbone diagram tools to analyze the issues related to inventory management. Thus, the elimination of the wastes by proper monitoring will help to maximize the full use of materials being present and then to order for the new lots. Hence, the scope of quality has been expanded to include the inventory within its boundary. In this paper, the raw materials and the sub-processed materials of a leading steel manufacturing industry is discussed and the main focus is on the inventory of raw materials by the quality tools to effectively monitor and to reduce its wastage.

Keywords Inventory · Seven tools of quality · Value advantage · Materials
Responsiveness

S. K. Nayak (✉) · D. K. Singh
Department of MPAE, NSIT, New Delhi, Delhi, India
e-mail: swadhinkumarnayak.1993@gmail.com

D. K. Singh
e-mail: dks662002@yahoo.com

© Springer Nature Singapore Pte Ltd. 2019
H. Malik et al. (eds.), *Applications of Artificial Intelligence Techniques
in Engineering*, Advances in Intelligent Systems and Computing 698,
https://doi.org/10.1007/978-981-13-1819-1_37

1   Introduction

TQM stands for "Total Quality Management" which explains that every corner or sphere is viewed to confirm that quality is maintained in the process, product confirmation, assembly and stock management. It is a coherent program or system of activities and policies which combine people, process, and technology in an institutional infrastructure that provides vision, incentive, and organizational support for achieving the goal of total quality. Whenever there lies a discussion about increasing the productivity, effective planning, quality control, process control, and inspection, it is seen that money plays a vital role in it which on the other terms is sensed as cost estimation, financing, budgeting, and profit maximization on the point of view of the manufacturer and then customer plays a vital role as one of the virtual stakeholder. Hence, it is a great need to bridge the gap between the customers' demand and the manufacturers' supply. Here, the importance of inventory can be visualized which tends to relate and equate the demand and the supply. Simply it can be explained as if inventory is kept in hand then only it can meet the future demand.

But on the contrary, if it is kept in excess there can be chance of spoilage, pilferage or direct damage due to external sources which not only increases the holding cost but also includes the cost of the waste material.

The theory of inventory control techniques classifies the various group of materials according to its cost and volume, which designates about how much to order and when to order so that the cost can be minimized and subsequently the demand can be met. Quality is Free [1] must be utilized which in turn increases the profit and proves to eliminate the wastage. The inefficient management of inventory of raw materials can turn to heavy loss in monetary terms which is more as compared to the qualitative management of raw materials. Quality aspects will clarify the certain key areas to be focused on so as to reduce the chances of wastage of inventory.

2   Seven Tools of Quality or the Magnificent Seven Used in Inventory Management

The magnificent seven tools of quality shows a fixed set of statistical techniques identified as being the most useful in troubleshooting the issues related to quality. These tools are basic in nature because they are suitable for employees with minimum training in statistics and also they can be used to monitor and analyze the important quality issues.

The seven quality tools include the following: (1) Pareto chart, (2) Cause and effect diagram, (3) Check sheet, (4) Control chart, (5) Scatter diagram, (6) Flowchart, and (7) Histogram.

Problem: A leading steel manufacturing industry analyzes the key areas due to which it finds unnecessary expenditure in the production process. It focuses only on production process earlier and had a tighter control on it. But still then, the cost did not reduce as expected. During analysis, it is found that the maximum cost is incurred in the raw materials department. The movement of materials is described below:

It has been observed that the raw materials inventory in the RMHP (Raw Materials Handling Plant) department is kept in an open environment after being received from mines, which tends to occupy a huge space making the area hazardous as well the materials are wasted. Ultimately during the urgent need, waste materials are being refined again and processed, which can have effect of producing low-quality steel. The raw materials get deposited due to environmental effect whose redressal incurs significant cost.

3 Cause and Effect Diagram and Its Use in the Analysis of Problem

The cause and effect diagram is widely also known as Fishbone Diagram or Ishikawa Diagram in Fig. 1. It identifies many possible causes for a particular problem. It can be used to structure a brainstorming session. It sorts the idea into useful clusters. The process has four major steps: identification of the problem; working out the major factors involved; identifying the possible causes in the corresponding factors and analyzing the cause and effect diagram, which are used to resolve numerous problems including risk management in production and services sector [2]. The layout provided by the diagram in the form of clusters helps team members to think in a very systematic approach. Some of the advantages of constructing a fishbone diagram are that it helps determine the root causes of a problem or quality characteristic using a structured approach, encourages brainstorming on the causes and concerns on group knowledge of the process, and identifies the areas where data should be collected for further study [2].

The application of cause and effect diagram in inventory management leads to define the causes of inventory wastage in an industry. Thus, it shows the alarming effect of the so-caused material loss which directly increases the lead time between the

Fig. 1 Cause and effect diagram [3]

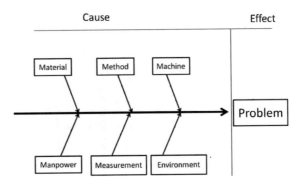

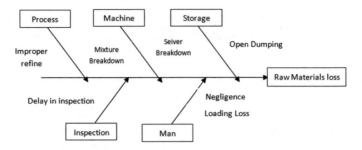

Fig. 2 Cause and Effect diagram of the inventory loss

departments resulting in lower productivity. A case has been analyzed below showing an industrial practice in which quality has been maintained in the production run process but due to mismanagement in inventory of raw materials, the cost increases. The causes of mismanagement in the raw material inventory are explained in the example below (Fig. 2).

4 Pareto Chart

It is named after Sir Vilfredo Pareto, although first described by the famous quality guru Sir Joseph M. Juran. This chart contains both bars and line graph, where individual values are represented in descending order by bars and the cumulative total is represented by the lines. The concept followed here is "Vital few and Trivial many" which shows that 80% of the important causes for any defect and 20% of minor causes.

Let from a monitoring report it is noted that there lies quality management issues. The causes of poor quality and its impact on the process are shown in Table 2 and Fig. 3.

The mentioned Table 1 and Fig. 3 highlights the important areas to be focused on and to take remedial steps to minimize the defects. The objective of a manufacturing

Fig. 3 A diagram for tallying the percentage of defects resulting from different causes to identify major quality problems from Table 1

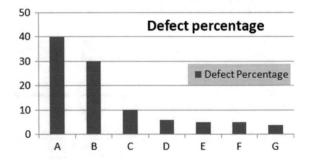

Table 1 Movement of materials within the departments of the plant

S.No.	Department	Materials		Transferred to
		In	Out	
1.	Raw materials handling plant (RMHP)	Iron ore, dolomite, quadrite, siderite	Refined raw materials	Sinter plant
2.	Sinter plant	Refined raw materials	Sinters	Blast furnace
3.	Coke oven	Coal	Coke	Blast furnace
4.	Blast furnace	Sinters	Molten iron	Basic oxygen furnace and continuous casting plant
5.	Bof/ccp	Molten iron	Steel billets, blooms, ingots	Section mill, wheel and axle plant
6.	Section mill	Steel billets, blooms	Rods, angles, channels	Customer
7.	Wheel and axle plant	Ingots	Wagon wheels, axle	Customer

Table 2 Causes of defect and their occurrence percentage

S.No.	Defects	Occurrence (percentage)	Priority
1.	A	40	Vital few causes
2.	B	30	
3.	C	10	
4.	D	6	Trivial many causes
5.	E	5	
6.	F	5	
7.	G	4	

firm is to focus on cost reduction which can lead to lean production system by reducing the wastages. With the help of Pareto analysis, the issues related to poor inventory in terms of keeping it in bad condition can be prevented. Pareto analysis being done on this issue: (Tables 2, 3 and Fig. 4).

5 Check Sheet and Its Use in Analysis of the Problem

A check sheet is a list in the form of a tabular format, prepared in advance to record data and is useful for later analysis. It is also represented as a tally sheet. There are five basic types of check sheets as given below [4]:

Table 3 Causes of defect in RMHP tending to regular inventory shortages of raw materials

S.No.	Defects	Occurrence (percentage)	Priority
1.	Open dumping	42	Vital few causes
2.	Improper refine	29	
3.	Loss due to loading	11	
4.	Employees negligence	4	Trivial many causes
5.	Sevier machine breakdown	7	
6.	Delay in sample testing	4	
7.	Mixture breakdown	3	

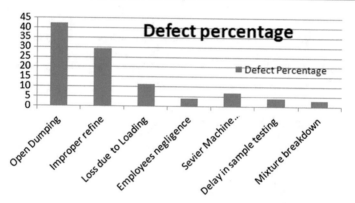

Fig. 4 Pareto chart showing the percentage of the causes of defect

(a) Classification type—It classifies the items under different categories
(b) Location type—It indicates the position or movement of an item
(c) Frequency type—It indicates the presence or absence of an item, and also the number of occurrences of that item
(d) Measurement scale type—It provides a measurement scale divided into certain intervals to enable easy marking
(e) Checklist type—It indicates the items or tasks to be performed to accomplish a task or process (Fig. 5).

6 Scatter Diagram and Its Use in Analysis of the Problem

While solving a problem or analyzing a situation, the relationship between two variables should be known (input variable is independent and output is dependent). A relationship may or may not exist between two variables. If a relationship exists,

CAUSES/WEEK	1ST	2ND	3RD	4TH
Raw materials handling plant (Monthly report on inventory depletion)				
Open Dumping	IIIIII	IIIII	IIIII	III
Improper refine	II	III	III	II
Loss due to Loading	III	II	II	
Employees negligence		II	I	
Sevier Machine breakdown	I	I		I
Delay in sample testing	I	II	II	I
Mixture breakdown	I	I	I	

Fig. 5 Checklist report for a month

it will highlight the impact of the input on the result which can be either positive or negative, it may be strong or weak. A tool that helps to study the relationship between two variables is known as Scatter Diagram. This diagram also determines the sensitivity analysis between the cause of wastage of materials in the problem mentioned above and its corresponding defect percentage rate. It consists of plotting a series of points representing the number of observations on a graph in which one variable is on X-axis (input) and its effect on the other variable in on Y-axis (output). If more than one set of values are identical, requiring more points at the same spot, a small circle is drawn around the original dot to indicate second point with the same values. The way the points lie scattered in the quadrant gives a better indication of the relationship between the two variables and better inferences can be drawn by the nature of curve [5]. It makes the operator and the shop floor manager aware about the trend and alarms to take a suitable corrective action.

The causes of inventory loss due to open dumping and improper refine have been shown using scatter diagram (Figs. 6 and 7).

Fig. 6 Scatter diagram showing the wastage due to open dumping

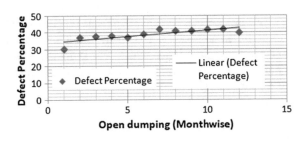

Fig. 7 Scatter diagram
showing the wastage due to
improper refine

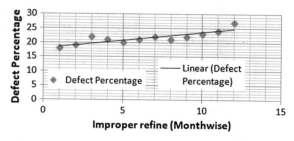

Activity	Department	Task Performed		Holding Time (days)
Raw Materials(Iron Ore) arrived	RMHP			7
Raw materials(Coal) arrived	Coke Oven			4
Sintering	Sinter Plant			3
Melting	Blast Furnace			2
Oxidizing	BOF			2

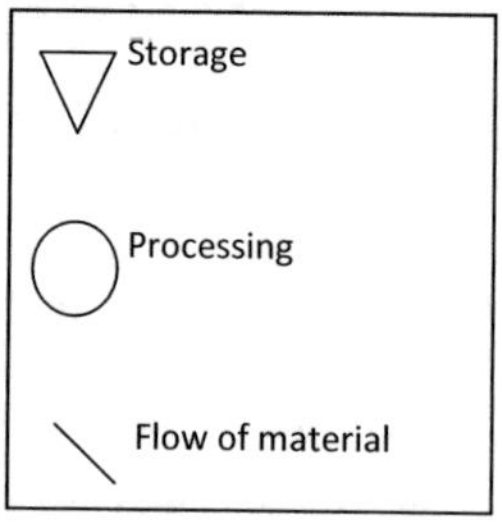

Fig. 8 Process flow chart

7 Process Flow Chart and Its Use in the Analysis of the Problem

A process flowchart is a diagram of the steps in a job, operation, or process. It enables all the employees involved in identifying and solving quality related problems to have a clear picture of how a specific operation works and a common frame of reference [5]. This process can be applied in managing the inventory by determining the pull and push demands of the customer. The above example can be analyzed by the flow chart in a descriptive which will show the holding time and the delivery of semifinished goods to the next department of the plant for next operational activity to be performed (Fig. 8).

8 Conclusion

Inventory is an important part of a business. Effective inventory management is essential for the successful operation of any business. Quality greatly affects inventory management because it is not only free but is a supreme source of profit. Poor quality products and services tend to increase inventory problems. The quality tools have great potential to improve quality that can be used to manage inventory effectively thereby reducing operational business costs.

References

1. P.B. Crosby, *Quality is Free: The art of Making Quality Certain* (Mc Graw Hill, 1979)
2. M. Hekmatpanah, Ardestan, The application of cause and effect diagram in the oil industry in Iran: the case of four liter oil canning process of Sepahan Oil Company. Afr. J. Bus. Manag. 5(26), 10900–10907 (2011)
3. Application of 7 QC Tools to Investigate the Rejection of Lathe Beds—Case Study of a Machine Tool Manufacturing Company
4. M. Magar, Dr. V.B. Shinde, Application of 7 quality control (7 QC) tools for continuous improvement of manufacturing processes. Int. J. Eng. Res. Gen. Sci. 2(4) (2014)
5. R.S. Russell, B.W. Taylor, *Operations Management: Creating Value in Supply Chains*, 5th edn. (Wiley)

Stock Prediction Using Machine Learning Algorithms

Pahul Preet Singh Kohli, Seerat Zargar, Shriya Arora and Parimal Gupta

Abstract Market systems are so complex that they overwhelm the ability of any individual to predict. But it is crucial for the investors to predict stock market price to generate notable profit. The ultimate aim of this project is to predict the behavior of Bombay Stock Exchange (BSE). We have taken into factors such as Commodity Prices (crude oil, gold, silver), Market History, and Foreign exchange rate (FEX) that influence the stock trend, as input attributes for various machine learning models to predict the behavior of Bombay Stock Exchange (BSE). The performances of the models are then compared against other benchmarks. A structured relationship was also determined among the different attributes used. The gold price attribute was found to have the highest positive correlation with market performance. The AdaBoost algorithm performed best as compared to other techniques.

Keywords Stock prediction · BSE index · Machine learning algorithms · Stock prediction classification

1 Introduction

Historically, high market prices often make the investors despondent from investing, while low market prices represent an opportunity. Predicting stock market price, therefore, becomes imperative for investors to yield a significant profit. Though predicting the financial markets and the stock movements is onerous [1], many

P. P. S. Kohli (✉) · S. Zargar · S. Arora · P. Gupta
Bharati Vidyapeeth's College of Engineering, New Delhi, India
e-mail: pahulpreet86@gmail.com

S. Zargar
e-mail: seeratzargar1996@gmail.com

S. Arora
e-mail: shriyaarora080696@gmail.com

P. Gupta
e-mail: guptaparimal1996@gmail.com

© Springer Nature Singapore Pte Ltd. 2019
H. Malik et al. (eds.), *Applications of Artificial Intelligence Techniques in Engineering*, Advances in Intelligent Systems and Computing 698,
https://doi.org/10.1007/978-981-13-1819-1_38

405

researchers from different fields have scrutinized and used many algorithms and different combination of attributes to predict the market movements. But these algorithms are all on the basis of stock price itself which has random property.

In this project, we have proposed the use of macroeconomic factors such as commodity price, market history, and foreign exchange rate to predict the Bombay Stock exchange (BSE). These are some of the vital factors [2, 3] that predict whether BSE will increase or decrease on a particular day. The project is implemented in Ipython Notebook.

The rest of the paper is organized as follows. Section 2 gives a brief overview of machine learning algorithms. Section 3 describes the method that is proposed for the implementation of models on the stock market data. Section 4 presents the simulation and test results of the paper. Section 5 concludes this paper.

2 Methodology

In this research, four machine learning algorithms are used and compared on the basis of their training accuracy. These models are as follows.

2.1 Support Vector Machines

SVM algorithm is based on statistical learning hypothesis. Both regression and classification can be done using this supervised machine learning algorithm. SVM can also be used for outlier's detection [2, 3]. The data points are plotted in the n-dimensional space. SVM performs classification by hyperplane which is a boundary constructed over the dataset. The hyperplane separates the cases of different class labels. This hyperplane is constructed in the multidimensional space as given in Fig. 1. To find the right hyperplane in the case of classification, the distance between the nearest data points called support vectors and the decision boundary is maximized.

Hence, minimization of the norm of the vector w is needed; where w define the separating decision boundary. This is analogous to maximizing the margin between the two classes [4]. Considering the above figure, if we assume u to be some unspecified data point and was a vector which is perpendicular to the hyperplane, then the decision rule in SVM is given by:

$$\vec{\omega} \cdot \vec{u} + b \geq 0 \qquad (1)$$

Fig. 1 Decision boundary in
support vector machine

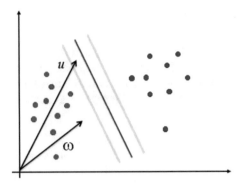

The maximization of width of the hyperplane is required for the expansion of the
spread

$$W = \left[\frac{2}{||\omega||} \right] \tag{2}$$

$$W = max \left[\frac{2}{||\omega||} \right] \tag{3}$$

2.2 Random Forest

Random forest [5] is a machine learning algorithm which uses ensemble method for
classification and regression problems. It uses the benefits of both Decision Trees and
Bagging (Bootstrap Aggregation). Thus, it overcomes the problem of overfitting in
decision trees. Bagging reduces the variance of high variance algorithms like decision
trees (CART). Random forest is a group of unpruned classification and regression
trees that are obtained from the subsamples of the training dataset.

In this algorithm, the process is as follows:

- The subsamples from the training dataset are created.
- CART model is applied to each of the subsamples to obtain a predicted output
  depending on the model.
- The bagging process is applied to get the ensemble of the predicted outputs of
  each model.
- Thus by this process, the predicted output has a lesser prediction error than each
  of the individual model.

2.3 Gradient Boosting

Gradient boosting is a kind of a boosting algorithm that trains many models sequentially. The loss function is gradually minimized by each new model. In Gradient Boosting, we assume a uniform distribution say D1 which is $1/n$ for all n observations. Then the algorithm progress according to the following steps:

- We assume an α_t.
- Calculate a weak classifier $h(t)$.
- Update the population distribution for the next step.
- The new population distribution is used again to find the next learner.
- Iterate Step 1–Step 4 until no hypothesis is found which can further improve the accuracy.
- Take the weighted average of the frontier using all the learners used till now.

2.4 Adaptive Boosting (AdaBoost)

Adaptive Boosting being [6] one of the first successful boosting algorithms can be used for both classification and regression processes. It involves ensemble of weak classifiers to build a strong one. Adaptive Boosting focuses on predicting current data set by giving equal weight to each attribute. If the prediction is incorrect, then it gives higher weight to the incorrect observation [7]. The iteration continues till almost no error is received. The basic prediction is made using a basic algorithm or decision stumps. The final output is predicted by calculating the weighted average of each weak classifier. Since the weights are calculated on the basis of the false predictions and changed accordingly, this algorithm is adaptive in nature, hence the name Adaboost.

3   Proposed Model

The idea implemented in this research is to use different attributes such as commodity prices, market history and foreign exchange rate as input attribute to predict the Bombay Stock Exchange (BSE) [8]. These input attributes were continuous numeric value of varied range, so in order to classify them, they were normalized as [−1, 1]. This is because all the input attributes can have either positive or negative values. The paper compares the outputs of all the four machine learning algorithms used. The output of each model is either 1 or −1 to describe either positive or negative impact on the market, respectively. All the factors are discussed separately (Fig. 2).

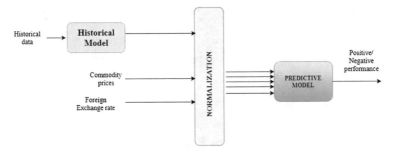

Fig. 2 Proposed model

3.1 Factors

The factors that affect the stock market prices are as follows:

- **Market History**

  A historical data of 2 years from January 2015 to December 2016 was collected. This historical data was not applied to the model directly but after applying a model based on historical data which was training on factors such as change in opening price, low price, change in volume of stocks, and high price.

- **Commodity Price**

  The price change of various different commodities (attributes) such as gold, silver, and crude oil has an impact on the overall change in the stock prices of the BSE.

- **Foreign Exchange**

  The foreign exchange rate change is known to play a vital role in the market performance. Foreign exchange rate between the INR (Indian National Rupee) and USD (United States dollar) is used as an input attribute in the model.

Based on the above factors, a total of 5 attributes are used as an input to the model and output as +1 (positive market) or −1 (negative market) is found.

3.2 Scope

A data spread of over 9 months from January 2017 to September 2017 is used in the research. All the data is taken from website https://www.investing.com (Fig. 3).

Fig. 3 Variation in input and output attributes

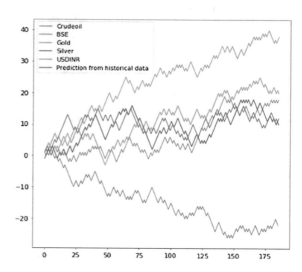

4   Simulation and Results

The data of different attributes was extracted and manipulated to convert it into a form that can be used as an input in model. The output of this model is +1 to represents positive market which describes the increase in BSE prices and −1 to represent negative market which describes decrease in BSE prices. The stimulation was performed using four different algorithms and accuracy for each was calculated from train/test split method. In this method, the data is split into training data and test data. The training set contains a known output and the model learns on this data in order to be generalized to other data later on.

4.1   Implementation of Different Algorithms

The abovementioned machine learning algorithms were separately and successfully implemented on the dataset in python and accuracy for each model was calculated using different values of training and test dataset size. It observed an accuracy of 68.4% for 90% training data and 73% for 70% training data for Random Forest. For SVM, the accuracy was 78.95% for 90% training data and 73% for 70% training data. Accuracy for Gradient Boosting being equal to 78.95 for 90% training data and 73.2% for 70% training data. For AdaBoost, the accuracy is 78.95% for 90% training data and 77% for 70% training. Figure 4 shows the comparison between accuracy of different algorithms in the form of bar graph.

Fig. 4 Accuracy
comparison of different
machine learning algorithms

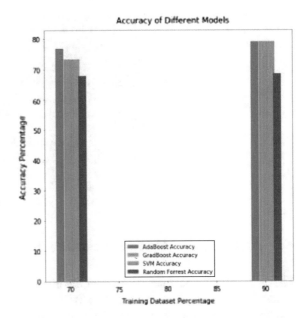

4.2 Dependency of Market Performance on the Attributes

Correlation and covariance between various attributes and market performance were
calculated. The resulting dependency of the attributes in the form of correlation and
covariance is described in Table 1.

It is evident from Table 1 that the attribute "Gold Rate" shows maximum positive correlation with market that is one variable decreases as the other variable
decreases, or one variable increases while the other increases. Also the attribute
"Foreign Exchange rate" shows maximum negative correlation with market that is
when one variable increases as the other decreases and vice versa. The attribute that
has the least impact was found to be "Silver Rate". The following Fig. 5 shows
scatter plot between different attributes for spotting structured relationship between
different attributes and market performance.

Table 1 Correlation and covariance between different attributes and market performance

S. No	Attributes	Correlation	Covariance
1	Oil rate	−0.7451	−102579.41
2	Gold rate	0.75293	16781.41
3	Silver rate	−0.2866	−114.30
4	Foreign exchange rate	−0.8961	−698.39

Fig. 5 Scatter plot between
market performance and
different attributes

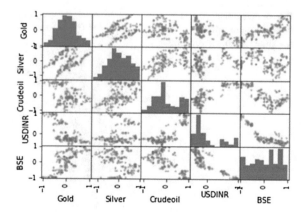

4.3 Evaluation of Machine Learning Algorithms

The accuracy of all four machine learning algorithms over different values of training
dataset and test dataset are compared in Table 2. From the given table, it is evident
that the accuracy of AdaBoost is the highest in all of the four algorithms.

The AdaBoost was selected as the best predicting algorithm and was tested on
untrained dataset and accuracy was found to be 75%. Figure 6 shows the scatter plot
of actual and predicted market performance by AdaBoost. The number of instances
is represented by the x-axis and the y-axis represents the class.

Table 2 Accuracy of different predictive models

Machine learning algorithms	Dataset used for verification		
	Training set (%)	Test set (%)	Accuracy (%)
AdaBoost	90	10	78.95
	70	30	76.79
Gradient boosting	90	10	78.95
	70	30	73.21
SVM	90	10	78.95
	70	30	73.00
Random forest	90	10	68.4
	70	30	67.8

Fig. 6 Actual and predicted market performance by AdaBoost

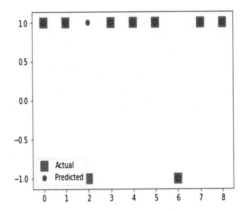

5   Conclusion and Future Scope

The outcome of this research concludes that the machine learning algorithms can be used to predict the increase or decrease in the stock market performance. It verifies the dependency of BSE on the factors taken in the study. Our findings confirm that the dependency of BSE is highest on the gold rate, since the correlation factor is highest. Also, the correlation factor is lowest for silver rate, showing least dependency of BSE on it. Of all the machine learning algorithms used, AdaBoost shows the highest accuracy of 76.79% for 70% training data and 75% for untrained data. There is still a scope of improvement in this project. The project can be further extended to include additional variables such as interest policy, political, and economic reforms to get more accurate results.

References

1. A. Nayak, M. Pai, R. Pai, Prediction models for Indian stock market. Proced. Comput. Sci. **89**, 441–449 (2016)
2. Z. Hu, J. Zhu, K. Tse, Stocks market prediction using support vector machine, in *2013 6th International Conference on Information Management, Innovation Management and Industrial Engineering* (2013)
3. Wei Huang, Yoshiteru Nakamori, Shou-Yang Wang, Forecasting stock market movement direction with support vector machine. Comput. Oper. Res. **32**(10), 2513–2522 (2005)
4. C. Cortes, V. Vapnik, Support vector networks. Mach. Learn. **20**, 273–297 (1995)
5. J. Ali, R. Khan, N. Ahmad, I. Maqsood, Random forests and decision trees. Int. J. Comput. Sci. Issues **9**(5), 272–278 (2012)
6. S. Yutong, H. Zhao, Stock selection model based on advanced AdaBoost algorithm, in *2015 7th International Conference on Modelling, Identification and Control (ICMIC)* (2015)
7. P. Wu, H. Zhao, Some analysis and research of the AdaBoost algorithm, in *Communications in Computer and Information Science* (2011), pp. 1–5
8. L. Zhao, L. Wang, Price trend prediction of stock market using outlier data mining algorithm, in *2015 IEEE Fifth International Conference on Big Data and Cloud Computing* (2015)

9. M. Usmani, S. Hasan Adil, K. Raza, S. Ali, Stock market prediction using machine learning techniques, *ICCOINS* (2016)
10. J. Ali, R. Khan, N. Ahmad, I. Maqsood, Random forest and decision trees. Int. J. Comput. Sci. Issues **9**(5), 272–278 (2012)
11. A. Bhargava, A. Bhargava, S. Jain, Factors affecting stock prices in India: a time series analysis. IOSR J. Econ. Finance **07**(04), 68–71 (2016)
12. N. Pahwa, N. Khalfay, V. Soni, D. Vora, Stock prediction using machine learning a review paper. Int. J. Comput. Appl. (0975–8887) **163**(5), 36–43 (2017)

Fuzzy Concept Map Generation from Academic Data Sources

Rafeeq Ahmed and Tanvir Ahmad

Abstract Increasing unstructured data in the internet by academics, digital document, and social media leads to task of mining and representing the hidden knowledge correctly. But in polysemic words, many morphologically similar terms, difficulty in human languages makes the task much difficult for the machine while humans understands and disambiguates the meaning of text by context. Text mining, being an important field of research, has many ground-level challenges because of no common platform, especially for free-flowing text or natural languages. Extracting hidden basic information in the text is clearly a major challenge. So, important topics or concepts, which user wants to learn, are mined with their fuzzy context. Taxonomies or concept hierarchies' plays an important role in any knowledge representation system like online learning or E-learning. Application of concept extraction is creation of concept maps, a knowledge representation technique, for domains like E-learning. In this paper, we have mined the important topics or concept. We have then applied windowing process to get context vector of these concepts. We have applied Mutual Information (MI) and Balanced Mutual Information (BMI) techniques to calculate fuzzy membership values to get fuzzy context vector. The final results have reflected the fuzzy distance between concepts with the aim to implement a concept map learning system with unsupervised learning from the academic data sources.

Keywords Text mining · Concept mining · Concept map

R. Ahmed (✉) · T. Ahmad
Computer Engineering Department, Jamia Millia Islamia, New Delhi, India
e-mail: rafeeq.amu@gmail.com

T. Ahmad
e-mail: tahmad2@jmi.ac.in

© Springer Nature Singapore Pte Ltd. 2019
H. Malik et al. (eds.), *Applications of Artificial Intelligence Techniques in Engineering*, Advances in Intelligent Systems and Computing 698,
https://doi.org/10.1007/978-981-13-1819-1_39

415

1   Introduction

Text mining or knowledge discovery from unstructured data is a source of very high commercial values. Unstructured data points to storage of information which is neither with any schema nor stored in a predefined manner. Text provides storage of information in the most natural form. A recent study showed that an information in a company stored in Digital form is greater than 80%. However, text mining, being more complex task than that of data mining, involves dealing with the underlying non-organized and unclear data, i.e., fuzzy values.

Getting deep insights into unstructured documents like research papers, textbooks, articles, news, and blogs are really very difficult task. The reason is that the content of document is free-flowing natural English language, semantics extraction of important concepts, which humans can understand by analyzing the contexts of the concept [1]. Knowledge retrieval or acquisition from knowledge-based system is not much efficient. To overcome this problem partially, we propose a novel framework for automatic learning of concept hierarchy from a text corpus. Work is being done for building up a knowledge base using existing documents by learning composite concepts. Extraction of concept is basically taken from key phrase extraction in information retrieval as well as area of text mining. A framework has been designed and proposed for concept mining from academic text (unstructured format). This framework can be generalized for different input format like ppt, html pages, pdf documents, and so on. We used only few documents to check the accuracy.

Because of the existence of ambiguity in mappings of words to concepts, several possible concepts can be derived from each word in a given language. When a single concept is pointed by all the terms in a query, then arriving at the solution is easy. But as we know the text mining contains fuzziness, so when separate concepts referred by many morphologically similar terms (showing fuzzy behavior), then arriving at a solution becomes difficult. Before applying any knowledge generation or representation techniques to such polysemic word in text mining, word sense disambiguation becomes imperative.

The paper is organized in the following sections: Sect. 2 contains related work. Methodology has been discussed in Sect. 3. In Sect. 4, we have shown the results of the experimental analysis. Conclusion and future work has been discussed in Sect. 5.

2   Related Work

Extraction of concept is basically taken from key phrase extraction in information retrieval as well as area of text mining. One of the techniques to analyze documents is key phrase extraction methods by finding relevance of a phrase or concept which can be a single word or multiword. Statistical features like Frequency of key phrase or linguistic features like Parts of Speech (POS) can be modeled to determine the relevance of a phrase, i.e., passing certain threshold will make them key phrase or concepts. As we know Term Frequency–Inverse Document Frequency (TF–IDF) has been used mostly in the areas of machine learning and information retrieval. The intention behind this procedure is the phrases that often appear in the document, but perhaps the document collectively, often plays a key role in distinguishing between the documents. In this way, document rating and document clustering in text-based applications are widely represented using a bag of words.

Efficient representation of knowledge can help deeper understanding of the content of the document [2]. Semantic Net, Conceptual graphs, Frames, Logic, and Rules establish the base of Knowledge Representation mechanisms and models. Some of the common technology behind the representation of knowledge is learning object and metadata based on ontology, attributing to rising learning standards like Shareable Content Object Reference Model (SCORM), IEEE Learning Object Metadata (IEE LOM) and binding with Web Ontology Language (OWL), Resource Description Framework (RDF), and Extended Markup Language (XML). In "*Ontology-based metadata*" knowledge representation scheme, it characterizes the formal general consent about meaning or sense of data, enabling metadata effective and more general. And thus, domain concepts can be represented using it [3]. "*Learning Object*" is basically a digital resource which can be used for learning [4]. Getting small modules by subdividing learning materials/content, learning object has many plus points like reusability, accessibility, etc. [3, 5]. "*Semantic Link Network*" (SLN) made up of nodes (entities, features, communities, schemas, or concepts) and semantic links between nodes. "*Domain Concept*" forms the basis for constructing domain knowledge structure, including domain concepts and relationships among them.

3 Methodology

We have made certain assumptions that frequent keyword found in the document represents one essential concept, a fundamental entity for construction of concept maps for e-learning. In general, keywords can be used to represent model, concept, theory, methodology, method [6]. Since the documents are unstructured, document preprocessing such as stop word removal, POS tagging, and word stemming are performed [7]. Next step is to find information form context distance and morphology. Context distance using fuzzy set is a measure that we use to find the meaning of the word unambiguously. Extracting the semantics of lexical elements or concepts can be done from the idea of collocational expressions from computational linguistics. Collocational expressions can be elaborated as a set of words similar in meaning, and most of the basic words in the expression unit are found in close proximity to some adjacent words in a textual unit [8].

3.1 Documents Preprocessing

After tokenization and stop word elimination, we get valuable set of words. Parts of Speech (POS) tagger can also be applied to extract noun-noun or noun adjectives, etc., but we have used frequency as a parameter to keep important concept as this will only slightly reduce the accuracy of the results. Stemming can be done using Porter stemmer. After this, frequent keywords are selected as keyword based on certain threshold value, taking into consideration various parameters as number of documents and total number of words after stemming and stop word removal.

3.2 Fuzzy Context/Semantics Vector Generation

In linguistics, distributional hypothesis says that terms (concepts) are semantically similar with the extent of sharing similar linguistic contexts, i.e., they collocational terms as chances of their co-occurrence probability is greater. For getting the context vector of a concept using windowing process [9], we have to obtain the statistical information among tokens, so start scanning the document with a δ words of virtual window to add all words in the vector if target concept is present in the window else next window is scanned. We obtained context vector as a result. For example, for the concept "ONTOLOGY", we have

ONTOLOGY {fuzzy, domain, concept, learn, extract, method, map, algorithm}

To obtain fuzzy membership values of elements in the context vector, potential effect of "close" words, "close" in context, can be calculated over all instances of the word in the data set. For each concept C, a context vector with fuzzy membership value is defined as

$$\text{Concept } C_i = \left\{ t_1(\mu_{c_i}(t_i)), \ t_2(\mu_2) \dots t_m(\mu_m) \right\} \tag{1}$$

where t_i is the term in the local context obtained through windowing process. μ_i is the weight of the term t_i for the corresponding concept, which is calculated through windowing process given below

$$\mu_i = \text{Fr}\left(T_i/C_j\right) / \text{Fr}\left(C_j\right) \tag{2}$$

Here, $\text{Fr}(T_i/C_j)$ is the number of window containing the term t_i and Concept C_j. $\text{Fr}(C_j)$ is the number of window containing the concept C_j

3.3 Mutual Information Between Two Terms

Mutual information is a technique to compute the dependency between two entities and is defined by [1]

$$MI(t_i, t_j) = \log_2\left(\frac{\mathbf{Pr}(t_i, t_j)}{\mathbf{Pr}(t_i) * \mathbf{Pr}(t_j)}\right) \tag{3}$$

Here, $MI(t_i, t_j)$ represents information shared mutually between term t_i and term t_j. Probability of appearing both terms together in a window is calculated by $Pr(t_i, t_j)$ and probability of getting a term in window is represented by $Pr(t_i)$. The probability $Pr(t_i)$ is estimated based on w_t/w, where w_t is the number of windows containing the term t and w is the total number of windows constructed from a corpus [9]. Simplifying the above equation, we get

$$R(t_i, t_j) = \log_2\left(\frac{w_{ij}(t_i, t_j) * w}{w(t_i) * w(t_j)}\right) \tag{4}$$

where $R(t_i, t_j)$ is the same as $MI(t_i, t_j)$ between term t_i and term t_j. $w(t_i, t_j)$ is the count of co-occurrence that both terms appear in a text window and $w(t_i)$ is the count that a term t_i appears in a text window.

The corpus relevance/mutual information of two words or membership function of term with its concept can be obtained by doing the windowing process over whole dataset, but work has been done to calculate the corpus relevance between two terms/concepts or words by many methods including Kullback–Leibler divergence (KL), Jaccard (JA), Normalized Google Distance (NGD), conditional probability (CP), Expected Cross Entropy (ECH), and Balanced Mutual Information (BMI) [3]. MI and BMI methods are used to calculate the membership values of terms with the concept and results are compared for each technique. BMI gives more effective results. Simplifying the BMI with window size and window frequency for $Pr(\neg t_i, \neg t_j)$ and others, we have Eq. (6).

$$
\begin{aligned}
BMI(t_i, t_j) = \beta \times & \left[Pr(t_i, t_j) \log_2(\frac{Pr(t_i, t_j) + 1}{\mathbf{Pr}(t_i)\mathbf{Pr}(t_j)}) + Pr(\neg t_i, \neg t_j) \log_2(\frac{Pr(\neg t_i, \neg t_j) + 1}{Pr(\neg t_i)\, Pr(\neg t_j)}) \right] \\
- (1-\beta) \times & \left[Pr(t_i, \neg t_j) \log_2(\frac{Pr(t_i, \neg t_j) + 1}{Pr(t_i)\, Pr(\neg t_j)}) + Pr(\neg t_i, t_j) \log_2(\frac{Pr(\neg t_i, t_j) + 1}{Pr(\neg t_i)\, Pr(t_j)}) \right]
\end{aligned} \tag{5}
$$

Simplifying the BMI with window size and window frequency for $Pr(\neg t_i, \neg t_j)$ and others, we have

$$
\begin{aligned}
BMI(t_i, t_j) = \beta \times & \left[\frac{w_{ij}}{w} \log_2(\frac{(w_{ij} + w) \times w}{w_i \times w_j}) + \frac{(w - w_{ij})}{w} \log_2(\frac{(2w - w_{ij}) \times w}{(w - w_i) \times (w - w_j)}) \right] \\
- (1-\beta) \times & \left[\frac{(w_i - w_{ij})}{w} \log_2(\frac{(w_i - w_{ij} + w) \times w}{w_i \times (w - w_j)}) + \frac{(w_j - w_{ij})}{w} \log_2(\frac{(w_j - w_{ij} + w) \times w}{(w - w_i) \times w_j}) \right]
\end{aligned} \tag{6}
$$

3.4   Concept-Concept Relationships

After generation of fuzzy context vector of important concept from the document, we have calculated the distances between the two concept using different techniques. The distance of the two concepts is based on their distance between context vectors which is ultimately based on the mutual information/corpus relevance between the words in the context vector. We have used windowing process for getting mutual information value between two words as a parameter of relatedness between two words. High value indicates strong relationship between words. This approach of semantic analysis extracts relevant and useful word correlations which might not be present in lexicon. In this approach of finding concept to concept distance or relationship, we first computed the mutual information between the corresponding relations between words in two context vectors.

Suppose context vector C_1 and C2 consists of n terms $t_{1,c_1} \ldots t_{n,c_1}$ and context vector C_2 consists of n terms: $t_{1,c_1} \ldots t_{n,c_2}$, for every term t_{i,c_1} from first vector, $i = 1$ to n, we calculate mutual information by doing windowing process with every term t_{j,c_1} from second vector, $j = 1$ to n. We then store the n max value with highest mutual information of every term from first vector to terms in second vector, i.e., store the n values from 1 to n from the following formula:

$$\{t_{i,c_1}, t_{m(i),c_2}\} = \max_{j=1 \text{ to } n} \{MI(t_i, c_1, t_{j c_2})\} \tag{7}$$

where m (i) means the match of i, for $i = 1 \ldots n$, we compute the distance between the two context vectors based on the matching in the first step and the precomputed mutual information. If we represent the two context vectors as

$$C_1 = \{t_1(\mu_{c_1}(t_1)), t_2(\mu_{c_1}(t_2)) \ldots t_n(\mu_{c_1}(t_n))\} \tag{8}$$
$$C_2 = \{t_1(\mu_{c_2}(t_1)), t_2(\mu_{c_2}(t_2)) \ldots t_n(\mu_{c_2}(t_n))\} \tag{9}$$

then context distance is computed as follows:

$$\text{Dist}(c_1, c_2) = \frac{1}{n} \sum_{i=1}^{n} R(t_i, t_{m(i)}) \times \mu_{c_i}(t_i) \times \mu_{c_2}(t_{m(i)}) \tag{10}$$

where $R(t_i, t_{m(i)})$ means mutual information of t_i and $t_{m(i)}$. Now, the average distance between two concepts is obtained by dividing $R(t_i, t_{m(i)})$ (distance) with the vector size n. We obtained a distance in the range of 0–1.

Algorithm:- Concept_Map_Generation
Input: n Input Document: ID
Output: Concept Map Generation
Procedure
1. For i=1 to n
    1.1 Doc_i =Read document d_i
    1.2 Pre process Doc_i
        1.2.1 Remove stop words, redundant data, etc & do stemming
        1.2.3 Select the word or term t_i
        1.2.4 for each $t_i \in Doc_i$, Calculate the frequency f_i for t_i
        1.2.5 Store terms for document d_i in t_{d_i} & corresponding frequencies in Fr_{d_i, t_i}
2. Selection of Concepts C_i
    2.1 for each document $d_i \in ID$
        2.1.1 If Frequency $Fr_{d_i, t_i} >$ threshold value $ConceptArr_{d_i} = ConceptArr_{d_i} \cup t_{d_i}$
3. for each concept $C_i \in$ Concept $Array_{d_i}$
        3.1 Construct text window $w \in d_i$
        3.2 Calculate the joint probability $Pr(c_i, t_j)$ of the term t_j with C_i, $\forall\, t_j \in C_i$
        3.3 Calculate the membership value using BMI, MI
        3.4 Construct the Context Vector, $C_i = \{t_1(\mu_{c_i}(t_1)), t_2 (\mu_{c_i}(t_2)) . t_n((\mu_{c_i}(t_n))\}$
4. for each concept $C_i \in ConceptArray_{d_i}$
    4.1 For each concept $C_j \in ConceptArray_{d_i}$
        4.1.1 $\{t_{k,c_i}, t_{m(k), c_j}\} = \max_{k=1\,to\,n} \{ MI(t_{k, c_i}, t_{j, c_j}) \}$
        4.1.2 $d(C_i, C_j) = \frac{1}{n}\sum_{k=1}^{n} R(t_k, t_{m(k)}) \times \mu_{c_i}(t_k) \times \mu_{c_j}(t_{m(k)})$
5. Display the semantic distance of C_i with C_j.

4 Experimental Analysis

4.1 Data Sets Used

We have taken the several research papers itself for finding out the important concept which can help in populating knowledge base. The article with reference number [3, 7, 10–16] has been used as data set for our work.

4.2 Performance Evaluation

Find the relevant semantics in the documents helps to represent the knowledge and provides deep insight into the data. We have compared the membership values of the terms with the concept using Balanced Mutual Information (BMI) keeping $\beta = 0.75$ with Mutual Information (MI) formula. Out of 9 documents, 11 concepts has been

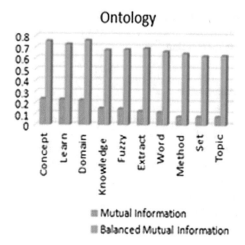

Fig. 1 CV for "Ontology"

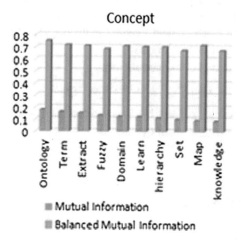

Fig. 2 CV for "Concept"

extracted with a threshold of 33%. Context vector has been generated for them with each techniques explained above. The context vector has been shown graphically in Figs. 1, 2 and 3.

The concept map generated for the given academic data sources is shown in Fig. 4. Important learning concepts in the document are those topics which the reader wants to learn or study online from the documents. Here, important concepts are represented by nodes and edges are relationships between them with 0 as min and 1 as max.

Fig. 3 CV for "Knowledge"

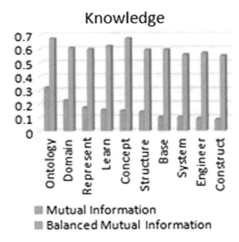

Fig. 4 Concept map for the nine journal papers with MI method

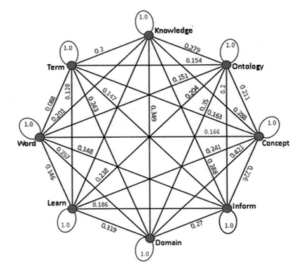

5 Conclusion and Future Work

The paper has demonstrated the technique of unsupervised concept map generation from academic research papers which can be extended for articles, documents, pdfs, ppts, blogs, emails, online learning forums, etc. This work provides a better semantic view of document and can helps us to give deep insights into the research articles by doing mathematical modeling of it, especially in the era of Big data where it is really very difficult to read all the data.

References

1. P. Gärdenfors, *Conceptual spaces: The Geometry of Thought* (MIT press, 2004)
2. S. Naidu, E-learning: a guidebook of principles, procedures and practices. Commonwealth Educational Media Centre for Asia (CEMCA) (2006)
3. R.Y.K. Lau, D. Song, Y. Li, T.C.H. Cheung, J.-X. Hao, Toward a fuzzy domain ontology extraction method for adaptive e-learning. IEEE Trans. Knowl. Data Eng. **21**(6), 800–813 (2009)
4. C.-H. Lee, G.-G. Lee, Y. Leu, Application of automatically constructed concept map of learning to conceptual diagnosis of e-learning. Expert Syst. Appl. **36**(2), 1675–1684 (2009)
5. G. Salton, M. McGill, *Introduction to Modern Information Retrieval* (McGraw-Hill, 1983)
6. A. Zadeh, Fuzzy sets. Inf. Control **8**(3), 338 353 (1965)
7. R.Y.K. Lau, J.X. Hao, M. Tang, X. Zhou, Towards context-sensitive domain ontology extraction, in *40th Annual Hawaii International Conference on System Sciences, 2007, HICSS 2007* (IEEE, 2007), pp. 60–60
8. N.-S. Chen, C.-W. Wei, H.-J. Chen, Mining e-Learning domain concept map from academic articles. Comput. Educ. **50**(3), 1009–1021 (2008)
9. H. Jing, E. Tzoukermann, Information retrieval based on context distance and morphology, in *Proceedings of the 22nd Annual International ACM SIGIR Conference on Research and Development in Information Retrieval* (ACM, 1999), pp. 90–96
10. Y. Wu, S. Zhang, W. Zhao, Towards learning domain ontology from legacy documents, in *Fourth International Conference on Digital Society, 2010, ICDS'10* (IEEE, 2010), pp. 164–171
11. X. Zhou, Y. Li, P. Bruza, Y. Xu, R.Y.K. Lau, Two-stage model for information filtering, in *IEEE/WIC/ACM International Conference on Web Intelligence and Intelligent Agent Technology, 2008, WI-IAT'08*, vol. 3 (IEEE, 2008), pp. 685–689
12. M.-F. Moens, R. Angheluta, Concept extraction from legal cases: the use of a statistic of coincidence, in *Proceedings of the 9th International Conference on Artificial Intelligence and Law* (ACM, 2003), pp. 142–146
13. B. Fortuna, M. Grobelnik, D. Mladenić, System for semi-automatic ontology construction (2006)
14. A. Maedche, V. Pekar, S. Staab, Ontology learning part one—on discovering taxonomic relations from the web. Web Intell. 301–319 (2003). Springer Berlin Heidelberg
15. K. Englmeier, F. Murtaghμ, J. Mothe, Domain ontology: automatically extracting and structuring community language from texts. IADIS Appl. Comput. 59–66 (2007). (Spain, Espagne)
16. L. Drumond, R. Girardi, Extracting ontology concept hierarchies from text using markov logic, in *Proceedings of the 2010 ACM Symposium on Applied Computing* (ACM, 2010), pp. 1354–1358

Rural Electrification Using Microgrids and Its Performance Analysis with the Perspective of Single-Phase Inverter as Its Main Constituent

Apoorva Saxena and Durg S. Chauhan

Abstract This paper explores the various changes that have occurred in traditional electricity grid which have led to the emergence of microgrids as a possible solution for rural electrification. The single-phase inverters constitute a major component of microgrids and keeping in view the requirement of a traditional rural household, design aspects of a 400 W H bridge inverter is discussed in this paper. The open-loop analysis of single phase full bridge inverter is performed and selection of design parameters like filter inductance and capacitance, DC bus capacitor, switching frequency, etc., is done in detail. The effect of varying these parameters on the output voltage and current of the inverter is also done with the help of PSIM software. Finally, the performance of inverter is analyzed under rated load conditions.

Keywords Electricity market · Microinverter · Design parameters
Total harmonic distortion (THD) · Inverter performance

1 Introduction

Traditionally in earlier days, the electricity market was based on the principles of economies of scale, cost-plus pricing, and monopoly franchise with fossil fuel-based centralized generation as the most popular method of electricity generation. The electrification of remote areas was done through extension of the existing grid. But post 1970, with increased focus on environment and sustainable development, the electricity market has undergone a paradigm shift. Renewable integration, two-way power flow, and competitive pricing have shifted the focus towards a smart distribution network.

The rural electricity supply, especially in India, is suffering both in terms of availability for measured number of hours and penetration level. More than 25% of their rural households are yet to have an access to electricity [1]. A major bottleneck in

A. Saxena (✉) · D. S. Chauhan
GLA University, Mathura, India
e-mail: apoorvatu@gmail.com

© Springer Nature Singapore Pte Ltd. 2019
H. Malik et al. (eds.), *Applications of Artificial Intelligence Techniques in Engineering*, Advances in Intelligent Systems and Computing 698,
https://doi.org/10.1007/978-981-13-1819-1_40

the development of the power sector is the poor financial state of the State electricity boards (SEBs), which can be attributed to the lack of adequate revenues, state subsidies for supply to the rural subscribers and high T&D losses to the tune of over 25%. Due to high T&D losses and low collection efficiency state utilities have very little incentive to provide electricity to rural areas, which in turn further adds to already poor financial status of utilities giving rise to a "vicious cycle". In this paper, the concept of microgrids has been explored as a possible solution for rural electrification and some of the challenges in its implementation are discussed. Its major constituent, the single-phase H bridge inverter, is also explained in detail, so that the practical designing aspects can be analyzed.

The paper is organized as follows: Sect. 2 outlines the major changes that have happened in traditional grid that supports the concept of microgrids as a possible solution for rural electrification. Section 3 explains the concepts of microgrids and main drivers for this technological evolution. Section 4 discusses the selection of various simulation parameters values of a single-phase H bridge inverter like DC bus capacitor, frequency modulation index and amplitude modulation index of gate driver circuit, filter inductance and capacitance. In Sect. 5, the effect of parasitic resistance of filter inductor, ESR of filter capacitor, and variation of load on inverter output are also analyzed using PSIM software. Section 6 consists of conclusion to the paper.

2 Major Changes in Traditional Grid

Post 1870, the evolution of traditional electricity grid was based on the concept of economies of scale which literally means the larger production directly implies cheaper rates. In late 1800s, big business houses realized the good business prospects in electric generation and distribution area. So, concept of monopoly franchise with virtually no competition came into existence. The cost-plus pricing model in which a small profit was added by the utility in addition to actual cost was practised.

Post 1970 due to labor laws, land acquisition for centralized power plant, inflation, etc., economies of scale concept started to be less relevant. With restructuring and deregulation of electric energy sector and competitive bidding of power blocks, monopoly franchise also started to wither away. Increased focus on reduction of carbon emissions and sustainable development has changed the way energy is generated and distributed around the world.

These concepts are summarized in Fig. 1, which clearly indicates that the electricity energy market scenario has undergone considerable change and so the conventional approach of fossil fuel based centralized generation and grid extension to remote rural areas may not be the way forward. These factors have led to the evolution of traditional grid by employing innovative products and services with intelligent monitoring and control. Microgrids essentially form the basic building block of these smart distribution networks.

Fig. 1 Change in electricity
energy scenario

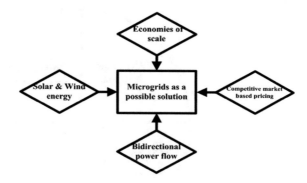

3   Concept of Microgrid

Microgrids comprises LV distributed systems with distributed energy sources (PV, wind) together with storage devices (flywheels, batteries and ultra-capacitors) and can operate in autonomous way, if disconnected from grid or in grid-connected mode [2]. The microgrid implementation for rural electrification has distinct advantages over the conventional grid extension approach.

The effective use of local resources like wind and solar, less T&D losses, and reduction in cost due to reduction in cost of heavy transmission lines and transformers makes microgrid an excellent viable option [3] as shown in Fig. 2. The simplified diagram of microinverter (less than 500 W) connected in a mircogrid working in autonomous mode is shown in Fig. 3. It mainly consists of a renewable energy source like PV array, dc bus capacitor to maintain the smooth dc, a single-phase DC/AC inverter. The output of this inverter is passed through the filter circuit to get sinusoidal AC output. The values of filter parameters are selected to minimize THD in the output. The voltage and current control loops are provided to get the feedback control which can be designed for analog mode or digital mode. The gate driver circuit is operated with Pulse Width Modulation techniques (PWM) in this paper to minimize the harmonics.

The output impedance of the inverters is considered to be mainly inductive as the filter inductor has predominant effect on the overall impedance of the inverter. However, the output impedance of the inverters also depends on the control strategy employed and it is possible to design Resistive output impedance (R inverters), Capacitive output impedance (C inverters), Inductive output impedance (L inverters), Resistive capacitive output impedance (RC inverters), and Resistive inductive impedance (RL inverters) by adding a virtual impedance loop and suitably selecting the values of feedback parameters [4, 5].

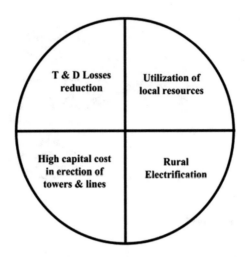

Fig. 2 Reasons for microgrid implementation

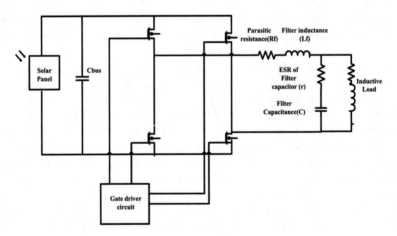

Fig. 3 Microinverter in autonomous mode as major component of microgrid

4 Selection of Design Parameters

A single-phase H bridge inverter is simulated on PSIM software with bipolar PWM switching. The DC side voltage obtained from solar panel module is taken as $V_{in} =$ 355.75 V and a resistive load of 50 Ω is connected at the load terminals. The value of switching frequency, amplitude modulation index, DC bus capacitor and filter inductance, and capacitance are calculated as per details given later in this section.

Table 1 Variation with switching frequency of triangular wave

Switching freq (kHz)	1.05	5.55	10.55	21.05	31.55	42.05	84.05
V_{out} (V)	162.5	61.4	61.6	61.7	61.6	161.5	161.6
I_{out} (A)	3.25	3.23	3.23	3.23	3.23	3.23	3.23
THD (%)	7.52	2.55	2.00	2.37	2.78	1.99	2.90

Table 2 Variation with amplitude modulation index

Modulation index (m_a)	0.2	0.4	0.6	0.617	0.7	0.8	1	1.2
V_{out} (V)	56.2	109.6	157.9	161.5	178.8	196.2	215.1	207.6
I_{out} (A)	1.12	2.19	3.15	3.23	3.57	3.92	4.30	4.15
THD (%)	0.50	1.70	2.53	1.99	3.42	3.85	5.84	17.6

4.1 Selection of Switching Frequency of Triangular Wave

It is desirable to use higher switching frequencies because it is easier to filter harmonic content in voltages at higher frequencies. The only limitation is that the switching losses across the MOSFET switch increases with the higher switching frequencies. In most of the applications, switching frequency is chosen to be either less than 6 kHz or more than 21 kHz [6]. In the case of this inverter, THD is also minimized at 10.55 kHz but to avoid the audible frequency range, the switching frequency will be selected as 42.05 kHz where the value of THD is only 1.99% as shown in Table 1. The value of carrier sinusoidal signal frequency is taken as 50 Hz.

4.2 Variation of Amplitude Modulation Index m_a

The amplitude modulation index was varied by changing the amplitude of sinusoidal control signal and results obtained are shown in Table 2. It can be observed that the fundamental frequency output voltage varies linearly with amplitude modulation index (m_a) for $m_a \leq 1$. If the value of modulation index $m_a \geq 1$, the value of fundamental component of output voltage is increased while the value of THD is also increased due to increased harmonics in sidebands.

In Figs. 4 and 5, it can be clearly seen that the sideband harmonics around the operating frequency of 50 Hz has increased as compared to case when $m_a = 0.617$ when the value of THD is minimum. If the value of m_a is increased beyond 1, to get higher output fundamental frequency voltage, the inverter is said to be operating in over-modulation region. So in this paper, the value of m_a is taken as 0.617 because at this value the THD is minimum and the power quality of the output will be better.

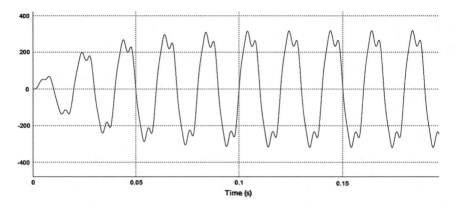

Fig. 4 V_{out} for $m_a = 1.2$

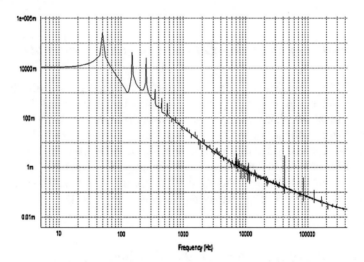

Fig. 5 FFT analysis of V_{out} for $m_a = 1.2$

FFT analysis of output load voltage as shown in Figs. 6 and 7 clearly indicates that for $m_a = 0.617$, the harmonic voltages around fundamental frequency is reduced, resulting in better power quality.

4.3 Varying the Frequency of Sinusoidal Control Signal Frequency

The effect of change in variation of the value of control signal frequency is also analyzed in this paper once the value of m_a is fixed at 0.617. It has been observed

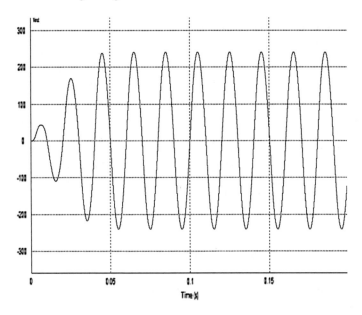

Fig. 6 V_{out} for $m_a = 0.617$

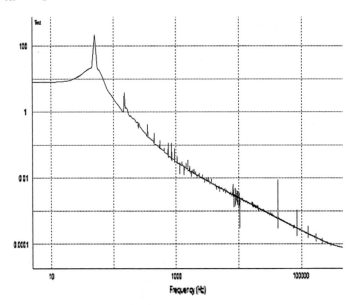

Fig. 7 FFT analysis of V_{out} for $m_a = 0.617$

Table 3 Variation of control signal frequency

Modulation index (m_a)	0.617 $f_{control}$ = 40 Hz	0.617 $f_{control}$ = 60 Hz
V_{out} (V)	158.9	165.2
I_{out} (A)	3.178	3.30
THD (%)	3.46	3.75

that in Table 3, variation of control signal frequency is the fundamental frequency voltage and current varies indirect proportion with the variation of control signal frequency. But the value of THD is higher for both 40 and 60 Hz control signal frequency as compared to the THD at 50 Hz.

4.4 Selection of DC Bus Capacitor

The presence of ripples in input current and voltage does not harm the solar PV panel but the output power extracted from the PV panel is reduced [7]. So, a dc capacitor is connected in parallel with the solar PV panel to minimize these ripples on input side. The variable output power from solar PV panel is also a point of concern while selecting the values of DC bus capacitor but in this paper, the output from solar panel is assumed to be constant for solar irradiation of 1000 W/m^2 operating at 25 °C. Hence, the major point of concern for selecting the value of C_{bus} in this paper is the magnitude of allowable voltage ripple magnitude, which is actually twice the inverter output supply frequency. Life expectancy, operating temperature range, and cost are some of the issues for the choice between electrolytic capacitor and film capacitor, once the value of C_{bus} is calculated [8]. The value of C_{bus} is calculated using Eq. 1 is explained in [7]

$$\Delta V_{dc(p-p)} \geq \frac{P}{\omega V_{dc} C_{bus}} \tag{1}$$

The variation of input parameters with the change in value of C_{bus} is represented in Table 4. This clearly shows that the DC bus voltage reduces with increase in the value of C_{bus}.

In Fig. 8, it is clearly evident that the current I_{dc} from the solar panel is increased with the value of C_{bus} and more power is extracted from the solar panel but value of V_{dc} is decreased. So an optimum value of C_{bus} using Eq. 1 and taking 5.51% allowable DC ripple. The value of C is selected as 290 μF in this paper.

Further, it can be observed that the variation of C_{bus} also effects the output voltage and current as shown in Table 5. In this paper, output THD of 1.99% corresponding to C_{bus} = 290 μF is selected.

Table 4 Variation of input power with Cbus

DC bus capacitor (C_{dc})	250u	270u	290u	310u	330u	350u	370u	390u
V_{dc} (V)	359	357.3	355.7	354.0	352.3	350.6	348.9	347.2
I_{dc} (A)	2.16	2.19	2.22	2.25	2.28	2.31	2.34	2.37
P_{dc} (W)	627.3	630.8	634.1	637.2	640.2	643.1	646.01	648.7
S (VA)	778.1	784.6	791.1	797.6	804.1	810.6	816.9	823.1

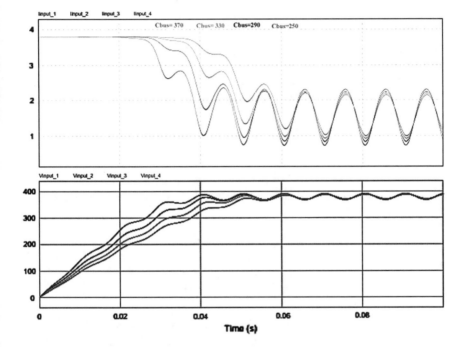

Fig. 8 Variation of input voltage and currents for different values of Cbus

4.5 Selection of Filter Inductance and Capacitance

A filter is required at the output of an inverter to reduce the harmonics around the fundamental frequency switching frequency and their multiples. A large value of inductor reduces the current harmonics around fundamental frequency but it reduces the dynamic response [9]. A smaller value of inductance can lead to resonance interactions. A cutoff frequency is desired to be around between power frequency and $0.1 f_s$ to reduce high-frequency harmonics of inverter output. If cutoff frequency is of low value, it will imply corresponding higher values of filter inductance and capacitance. This will directly affect the cost and losses occurring in filter circuit. The

Table 5 Variation of output parameters with C_{bus}

DC bus capacitor (C_{dc})	250u	270u	290u	310u	330u	350u	370u	390u
V_{ac} (V)	162.7	162.2	161.5	160.9	160.3	159.9	158.9	158.2
I_{ac} (A)	3.25	3.24	3.23	3.21	3.2	3.19	3.17	3.16
P_{ac} (W)	530	526.2	522.2	518.1	513.9	509.7	505.3	500.8
S (VA)	530	526.2	522.2	518.1	513.9	509.7	505.3	500.8
THD (%)	2.30	2.14	1.99	1.86	1.75	1.64	1.54	1.46

Table 6 Variation of output parameters with R

Load resistance (R)	200 Ω	100 Ω
V_{ac} (V)	170.05	167.3
I_{ac} (A)	0.85	1.67
P_{ac} (W)	144.5	280.15
S (VA)	144.5	280.15
THD (%)	1.64	1.67

value of filter inductance and capacitance are calculated with the help of equations given in [10]

$$L = \frac{V_{dc} - V_{o/rms}}{2\Delta i_{o/rms}} DT_s \qquad (2)$$

$$C = \frac{V_{dc} - V_{o/rms}}{16L\,\Delta v_{c/rms}f_s^2} D \qquad (3)$$

Using these formulas and taking a allowance of 15.78% as ripple in the dc component of output current, the value of L is calculated as 8.42 mH. If the value of ripple is allowed beyond 20%, it will increase the peak current ratings of inductor and switching devices thus increasing the overall cost and size [11].

The small ripple approximation or linear ripple approximation can be assumed for output voltage ripple [11], in this paper it is taken as 0.78% corresponding to which the value of filter capacitor comes out to be 64.1 µF. It is evident from the results in Table 6 that the output power and output current decreases with increase in load resistance R. One important point to note from Table 7 is that the value of DC bus voltage drops very sharply with the increase in the value of load.

Table 7 Variation of input parameters with R

Load resistance (R)	200 Ω	100 Ω
V_{dc} (V)	172.8	367.4
I_{dc} (A)	1.61	1.77
P_{dc} (W)	264.5	397.6
S (VA)	600.2	651.6

Table 8 Various parameters with R_L load

V_{out}	I_{out}	THD	V_{dc}	I_{dc}	P_o	P_{in}	S_o	S_{in}
158.3	3.16 A	0.54% (I); 1.08% (V)	362.1	1.95 A	403.3	521.3	501.45	708.07

5 Effect of Load Variation and Parasitic Parameters on Inverter

5.1 Effect of Parasitic Filter Components

The value of filter capacitor ESR is taken as 1 mΩ and filter inductor resistance is taken as 740 Ω [12]. It has been observed that the value of filter inductor resistance decreases the magnitude of output voltage, output current. The ESR of filter capacitor has negligible effect on output parameters of inverter.

5.2 Effect of Resistive-Inductive Load

In place of pure resistance of 50 Ω, now a more practical condition of inductive load is taken into consideration. The value of R is now taken as 40 Ω for load resistance and load inductance as 95.5 mH such that overall impedance at 50 Hz is 50 Ω with a power factor of 0.8 lagging. The final results are summarized in Table 8 and output and input voltages waveforms are shown in Fig. 9.

It can be clearly observed that system is settling to steady state in 0.05 s. The response can be further improved by providing a feedback loop to modulate the reference voltage and duty cycle of switches [13].

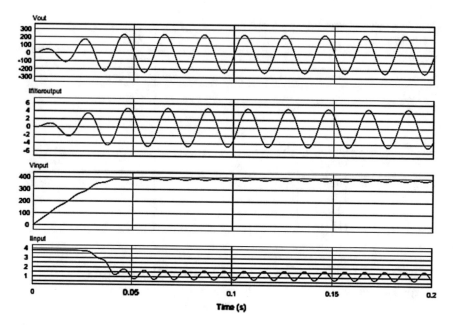

Fig. 9 Output and input parameter variation

6 Conclusion

In this paper, the detailed analysis of single-phase inverter is performed along with the systematic explanations for choice of different inverter parameters like DC bus capacitance, filter parameters. etc. Since these inverters are designed to operate in standalone mode for low power applications these are sometimes referred as microinverters in this paper. It has been observed that rms value of inverter output voltage and current virtually remains constant with the variation of switching frequency. The output THD is effected by switching frequency and for the microinverter discussed in this paper. The THD was minimized at 42.05 kHz. It can be observed that the fundamental frequency output voltage varies linearly with amplitude modulation index (m_a) for $m_a \leq 1$. At $m_a = 0.617$, the value of THD is reduced. Whereas for higher values of modulation index the fundamental component of output voltage is increased but THD also increases correspondingly. The values of filter inductance and capacitance are selected keeping in view the allowable ripple attenuation and dynamic response of the system.

With the perspective that the inverters forms main component of this standalone system, the detailed open-loop analysis done will help to design the inverter in real-time applications. The performance and stability aspects of these inverters can be improved by employing voltage and current inner loops. The performance, in terms of lower THD and better reactive power support, can be further improved by making the output impedance of these inverters as capacitive or resistive—capacitive through virtual feedback control loop.

References

1. ERNST & YOUNG, Models of rural electrification—report to forum of Indian regulators, p. 16
2. N. Hatziargyriou, *Microgrids Architecture and Control,* vol. 1 (IEEE press, 2014), p. 4
3. R.H. Lassester, Microgrids, in *Power Engineering Society Winter Meeting,* IEEE 2002
4. Y. Sun, X. Hou, J. Yang, G. Han, M. Su, J.M. Guerreo, New perspective on droop control in AC microgrids. IEEE Trans. Ind. Electron. **64**(7), 5741–5745 (2017)
5. Q.-C. Zhong, Y. Zeng, Universal droop control of inverters with different output impedance. IEEE Access **4**, 702–712 (2016)
6. N. Mohan, T.M. Undelands, W.P. Robbins, in *Power Electronics: Converters, Applications & Design,* vol. 3 (Wiley, 2003), p. 158
7. S.K. Yarlagadda, A maximum power point tracking technique for single phase photovoltaic systems with reduced DC link capacitor, M.S. dissertation, Department Electrical & Computer Engineering, University of Houston, 2012
8. X. Zong, A single phase grid connected DC/AC inverter with reactive power control for residential PV application, M.Tech. Department Electrical Engineering, University of Toronto, 2011
9. J. Bauer, Single phase voltage source inverter photovoltaic application. Acta Ploytechnica **50**(4) (2010)
10. A. Roshan, A dq rotating frame controller for single phase full bridge inverters used in small distributed generation systems, M.S. dissertation, Department Electrical Engineering, Virginia polytechnic institute and state university, 2006
11. R.W. Erickson, D. Maksimovic, *Fundamentals of Power Electronics*, 2nd edn. (Kluwer academic publishers, 2004), pp. 18–22
12. Shurter datasheet, 1-phase line filters, AC filter high symmetrical attenuation, FMAB NEO
13. Y. Chen, J.M. Guerrero, Z. Shuai, Z. Chen, L. Zhou, A. Luo, Fast reactive power sharing, circulating currents and resonance suppressions for parallel inverters using resistive—capacitive output impedance. IEEE Trans. Power Electron. **31**(8), 5524–5537 (2016)

Speed Control of PMBLDC Motor Using Fuzzy Logic Controller in Sensorless Mode with Back-EMF Detection

Mohammad Zaid, Zeeshan Sarwer, Farhan Ahmad and Mukul Pandey

Abstract Recent advances in the field of power electronics have made PMBLDC motors very popular. They are being used in many applications because they possess certain desirable features as compared to brushed DC motor and servo motors. This paper presents speed control of PMBLDC motor in sensorless mode of operation using indirect Back-EMF detection. The Back-EMF estimation is done indirectly using the line voltage difference method to obtain the exact commutation instants of current. In this paper, a Fuzzy Logic Controller (FLC) is employed for speed control of PMBLDC motor which is running in sensorless mode of operation is becoming increasingly popular due to various problems associated with Hall sensors. The proposed Fuzzy Logic-based speed controller allows better speed control of motor as compared to PI speed controller, especially with varying load conditions.

Keywords Fuzzy logic controller (FLC) · PMBLDC motor · PI controller
DC motor

1 Introduction

The first electronically commutated brushless DC motor was developed with the help of Hall elements in 1962 [1] since then tremendous development has been made in the field of drives and permanent magnet materials. Today, Permanent Magnet Brushless DC motor is used in many applications from aerospace, automobile industry to household appliances. The PMBLDC motor possesses certain features such as high efficiency, good power factor, lower maintenance, precise and accurate control, and high-power density which are desirable in any machine. The PMBLDC motor generally has a trapezoidal Back EMF. Unlike brushed DC motors, the commutation

M. Zaid (✉) · Z. Sarwer · F. Ahmad · M. Pandey
Zakir Husain College of Engineering and Technology, AMU, Aligarh, India
e-mail: zaidahmadzhcet@gmail.com

Z. Sarwer
e-mail: z.sarwer@gmail.com

© Springer Nature Singapore Pte Ltd. 2019
H. Malik et al. (eds.), *Applications of Artificial Intelligence Techniques in Engineering*, Advances in Intelligent Systems and Computing 698,
https://doi.org/10.1007/978-981-13-1819-1_41

is electronically controlled and Hall sensors are used for knowing the exact position of rotor. The output of Hall sensors is used for generation of switching signals for the inverter. Hall sensors are costly and less reliable, especially in space application. Due to these reasons, various sensorless techniques have been developed. Each of the sensorless techniques employed has their own advantages and disadvantages. The main objective of sensorless techniques is used to find the position of rotor indirectly. Most popular and widely used technique is Back EMF detection using line voltage difference method [2]. Actually, in any sensorless scheme, we need to identify and process exact commutation instants of current for the generation switching signals. In the scheme used in this paper, the difference of two-line voltages gives the zero crossing of Back EMF of any one phase [3, 4]. The zero crossing instants of that phase EMF waveform gives the approximate commutation instants of the current of that phase. The zero-crossing points (ZCPs) need to be phase shifted to get the exact commutation point. A low-pass filter generally introduces the delay required for the operation. Most of the Back-EMF detection techniques suffer from serious drawbacks that during starting EMF cannot be detected because of low speed, hence some starting methods need to be employed before motor accelerates to minimum threshold speed [5]. Another improved method for detection of rotor position is by using of third harmonic voltage in the EMF waveform of the motor. The voltage between the artificially made neutral and motor neutral gives the third harmonic voltage component which contains the ZCPs of Back EMFs of the three phases [6]. Generally, motor neutral is not accessible hence midpoint of DC link can also be used for generation of third harmonic voltage, but this signal is noisier as compared to the previous signal obtained between the two neutrals [7, 8].

This paper is divided into six sections. Section 2 of the paper describes the overall system configuration of PMBLDC motor drive. In Sect. 3 of this paper, sensorless technique for estimation Back EMF using line voltage difference method are discussed. In Sect. 4, the design of a Fuzzy Logic Controller is discussed. The problem with conventional controllers comes when either plant structure is unknown or if known is so complex that design of controller by classical approach would be impractical and cumbersome. In Sect. 5, comparison is made between the performance of motor with proposed FLC and conventional PI speed controller when the load is changed and step change in reference speed of motor takes place. Conclusion is discussed in Sect. 6.

2 System Configuration

The PMBLDC stator is fed by a three-phase inverter which is operated in 120° mode. The rotor is made of permanent magnet. Only two-phase conduct at a time and third phase is floating. The rotor position is detected knowing the ZCPs of Back EMF from the difference of line voltages. The speed controller used can be a PI controller or a FLC. The control loop has outer speed controller and inner current controller.

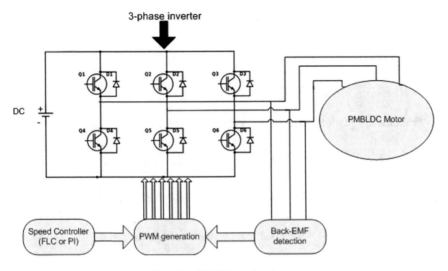

Fig. 1 PMBLDC motor with indirect BackEMF detection

Table 1 The PMBLDC motor specifications

Parameters	Value	Units
No. of phases	3	–
Torque constant	0.042	Nm/A
Poles	4	–
Back-EMF constant	0.42	V/rad/s
Resistance per phase (R)	2.90	ohm
Inductance per phase (L)	2.80	mH
Viscous damping (B)	0.00009	Nm (rad/s)
Rotor inertia (J)	0.0005	kg m^2
DC link voltage	300	V

Figure 1 shows the overall configuration for speed control of motor. Table 1 gives details of PMBLDC motor specifications.

3 Proposed Technique

The ZCP determined needs to be phase shifted by 30° to get real commutation instants. In the proposed method, the ZCP of EMF is estimated using the difference in line voltage. With respect to neutral points, the phase a voltage of motor is

$$u_{an} = Ri_a + (L - M)\frac{d}{dt}i_a + E_a \tag{1}$$

Similar equations can be written for phases b and c.

$$u_{bn} = Ri_b + (L - M)\frac{d}{dt}i_b + E_b \tag{2}$$

$$u_{cn} = Ri_c + (L - M)\frac{d}{dt}i_c + E_c \tag{3}$$

From these equations, line-to-line voltages can be found

$$u_{ab} = R(i_a - i_b) + L\left[\frac{d}{dt}(i_a - i_b) + e_{an} - e_{bn}\right] \tag{4}$$

$$u_{bc} = R(i_b - i_c) + L\left[\frac{d}{dt}(i_b - i_c) + e_{bn} - e_{cn}\right] \tag{5}$$

$$u_{ca} = R(i_c - i_a) + L\left[\frac{d}{dt}(i_c - i_a)\right] + e_{cn} - e_{an} \tag{6}$$

The relationship between currents is given by

$$i_a + i_b + i_c = 0 \tag{7}$$

Above equation is rewritten as

$$i_a = -(i_b + i_c) \tag{8}$$

Substituting Eq. (8) in Eqs. (4) and (5), we get

$$u_{ab} = R(i_a - i_b) + L\frac{d}{dt}(i_a - i_b) + e_a - e_b \tag{9}$$

$$u_{bc} = R(i_a - 2i_b) + L)\frac{d}{dt}(i_b - i_c) + e_b - e_{cb} \tag{10}$$

Now to find the difference in line voltage, subtract Eq. 5 from 4. In this technique, no neutral point is required for estimation of line voltages.

$$u_{abbc} = R(i_a - 2i_b + i_c) + L\frac{d}{dt}(i_a - 2i_b + i_c) + e_{an} - 2e_{bn} + e_{cn} \tag{11}$$

Now consider a situation in which phase a and phase c is conducting and phase b is open. In this situation $e_{an} = -e_{cn}$. Therefore, in that interval (11) may be simplified as

$$u_{abbc} = e_{an} - 2e_{bn} + e_{cn} = -2e_{bn} \tag{12}$$

The above result shows that the difference in line voltage u_{abbc} gives the inverted and magnified waveform of Back EMF of phase b. Similarly, u_{bcca} and u_{caab} gives the inverted and magnified Back-EMF waveforms of phase c and a. The above derivation

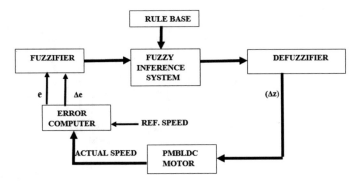

Fig. 2 Basic structure of FLC

shows that ZCP of Back EMF can be found indirectly by proper processing of three stator voltages. From the virtual hall, signals switching signals are generated for inverter. The ZCP determined needs to be phase shifted by 30° to get real commutation instants.

4   Implementation of FLC

The basic structure of a FLC for speed control of motor is as shown below in Fig. 2.

For the design of a FLC the error in the speed in speed loop is utilized by collating the actual speed with the set point speed. The triangular membership function with 5 linguistic variables and 25 rules are used in the FLC design. The common interval for three membership functions, namely error (e), change in error (Δe), and change in output of FLC controller (Δz) is between $[-1, 1]$. Scaling factors are used for tuning FLC.

5   Simulation Results and Discussion

The PMBLDC motor is first started with sensors and is then switched over to sensorless control at 200 ms. The other problem is to determine the instant when the control is shifted from sensor control to sensorless control. Practically this is done by first exciting two phases out of three for a fixed period. At the end of this period, motor has moved from an unknown position to a predetermined position. Filtering helps in acquiring accurate commutation instants. Exact commutation instant will be 30° phase shifted from zero-crossing point for which a delay circuit is used.

Figures 3 and 4 gives comparison between real hall signal originally generated by sensors and virtual hall signals generated by detecting exact commutation instants,

Fig. 3 Real hall signal

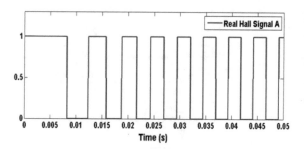

Fig. 4 Virtual hall signal

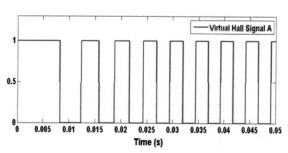

both signal exactly matches which is necessary for satisfactory operation of motor in sensorless control.

The performance of motor with PI controller in sensorless mode is evaluated in this section. Load torque of 1 Nm is applied at 0.2 s. From Fig. 5, it can be seen that there is a significant reduction in setpoint speed with application of load torque. Speed is reduced to 194 rad/s from set point speed of 200 rad/s. Motor speed again reaches to setpoint speed after significant delay of 0.3 s.

After PI controller, FLC with sensorless control is employed. The load torque of 1Nm is applied at $t=0.2$ s. The speed response in Figs. 6 and 7 shows that there is almost no reduction in speed of the motor when load torque is applied at $t=0.2$ s which shows the superiority of fuzzy logic controller if properly tuned over PI controller. From Fig. 7, it can be observed that the reduction in speed after application of load torque is less than 1 rad/s, and motor gets back to set point speed almost instantaneously. This is great improvement over PI controller which takes almost 0.3 s to get back at same speed reference for same amount of load torque applied.

The trapezoidal waveform of Back EMF and current of motor running in sensorless mode are shown in Figs. 8 and 9.

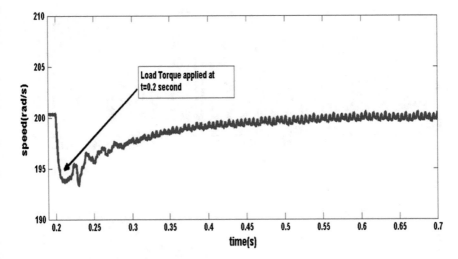

Fig. 5 Motor speed with PI controller

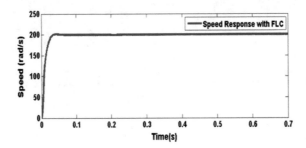

Fig. 6 Motor speed with FLC

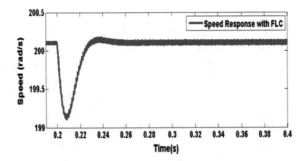

Fig. 7 Negligible reduction in speed with FLC

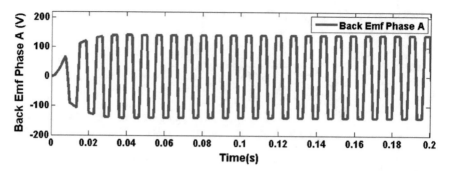

Fig. 8 Back EMF (V)

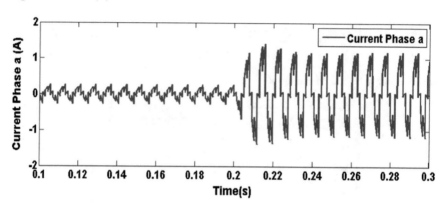

Fig. 9 Phase current (A)

6 Conclusion

A Simulink model of PMBLDC motor without Hall sensors employing indirect Back-EMF detection technique has been developed and its speed is controlled by using both FLC and PI controller. Motor is found to be running smoothly in sensorless mode of operation and all the waveforms, i.e., motor phase currents, Back EMF, and rotor position has been obtained from Simulink model. The use of Fuzzy controller has generally reduced the rise time and settling time of the speed response of the motor. Hence, a tuned Fuzzy controller has outperformed conventional PI controller. However, the main advantage of using Hall sensors is that motor design remains simple, and no extra circuitry is needed.

References

1. C.L. Xia, *Permanent Magnet Brushless DC Motor Drives and Controls* (Wiley Press, Beijing, 2012)
2. A. Girolkar, G. Bhuvaneswari, Control of PMBLDC motor using third harmonic back EMF sensing with zigzag transformer, in *2nd IEEE Conference on Electrical Energy Systems*, 2014, pp. 110–115. https://doi.org/10.1109/icees.2014.6924151
3. P. Damodharan, K. Vasudevan, Sensorless brushless DC motor drive based on the zero-crossing detection of back electromotive force (EMF) from the line voltage difference. IEEE Trans. Energy Convers. **25**(3) (2010)
4. S. Tara, Md. Syfullah Khan, Simulation of sensorless operation of BLDC motor based on the zero-cross detection from the line voltage. Int. J. Adv. Res. Electr. Electron. Instrum. Eng. **2**(12) (2013). ISSN 2320-3765
5. P.P. Acarnley, J.F. Watson, Review of position sensorless operation of brushless permanent magnet machines. IEEE Trans. Industr. Electron. **53**(2), 2006 (2006)
6. S. Ogasawara, H. Akagi, An approach to position sensorless drive for brushless DC motors. IEEE Trans. Ind. Appl. **27**, 928–933 (1991)
7. K.M.A. Prasad, U. Nair, An intelligent fuzzy sliding mode controller for a BLDC motor, in *2017 International Conference on Innovative Mechanisms for Industry Applications (ICIMIA)*, 2017, pp. 274–278
8. R.M. Pindoriyal, A.K. Mishra, B.S. Rajpurohie, R. Kumar, Analysis of position and speed control of sensorless BLDC motor using zero crossing back-EMF technique, in *1st IEEE International Conference on Power Electronics. Intelligent Control and Energy Systems, ICPEICES-2016*, pp. 1–6. https://doi.org/10.1109/icpeices.2016.7853072

Modeling and Analysis of the Photovoltaic Array Feeding a SPWM Inverter

Shirazul Islam, Farhad Ilahi Bakhsh and Qamar Ul Islam

Abstract This chapter deals with the advanced modeling and analysis of the photovoltaic cell when developed in MATLAB/Simulink environment. The modeling of the solar cell has been carried on the basis of the performance equation of the photovoltaic cell. The effect of the change in solar radiation and the temperature change has been successfully incorporated in this model. The major contribution made by the chapter is development of single-phase Sinusoidal Pulse Width Modulated Inverter (SPWM) which is being fed by the developed photovoltaic cell. The THD has been investigated by FFT analysis at the output of the inverter using power GUI block. The introduction of the inductive load at the output of the inverter shows the steep decrease in the value of THD in output current and voltage waveforms of the inverter and leading to improved performance of the SPWM inverter.

Keywords Modeling · Analysis · Solar photovoltaic (SPV) · MATLAB
Sinusoidal pulse width modulated inverter (SPWM)

1 Introduction

Nowadays, due to continuous reduction in the cost of SPV system and steady growth in SPV technology; the SPV generation has added substantial social and economic profits. Large-scale SPV generation is a vital novel alternative source of energy for twenty-first century [1]. Since the output of a single cell is very small, therefore, large number of cells is connected in parallel and/or series to form panels or modules or arrays [2]. The performance of a SPV system (output power, current, and voltage) depends on the solar cell and array design quality as well as on the operating

S. Islam
Department of Electrical Engineering, Indian Institute of Technology Kanpur, Kanpur, India

F. I. Bakhsh (✉) · Q. U. Islam
Department of EREE, SOET, Baba Ghulam Shah Badshah University, Rajouri, Jammu and Kashmir, India
e-mail: farhad.ilahi@coetbgsbu.org

© Springer Nature Singapore Pte Ltd. 2019
H. Malik et al. (eds.), *Applications of Artificial Intelligence Techniques in Engineering*, Advances in Intelligent Systems and Computing 698,
https://doi.org/10.1007/978-981-13-1819-1_42

conditions (temperature, load current, and solar irradiation level). Hence, the effects of these variables should be considered while designing the SPV array [3, 4]. Various auxiliary power circuit configurations using SPV arrays has been proposed by different researchers [5–11].

The major contribution made by this chapter is development of the mathematical model of the solar cell comprising the effects of the change in operating temperature and solar radiation in the MATLAB/Simulink environment and its connection to single-phase SPWM for harnessing the solar energy. The research works has been divided into two parts. The first part deals with the effect of change in solar radiation and operating temperature and the second part deals with the development of the SPWM inverter and harmonic analysis for the output voltage of the proposed inverter. The output of the developed inverter shows drastic decrease in the harmonics with inductive load which faithfully maintains the power quality issues.

2 Mathematical Modeling of the Solar Cell

The equivalent circuit of the solar cell available in the literature is shown in Fig. 1 below. From the circuit diagram, it is clear that the single unit of the solar cell has been replaced by an equivalent current source bypassed by a diode circuit. Therefore, the current flowing through the load is given by Eq. (1).

$$I = I_{ph} - I_s (e^{\frac{q(V+IR_s)}{NkT}} - 1) - \frac{(V + IR_s)}{R_{sh}} \tag{1}$$

The Eq. (1) takes into account the effect of the solar radiation and the effect of change in temperature due to the change in temperature.

Fig. 1 PV cell equivalent circuit

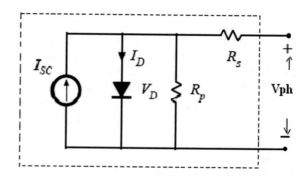

2.1 Effect of Change of Solar Radiation on the Solar Cell Current

There is a direct effect of the solar radiation or solar energy on the output solar cell current. The solar energy consists of photons of energy which when striking with the surface of the crystalline silicon material of the solar cell emits energy which is given by Eq. (2)

$$I = [I_{sc} - K_i(T - 298)]\frac{\beta}{100} \tag{2}$$

2.2 Effect of Change in Operating Temperature on the Solar Cell Current

There is a drastic change in the value of the reverse saturation of the solar cell when the operating temperature changes abruptly. The following mathematical equation explores the effect of the change in the value of the current I_s due to the change in operating temperature. The following mathematical equation explores the relation between I_s and temperature T.

$$I_s(T) = I_s\left(\frac{T}{T_{nom}}\right)^3 (e^{\frac{qEg}{NkT}(\frac{T}{T_{nom}}-1)} - 1) \tag{3}$$

2.3 Derivation for the Output Voltage of the Solar Cell

Now, from the Fig. 1, the current through the load connected across the output of the inverter can also be expressed by Eq. (4)

$$V_c = \frac{AkT}{q}\ln(\frac{I_{ph} + I_s - I}{I_s}) - R_s I \tag{4}$$

3 Modeling of the Solar Cell with Simulink

For carrying out the mathematical modeling of the solar cell, consider the performance equation of the solar cells which are given by the Eqs. (1)–(4). These three equations are combined together and the output of the solar cell appears in the form of the output voltage which is dc and having constant magnitude. The modeling of the solar cell has been carried in Simulink environment. The Simulink diagram is

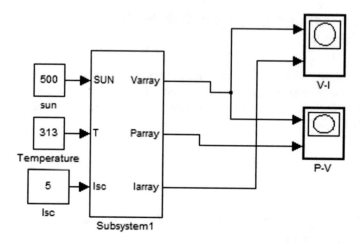

Fig. 2 The Simulink model

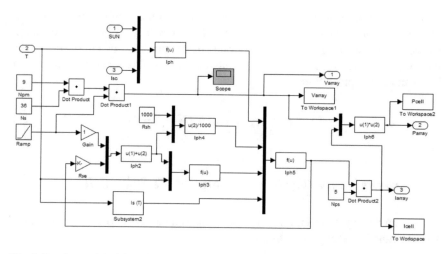

Fig. 3 Interior model of the subsystem 1 (single solar cell unit)

presented in Fig. 2. Figure 3 shows the subsystem 1 of the Fig. 2 which is the model
of the solar cell in MATLAB/Simulink. Figure 4 shows the subsystem 2 of the Fig. 3.

3.1 Simulation Results of Solar Cell

The performance analysis of the solar cell has been carried for the four operating
temperature ranges of 40, 50, 60, and 70 °C. The values of the solar radiations are
assumed to be 500, 700, 900, and 1100 W/m^2.

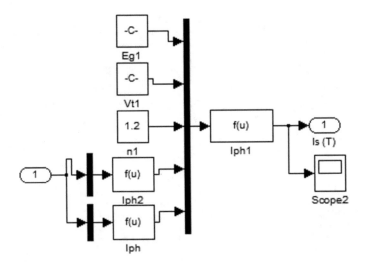

Fig. 4 Interior model of the subsystem 2

Figure 5 shows the I-V and P-V output characteristics of PV module with varying irradiance at the constant temperatures and varying temperature and constant irradiance. Figure 5a shows the I-V characteristics at constant temperature and different insolation levels. It is depicted that with increase in insulation level, the P-V output current increases rapidly and it delivers its maximum power to the load at an optimum operating point.

Figure 5b shows the P-V characteristics of the PV module at different irradiance and at the constant temperature. From the graphs, it can be illustrated that with the increase in the solar irradiance, the voltage and current output also increases. This lead to the increment in power output with an increase in irradiance at the constant temperature. The I-V and P-V characteristics under constant irradiance with varying temperature are presented in Fig. 5c and d, respectively. When the operating temperature increases, the current output increases marginally but the voltage output decreases drastically, which results in gross reduction in power output.

3.2 Simulation of Solar Array

A single solar cell produces an open circuit voltage of 0.7 V. In the given circuit, 36 cells are connected in a string to produce a typical voltage of 25.2 V and 5 strings are connected in parallel, so as to produce a typical output current of 12.5 A in the form of solar module. Ultimately, nine solar modules are connected in series to produce an output voltage of 226.8 V = 230 V in the form of solar array. The Fig. 6a, b shows the I-V curve and P-V curve of the solar array at the temperature of 40 °C and solar irradiance of 500 W/m^2.

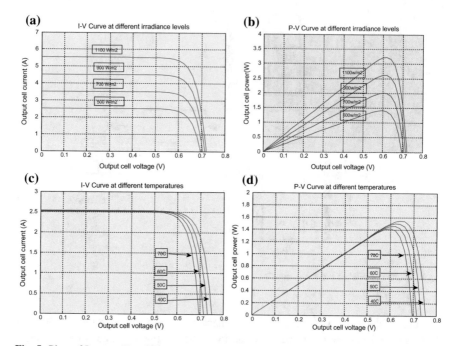

Fig. 5 Plots of I versus V and P versus V at different insolation levels and temperatures

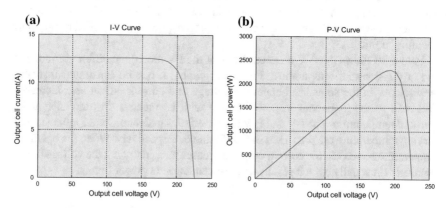

Fig. 6 I-V and P-V characteristics plot of a typical solar array

3.3 Simulation of Single-Phase PWM-Based VSI Inverter

Figure 7 shows the Simulink model of the single-phase solar array-based sinusoidal pulse width modulated VSI. The output of solar array is fed to the subsystem 3 which is the SPWM inverter. The solar cell output consists of the dc voltage and the harmonics are contained in it. Therefore, filter circuits are needed to make the

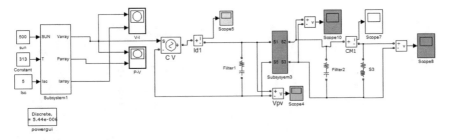

Fig. 7 Single-phase solar array fed SPWM voltage source inverter

output of the solar cell steady and ripple-free. The filter circuit consists of the parallel combination of R and C connected across the subsystem 3. The typical values of the R and C for the filter circuit are chosen as $R = 10\ \Omega$ and $C = 400$ MFD. A diode is also connected in series with solar cell output, so as to prevent the reverse flow of current.

4 Simulation Results of Single SPWM Inverter

In this case, the value of modulation index m $= 0.8$. The modulation technique used for the simulation is the sinusoidal pulse width modulation technique. The load connected at the output of the inverter is considered to be inductive load having the value of R $= 10\ \Omega$ and L $= 70$ mH. Figure 8a, b shows the shape of the output load voltage and the load currents.

Similarly, Fig. 8c, d shows the harmonic spectrum of the load voltage and the load current. The THD of the load voltage is 56% and the THD of the output current is only 1.62% which is very small and can be safely neglected. The waveform of the load current shows the purely sinusoidal waveform. The above values of the current and voltage have been calculated for an operating temperature of 40 °C and solar radiation of 500 W/m². The load current is almost sinusoidal because of the low harmonic content.

4.1 Simulation Results at Different Radiation and Temperature

Since the inverter is fed by solar cell, the output of the inverter will be affected by variation in temperature and solar radiation. First, the performance of the inverter is studied at constant temperature and at varying insolation levels. The corresponding values of the output voltage, load current and their harmonic contents are also noted down. From the results, it is noted that the magnitude of the harmonics remains almost

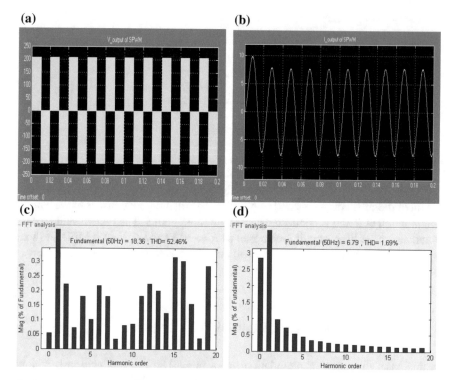

Fig. 8 **a** Inverter output voltage. **b** Inverter output current. **c** Inverter voltage harmonic spectrum. **d** Load current harmonic spectrum

constant only the magnitude of the output voltage and current is affected (following the same trend as in Fig. 8) for different temperature and insolation values.

5 Conclusion

In this chapter, a solar cell driving a PWM inverter has been proposed which utilizes Sinusoidal pulse width modulation technique for its implementation. The effects of the different operating variables like the operating temperature and the solar radiation of the photovoltaic cell and its impact on the output voltage, current, and their harmonic contents has been elaborated in this chapter. The advantages of the proposed solar cell based inverter are the low harmonic contents in the load current and the output voltage is almost sinusoidal. Extensive simulations in different conditions have been made in the MATLAB/Simulink.

References

1. H. Xue-hao, Z. Xiao-xin, B. Xiao-min et al., Development prospects for the very large-scale photovoltaic power generation and its electric power systems in China. Sci. Technol. Rev. **11**, 4–8 (2004)
2. J.S. Kumari, C.S. Babu, Mathematical modeling and simulation of photovoltaic cell using MATLAB-Simulink environment. Int. J. Electr. Comput. Eng. **2**(1), 26–34 (2012)
3. Y.-C. Kuo, T.-J. Liang, J.-F. Chen, Novel maximum-power-point-tracking controller for photovoltaic energy conversion system. IEEE Trans. Ind. Electron. **48**(3), 594–601 (2001)
4. T. Noguchi, S. Togashi, R. Nakamoto, Short-current pulse-based maximum-power-point tracking method for multiple photovoltaic-and-converter module system. IEEE Trans. Industr. Electron. **49**(1), 217–223 (2002)
5. A. Kulkarni, V. John, Mitigation of lower order harmonics in a grid connected single phase PV inverter. IEEE Trans. Power Electron. **28**(11), 5024–5037 (2013)
6. F. Filho, Y. Cao, L.M. Tolbert, 11 level cascaded H bridge grid tied inverter interface with solar panels, in *2010 Twenty-Fifth Annual IEEE Applied Power Electronics Conference and Exposition (APEC)* (2010)
7. J. Selvaraj, N.A. Rahim, Multilevel inverter for grid connected PV system employing digital PI controller. IEEE Trans. Industr. Electron. **56**(1), 149–158 (2009)
8. S. Mekhilef, N.A. Rahim, A.M. Omar, A new solar energy conversion scheme implemented using grid-tied single phase inverter, in *2000 TENCON Proceedings. Intelligent Systems and Technologies for the New Millennium*, Kuala Lumpur (2000), pp. 524–527
9. H. Dehbonei, C. Nayar, L. Borle, M. Malengret, A solar photovoltaic in-line UPS system using space vector modulation technique, in *Power Engineering Society Summer Meeting*, Vancouver, Canada (2001), pp. 632–637
10. Y. Sozer, D.A. Torrey, Modeling and control of utility interactive inverters. IEEE Trans. Power Electron. **24**(11), 2475–2483 (2009)
11. Z. Yao, L. Xiao, Two-switch dual-buck grid-connected inverter with hysteresis current control. IEEE Trans. Power Electron. **27**(7), 3310–3318 (2012)

Optimal Material Selection on Designed Chassis

Manas Desai, Devendra Somwanshi and Shubham Jagetia

Abstract This paper deals with the material selection on a designed chassis on basis of the detailed mechanical properties of the materials. The paper also deals with the static structural analysis of an automobile chassis—monocoque chassis and design modification to improve the strength of chassis. The chassis is the backbone of the vehicle, so safety is the utmost priority. Chassis should be capable enough to withstand the loads—fluctuating loads, loads due to vibration, loads due to torsion, and other loads. Chassis is also considered to have adequate bending stiffness for better handling. So, maximum stress, maximum equivalent strain, maximum equilateral stress, and deflection are important criteria for the design of the chassis. Weight factor also plays an important role in manufacturing the chassis; hence, a group of materials with their detailed mechanical properties were considered, and the static structural analysis was performed considering the properties of the materials in order to identify the optimal material.

Keywords Monocoque chassis · Material selection
Design modification of chassis · Roll-over index

M. Desai (✉) · S. Jagetia
Department of ME, Poornima University, Jaipur 303905, India
e-mail: manasme1514@poornima.edu.in

D. Somwanshi
Department of EEE, Poornima University, Jaipur 303905, India
e-mail: imdev.som@gmail.com

© Springer Nature Singapore Pte Ltd. 2019
H. Malik et al. (eds.), *Applications of Artificial Intelligence Techniques
in Engineering*, Advances in Intelligent Systems and Computing 698,
https://doi.org/10.1007/978-981-13-1819-1_43

459

1 Introduction

Chassis is the foundational structure of an automobile on which all the other parts of the vehicle are mounted. The different parts that are mounted on the vehicle are suspensions, wheels, tires, engine, transmission, body, and other parts. Chassis undergo various forces due to self-load (static), vibrations, twist (dynamic) forces.

Out of many types of chassis, monocoque chassis is one which plays the role of both body surface and the structure. The typical monocoque chassis is very weight efficient and rare. It is the most widely used type of modern car structure and is well suited for mass production.

2 Literature Survey

Device to calculate the risk of rolling over has been designed and a lateral acceleration is produced in a direction opposite to that of rolling over to prevent the effect. But this was more prone to accidents as the car went in opposite direction as intended by the driver. So, a unified chassis control (UCC) system was designed, which calculates the risk of rolling over and then calculates the braking force required to stop the vehicle, and the brakes are applied to prevent rolling over (Fig. 1).

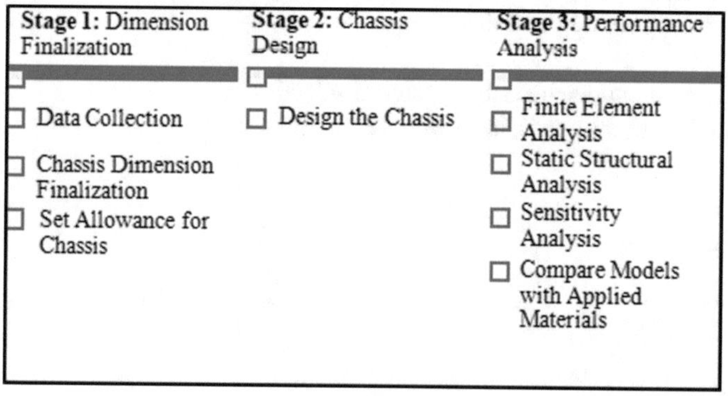

Stage 1: Dimension Finalization	Stage 2: Chassis Design	Stage 3: Performance Analysis
☐ Data Collection	☐ Design the Chassis	☐ Finite Element Analysis
☐ Chassis Dimension Finalization		☐ Static Structural Analysis
☐ Set Allowance for Chassis		☐ Sensitivity Analysis
		☐ Compare Models with Applied Materials

Fig. 1 Architectural diagram for designing and analysis

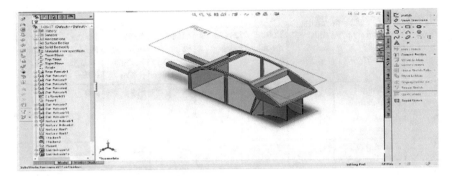

Fig. 2 Chassis model designed

2.1 Structural Analysis of the Monocoque Chassis

2.1.1 Chassis Specifications

Length: 3676.5 mm   Width: 1539 mm   Height: 1150 mm

Front Overhang (a) = 450 mm   Rear Overhang (c) = 671.5 mm   Wheelbase (b): 2555 mm

2.1.2 Modeling of the Chassis

The chassis is modeled on the designing software Solid Works 2014. The feature commands have been majorly used to design the chassis of the mentioned chassis specification. The method for designing the chassis was subtractive and eased the designing procedure. The overlook of the chassis of the abovementioned dimensions is as shown in Fig. 2.

The model designed was having the average thickness was **72.33 mm**. The detailed properties of the various materials to be used for analysis on the designed chassis are as follows.

2.1.3 Analysis of the Chassis

The chassis designed in Solid Works 2014 was analyzed in the ANSYS Workbench 15.0. The design file from Solid Works was made portable by saving it with the .IGS extension. The structural analysis component system was used for the purpose. The body was provided a fixed support at the back of the chassis and a force of 2000 N was applied at the very front of the chassis to assess the performance. Also, the chassis was analyzed by applying a total of 13 materials as mentioned in Table 1. The results of the structural analysis are shown in the following text.

Table 1 Material survey table

S. no	Name of material	Paper Ref.	Bulk modulus (GPa)	Poisson's ratio	Modulus of elasticity (GPa)	Tensile ultimate strength (MPa)	Tensile yield strength (MPa)	Shear modulus (GPa)
1	Austenitic (Steel Alloy)		166	0.25	193	579	290	77
2	ASTM A710 Steel Grade B	[1]	160	0.29	205	620	585	80
3	ASTM A710 Steel Grade A	[1]	160	0.29	205	620	585	80
4	ASTM A302 Alloy Steel	[1]	160	0.29	200	690	345	80
5	Aluminum Alloy 6063 T6	[1]		0.33	69	240	215	25.8
6	Al 360	N	62	0.33	71	317	170	26.5
7	Cast Iron		41	0.26	190	448	241	78
8	AISI 1018	[2]	159	0.29	205	440	370	80
9	AISI 1020		140	0.29	200	420.5	351	77
10	St 52	[3–5]		0.31	210	500	355	
11	St. 37	[6]		0.29	210	490	350	
12	AISI 1118	[7]	140	0.27–0.30	200	525	315	80
13	AISI 4130 (Quenched and Tempered)	[8–10]	140	0.285	205	731	460	80

2.1.4 Results for Static Structural Analysis

The results are calculated on bases of the **equivalent stress generated**, the **equivalent strain generated,** and the **total deformation indented** on the body when a total force of **2000 N** is applied from the front side of the chassis, and the fixed supports are applied on the backend side of the chassis considering the 13 materials along with their detailed mechanical properties are as follows:

The static structural analysis results were further showcased in the form of graphical representation as follows (Figs. 3, 4, 5 and 6):

The **total deformation generated** lied in the range of **0–3.31 × 10^{-004} m**.

The **equivalent stress generated** lied in the range of **2.41 × 10^{+006} to 1.22 × 10^{+007} Pa.**

The **equivalent strain generated** lied in the range of **1.20 × 10^{-005} to 7.8 × 10^{-005} m/m**. According to the static structural analysis, the least equivalent stress generated on the chassis was by using the material Aluminum Alloy 6063 T6 and Al 360. Weight also plays an important role in the manufacturing of the chassis manufacturing. According to the survey of the recent sedan cars trending in the market, it can be concluded that an average weight of a monocoque chassis should not exceed 1200 kg so as to avoid the weight of the car to exceed 1500 kg. Hence,

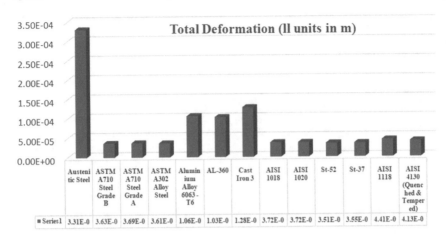

Fig. 3 Graphical representation of total deformation generated

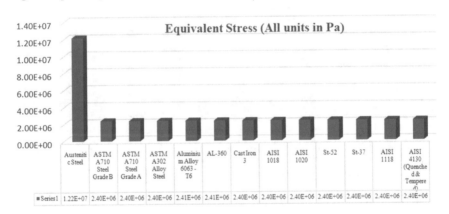

Fig. 4 Graphical representation of equivalent stress generated

weight analysis is performed by applying the densities of all 13 materials, and the materials falling under 1200 kg were further taken into consideration.

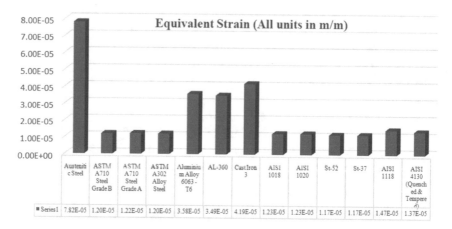

Fig. 5 Graphical representation of equivalent strain generated

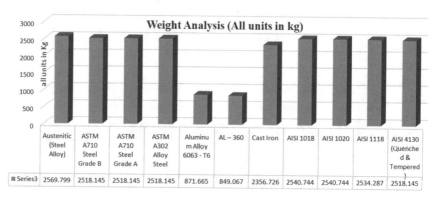

Fig. 6 Graphical representation of weight analysis

3 Results

From the above static structural analysis based on the consideration of the 13 materials along with their detailed mechanical properties, it can be clearly seen that:

The maximum total deformation generated was on austenitic steel having the value of 3.31×10^{-004} m and the minimum total deformation generated was on AISI 1118 having the value of 4.41×10^{-005} m.

The maximum equivalent strain generated was on austenitic steel having the value of 7.82×10^{-005} m/m, and the minimum equivalent strain generated was on ASTM A710 Steel Grade B having thickness less than 6.4 mm and ASTM A302 Alloy Steel having the value of 1.20×10^{-005} m/m.

Table 2 The results of structural analysis

Name of material	Total deformation	Equivalent strain	Equivalent stress	Name of material	Total deformation	Equivalent strain	Equivalent stress
M1: Austenitic Steel				M8: AISI 1018			
M2: ASTM A710 Steel Grade B				M9: AISI 1020			
M3: ASTM A710 Steel Grade A				M10: St. 52			
M4: ASTM A302 Alloy Steel				M11: St. 37			
M5: Aluminium Alloy 6063 T6				M12: AISI 1118			
M6: AL 360				M13: AISI 4130 (Q & T)			
M7: Cast Iron				Where, Q & T = Quenched & Tempered, M = material			

The maximum equivalent stress generated was on austenitic steel having the value of $1.22 \times 10^{+007}$ Pa, and the minimum equivalent stress generated was on Aluminum Alloy 6063 T6 and Al 360 having value of $2.41 \times 10^{+006}$ Pa.

The results for the weight analysis stated that Aluminum Alloy 6063 T6 and Al 360 having their densities as 2.70 and 2.63 g/cc and the chassis weight being 871.665 kg and 843.067 kgs are the only two materials which are best suitable for the manufacturing of the chassis as they fall under the weight constraints (Tables 2, 3 and 4).

The results for the cost analysis stated that Aluminum Alloy 6063 T6 is cheaper having the rate of 6000 per kg compared to Al 360 having the rate of 11,500 per kg.

Table 3 Results for structural analysis

Sr. no	Name of material	Total Deformation generated (m)	Equivalent Strain generated (m/m)	Equivalent Stress generated (Pa)
1	Austenitic Steel	3.31×10^{-004}	7.82×10^{-005}	$1.22 \times 10^{+007}$
2	ASTM A710 Steel Grade B	3.63×10^{-005}	1.20×10^{-005}	$2.40 \times 10^{+006}$
3	ASTM A710 Steel Grade A	3.69×10^{-005}	1.22×10^{-005}	$2.40 \times 10^{+006}$
4	ASTM A302 Alloy Steel	3.61×10^{-005}	1.20×10^{-005}	$2.40 \times 10^{+006}$
5	Aluminium Alloy 6063-T6	1.06×10^{-004}	3.58×10^{-005}	$2.41 \times 10^{+006}$
6	AL-360	1.03×10^{-004}	3.49×10^{-005}	$2.41 \times 10^{+006}$
7	Cast Iron 3	1.28×10^{-004}	4.19×10^{-005}	$2.40 \times 10^{+006}$
8	AISI 1018	3.72×10^{-005}	1.23×10^{-005}	$2.40 \times 10^{+006}$
9	AISI 1020	3.72×10^{-005}	1.23×10^{-005}	$2.40 \times 10^{+006}$
10	St-52	3.51×10^{-005}	1.17×10^{-005}	$2.40 \times 10^{+006}$
11	St-37	3.55×10^{-005}	1.17×10^{-005}	$2.40 \times 10^{+006}$
12	AISI 1118	4.41×10^{-005}	1.47×10^{-005}	$2.40 \times 10^{+006}$
13	AISI 4130 (Quenched & Tempered)	4.13×10^{-005}	1.37×10^{-005}	$2.40 \times 10^{+006}$

Table 4 Material density table for weight analysis

Sr. no.	Name of material	Density (g/cc)	Weight of chassis (kg)
1	Austenitic (Steel Alloy)	7.96	2569.799
2	ASTM A710 Steel Grade B	7.80	2518.145
3	ASTM A710 Steel Grade A	7.80	2518.145
4	ASTM A302 Alloy Steel	7.80	2518.145
5	Aluminium Alloy 6063 - T6	2.70	871.665
6	AL – 360	2.63	849.067
7	Cast Iron	7.30	2356.726
8	AISI 1018	7.87	2540.744
9	AISI 1020	7.87	2540.744
12	AISI 1118	7.85	2534.287
13	AISI 4130 (Quenched & Tempered)	7.80	2518.145

4 Conclusion

Based on the analysis and simulation, it can be concluded that ASTM A710 Steel Grade B showed the least equivalent strain generated, hence possessed the best strength. Aluminum Alloy 6063 T6 and Al 360 showed the least equivalent stress generated and had the least raw material cost per kg; hence, it can be concluded that Aluminum Alloy 6063 T6 is the best suitable material for manufacturing a chassis.

References

1. A.H. Kumar, V. Deepanjali, Design & analysis of automobile chassis. Int. J. Eng. Sci. Innov. Technol. (IJESIT) **5**, 187–196 (2016)
2. R. Thavai, Q. Shahezad, M. Shahrukh, M. Arman, K. Imran, Static analysis of go-kart chassis by analytical and solid works simulation. Int. J. Multidiscip. Educ. Res. (IJMER) **5**, 64–68 (2015)
3. J. Rajpal, R.S. Bhirud, A.K. Singh, A.V. Hotkar, S.G. Thorat, Finite element analysis and optimization of an automobile chassis. Int. J. Eng. Res. Technol. (IJERT) **3**, 2075–2082 (2015)
4. J. Rajpal, S.G. Thorat, B.S. Basavaraj, S. Kothavale, S.S. Hatwalane, Design considerations for automobile chassis for prevention of rolling over of a vehicle. Appl. Mech. Mater. **6**, 41–49 (2014)
5. V.V. Patel, R.I. Patel, Structural analysis of automotive chassis frame and design modification for weight reduction. Int. J. Eng. Res. Technol. (IJERT) **1**, 1–6 (2012)
6. R.L. Patel, K.R. Gawand, D.B. Morabiya, Design and Analysis of Chassis Frame of Tata 2516tc. Int. J. Res. Appl. Sci. Eng. Trends (IJRASET) **2**, 115–119 (2014)
7. A. Makhrojan, S.S. Budi, J. Jamari, A. Suprihadi, R. Ismail, Strength analysis of monocoque frame construction in an electric city car using finite element method, in *Electric Vehicular Technology and Industrial, Mechanical. Electrical and Chemical Engineering, Joint International Conference, IEEE, 2015*, pp. 275–279
8. N.L. Rakesh, K.G. Kumar, J.H. Hussain, Design and analysis of Ashok Leyland chassis frame under 25 ton loading condition. Int. J. Innov. Res. Sci. Eng. Technol. (IJIRSET) **3**, 17546–17551 (2014)
9. R. Rajappan, M. Vivekanandhan, Static and modal analysis of chassis by using fea. Int. J. Eng. Sci. (IJES) **2**, 63–73 (2013)
10. D. Lavanya, G.G. Mahesh, V. Ajay, C. Yuvaraj, Design and analysis of a single seater race car chassis frame. Int. J. Res. Aeronaut. Mech. Eng. (IJRAME) **2**, 12–23 (2014)

Optimization and Comparison of Deformation During Closed Die Forging of Different Parts

Jasleen Kaur, B. S. Pabla and S. S. Dhami

Abstract The key requirement of forging industry is to obtain a defect-free part with minimal trials on the shop floor. But the hit and trial method used by the industries does not give optimum results. The use of FEA for the optimization of parts is a better option. The definition of input parameters basically defines the metal flow within the die cavity during the forging process. In this paper, the process parameters have been optimized for better metal flow in two different parts. The simulated study was done using the input parameters of billet size, temperature, and coefficient of friction. The Response Surface Method (RSM) was used for the design of experiment, the results of which revealed the significant responses affecting the forging process. The optimization of the process was done for minimum effective strain rate, minimum die wear, and maximum material flow rate. The predictive models of the responses were obtained. Then, the comparison of the parts was done in terms of the responses. It was observed that the biggest part had the largest inhomogeneity, while the smaller part with higher shape complexity factor had larger homogeneity during deformation. It was easy to deform the bigger, yet simpler part. The material flow rate gave the opposite results. The bigger and simpler shape had minimum variation in the material flow rate, while the complex shapes had larger variation. The smallest part with complex contour had the highest associated die wear. The results could be used as an empirical form to frame the forging of new parts with minimum trials. The novelty of the research lies in the quantitative analysis of the responses in relation to metal flow, which has not been reported in the literature.

Keywords Simulation · Forging · Material flow rate · Effective strain rate
Die wear

J. Kaur (✉) · B. S. Pabla · S. S. Dhami
Department of Mechanical Engineering, NITTTR Chandigarh, Chandigarh, India
e-mail: jasleenkaur12636@gmail.com

© Springer Nature Singapore Pte Ltd. 2019
H. Malik et al. (eds.), *Applications of Artificial Intelligence Techniques in Engineering*, Advances in Intelligent Systems and Computing 698,
https://doi.org/10.1007/978-981-13-1819-1_44

469

1 Introduction

The modeling and simulation techniques [1] for closed die forging process have been used widely by the researchers owing to the ease of use through simulation software and platform for detailed research. The plastic deformation [2] during the hot forging operation is analyzed critically using the numerical simulation [3] technique. The simulation software provide certain features which have been used by the researchers to study the metal flow [4] during deformation, analyze the split flow design [5] and study the strain distribution pattern [6]. The other responses which have been studied include the deformation speed [7], preform design [8, 9], upset ratio [7], forming load [10], etc. The shape complexity factor [11] of the part has also been accounted in the literature. The optimization of the forging process [12] to improve the quality of the part and deformation homogeneity [13] is a crucial requirement of the manufacturing industry and has been used widely by the researchers. The various case studies taken up for the optimization include turbine blade [14], yoke [15], impeller [8] and spiral bevel gear [16]. The forming of a part with different metals like stainless steel [17], alloy [4, 17] and bimetal components [10] has also been reported. In this paper, the optimization of two components has been taken up, and the deformation among them has been compared using various responses. The following sections include the methodology used for experimentation and results and discussions. Graphical results for each part have been discussed in detail for the responses, along with the predictive models.

2 Methodology

The reported research has revealed that the round billet at 10% flash allowance of the finished part was ideal for forging a gear blank [18]. Two new forged parts were used for similar analysis, and then, the comparison of three parts was done to generalize the results. The input parameters of billet size, temperature, and coefficient of friction were varied according to experiments designed by the response surface method. The results of Analysis of Variance (ANOVA) in RSM clarified the significant parameters having an effect on the forging process. Further, the predictive models of the responses of minimum effective strain rate, minimum die wear, and maximum material flow rate were obtained. The numerical optimization of the parameters was done and validated as well.

3 Results and Discussions

3.1 Part I

The methodology given by Kaur et al. [18] was followed, and the deformation pattern of the resulting part is shown in Fig. 1. From the ANOVA analysis, it was observed that out of the three input parameters, the dimension of billet was found to be the most dominant parameter, followed by temperature of billet. The coefficient of friction was found to have minimal effect on the responses. So, it was concluded that the effect of the dimension of round billet had a major role in determining the normal and shear stresses, i.e., the deformation pattern of billet along with the other responses of effective stress, effective plastic strain, effective strain rate, die wear, and material flow rate. The load was not found to be significantly affected because the volume of material in all three billet sizes was kept same.

Graphical analysis of results: The graphical variation of responses with the input parameters can be observed from Figs. 2, 3 and 4. The details are given below.

1. *Effective strain rate*: The effective strain rate is the response which defines the homogeneity of deformation. Less variation in the effective strain rate is the requisite of the deformation process. The perturbation, Fig. 2a, shows that maximum slope is observed for billet dimension, indicating that it is the most dominant parameter affecting the process, followed by coefficient of friction and temperature of billet. The temperature of billet is having negligible effect. The detailed interaction effects can be seen from the 3D graphs (Fig. 2b–d). As the billet dimension is increased, the effective strain rate decreases significantly. However, it increases as the coefficient of friction increases. As the temperature of billet is increased, it decreases. The regression analysis gave the predictive model for the response as shown in Eq. (1).

$$\text{Effective strain rate} = +254.09 + 17.95 * A - 13.80 * B - 54.61 * C$$
$$+ 27.85 * A * B - 23.15 * A * C$$
$$+ 26.10 * B * C - 34.92 * A^\wedge 2 - 36.85 * B^\wedge 2 + 96.02 * C^\wedge 2 \qquad (1)$$

Fig. 1 Flowlines showing deformation in defect-free part 1

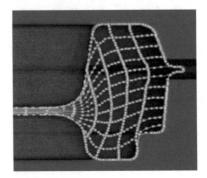

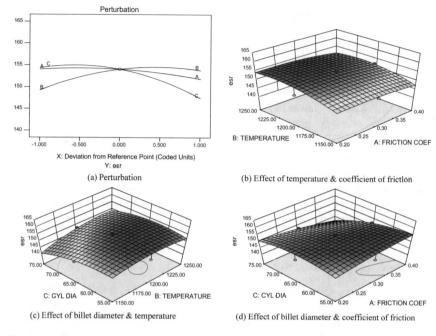

(a) Perturbation

(b) Effect of temperature & coefficient of friction

(c) Effect of billet diameter & temperature

(d) Effect of billet diameter & coefficient of friction

Fig. 2 Results of analysis for part 1 for effective strain rate

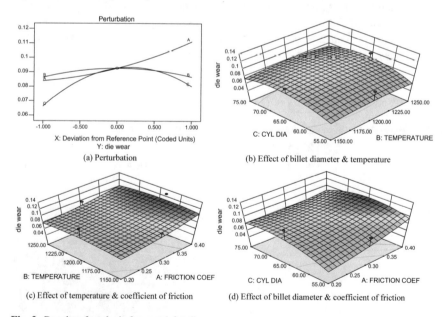

(a) Perturbation

(b) Effect of billet diameter & temperature

(c) Effect of temperature & coefficient of friction

(d) Effect of billet diameter & coefficient of friction

Fig. 3 Results of analysis for part 1 for die wear

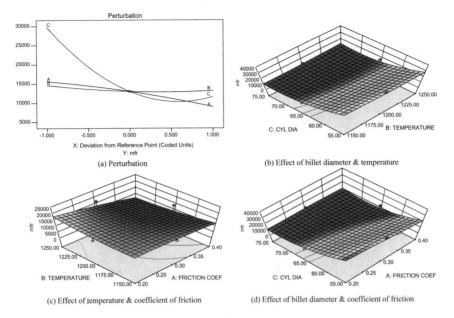

(a) Perturbation

(b) Effect of billet diameter & temperature

(c) Effect of temperature & coefficient of friction

(d) Effect of billet diameter & coefficient of friction

Fig. 4 Results of analysis for part 1 for material flow rate

where A = Coefficient of friction, B = Temperature of billet, and C = Billet size.

2. *Die wear*: The perturbation, Fig. 3a, shows that maximum slope is observed for billet dimension, indicating that it is the most dominant parameter affecting the process, followed by temperature of billet and coefficient of friction. The detailed interaction effects can be seen from the 3D graphs (Fig. 3b–d). As the billet dimension and temperature are increased, the die wear decreases. However, it increases as the coefficient of friction increases. The regression analysis gave the predictive model for die wear as given in Eq. (2).

$$
\begin{aligned}
\text{Die wear} = {}&+0.45 + 0.063 * A - 0.089 * B - 0.078 * C \\
&+ 0.071 * A * B - 0.068 * A * C \\
&+ 0.071 * B * C + 0.14 * A^2 + 0.24 * B^2 + 0.32 * C^2
\end{aligned} \tag{2}
$$

where A = Coefficient of friction, B = Temperature of billet, and C = Billet size.

3. *Material flow rate*: The perturbation, Fig. 4a, shows that maximum slope is observed for billet dimension, indicating that it is the most dominant parameter affecting the process, followed by coefficient of friction and temperature of billet. The detailed interaction effects can be seen from the 3D graphs (Fig. 4b–d). For smaller billet size, the material flow rate decreases as temperature and coefficient of friction are increased. However, for large billet size, it increases with the

Table 1 Numerical optimization of parameters for part I

Constraints		Part 1			Part 2		
Name	Goal	Predicted solution	Simulated solution	Error (%)	Predicted solution	Simulated solution	Error (%)
A: Coefficient of friction	Is in range	0.3			0.2		
B: Temperature of billet (°C)	Is in range	1200			1150		
C: Billet size (mm)	Is in range	65			55		
Effective strain rate	Minimize	208	220	5	147.926	149.8	1
Die wear (mm)	Minimize	0.338	0.345	2	0.037	0.040	8
Material flow rate (mm/s)	Maximize	25500	26560	4	37570	37230	0.9

increase in billet temperature. The regression analysis gave the predictive model for material flow rate as given in Eq. (3).

$$\text{Material flow rate} = 35763.09 - 317 * A - 123 * B - 4561 * C$$
$$+ 80 * A * B - 125 * A * C$$
$$+ 275 * B * C - 4327.73 * A^2 - 3747.73 * B^2 - 4297.73 * C^2 \quad (3)$$

where A = Coefficient of friction, B = Temperature of billet, and C = Billet size.

Optimization of parameters for part I: RSM was used to optimize the input parameters which would yield the desired responses. The numerical optimization and comparison of predicted and simulated results are shown in Table 1.

3.2 Part II

This part was analyzed on the same pattern and the deformation in the resulting part is shown in Fig. 5. From the ANOVA analysis, it was observed that out of the three input parameters, the dimension of billet was found to be the most dominant parameter, followed by temperature of billet and then the coefficient of friction.

Graphical analysis of results: The graphical variation of responses with the input parameters can be observed from Figs. 6, 7 and 8. The details are given below.

1. *Effective strain rate*: The effective strain rate is the response which defines the homogeneity of deformation. Less variation in the effective strain rate is the

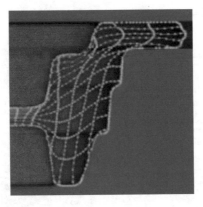

Fig. 5 Flowlines showing deformation in defect-free part II

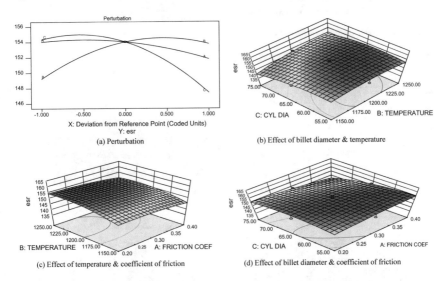

(a) Perturbation

(b) Effect of billet diameter & temperature

(c) Effect of temperature & coefficient of friction

(d) Effect of billet diameter & coefficient of friction

Fig. 6 Results of analysis for part II for effective strain rate

requisite of the deformation process. The perturbation, Fig. 6a, shows that maximum slope is observed for billet dimension, indicating that it is the most dominant parameter affecting the process, followed by temperature of billet and then coefficient of friction. The detailed interaction effects can be seen from the 3D graphs (Fig. 6b–d). As the billet dimension is increased, the effective strain rate decreases. When the coefficient of friction is increased, the effective strain rate increases at low temperature and decreases at high temperature. With the increase in the temperature of billet, there is a slight increase in the value of effective strain rate. The regression analysis gave the predictive model for the effective strain rate as given in Eq. (4).

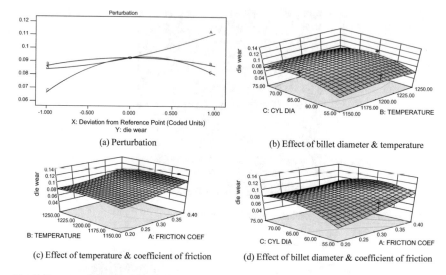

(a) Perturbation

(b) Effect of billet diameter & temperature

(c) Effect of temperature & coefficient of friction

(d) Effect of billet diameter & coefficient of friction

Fig. 7 Results of analysis for part II for die wear

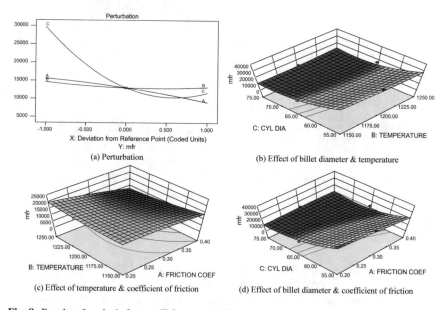

(a) Perturbation

(b) Effect of billet diameter & temperature

(c) Effect of temperature & coefficient of friction

(d) Effect of billet diameter & coefficient of friction

Fig. 8 Results of analysis for part II for material flow rate

$$\text{Effective strain rate} = 154.03 - 1.11 * A + 2.27 * B - 3.46 * C - 0.31 * A * B - 3.17 * A * C$$
$$+ 1.84 * B * C - 1.11 * A^2 - 2.52 * B^2 - 3.14 * C^2 \qquad (4)$$

where A = Coefficient of friction, B = Temperature of billet, and C = Billet size.

2. *Die wear*: The perturbation, Fig. 7a, shows that maximum slope is observed for coefficient of friction, indicating that it is the most dominant parameter affecting the process, followed by billet dimension and temperature. The detailed interaction effects can be seen from the 3D graphs (Fig. 7b–d). As the billet dimension and temperature are increased, the die wear decreases at higher values of coefficient of friction, while it increases at low values. The regression analysis gave the following predictive model of die wear.

$$Die\ wear = 0.093 + 0.014 * A + 1.900E - 004 * B$$
$$+ 6.950E - 003 * C - 1.963E - 003 * A * B - 0.011 * A * C - 9.375E - 004 * B * C$$
$$+ 4.832E - 003 * A^2 - 6.168E - 003 * B^2 - 0.019 * C^2 \qquad (5)$$

where A = Coefficient of friction, B = Temperature of billet, and C = Billet size.

3. *Material flow rate*: The perturbation, Fig. 8a, shows that maximum slope is observed for billet dimension, indicating that it is the most dominant parameter affecting the process, followed by coefficient of friction and temperature of billet. The detailed interaction effects can be seen from the 3D graphs (Fig. 8b–d). As the billet dimension and the temperature are increased, the material flow rate decreases. For the smaller value of coefficient of friction, the material flow rate decreases, while it increases with the larger billet size. The regression analysis gave the following predictive model for material flow rate.

$$Material\ flowrate = 13020 - 3304 * A - 710 * B - 9066 * C - 1953.75 * A * B$$
$$+ 3716.25 * A * C + 1823.75 * B * C - 710 * A^2$$
$$+ 910 * B^2 + 7670 * C^2 \qquad (6)$$

where A = Coefficient of friction, B = Temperature of billet, and C = Billet size.

Optimization of parameters for part II: The numerical optimization of parameters was done using RSM, and the predicted values of the responses and the simulated responses were compared as given in Table 1.

4 Comparison

The two parts were compared with the third part [18] as shown in Fig. 9, and the following interpretations were obtained. The minimum variation of effective plastic strain is for part 3 and maximum for part 2. This implies part 3 is easier to deform than part 2, even though the weight of part 3 is more than part 2. The same result is observed when the shape complexity factor is calculated. It can be observed that the shape of the part is more critical than the size of the part. The minimum variation of effective stress is found in part 3, while highest in part 2. The die cavity for part 2 is deep with multiple cross sections which increase the variation of effective stresses. The minimum variation of effective strain rate is found in part 2, while the maximum

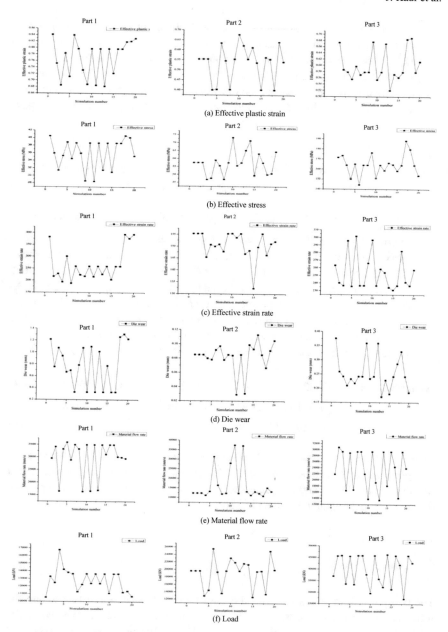

(a) Effective plastic strain

(b) Effective stress

(c) Effective strain rate

(d) Die wear

(e) Material flow rate

(f) Load

Fig. 9 Comparison of responses for three parts

is found in part 3. It implies that the deformation is most homogenous for part 2, while minimum homogeneity is found in part 3. This implies that the larger weight of the part is difficult to deform uniformly. The maximum variation of material flow rate is found with part 2, while minimum with part 3. It is because of the multiple cross sections of part 2, which requires different metal flow rates at different cross sections. Part 3 mainly requires lateral flow of material since the size of the part is large.

5 Conclusions

It was concluded that the weight of the component is important in deciding the deformation energy, homogeneity of deformation, and die life. The rate of the change of effective strain during deformation must be uniform with minimum variation. It requires a billet shape which spreads out evenly in all directions into the die cavity. Further, the size of cavity decides the uniformity of flow. A large size cavity will not give a uniform flow, while a medium size and small size cavities get filled up quickly at a steady rate. The square billet got sheared on the corners while filling the die cavity, which increased die wear; hence, the round billet gave better results. A part with larger surface area is affected significantly by the value of coefficient of friction, while that with a smaller surface area is affected by the temperature of billet, rather than the coefficient of friction. The deformation is homogenous with round billet. The die wear is also less because the material of billet flows out uniformly in all the directions. The contoured cavity requires more forces for deformation with round billet than with square billet. For a proper die design, ISO recommended flash allowance of 10% is sufficient to manufacture the part without any defects. The predicted and simulated results for the parts have an error less than 10%, which goes well with observed results for part 3 [18]. The scope of the paper lies in the use of results in intelligent techniques which are self-leaning and can be used to design the forging process in a better manner.

References

1. G.E. Dieter, H.A. Kuhn, S.L. Semiatin, *Handbook of Workability and Process Design*, ASM International
2. P. Veena, D. Maheshwari Yadav, C. Naga Kumar, A critical review on severe plastic deformation. IJSRSET 3(2), 336–343 (2017)
3. D. Luca, A numerical solution for a closed die forging process, in *MATEC Web of Conferences*, 2017. https://doi.org/10.1051/matecconf/201711202008
4. M.J. Zhao, Z.L. Wu, Z.R. Chen, X.B. Huang, Analysis on flow control forming of magnesium alloy wheel, in *3rd International Conference on Advanced Materials Research and Applications*, 2017, pp. 1–6. https://doi.org/10.1088/1757-899x/170/1/012006

5. L.I. Xu Bin, Z. Zhi Mi, W. Qiang, L.I. Guo Jun, Mechanism of split-flow forming characteristics in plastic forming processing. Int. J. Adv. Manuf. Technol. **89**, 3187–3194 (2017). https://doi.org/10.1007/s00170-016-9264-2

6. D.B. Gohil, The simulation and experimental investigation to study the strain distribution pattern during the closed die forging process. Int. J. Mech. Mech. Eng. **10**(10), 1788–1793 (2016)

7. T. Ram Prabhu, Modelling studies on effects of deformation speed, preform shape, and upset ratio on the forging characteristics of the aerospace structural Al alloys. Int. J. Mater. Prod. Technol. **54**(4) (2017). https://doi.org/10.1504/IJMPT.2017.082621

8. T.R. Prabhu, Simulations and experiments of the non-isothermal forging process of a Ti-6Al-4 V impeller. J. Mater. Eng. Perform **25**(9), 3627–3637 (2016). https://doi.org/10.1007/s11665-016-2186-1

9. M.C. Song, C.J. VanTyne, J.R. Cho, Y.H. Moon, Optimization of preform design in forging of heavy crankshafts. J. Manuf. Sci. Eng. **139** (2017)

10. P. Wu, B. Wang, J. Lin, B. Zuo, Z. Li, J. Zhou, Investigation on metal flow and forming load of bi-metal gear hot forging process. Int. J. Adv. Manuf. Technol. **88**, 2835–2847 (2017). https://doi.org/10.1007/s00170-016-8973-x

11. R. Hosseini-Ara, P. Yavari, R. Hosseini-Ara, P. Yavari, A new criterion for preform design of H-shaped hot die forging based on shape complexity factor. Int. J. Mater. Form. 1–6 (2017). https://doi.org/10.1007/s12289-017-1345-8

12. F. Zhu, Z. Wang, M. Lv, Multi-objective optimization method of precision forging process parameters to control the forming quality. Int. J. Adv. Manuf. Technol. **83**, 1763–1771 (2016). https://doi.org/10.1007/s00170-015-7682-1

13. K. Wei, X. Fan, M. Zhan, H. Yang, P. Gao, Improving the deformation homogeneity of the transitional region in local loading forming of Ti-alloy rib-web component by optimizing unequal-thickness billet. Int. J. Adv. Manuf. Technol. **92**, 4017–4029 (2017). https://doi.org/10.1007/s00170-017-0477-9

14. F. Chen, F. Ren, J. Chen, Z. Cui, O. Hengan, Microstructural modeling and numerical simulation of multi-physical fields for martensitic stainless steel during hot forging process of turbine blade. Int. J. Adv. Manuf. Technol. **82**, 85–98 (2016). https://doi.org/10.1007/s00170-015-7368-8

15. Kumbhar, J. Paranjpe, N. Karanth, Optimization in tube yoke forging process using computer simulation, SAE Technical Paper (2017). https://doi.org/10.4271/2017-26-0238

16. Z.-S. Gao, J.-B. Li, X.-Z. Deng, J.-J. Yang, F.-X. Chen, X. Ai-Jun, L. Li, Research on gear tooth forming control in the closed die hot forging of spiral bevel gear. Int. J. Adv. Manuf. Technol. (2017). https://doi.org/10.1007/s00170-017-1116-1

17. J.H. Chen, Numerical simulation of hot die forging process of Ti-6Al-4 V alloy blade. Mater. Sci. Forum **898**, 1325–1331 (2017). https://doi.org/10.4028/www.scientific.net/MSF.898.1325

18. J. Kaur, B.S. Pabla, S.S. Dhami, Parametric analysis and optimization of closed die forging of gear blank. Indian J. Sci. Technol. **10** (2017)

Material Removal Rate Prediction of EDDG of PCBN Cutting Tool Inserts Based on Experimental Design and Artificial Neural Networks

Rohit Shandley, Siddharth Choudhary and S. K. Jha

Abstract Electric discharge diamond grinding (EDDG) is a hybrid process, in which material removal occurs by the combined action of electric discharge machining (EDM) and grinding. In this paper, the influence of material removal rate on the process parameters of a bi-plate PCBN cutting tool insert ground with the help of EDDG is studied. Design of experiments is used to identify the most influential parameters, and a second-order regression model is developed, denoting a relationship between the response and the process variables. An artificial neural network (ANN) model is employed to compare and validate the regression model, which has shown a closer proximity to the experimental data.

Keywords Electric discharge diamond grinding (EDDG) · Design of experiments (DOE) · Artificial neural networks (ANN) · Bi-plate PCBN · Advanced manufacturing

1 Introduction

PCBN belongs to a class of ultrahard cutting tool materials, capable of machining difficult to cut materials such as hardened and high strength alloy steels, Ni super alloys, and some varieties of cast irons [1, 2]. PCBN cutting inserts with their chemical inertness and high hot hardness have been widely adopted to produce components possessing excellent surface finish and precise geometry at high cutting speeds in a cost-effective manner. Besides being a better candidate material than artificial diamond when machining ferrous alloys, PCBN can be used for hard turning [3, 4],

R. Shandley · S. Choudhary · S. K. Jha (✉)
Division of MPAE, Netaji Subhas Institute of Technology, New Delhi, India
e-mail: skjha63@rediffmail.com; skjha63@gmail.com

R. Shandley
e-mail: rohit_shandley@yahoo.in

S. Choudhary
e-mail: sid.choudhary007@gmail.com

© Springer Nature Singapore Pte Ltd. 2019
H. Malik et al. (eds.), *Applications of Artificial Intelligence Techniques in Engineering*, Advances in Intelligent Systems and Computing 698,
https://doi.org/10.1007/978-981-13-1819-1_45

481

wherein an additional grinding step can be eliminated without compromising the surface integrity of the work material.

At present, the market offers many registered trademarks of PCBN-based cutting inserts whose composition and properties are tailored to optimize the performance in the targeted areas. One of them is Tomal (a Russian trademark), belonging to a category of synthetic polycrystalline bi-plate composite material [5]. Like other commercially available trademarks, Tomal is available in various grades such as Tomal-10 (hereafter termed T10). Such superhard materials are often ground using diamond wheels, which when performed under the conventional grinding conditions has been proven to be uneconomical and slow [1].

In the recent times, various hybrid machining processes are available to solve the aforementioned problem. In 1943, Lazarenko and Lazarenko exploited the energy of the electric arc for material removal in metals, the process now known as electric discharge machining (EDM) [6]. EDM combined with diamond grinding has given rise to a new process called Electric Discharge Diamond Grinding (EDDG), first conceived by Grodzinskii and Zubotova [7] in the late 70s.

EDDG is a hybrid process where material removal takes place by the mechanical action of the metal bonded diamond grinding wheel in conjunction with erosion due to the electric discharge. The discharge helps maintain the wheel in an optimum state of sharpness with the help of in-process dressing [8–10], mitigating the chances of wheel glazing, clogging, and unwanted wheel wear when grinding super hard materials. Additionally, the heat generated by the electric discharge also helps in softening of the work material, and hence, mechanical grinding by diamond grains becomes easier. Thus, the use of EDDG translates to minimization of grinding forces, uninterrupted cutting and higher uptime, excellent surface finish, and dimensional accuracy of the machined part at a fraction of the cost compared to conventional grinding.

Although a large number of researchers have used this concept to machine various hard to machine materials efficiently and economically with different labels [10, 11] under different names, however, there still exists a handful of research data in the open literature on the grinding of Tomal.

The aim of this paper is to establish some of the most influential factors, considering the material removal rate (MRR) as the response parameter in the EDDG of T10. For this, an RSM-based systematic design approach has been employed to establish the optimum settings of process variables. Using a statistical technique called analysis of variance, a second-order quadratic polynomial regression model is established between the process and response parameters. Furthermore, artificial neural networks are used to construct a model which can validate and compare the results of the regression model.

2   Materials and Experimentation

2.1   Tomal-10 (T10)

Tomal-10 which was chosen as the work material has a 1:1 ratio of a working layer of cubic boron nitride (CBN) on a substrate consisting of BN along with Ti and Cu. An interesting fact regarding the success of Tomal is that it closely resembles other bi-plate PCBN products such as Sumiboron (Sumitomo, Japan) and Amborite (Element Six, South Africa) [5].

2.2   EDDG Setup

A universal tool and cutter grinder, model 3B642 (Russian made), was redesigned to perform EDDG. For this purpose, special modifications were made to fix brushes to the spindle head and the work table to supply power as shown in Fig. 1. Bronze alloy bonded diamond flaring cup type grinding wheels were used to grind T10. For the power source, a specially fabricated small sized generator was used whose characteristics could be adjusted to provide controlled electrical parameters, suitable for machining.

2.3   Design of Experiments

RSM is one of the effective strategies which can be applied to conduct experiments. RSM is based on statistical techniques used for establishing, enhancing and optimizing a process. With RSM, a relationship between the responses and the most influencing input variables is obtained in the form of 2D plots or a 3D surface.

In the present study, four input parameters, namely, depth of cut (μm/double stroke), wheel speed (m/s), grit size (μm), and diamond concentration (%) with three levels each as shown in Table 1, were selected. Here, Level 1 refers to the lowest level and Level 3 refers to the highest level. The range of factors was determined

Fig. 1 Sectional view of the modified EDDG setup

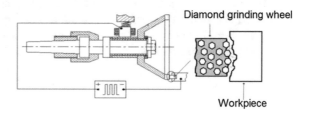

Diamond grinding wheel

Workpiece

484

R. Shandley et al.

Table 1 Process parameters and their levels

Parameters	Symbol	Factor levels			
		Level 1 (−1)	Level 2 (0)	Level 3 (+1)	Level interval
Depth of cut (μm/double stroke)	d	16	32	48	16
Wheel speed (m/s)	V	15	25	35	10
Grit size (μm)	z	40	120	200	80
Diamond concentration (%)	K	50	100	150	50

by prior experimentation with one factor at a time approach. Table 2 represents the design matrix with the corresponding MRR values.

Attributing to the ease of formation of a second-order polynomial model, a Box–Behnken design was chosen. For the present design, the regression equation obtained is of the following form:

$$\eta = a_0 + \sum a_1 X_i + \sum a_{11} X_i^2 + \sum a_{12} X_i X_j \tag{1}$$

where η is the response (area); a_o is the intercept; a_1, a_{11}, and a_{12} are regression coefficients; and X_i, X_j are input variables.

2.3.1 Analysis of Variance (ANOVA)

ANOVA uses simple algebra calibration which compared to complex matrix calculation is easy and hence can be calculated manually. Variance which stands for square of standard deviation depicts the scattering of data from the mean. ANOVA makes use of f-test to know whether the variation between groups is greater than variation of observations within the groups. If this ratio is significantly high, then not all means are equal. By using f-statistics and f-tests, fits of different models are compared, specific regression terms tested, and equality of means tested. Low f-value is the case of low variation, i.e., closer group means w.r.t. variation within groups, and high f-value is case of high variation, i.e., wide group means w.r.t. variation within groups. High f-value is needed for rejection of null hypothesis (i.e., group means are equal). F-value sometimes is difficult to interpret, and therefore, f-distribution is used to calibrate probabilities.

Table 2 Design matrix

Run	d (μm/double stroke)	V (m/s)	z (μm)	K (%)	MRR (mg/mm^3)
1	−1	−1	−1	−1	0.29
2	−1	−1	−1	−1	0.28
3	−1	−1	−1	1	0.22
4	−1	−1	1	−1	0.52
5	−1	−1	1	1	0.81
6	−1	1	−1	−1	0.67
7	−1	1	−1	1	0.28
8	−1	1	1	−1	0.76
9	−1	1	1	1	0.685
10	1	−1	−1	−1	0.17
11	1	−1	−1	1	0.13
12	1	−1	1	−1	0.4
13	1	−1	1	1	0.325
14	1	1	−1	−1	0.35
15	1	1	−1	1	0.26
16	1	1	1	−1	0.64
17	1	1	1	1	0.319
18	1	0	0	0	0.18
19	−1	0	0	0	0.32
20	0	1	0	0	0.37
21	0	−1	0	0	0.28
22	0	0	1	0	0.59
23	0	0	−1	0	0.22
24	0	0	0	1	0.3
25	0	0	0	−1	0.43
26	1	1	1	1	0.322
27	−1	−1	−1	−1	0.282
28	−1	1	−1	1	0.278
29	0	0	0	−1	0.43

2.4 Artificial Neural Network

Artificial neural networks (ANN) are nonlinear signal processing blocks made up of interconnected elementary processing units called neurons. ANN is inspired from the structure and functions of neural networks found in animals. Such a network learns by examples to solve a specific problem, mostly in the absence of problem-specific programming. Artificial neural networks are made up of neuron network architecture, designated weights, and activation function (refer Fig. 2). The neuron network architecture generally means the type of connection between neurons. There

Fig. 2 Schematic of a
simplified artificial neural
network

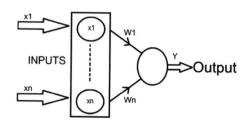

are different types of neural network architecture; feed forward network, feedback
network, fully interconnected network, competitive network, etc. The designation of
weights is also termed as learning or training. In this learning or training of ANN,
the weights in accordance to the priority are allotted on such neural networks.

Backpropagation neural networks are often considered as feed forwarded where a
gradient descent technique is used to train a number of hidden layers in a multilayered
network. These algorithms comprise three layers, namely, input layer, hidden layer,
and output layer. Feed forward backpropagation algorithm-based neural networks
have been applied in combination with design of experiments (DOE) to study the
grinding behavior by a few researchers [12–14].

The present study makes the use of a forward backpropagation algorithm to
develop a neural network model to compare the results with the actual experimentation. A network consisting of 10 hidden neurons and 1 output neuron was sufficient
to give satisfactory results. An array of 20 randomly selected data points of 4 inputs
each was fed into the algorithm for the purpose of training.

3   Results and Discussion

3.1   Estimation of Critical Process Parameters Using Design
of Experiments

Table 3 shows the analyzed ANOVA for material removal rate (MRR). P-values
of the second-order polynomial model (Model), process parameters which include
depth of cut (d), wheel speed (V), grit size (z) and the interaction of wheel speed
(V) and diamond concentration (K) are less than 0.005 (α level). This implies that
the abovementioned factors play a pivotal role in determining the behavior of the
grinding process and are thus considered significant.

Among the significant parameters, grit size (z) of the diamond grains has the
largest contribution in the grinding of T10, as indicated by the largest F-value. This
is followed by the depth of cut (d), which has the second highest F-value. Finally,
wheel speed (V) and its interaction (VK) with diamond concentration (K) have a
similar influence on the material removal rate (MRR).

Table 3 ANOVA table for MRR of T10

Source	Sum of squares	Degree of freedom	Mean square	F-value	P-value
Model	0.8484	14	0.0606	9.881470235	<0.0001
d	0.208003	1	0.208003	33.91702425	<0.0001
V	0.078031	1	0.078031	12.72371241	0.0031
z	0.37941	1	0.37941	61.86686033	<0.0001
VK	0.074276	1	0.074276	12.11148813	0.0037
R-squared: 0.9081			Adj. R-squared 0.8162		

Fig. 3 Normal probability plot of residuals for MRR during EDDG of T10

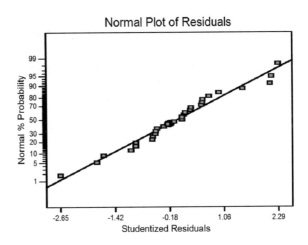

The model can be described by the following second-order regression Eq. (1):

$$MRR = 0.31 - 0.1d + 0.061V + 0.14z - 0.055K - 0.058d^2 + 0.017V^2$$
$$+ 0.097z^2 + 0.06K^2 + (6.026 \times 10^{-4})dV - 0.036dz$$
$$- 0.016dK - 0.024Vz - 0.064VK + 0.027zK$$

The R^2 value of the model obtained is 0.9081, which suggests that 90.81% of the variance in MRR can be explained by the grinding parameters considered in this study, thus rendering the model adequate. The model adequacy can also be verified by the normal probability plot of residuals, as shown in Fig. 3. Here, points lie in close proximity of the normal probability line, which confirms the adequacy of the model.

3.2 Assessment of MRR Using ANN

The predicted results of MRR using the 4-10-1 network, compared with the observed experimental findings, are shown in Fig. 4. From the figure, it can be easily seen that the neural network model is in agreement with the experimental data. The trained

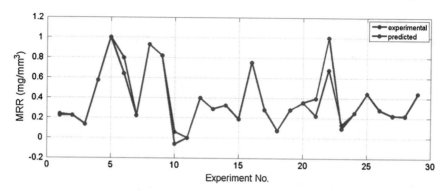

Fig. 4 Comparison of experimental and ANN results plotted against normalized MRR

Table 4 A comparison of regression and ANN model predictions of MRR with the experiment

Experiment no.	Experimental measurement	Regression model	ANN model
4	0.760	0.736	0.760
14	0.430	0.438	0.429
15	0.300	0.330	0.300
17	0.687	0.625	0.665
24	0.220	0.224	0.220
26	0.400	0.421	0.399

network has an RMS error percentage of 2.13, which is well below the permissible value of 10%. A mean square error (MSE) of 4.55×10^{-4} was also realized. Table 4 shows a comparison of regression and ANN model predictions of MRR with the experiment. The comparison shown in Table 4 elucidates that the predictions of the ANN model are more accurate compared to the regression model which was developed using statistical techniques. The maximum test error in the Regression model is 9% compared to the ANN model which has a maximum error of 2.7%.

4 Conclusions

The influential process parameters governing the material removal of Tomal-10, using electric discharge diamond grinding (EDDG) were studied in this paper. A design of experiments (DOE) based approach, involving the use of Box-Behnken design was employed to identify significant process parameters, and their relationship with material removal rate (MRR) was developed in the form of a quadratic regression equation. Analysis of variance (ANOVA) of the design confirmed the quadratic model to be significant. From the F-test, grit size (z) of the diamond particles was identified as the most significant parameter in material removal of T10, followed by depth of cut (d).

Lastly, wheel speed (V) and the interaction of wheel speed in conjunction with diamond concentration were found to have a crucial say in the material removal process. In order to compare and validate the results of the regression model, a multilayered ANN model was developed. The predicted results of the ANN model were in close proximity to the actual experiment as compared to the regression model. Thus, it can be said that the ANN model is reliable to a higher degree. But overall, both the models predicted the values of MRR within the permissible error range of 10%.

Acknowledgements The authors would like to thank National Technical University (Kharkiv Polytechnic Institute), Ukraine, for carrying out the experiment and interpretation of results.

References

1. B. Denkena, J. Köhler, C. Ventura, Grinding of PCBN cutting inserts. Int. J. Refract. Metal Hard Mater. **42**, 91–96 (2014). https://doi.org/10.1016/j.ijrmhm.2013.08.007
2. A. McKie, J. Winzer, I. Sigalas, M. Herrmann, L. Weiler, J. Rödel, N. Can, Mechanical properties of cBN–Al composite materials. Ceram. Int. **37**(1), 1–8 (2011). https://doi.org/10.1016/j.ceramint.2010.07.034
3. G. Poulachon, A. Moisan, I. Jawahir, Tool-wear mechanisms in hard turning with polycrystalline cubic boron nitride tools. Wear **250**(1), 576–586 (2001). https://doi.org/10.1016/S0043-1648(01)00609-3
4. C. Ventura, J. Köhler, B. Denkena, Cutting edge preparation of PCBN inserts by means of grinding and its application in hard turning. CIRP J. Manuf. Sci. Technol. **6**(4), 246–253 (2013). https://doi.org/10.1016/j.cirpj.2013.07.005
5. A. Mamalis, M. Horvath, A. Grabchenko, Diamond grinding of super-hard materials. J. Mater. Process. Technol. **97**(1), 120–125 (2000). https://doi.org/10.1016/S0924-0136(99)00358-1
6. B. Schumacher, R. Krampitz, J.-P. Kruth, Historical phases of EDM development driven by the dual influence of "Market Pull" and "Science Push". Procedia CIRP **6**, 5–12 (2013). https://doi.org/10.1016/j.procir.2013.03.001
7. E.Y. Grodzinskii, L. Zubatova, Electrochemical and electrical-discharge abrasive machining. Sov. Eng. Res. **2**(3), 90–92 (1982)
8. P. Koshy, V. Jain, G. Lal, Grinding of cemented carbide with electrical spark assistance. J. Mater. Process. Technol. **72**(1), 61–68 (1997). https://doi.org/10.1016/S0924-0136(97)00130-1
9. E.-S. Lee, Surface characteristics in the precision grinding of Mn–Zn ferrite with in-process electro-discharge dressing. J. Mater. Process. Technol. **104**(3), 215–225 (2000). https://doi.org/10.1016/S0924-0136(00)00562-8
10. J. Kozak, Abrasive electrodischarge grinding (AEDG) of advanced materials. Arch. Civil Mech. Eng. **2**(1), 83–101 (2002)
11. J. Xie, J. Tamaki, An experimental study on discharge mediums used for electro-contact discharge dressing of metal-bonded diamond grinding wheel. J. Mater. Process. Technol. **208**(1), 239–244 (2008). https://doi.org/10.1016/j.jmatprotec.2007.12.115
12. M. Rahman, K. Kadirgama, A.S. Ab Aziz, Artificial neural network modeling of grinding of ductile cast iron using water based Sio^ Sub 2^ nanocoolant. Int. J. Automot. Mech. Eng. **9**, 1649 (2014). https://doi.org/10.15282/ijame.9.2013.15.0137
13. S. Kumar, S. Choudhury, Prediction of wear and surface roughness in electro-discharge diamond grinding. J. Mater. Process. Technol. **191**(1), 206–209 (2007). https://doi.org/10.1016/j.jmatprotec.2007.03.032
14. N.B. Fredj, R. Amamou, Ground surface roughness prediction based upon experimental design and neural network models. Int. J. Adv. Manuf. Technol. **31**(1), 24–36 (2006). https://doi.org/10.1007/s00170-005-0169-8

Adaptive Control of a Nonlinear Surge Tank-Level System Using Neural Network-Based PID Controller

Alka Agrawal, Vishal Goyal and Puneet Mishra

Abstract A conventional Proportional–Integral–Derivative (PID) controller is not able to adapt to the changes in the system of a plant having nonlinear dynamics. In this paper, a Neural Network PID (NN-PID) controller is designed based on Multi-layer Neural Network (MLN) technique for controlling of the liquid level in a nonlinear surge tank system. A separate MLN identifier is implemented to approximate the plant's dynamics which operates in parallel to the controller with disturbance and parametric uncertainties in the system. The NN-PID controller works using backpropagation algorithm and weights are updated according to the gradient descent based learning rule. The simulation results show that as the variations occur in the plant, MLN identifier follows the plant's dynamics by adjusting its parameters and the controller reads just the plant's output to the desired level by adjusting its own parameters. In addition to this, NN-PID controller response is much accurate and faster than the conventional PID controller and an improvement of 97.35% was achieved in terms of ISE.

Keywords PID controller · Surge tank · Liquid level
Multilayered neural network · Gradient descent rule

A. Agrawal (✉) · V. Goyal · P. Mishra
Department of ECE, GLA University, 17km Stone, NH-2, Mathura, India
e-mail: alka.agrawal@gla.ac.in

V. Goyal
e-mail: vishal.glaitm@gmail.com

P. Mishra
e-mail: puneet.mishra@ymail.com

P. Mishra
Department of EEE, Birla Institute of Technology and Science, Pilani, Pilani, Rajasthan, India

© Springer Nature Singapore Pte Ltd. 2019
H. Malik et al. (eds.), *Applications of Artificial Intelligence Techniques in Engineering*, Advances in Intelligent Systems and Computing 698,
https://doi.org/10.1007/978-981-13-1819-1_46

491

1 Introduction

In the past, conventional Proportional–Integral–Derivative (PID) controllers have been the most preferred controllers as compared to others due to simplicity in their structure [1]. But, as the complexity of the plant grows and variations occur in the system parameters, the PID gains have to be adjusted to give the desired performance. Many tuning methods are proposed in the literature [2] for the parameters tuning. Adaptive control is often most useful in a practical environment, as it requires only a limited knowledge of the plant structure and parameters. In the literature, different techniques have been presented, which exhibit the advantages and features of the adaptive control [3]. In 1958, Whitaker introduced the first Model Reference Adaptive control design (MRAC) which was implemented in flight control [4]. After that in 1970s, K. Astrom and B. Egardt gave the theory of self-tuning regulators with many successful applications [5, 6]. In 1990, Parthasarthy proposed that a neural network is a tool which should be used for efficiently identifying and controlling dynamical systems [7]. To identify and control the systems using neural networks, a unified framework of systems theory is given in [8]. A massive amount of work has been presented in literature depicting the dynamics of the neural network and its robustness for many plants [9, 10]. In [11], Hayakawa gave the remarkable advantage of Neural Network adaptive control over adaptive control. It states that the best feature of an NN network is its parallel structure which makes it to effectively update the parameters online. In [12–15], the authors designed a PID controller using neural network technique for a nonlinear system and verified that the PID controller using neural network has greater robustness against parameter variations and disturbance as compared to conventional PID.

In this paper, a PID controller using a Multilayer Neural Network (MLN) technique is proposed for controlling the liquid level of a surge tank system. The conventional PID controllers are not able to control the liquid level accurately under the disturbance effect and the parametric uncertainties. Hence to overcome the limitation of the conventional PID, a neural network based PID (NN-PID) using MLN technique is employed in this paper. The training of the weights occur using backpropagation algorithm and the weights updating law has been taken as gradient descent law.

Further, this paper has been categorized as follows: In Sect. 2, a mathematical model describing the dynamics of a surge tank system is represented. Section 3 consists of two subsections as 3.1 and 3.2 in which recursive weight update equations are derived for MLN controller and identifier, respectively, using backpropagation algorithm. Section 4 consists of the simulation results which shows the performance of MLN PID controller and compared it with the conventional PID controller and Sect. 5 gives the conclusion.

2 Surge Tank System

The differential equation for determining the level of liquid in a surge tank system is as follows [16]:

$$\frac{dh(t)}{dt} = \frac{-\sqrt{2gh(t)}}{A(h(t))} + \frac{1}{A(h(t))}u(t) \tag{1}$$

where $u(t)$ represents the flow of liquid at the input, $h(t)$ represents the height of liquid level in the tank (output of the system), and $A(h(t)) = \sqrt{ah^2(t) + b}$ is the cross-sectional area of the tank. The input $u(t)$ can be both positive and negative. The parameters 'a' and 'b' are taken as 1.

3 PID Controller Using Neural Network (NN-PID Controller)

Figure 1 shows the structure of NN-PID. It can be observed from Fig. 1 that the developed controller employs two MLNs, one for identification, and the other for implementation of PID action.

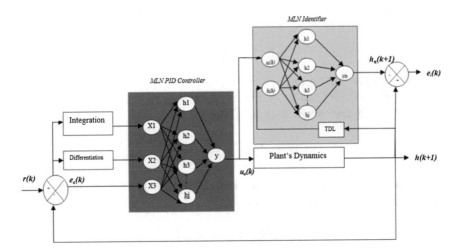

Fig. 1 Structure of NN-PID controller used for identification and control of plant's output [14]

3.1 Implementation of MLN PID Controller

In this work, the online training of the nodes is performed through backpropagation and synaptic weights are adjusted according to gradient descent rule [16]. Based on this technique, the weight update equations for the PID controller are derived as follows.

The discrete form of PID controller output is given by

$$u_c(k) = K_p e_c(k) + K_i T \sum_{i=0}^{k} e_c(k) + K_d \left(\frac{e_c(k) - e_c(k-1)}{T} \right) \tag{2}$$

$$\text{where} \quad e_c(k) = y_d(k) - y(k) \tag{3}$$

Here, $y_d(k)$ is the desired output and $y(k)$ is the plant's output at any instant. The three inputs to the NN-PID controller's input nodes are given as

$$X_1(k) = e_c(k); \quad X_2(k) = T \sum_{i=0}^{k} e_c(k) \text{ and } X_3(k) = \frac{e_c(k) - e_c(k-1)}{T} \tag{4}$$

and K_p, K_i and K_d are the parameters of the controller which will be updated during the online control and identification process.

The objective of backpropagation theorem is to minimize the cost function which is defined as

$$E_c(k) = \frac{1}{2} e_c(k)^2 \tag{5}$$

where $e_c(k)$ is given in Eq. (3).

The gradient descent rule states that for minimizing the cost function, the update equation for the weight vector $\mathbf{W}$ should be

$$W(t+1) = W(t) - \eta \frac{\partial E}{\partial W} \tag{6}$$

where η is the learning rate whose value ranges from 0 to 1. As shown in Fig. 1, the input to the hidden layer neurons is given by:

$$V(k) = W_h^c(k)^T X(k) \tag{7}$$

where W_h^c is the weight vector from the input to the hidden layer.

The output of the layer is given by

$$S(k) = f_h(V(k)) \tag{8}$$

where f_h denotes the tangent hyperbolic activation function activating each neuron.

The output of each hidden layer neuron goes to the input of output neuron by multiplying by their respective weights.

$$y(k) = W_o^c(k)^T S(k) \tag{9}$$

where W_o^c is the weight vector from hidden to the output layer.

As the output layer is activated by a linear activation function, the output of the output layer will be:

$$u_c(k) = y(k) \tag{10}$$

From Eq. (6), it is clear that to find the weight update equation for the output layer, we have to calculate

$$\Delta W_o^c(k) = -\eta \frac{\partial E_c(k)}{\partial W_o^c(k)} \tag{11}$$

To find the value of $\frac{\partial E_c(k)}{\partial W_o^c(k)}$ chain rule will be applied as

$$\frac{\partial E_c(k)}{\partial W_o^c(k)} = \frac{\partial E_c(k)}{\partial h(k)} \frac{\partial h(k)}{\partial u_c(k)} \frac{\partial u_c(k)}{\partial W_o^c(k)} \tag{12}$$

$$\text{where} \quad \frac{\partial E_c(k)}{\partial h(k)} = -(r(k) - h(k)) = -e_c(k) \tag{13}$$

$$\text{and} \quad \frac{\partial u_c(k)}{\partial W_o^c(k)} = S(k) \tag{14}$$

The update equation for the output layer's weights is found by putting the value calculated in Eq. (11) into Eq. (6) as

$$W_o^c(k+1) = W_o^c(k) + \eta e_c(k) J(k) S(k) \tag{15}$$

Similarly, to find the input weight equation, we have to find

$$\Delta W_h^c(k) = -\eta \frac{\partial E_c(k)}{\partial W_h^c(k)} \tag{16}$$

$$\text{where} \quad \frac{\partial E_c(k)}{\partial W_h^c(k)} = \frac{\partial E_c(k)}{\partial h(k)} \frac{\partial h(k)}{\partial u_c(k)} \frac{\partial u_c(k)}{\partial S(k)} \frac{\partial S(k)}{\partial V(k)} \frac{\partial V(k)}{\partial W_h^c(k)} \tag{17}$$

Putting values of each term, update equation for the hidden layer is given as

$$W_h^c(k+1) = W_h^c(k) + \eta e_c(k) J(k) W_o^c(k) (1 - S(k)^2) X(k) \tag{18}$$

3.2 Implementation of MLN Identifier

$$W_o^i(k+1) = W_o^i(k) + \eta e_i(k)R(k) \tag{19}$$

where W_o^i is the output weight vector of the output layer.

$$\text{and} \quad W_h^i(k+1) = W_h^i(k) + \eta e_i(k)W_o^i(k)\left[1 - R(k)^2\right]M(k) \tag{20}$$

where W_h^i is the input to the hidden layer weight vector and $R(k)$ is the output of hidden layer neurons of MLN identifier. $M(k)$ is the input applied to the input neurons of the identifier.

4 Simulation Results

The numbers of neurons are taken as 30 in the hidden layer whereas the numbers of input layer neurons are 3 and the values of η are set to be 0.009 and 0.005 for the identification model and the PID controller respectively. The output is simulated for 5000 samples with the time period of 0.1 s and the desired value of the liquid level is taken as 10 m. A conventional PID controller is employed to compare the results of the proposed NN-PID whose gains are considered as $K_p = 0.8$, $K_i = K_d = 0.5$.

Figure 2 represents the response of MLN identifier during online training with plant output for initial phase via error convergence plot. It is clear from the figure that initially the learning is poor as the error is maximum but as the training of the neural network goes on, it is updating its parameters with the help of weight update Eqs. (19) and (20) and the error between the NN-identifier response and plant's response is decreasing as shown in Fig. 2. As the weight updating goes on according to gradient descent rule and after sufficient learning, the error becomes zero and the NN-identifier acquired the plant's dynamics successfully as shown in Fig. 3. The same action is repeated for the NN-PID controller and it is clear from the Fig. 4 that as the training continues the plant's output is reaching to the desired level and after some iteration, the plant's output controlled with NN-PID gets superimposed with the desired level showing the successful training of the NN controller. In addition to this, in Fig. 4 the plant's response with NN-PID is compared with that of the conventional PID and it can be clearly inferred from the figure that the action of the NN-PID controller is more speedy and precise. Figure 5 shows the error convergence as the error between the output of the plant and the desired value decreases as training goes on.

Table 1 shows the comparative analysis of NN-PID and conventional PID on the basis of their performance indices. It is found that the NN-PID controller is superior as the Integral of Squared Error (ISE) of the NN-PID controller is 32.95, which is about 97% less than the ISE of conventional PID controller and the integral of absolute error (IAE) of NN-PID is also about 94% less than the conventional PID.

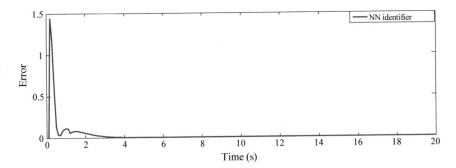

Fig. 2 Error convergence plot for NN-identifier

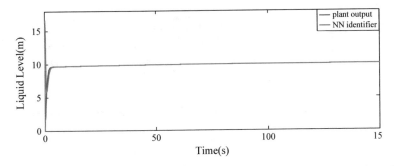

Fig. 3 NN-identifier response as the training occurs sufficiently

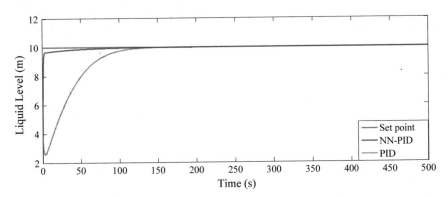

Fig. 4 Response of plant using NN-PID controller after sufficient training and comparison with conventional PID controller for nominal case

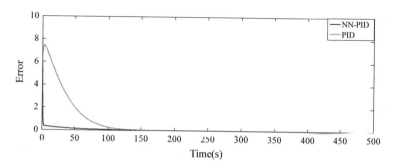

Fig. 5 Comparison of error between setpoint and plant's output with NN-PID controller and conventional PID controller for the nominal case

Table 1 Comparison of both controllers on the basis of the performance index

Type of controller	ISE	IAE
PID	1.246E3	379.9953
NN-PID	32.95	21.59

4.1 Testing of NN-PID Controller When Disturbance Occur in the Plant

A disturbance signal of amplitude 5 is added in controller's output at $t = 250$ s. Figure 6 shows the behavior of NN-PID controller which depicts that it is capable of rejecting the disturbance more efficiently than the conventional PID controller. It is observed that the response of NN-PID controller is better as the ISE of NN-PID controller during the disturbance effect is only 39.0484 while the conventional PID controller is having the ISE of around 1.43E3.

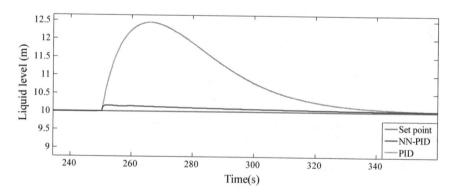

Fig. 6 Comparison of plant's output with NN-PID controller and conventional PID controller when a disturbance signal is added in the controller's output at $t = 250$ s

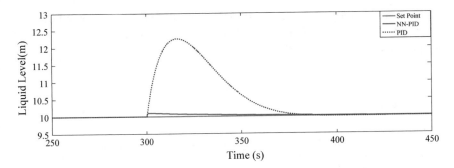

Fig. 7 Response of the plant with NN-PID controller and conventional PID controller when 30% variation in α occurs at t = 300 s

4.2 Testing of the NN-PID Controller Under Parametric Uncertainties

To test the parametric uncertainties in the system, the system equation is taken as

$$h(k + 1) = h(k) + \alpha T \left(-\sqrt{2g} \sqrt{\frac{h(k)}{h(k)^2 + 1}} \right) + \beta T \left(\frac{u(k)}{\sqrt{h(k)^2 + 1}} \right) \qquad (21)$$

To check the robustness of the NN-PID controller, the parameter α is changed by 30% at t = 300 s. It is clear from the Fig. 7 that due to controller action the output of the plant reaches to the desired level again and the NN-PID controller is more robust and performs better than the conventional PID controller as the ISE of NN-PID is 38.9697 while the ISE of conventional PID controller is 1.4024E3.

5 Conclusion

An NN-identifier using MLN technique is designed which is able to successfully learn the parameters of the plant using gradient descent rule and readjust the parameters of the neural network based PID controller when the dynamics of the plant gets changed due to parametric uncertainties and disturbance. Various simulation studies are done to check the performance of the NN-PID controller and compare with the conventional PID controller. On comparing the performance indices of the NN-PID controller and the conventional PID controller, it is found that for the nominal case the NN-PID has integral of squared error (ISE) of 32.965 which is 97.35% less than the conventional PID controller having ISE of 1.246E3. Similarly, on comparing the ISE of both controllers during the disturbance and parametric variations, it is found that the NN-PID controller is more accurate and robust as ISE of NN-PID controller is around 97% lesser than the conventional PID controller in both cases.

On the basis of comparative results, it can be concluded that the proposed NN-PID controller gives better performance and is believed to be more useful to be used in the industrial processes having nonlinear and uncertain dynamics.

References

1. K.J. Astrom, T. Hagglund, *Advanced PID Control* (Research T. Park, NC, ISA, 2005)
2. I.-L. Chien, P.S. Fruehauf, Consider IMC tuning to improve controller performance. Chem. Eng. Prog. **86**, 33–41 (1990)
3. K.J. Astrom, B. Wittenmark, *Adaptive Control* (Dover Publications, Inc., Mineola, NY, USA, 2008)
4. P. Osburn, H. Whitaker, A. Kezer, *New Developments in the Design of Model Reference Adaptive Control Systems* (Institute of the Aerospace Sciences, Easton, PA, USA, 1961)
5. K.J. Astrom, B. Wittenmark, On self-tuning regulators. Automatica **9**, 185–199 (1973)
6. B. Egardt, *Stability of Adaptive Controllers* (Springer, Berlin, Germany, 1979)
7. K.S. Narendra, K. Parthasarathy, Identification and control of dynamical systems using neural networks. IEEE Trans. Neural Netw. **1**, 4–27 (1990). https://doi.org/10.1109/72.80202
8. L. Chen, K.S. Narendra, Identification and control of a nonlinear discrete-time system based on its linearization: a unified framework. IEEE Trans. Neural Netw. **15**(3), 663–673 (2004). https://doi.org/10.1109/TNN.2004.826206
9. F.C. Chen, H.K. Khalil, Adaptive control of nonlinear systems using neural networks. Int. J. Control **55**(6), 1299–1317 (1992). https://doi.org/10.1080/00207179208934286
10. S.L. Dai, C. Wang, F. Luo, Identification and learning control of ocean surface ship using neural networks. IEEE Trans. Industr. Inf. **8**(4), 801–810 (2012). https://doi.org/10.1109/TII.2012.2 205584
11. T. Hayakawa, W.M. Haddad, Neural network adaptive control for a class of nonlinear uncertain dynamical system with asymptotic stability guarantees. IEEE Trans. Neural Netw. **19**(1), 80–89 (2008). https://doi.org/10.1109/tnn.2007.902704
12. V. Kumar, P. Gaur, A.P. Mittal, ANN based self tuned PID like adaptive controller design for high performance PMSM position control. Expert Syst. Appl. **41**(17), 7995–8002 (2014). https://doi.org/10.1016/j.eswa.2014.06.040
13. Y. Zhao, X. Du, G. Xia, L. Wu, A novel algorithm for wavelet neural networks with application to enhanced PID controller design. Neuro Comput. **158**(c), 257–267 (2015). https://doi.org/1 0.1016/j.neucom.2015.01.015
14. R. Kumar, S. Srivastava, J.R.P. Gupta, Modeling and control of one-link robotic manipulator using neural network based PID controller, in *Proceedings of International Conference on Advances in Computing, Communications and Informatics (ICACCI)*, 21–24 Sept 2016, Jaipur, India
15. R. Kumar, S. Srivastava, J.R.P. Gupta, Modeling and adaptive control of nonlinear dynamical systems using radial basis function network. Soft. Comput. **21**(15), 4447–4463 (2017). https://doi.org/10.1007/s00500-016-2447-9
16. L. Behera, I. Kar, *Intelligent Systems and Control Principles and Applications* (Oxford University Press Inc., Oxford, 2010)

Gray Wolf Optimization-Based Controller Design for Two-Tank System

Tapan Prakash, V. P. Singh and Naresh Patnana

Abstract The tuning of parameters of controller is crucial while designing a controller for a system. In this brief, a new meta-heuristic optimization, i.e., gray wolf optimization (GWO), is utilized for tuning of proportional–integral (PI) controller for two-tank system. The integral of squared error (ISE) is minimized for deriving controller's parameters. ISE is formulated in terms of alpha and beta parameters. Thus, the formulated ISE is minimized using GWO and other state-of-the-art algorithms. A comparative study is also performed in which time-domain specifications and responses of system are compared. Detailed study proves that GWO provides better results when compared to other algorithms.

1 Introduction

The proportional–integral–derivative (PID) controllers are widely used controllers in industrial applications due to its simplicity. Many tuning rules are reported in literature for deriving PID controller's parameters. The tuning rules can broadly be categorized into two: rule-based tuning or artificial intelligence techniques-based tuning. The rule-based tuning is based on mathematical formulations while some error criterion is generally minimized in case of artificial intelligence techniques-based tuning. Methods based on rule-based tuning include Ziegler–Nichols (ZN) criterion, Kessler Landau Voda (KLV), integral of square time-weighted error (ISTE), no-overshoot (NOOV) rule, some-overshoot (SOOV) rule, Mantz–Tacconi Ziegler–Nichols (MT-ZN) criterion, refined Ziegler–Nichols (R-ZN) criterion, Pessen integral of absolute error (PIAE), etc.

The rule-based tuning methods do not provide universal formulations for all applications. Additionally, PID controllers designed using rule-based tuning do not always provide satisfactory performance. To overcome these limitations of PID controllers, artificial intelligence-based tuning is evolved. Many optimization algorithms such

T. Prakash · V. P. Singh (✉) · N. Patnana
EED, NIT, Raipur, Raipur 492010, India
e-mail: vinaymnnit@gmail.com

© Springer Nature Singapore Pte Ltd. 2019
H. Malik et al. (eds.), *Applications of Artificial Intelligence Techniques in Engineering*, Advances in Intelligent Systems and Computing 698,
https://doi.org/10.1007/978-981-13-1819-1_47

as bat algorithm [1], particle swarm optimization (PSO) [2], genetic algorithm (GA) [3], teacher–learner-based optimization (TLBO) [4], elephant herding optimization (EHO) [5], and colonial competitive algorithm (CCA) [6] are proposed for tuning the PID controller parameters.

In this brief, proportional–integral (PI) controller is designed for two-tank system using a recently proposed algorithm, namely, gray wolf optimization (GWO) [7]. The GWO uses the hunting behavior of gray wolves for finding the optimum. Another reason for utilizing this algorithm for PI controller tuning is being free from algorithm-specific parameters. This eases the implementation of GWO for controller design. The integral-square-error (ISE) of unit step input is minimized to tune the PI controller parameters. Quantitative and qualitative results are obtained for GWO-based PI controller design. To further support the supremacy of GWO-based PI controller, controllers based on other state-of-the-art algorithms are also designed. The results conclude that the GWO-based PI controller performs better than others.

The section "The Feedback Structure" discusses considered system. Descriptions of plant along with controller are also given in this section. Section "The Tuning Methodology" deals with the proposed tuning method. Section "Gray Wolf Optimization" provides the description of GWO. Section "Simulation Results" discusses the obtained results. The concluding remarks are carried out in section "Conclusion".

2 The Feedback Structure

The closed-loop feedback structure for two-tank system [8] is provided in Fig. 1 where $r(s)$, $e(s)$, and $y(s)$ denote reference input, error, and actual output in Laplace domain. The transfer functions $G_p(s)$ and $G_c(s)$ describe the plant and controller, respectively. The plant transfer function, $G_p(s)$, is given as

$$G_p(s) = \frac{K}{(\tau_a s + 1)(\tau_b s + 1)} \tag{1}$$

where K, τ_a, and τ_b are gain and time constants of two-tank system, respectively.

The controller transfer function, $G_c(s)$, is

$$G_c(s) = K_p\left(1 + \frac{1}{T_i s}\right) \tag{2}$$

Fig. 1 Closed-loop feedback system

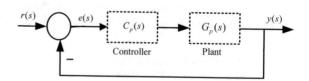

where K_p and T_i, respectively, denote proportional gain and integral time constant.

3 The Tuning Methodology

In this work, integral-square-error (ISE) is minimized to obtain PI controller's parameters. The ISE can be written as

$$J = \int\limits_{t=0}^{t=\infty} e^2(t)\, dt \tag{3}$$

where $e(t)$ is error function in time domain (Fig. 1) for unit step input. Laplace transform of $e(t)$ can, in general, be given as

$$e(s) = \frac{c_{p-1}s^{p-1} + \cdots + c_1 s + c_0}{d_p s^p + d_{p-1}s^{p-1} + \cdots + d_1 s + d_0} \tag{4}$$

where p is the order of error function. For (4), the ISE in terms of alpha and beta parameters [9] is given as

$$J = \sum_{i-1}^{p} \frac{1}{2} \frac{\beta_i^2}{\alpha_i} \tag{5}$$

where α_i and β_i for $i = 1, 2, \ldots, p$ are the alpha and beta parameters obtained from the coefficients of numerator and denominator of (4).

The ISE obtained in (5) is minimized using GWO algorithm to obtain the parameters of controller. Details of GWO algorithm are provided in the following section.

4 Gray Wolf Optimization

Mirjalili et al. [7], recently in 2014, proposed gray wolf optimization (GWO) algorithm. The GWO is based on leadership hierarchy and hunting behavior of gray wolves. The gray wolves are categorized as alpha, beta, delta, and omega. The alpha is mostly responsible for taking decisions regarding hunting and other activities. Hence, alpha is known as leader of the group. The alpha dictates other wolves, i.e., beta, delta, and omega. The second level in hierarchy is beta. The beta is subordinate wolf and helps alpha in taking decisions. The beta wolf receives the instruction from alpha and passes to lower levels. In the absence of alpha, beta takes care of the charge of alpha. The third level, i.e., the delta has to submit to alpha and beta and dominates the omega. The last level of hierarchy is omega. Omega has to submit to alpha, beta, and delta.

Fig. 2 Algorithm

1.	/* Initialization */
2.	Find alpha, beta, and delta
3.	/* while (termination criterion not met)
4.	Update solutions in population.
5.	Update alpha, beta, and delta.
6.	end */
7.	Print solution.

The GWO is easy to understand and implement. Additionally, there is no algorithm-specific parameter to tune in this algorithm. All the steps of GWO algorithm are summarized in Fig. 2.

In detail, the steps of GWO are discussed as follows:

1. Suppose the population contains a total of R search agents with dimension C. The jth dimension, $j = 1, 2, \ldots, C$ of ith search agent, $i = 1, 2, \ldots, R$, is given as $X_{i,j}$. Choose the initial population as

$$X_{i,j} = X_{i,j}^{\min} + r_{i,j}\left(X_{i,j}^{\max} - X_{i,j}^{\min}\right) \tag{6}$$

where $i = 1, 2, \ldots, R$ and $j = 1, 2, \ldots, C$; $X_{i,j}^{\min}$ and $X_{i,j}^{\max}$, respectively, represent the minimum and maximum values of $X_{i,j}$; and $r_{i,j}$ denote a vector of random numbers in the range [0, 1].

2. Evaluate entire population and find the best agent (i.e., alpha), second best agent (i.e., beta) and third best agent (i.e., delta). Denote alpha, beta, and delta as X_α, X_β, and X_δ.

3. Update the population by

$$nX_{i,j} = (X_1 + X_2 + X_3)/3 \tag{7}$$

where $nX_{i,j}$ is updated population and

$$X_1 = X_\alpha - A_1 D_\alpha, \quad D_\alpha = |C_1 X_\alpha - X| \tag{8}$$
$$X_2 = X_\beta - A_2 D_\beta, \quad D_\beta = |C_2 X_\beta - X| \tag{9}$$
$$X_3 = X_\delta - A_3 D_\delta, \quad D_\delta = |C_3 X_\delta - X| \tag{10}$$

The vectors A and C are obtained as

$$A = 2ar_1 - a \tag{11}$$
$$C = 2r_2 \tag{12}$$

where r_1 and r_2 are vectors having random numbers in the range $[0, 1]$ and a is decreased from 2 to 0 as the number of iterations increases.

4. Update the values of a, A, and C.
5. Evaluate the entire population and update X_α, X_β, and X_δ.
6. Go to step 3 and repeat until the termination is met.

5 Simulation Results

All the simulations are performed using MATLAB R2015a on a computer having configuration of 3.4 GHz, 4 GB, and 500 GB. Initially, error function, written in (5), is minimized using GWO algorithm. To further support the analysis, three other algorithms, namely, differential evolution (DE), teacher–learner-based optimization (TLBO), and particle swarm optimization (PSO) are also used to obtain controller's parameters.

The parameters of two-tank system are given as

$$K = 6, \ \tau_a = 2 \text{ and } \tau_b = 4 \tag{13}$$

The error function (5) for two-tank system takes the following form:

$$J = \frac{1}{2}\left\{\frac{y_1^2}{x_0 x_1} + \frac{y_2^2}{x_1 x_2 - x_0 x_3} + \frac{(x_1 y_3 - y_1 x_3)^2}{x_1 x_3 (x_1 x_2 - x_0 x_3)}\right\} \tag{14}$$

$$x_0 = T_i \, \tau_a \, \tau_b \tag{15}$$

$$x_1 = T_i (\tau_a + \tau_b) \tag{16}$$

$$x_2 = T_i \left(1 + K \, K_p\right) \tag{17}$$

$$x_3 = K \, K_p \tag{18}$$

$$y_1 = T_i \, \tau_a \, \tau_b \tag{19}$$

$$y_2 = T_i (\tau_a + \tau_b) \tag{20}$$

$$y_3 = T_i \tag{21}$$

The values of controller's parameters are considered in the range

$$600 \le K_p \le 700 \tag{22}$$

$$600 \le T_i \le 1000 \tag{23}$$

Table 1 shows the values of controller's parameters calculated by minimizing (14) using GWO algorithm, subject to constraints given in (22) and (23). The same table also shows the controller parameters obtained using DE, TLBO, and PSO. The qualitative results in the form of step response for different controllers are plotted in

Table 1 Values of controller parameters

	Proposed PI controller	DE-based PI controller	TLBO-based PI controller	PSO-based PI controller
K_p	600.21	600	657.62	700
T_i	1000	732.17885	989.02	609.2328

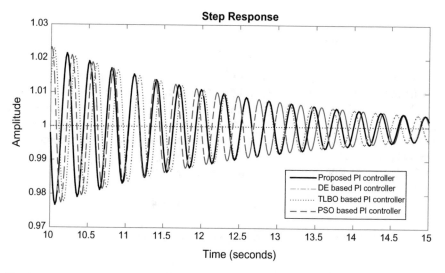

Fig. 3 Step response with different controllers

Table 2 Different time-domain specifications

	Proposed PI controller	DE-based PI controller	TLBO-based PI controller	PSO-based PI controller
Settling time (s)	10.3787	10.3807	10.4680	10.4274
Peak overshoot (%)	94.5519	94.5537	94.7911	94.9530

Fig. 3. Table 2 presents different time-domain specifications for different controllers considered in this work.

From the time-domain specifications as given in Table 2, it is clear that the value of settling time is minimum for GWO-based PI controller. Same is true for the value of peak overshoot. This proves that the GWO-based PI controller is providing better time response in comparison to others. Same can be inferred from Fig. 3. Hence, it is successfully confirmed that the performance of GWO is better for PI controller design in comparison to other state-of-the-art algorithms.

6 Conclusion

Proportional–integral (PI) controller is designed for two-tank system using gray wolf optimization (GWO). The main reason for adopting GWO for tuning lies in being free from algorithm-specific parameters. This simplifies implementation of GWO for controller design. The integral-square-error is treated as design criterion for tuning the PI controller's parameters. To support the analysis, quantitative and qualitative results are derived for GWO-based PI controller. To further support the efficacy of GWO-based PI controller, other controllers based on other state-of-the-art algorithms are also designed. The further extension of this work lies in controller design for three-tank system using GWO algorithm.

References

1. Y. Abdel-Magid, M. Dawoud, Optimal AGC tuning with genetic algorithms. Electr. Power Syst. Res. **38**(3), 231–238 (1996)
2. M.I. Solihin, L.F. Tack, M.L. Kean, Tuning of PID controller using particle swarm optimization (PSO). Int. J. Adv. Sci. Eng. Inf. Technol. **1**(4), 458–461 (2011)
3. A.K. Wong, W.W. Lin, M.T. Ip, T.S. Dillon, Genetic algorithm and PID Control together for dynamic anticipative marginal buffer management: an effective approach to enhance dependability and performance for distributed mobile object-based real-time computing over the internet. J. Parallel Distrib. Comput. **62**(9), 1433–1453 (2002)
4. R.K. Sahu, S. Panda, U.K. Rout, D.K. Sahoo, Teaching learning based optimization algorithm for automatic generation control of power system using 2-DOF PID controller. Int. J. Electr. Power Energy Syst. **77**, 287–301 (2016)
5. D. Sambariya, R. Fagna, A novel elephant herding optimization based PID controller design for Load frequency control in power system, in *2017 International Conference on Computer, Communications and Electronics (Comptelix)* (2017), pp. 595–600. IEEE
6. E. Atashpaz Gargari, F. Hashemzadeh, R. Rajabioun, C. Lucas, Colonial competitive algorithm: a novel approach for PID controller design in MIMO distillation column process. Int. J. Intell. Comput. Cybern. **1**(3), 337–355 (2008)
7. S. Mirjalili, S.M. Mirjalili, A. Lewis, Grey wolf optimizer. Adv. Eng. Softw. **69**, 46–61 (2014)
8. S. Bhanot, *Process Control: Principles and Application* (Oxford University Press, 2008)
9. S. Shrivastava, V. Singh, R. Dohare, S. Singh, D. Chauhan, PID tuning for position control of DC servo-motor using TLBO. Int. J. Adv. Technol. Eng. Explor. **4**(27), 23 (2017)

Comparative Performance Analysis of PID Controller with Filter for Automatic Generation Control with Moth-Flame Optimization Algorithm

B. V. S. Acharyulu, Banaja Mohanty and P. K. Hota

Abstract This paper introduces the comprehensive study of automatic generation control (AGC) system with proportional integral derivative controller with filter (PIDF). A nature-inspired optimization algorithm that is moth-flame optimization (MFO) algorithm is employed for controller gains concurrent optimization. First, a two-area non-reheat interconnected thermal power system is investigated and obtained superior performance with MFO-tuned PI controller as equated to recent competitive optimization techniques, but major improvement is investigated with PIDF controller. Then, the work is continued to a three-area interconnected hybrid system with proper generation rate constraint (GRC). PIDF controller can be implemented as secondary controller in the areas, whose performance is equated with conventional controllers. Result analysis divulges that MFO-tuned PIDF controller performs effective than all other controllers considered in this article. Robustness of the proposed technique is evaluated using sensitivity analysis.

Keywords Automatic generation control (AGC) · Generation rate constraint (GRC) · Moth-flame optimization (MFO) · Solar thermal power plant (STPP)

1 Introduction

Modern power plants are operated with interconnection to form a large system. The system frequency will depend upon the generation of real power and consumption of real power [1]. Major activity of AGC is to standardize power system frequency [2, 3]. Previously, AGC issue has been performed with conventional sources of thermal and hydro generation [4, 5]. Nowadays, works have been reported considering diversity of sources of generations like hydro–thermal–gas [6], hydro–thermal–gas–nuclear [7], and hydro–thermal–wind–diesel [8]. Therefore, for real-time design of AGC

B. V. S. Acharyulu (✉) · B. Mohanty · P. K. Hota
Department of Electrical Engineering, Veer Surendra Sai University
of Technology (VSSUT), Burla 768018, Odisha, India
e-mail: acharyulu201@yahoo.com

© Springer Nature Singapore Pte Ltd. 2019
H. Malik et al. (eds.), *Applications of Artificial Intelligence Techniques in Engineering*, Advances in Intelligent Systems and Computing 698,
https://doi.org/10.1007/978-981-13-1819-1_48

system, renewable sources along with conventional sources should be considered for analysis purpose. Bevrani [9] introduced renewable energy sources into AGC system. Sharma et al. [10] proposed a three-area AGC system including STTP in one area and only thermal power plants in the remaining two areas. So, further investigation is required to observe the system behavior with various sources along with STTP and also with other controllers.

Comparison of different conventional controllers like I, PI, and PID controllers is studied for two-area hydrothermal system [11]. Two-degree-of-freedom PID (2DOF-PID) controller is studied for the first time for AGC system by Sahu et al. [12]. Three-degree-of-freedom PID (3DOFPID) controller is studied for three-area thermal system in [13]. Many optimization algorithms have been successfully implemented for AGC system [14–18]. Every algorithm has its own advantages and disadvantages. Hence, new and promising optimization technique is always welcomed by researchers worldwide. Recently, a new optimizing algorithm named MFO is implemented by Mirjaili [19].

The main intention of current paper is

I. PIDF controller is designed for two-area and hybrid power system.
II. MFO technique is employed to tune PI/PIDF controller for a two-area thermal system, and dynamic performance of the system is compared with recent optimization algorithms.
III. Three unequal areas of interlinked power system are modeled with diverse sources of generation incorporating the STPP, and dynamic performances of the system with PIDF controller are compared with PI and PID controllers.

2 System Considered

To check the efficacy and superiority of proposed technique, two diverse interlinked power system networks are mentioned in the current investigation. Initially, the two-area interlinked power system and the details of nominal system parameters from Appendix-A are considered [4, 18].

2.1 Controller Design and Objective Function

For a given response, the use of a derivative filter coefficient leads to a much smoother control action, and the details are explained in [20]. Considering the advantages of PIDF controller, comparisons are made with classical controllers for a hybrid power system. ITAE is considered as objective function for optimization purpose and given by Eq. (1) [21]:

$$J = ITAE = \int_0^t \left(|\Delta f_i| + |\Delta P_{tiei-j}| \right) t\,dt \tag{1}$$

3 Moth-Flame Optimization Algorithm

Moth-flame optimization algorithm (MFO) is a nature-motivated algorithm developed by Mirjalili [19]. This algorithm is motivated by navigation of moth in nature called transverse direction. Moth travels a long distance in straight line effectively. It is observed that not only it moves in straight line but also moves in spirally around the lights.

Superiority of MFO algorithm is shown by solving 27 benchmark functions and 7 engineering problems and also applied to marine engineering problems [19]. MFO is also applied to find out weights and biases of multilayer perceptrons [22]. In that paper, MFO is compared with genetic algorithm, PSO, ant colony optimization, and evolution strategy. The steps of MFO algorithm are given below:

Step 1: Initialize the parameters of MFO: number of moths (n), tuning parameter (b), maximum number of iterations (T), dimension (d), lower bound, and upper bound of variables.

Step 2: Estimate fitness values of all moths and calculate flame number using Eq. (2):

$$\text{Flame number} = round\left(N - l * \frac{N-1}{T} \right) \tag{2}$$

where $N =$ maximum no of flames, $l =$ current number of iteration.

Step 3: If the iteration $= 1$, then go to step 4, otherwise go to step 5.

Step 4: Arrange the first population of mouths according to their fitness values and find the best flame in it.

Step 5: Select the n moths based on fitness values among the 2n moths and find the best flame in it.

Step 6: *for i = 1: n,*
for j = 1: d,
update r and t using Eqs. (2) and (3).
t is a random number in $[r, 1]$,
where r is linearly decreased from –1 to –2 over the course of iteration

$$t = (a - 1) * (r + 1) \tag{3}$$

where

$$a = 1 + l\left(\frac{-1}{T} \right) \tag{4}$$

Step 7:   Calculate distance (D) using Eq. (5) with related moth and update $M(i,j)$ using Eqs. (6), (7)

$$D_i = |F_j - M_i| \tag{5}$$
$$M_i = S(M_i, F_j) \tag{6}$$
$$S(M_i, F_j) = D_i \cdot e^{bt} \cos(2\pi t) + F_j \tag{7}$$

Step 8:   If the iteration number is equal to maximum number of iteration, then stop and show the results. If not next go to step 2.

4   Simulation Results and Discussion

4.1   Transient Analysis of Test System-1

The algorithm performance measures are scheduled in Table 1. At the first instant of study, MFO-tuned PI controller incorporated to the test system-1 considered from [4]. For step load change of 10% in area-1, the optimum gains of PI controller using MFO algorithm are provided in Table 2. The proposed MFO technique performance is equated with other newly published metaheuristic techniques such as CLPSO [21], EPSDE [21], DE [23], BFOA [23], GA [23], hBFOA–PSO [24], PSO [24], and ZN [25] tuned PI controller for the identical power system with similar fitness function. The comparative system performances are tabulated in Table 3. It is clearly noted from Table 3 and Fig. 1 that it is clear that transient performance of MFO is superior than recent optimization techniques considered in this paper.

For significant improvement in dynamic performances, PIDF controller is also introduced in the system. From Fig. 1 and Table 3, it is clearly noted that the pro-

Table 1 Performance measures of the proposed MFO-based methodology

Measuring factor	Value of ITAE
Minimum	0.1112
Maximum	0.1231
Average	0.11842
Median	0.1188
Standard deviation	0.00313
Variance	$9.8 * 10^{-6}$

Table 2 Optimum values of controller parameters for test system-1

MFO:PIDF	−0.499	−0.14	−0.49	−0.239	0.144	0.499	71.2	56.3
MFO:PI	0.416	0.079	−0.99	0.152	–	–	–	–

Table 3 Comparative performance of ITAE value and settling times for test system-1

Techniques/parameters	ITAE value	Settling time		
		ΔF_1	ΔF_2	ΔP_{tie}
MFO-tuned PIDF	0.1209	3.079	2.94	3.88
MFO-tuned PI	0.1422	3.13	3.99	4.01
EPSDE-tuned PI [21]	0.1539	6.22	7.8	6.67
CLPSO: PI [21]	0.1949	7.19	8.8	7.64
DE-tuned PI [23]	0.9911	8.96	8.16	5.75
BFOA-tuned PI [23]	1.7975	5.52	7.09	6.35
GA-tuned PI [23]	2.7475	10.59	11.39	9.37
Conventional PI [23]	3.7568	45	45	28
hBFOA–PSO-tuned PI [24]	1.1865	7.39	7.65	5.73
PSO-tuned PI [24]	1.2142	7.37	7.82	5
ZN-tuned PI [25]	3.7568	8.1	9.2	6.7

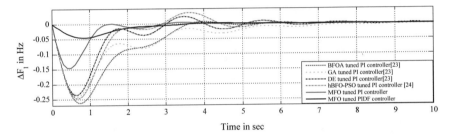

Fig. 1 Comparison of dynamic responses for test system -I

posed MFO-tuned PIDF controller gives better transient performances compared to other optimization tuned PI controller as reported in the literature. For this system, PIDF controller is only compared with PI controller. So, in succeeding section this, controller is compared with PI, PID controllers.

4.2 Transient Analysis of Test System-2

The supremacy of MFO algorithm is established in the previous section; in this section, PIDF controller is compared with other classical controllers for a hybrid three are power system with STTP thermal–hydro–gas as generating unit. In three unequal areas of power system, each area consists of up to two generating units. Area-1 having solar thermal–thermal power (STTP) stations, area-2 having thermal–hydropower stations, and area-3 having thermal–gas power stations. The linearized model of solar thermal, reheat thermal, hydro, and gas unit of three unequal areas (test system-2)

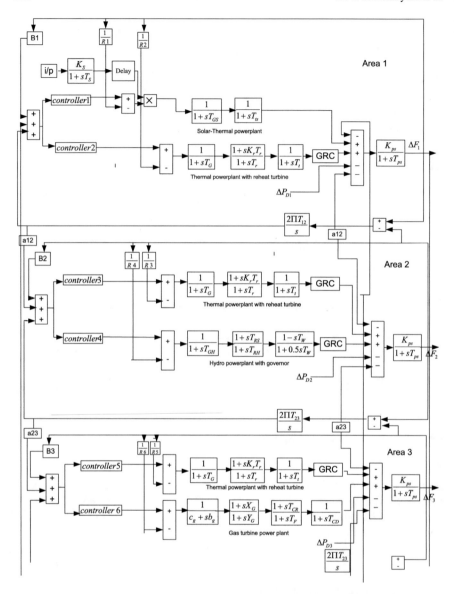

Fig. 2 Transfer function model of the multi-area (three) multisource system (Test system-2)

are shown in Fig. 2. MFO optimization technique is employed to optimize PI, PID, and PIDF controller parameters for the concerned system separately with 1% SLP in area-1. The dynamic responses of each controller associated with these optimum values are compared in Fig. 3. It is clear from Fig. 3 and Table 4 that PIDF controller provides superior dynamic responses compared to other conventional controllers.

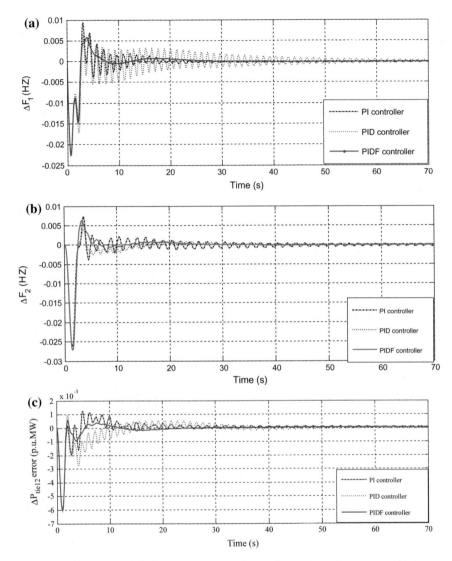

Fig. 3 Comparison of dynamic responses for different classical controllers (Test system-2) **a** frequency deviations in area-1 versus time, **b** frequency deviation in area-2 versus time, **c** Tie-line power deviations in the line connecting area-1 and area-2 versus time

4.3 Sensitivity Analysis

To show the superiority of PIDF controller optimum gains obtained at nominal loading conditions, the system loading condition and the synchronizing coefficients (T_{12}, T_{23}, T_{31}) are changed ±25% from nominal values, which are the two mainly disturb-

Table 4 The performance index, settling times, and peak overshoots of the responses of Fig. 3

Performance index	ITAE	PI	PID	PIDF
		3.3312	2.7922	1.7146
ΔF_1	ST	23.22	61.32	22.14
	OS	0.0095	0.0068	0.0059
ΔF_2	ST	68.14	31.33	21.99
	OS	0.0074	0.0040	0.0062
ΔF_3	ST	53.81	28.93	22.14
	OS	0.0042	0.0042	0.0058
ΔP_{tie12}	ST	15.24	25.44	4.63
	OS	0.0013	$9.7 * 10^{-4}$	$5.7 * 10^{-4}$
ΔP_{tie23}	ST	13.36	14.34	12.13
	OS	$9.1 * 10^{-4}$	$8.4 * 10^{-4}$	$9.11 * 10^{-4}$
ΔP_{tie31}	ST	11.23	19.32	12.15
	OS	0.00641	0.00641	0.00641

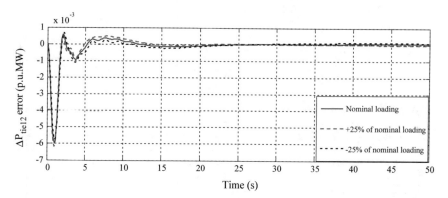

Fig. 4 Dynamic responses of the test system-2 equipped with the proposed controller under the uncertainties in loading

ing parameters of the AGC performance. The system dynamic performances with this variation are compared with nominal condition in Fig. 4. It is clear from Fig. 4 that dynamic responses of the system with this variation almost remain same or has negligible effect on system performance.

5 Conclusion

This article presents the design of moth-flame optimization (MFO) algorithm-tuned PIDF controller for hybrid AGC system. To illustrate the supremacy of the proposed method, the results of simulation are compared with some recent metaheuristic optimization algorithms for the test system-1 and major improvement is observed with MFO-optimized PI/PIDF controllers. Further, it is extended to a three-area unequal interconnected hybrid power system. Dynamic performances of MFO-optimized PIDF are compared with classical controllers and improved system performances are obtained with PIDF controller in terms of settling time and peak overshoots. Sensitivity analysis reveals that MFO-optimized PIDF controller obtained at nominal condition need not required to be reset for large variation in system parameters and loading conditions.

References

1. O.I. Elgard, *Electric Energy Systems Theory* (Mc Graw-Hill, New York, 1982)
2. J. Nanda, S. Mishra, L.C. Saikia, Maiden application of bacterial foraging based optimization technique in multi-area automatic generation control. IEEE Trans. Power Syst. **24**, 602–609 (2009)
3. B. Mohanty, S. Panda, P. Hota, Controller parameters tuning of differential evolution algorithm and its application to load frequency control of multi-source power system. Int. J. Electr. Power Energy Syst. **54**, 77–85 (2014)
4. R. Sahu, S. Panda, G. Chandra Sekhar, A novel hybrid PSO-PS optimized fuzzy PI controller for AGC in multi area interconnected power systems. Int. J. Electr. Power Energy Syst. **64**, 880–893 (2015)
5. S. Dhillon, J. Lather, S. Marwaha, Multi objective load frequency control using hybrid bacterial foraging and particle swarm optimized PI controller. Int. J. Electr. Power Energy Syst. **79**, 196–209 (2016)
6. K.P.S. Parmar, S. Majhi, D.P. Kothari, Load frequency control of a realistic power system with multi-source power generation. Electr. Power Energy Syst. **42**, 426–433 (2012)
7. B. Mohanty, TLBO optimized sliding mode controller for multi-area multi-source nonlinear interconnected AGC system. Electr. Power Energy Syst. **73**, 872–881 (2015)
8. B. Mohanty, S. Panda, P.K. Hota, Differential evolution algorithm based automatic generation control for interconnected power systems with non-linearity. Alex. Eng. J. **53**, 537–552 (2014)
9. H. Bevrani, A. Ghosh, G. Ledwich, Renewable energy sources and frequency regulation: survey and new perspectives. IET Renew. Power Gener. **4**(5), 438–457 (2010)
10. Y. Sharma, L.C. Saikia, Automatic generation control of a multi-area ST-Thermal power system using Grey Wolf Optimizer algorithm based classical controllers. Electr. Power Energy Syst. **73**, 853–862 (2015)
11. L.C. Saikia, J. Nanda, S. Mishra, Performance comparison of several classical controllers in AGC for multi-area interconnected thermal system. Electr. Power Energy Syst. **33**(3), 394–401 (2011)
12. R.K. Sahu, S. Panda, U.K. Rout, DE optimized parallel 2-DOF PID controller for load frequency control of power system with governor dead-band nonlinearity. Electr. Power Energy Syst. **49**, 19–33 (2013)
13. A. Rahman, L.C. Saikia, N. Sinha, Load frequency control of a hydro-thermal system under deregulated environment using biogeography-based optimised three degree- of-freedom integral-derivative controller. IET Gener. Transm. Distrib. **9**(15), 2284–2293 (2015)

14. Y. Arya, AGC performance enrichment of multi-source hydrothermal gas power systems using new optimized FOFPID controller and redox flow batteries. Energy **127**, 704–715 (2017)
15. Y. Arya, N. Kumar, Design and analysis of BFOA-optimized fuzzy PI/PID controller for AGC of multi-area traditional/restructured electrical power systems. Soft. Comput. **21**, 6435–6452 (2017)
16. D. Guha, P. Roy, S. Banerjee, Quasi-oppositional symbiotic organism search algorithm applied to load frequency control. Swarm Evol. Comput. **33**, 46–67 (2017)
17. S. Madasu, M. Kumar, A. Singh, Comparable investigation of backtracking search algorithm in automatic generation control for two area reheat interconnected thermal power system. Appl. Soft Comput. **55**, 197–210 (2017)
18. R.K. Sahu, S. Panda, S. Padhan, A hybrid firefly algorithm and pattern search technique for automatic generation control of multi-area power systems. Int. J. Electr. Power Energy Syst. **64**, 9–23 (2015)
19. S. Mirjalili, Moth-flame optimization algorithm: a novel nature-inspired heuristic paradigm. Knowl.-Based Syst. **89**, 228–249 (2015)
20. J.A. Romero, R. Sanchis, E. Arrebola, Experimental study of event based PID controllers with different sampling strategies. Application to brushless DC motor networked control system, in *International Conference on ICAT* (2015), pp. 1–6
21. D. Guha, P.K. Roy, S. Banerjee, Load frequency control of interconnected power system using grey wolf optimization. Swarm Evol. Comput. **27**, 97–115 (2016)
22. W. Yamanya, M. Fawzya, A. Tharwatb, A.E. Hassaniend, Moth-flame optimization for training multi-layer perceptrons, in *Computer Engineering Conference (ICENCO) 2015*
23. U.K. Rout, R.K. Sahu, S. Panda, Design and analysis of differential evolution algorithm based automatic generation control for interconnected power system. AinShamsEng **4**(3), 409–421 (2013)
24. S. Panda, B. Mohanty, P.K. Hota, Hybrid BFOA–PSO algorithm for automatic generation control of linear and non linear interconnected power system. Appl. Soft Comput. **13**(12), 4718–4730 (2013)
25. E.S. Ali, S.M. Abd-Elazi, BFOA based design of PID controller for two area load frequency control with nonlinearities. Int. J. Electr. Power Energy Syst. **51**, 224–231 (2013)

Analysis of Electrocardiogram Signal Using Computational Intelligence Technique

Papia Ray, Kishan Kumar Mandal and Biplab Kumar Mohanty

Abstract Amid various computational techniques, wavelet technique has taken universal place in biomedical signal processing area. Wavelet transform (WT) is an effective method for analysis of adaptable signals where both time and frequency information are vital. The most important role of wavelet transform is the elimination of noise from the biomedical signal. Analysis of ECG signals using computational intelligence techniques like discrete wavelet transform (DWT), principle component analysis (PCA), and independent component analysis (ICA) is presented vividly in this paper. A new modified wavelet transform called bionic wavelet transform (BWT) has been applied here for analysis of biomedical signals. By adapting the value of scales, T-function of BWT is varied and its effects on the value of the threshold is observed. This is called the multi-adaptive technique which is used for denoising purpose of electrocardiogram signal (ECG). The efficiency of various methods used by existing techniques for denoising of ECG signals has been evaluated and compared in terms of signal-to-noise ratio (SNR). From the proposed algorithm, i.e., with BWT, it is observed that there is an improvement in SNR than other conventional techniques.

Keywords Electrocardiogram · Discrete wavelet transform · Multi-adaptive bionic wavelet transform · Baseline wander reduction

P. Ray (✉) · K. K. Mandal · B. K. Mohanty
Department of Electrical Engineering, VSSUT, Burla, Odisha, India
e-mail: papia_ray@yahoo.co.in

K. K. Mandal
e-mail: kishan64@gmail.com

B. K. Mohanty
e-mail: biplabkumar123@gmail.com

© Springer Nature Singapore Pte Ltd. 2019
H. Malik et al. (eds.), *Applications of Artificial Intelligence Techniques in Engineering*, Advances in Intelligent Systems and Computing 698,
https://doi.org/10.1007/978-981-13-1819-1_49

1 Introduction

Biomedical signal processing is an interdisciplinary field which studies extract significant information from biomedical signals. The noises which is present in the biomedical signals is mostly interference due to power line and disturbances caused due to recording electrode movement. Often SNR of biomedical signals is low. Therefore removing noise and interference using filters are more important in medical practice. Electrocardiogram (ECG) signals have been taken for analysis purpose using different computational techniques in this paper. ECG signals are contaminated owing to electrical heart activities. An ECG signal consists of different types of interferences like electrode contact noise, motion artifacts, power line interference, electromyography (EMG) noise, and instrumentation noise.

The proposed work focuses on denoising of ECG signals and baseline wander reduction. Wavelets provide fast computational algorithms in which wavelet thresholding describes that the signal magnitude dominates the magnitude of noise. For noise cancelation of ECG signal using an adaptive filter which is discussed in [1].

For noise elimination, another method, i.e., adaptive impulse correlated filter (AICF), is described in [2]. It says about the passivity component part to the signal and noise. The AICF using algorithm shows that filter is linear. Using adaptive filter in the case of the least mean square (LMS) algorithm, it reaches to better steady-state results than modified character-weight algorithm (MWA). ICA is applied for noise removal of ECG signal in [3, 4]. But the SNR value of ICA is lower than the BWT and WT methods in [3, 4]. DWT along with Wiener filter (WF) is deliberated into [5].

WT is effectively used in the suppression of EMG noise when compared to linear filtering due to its time and frequency representation which is observed in [5]. While using DWT technique, the wideband EMG noise is more effectively suppressed in ECG signal when compared to linear filter as shown in [5]. The process of denoising is carried out using wavelet transform, as well as disintegration and further thresholding. Denoising of ECG signal by means of WT with soft thresholding is described in [6–8]. BWT is instigated profoundly in [9], and it is mainly used in areas relating to human biosystem and has given encouraging results about speech and its processing. From speech signal processing, it is concluded that the results of BWT are far better than WT for a cochlear implant. For calculation of T-function which is used in BWT technique, a detailed analysis is given in [10]. BWT which is used for speech processing in cochlear implants (CIs) is described in [11]. The advantages of BWT technique are better than the neural network simulation, i.e., reduction of necessary channel number, high noise resilience, and identification of average simulation duration for words [12]. For extermination of baseline wandering in ECG signal, wavelet adaptive filter (WAF) is used in [13]. In [13], two different type of filters are used i.e. a filter having 0.5 Hz cut-off frequency and an adaptive filter. Baseline reduction of ECG signal using DWT technique is described in [14]. Baseline reduction and power line interference using FIR (finite impulse response) filter are described in [15]. Application of wavelet transform and comparison study with Fourier transform is described in [16]. The analysis of biomedical signals using ICA is described in [18]. Based on the findings during the period of research work, it can be concluded that Biomedical signals are non-stationary signals whose analysis requires better time and frequency resolution. Such analysis includes denoising and compression. Analysis of ECG signals, i.e., denoising and baseline elimination using WT, principal component analysis, ICA, and BWT, has been presented in this paper. In this paper, from simulation results, it is observed that wavelet transform method and bionic wavelet transform method gives better accurate results than other known techniques. SNR improvements in case of BWT are also more than that of other conventional techniques like PCA and ICA.

The rest of the paper is organized in the following ways as described. Section 2 illuminates computational intelligence techniques alike discrete wavelet transform, principal component analysis, independent component analysis, and multi-adaptive bionic wavelet transform. In Sect. 3, the focus is made on baseline wander reduction using different computational intelligence technique. Section 4 discusses simulation outcome and Sect. 5 draws the conclusion.

2 Computational Intelligence Techniques

2.1 Discrete Wavelet Transform (DWT)

Wavelet takes two arguments: time and scale. The original wavelet transform is called mother wavelet h(t) and is used for generating all basic functions. Those functions are described as follows:

$$h_{a,b}(t) = \frac{1}{\sqrt{a}} h\left(\frac{t-b}{a}\right) \tag{1}$$

where $a \in R^+ - \{0\}, b \in R$. Using wavelet transform technique, the impulse component and the unwanted high frequency can be effectively eliminated. It uses fast pyramid algorithm, i.e., in the forward algorithm, the required signal is handled by WT and in the backward algorithm, the signal is processed by inverse discrete wavelet transform (IDWT) method. Wavelet transform of given sampled output waveform can be obtained by implementing DWT which is given by:

$$DWT(m, n) = 2^{-m/2} \sum m \sum nx(n)h^*\left(\frac{t - n2^m}{2^m}\right) \tag{2}$$

Where the discretized parent wavelet is

$$h_{m,n}(t) = \frac{1}{\sqrt{a_0 m}} h\left(\frac{t - nb_0 a_0{}^m}{a_0{}^m}\right) \tag{3}$$

Where, $a = a_0{}^m$ and $b = nb_0a_0{}^m$, $a_0, b_0 > 1$, $m, n \in Z$, where Z belongs to all integers.

Wavelet thresholding is a nonlinear technique in which each coefficient of WT is compared with a threshold value. There are two type of thresholding namely : hard and soft thresholding respectively.

$$\text{Thr}_{\text{hard}}(P, T) = \begin{cases} |P|, & |P| > T \\ 0, & |P| < T \end{cases} \tag{4}$$

$$\text{Thr}_{\text{soft}}(P, T) = \begin{cases} \text{sgn}(P)(|P| - T), & |P| > T \\ 0, & |P| < T \end{cases} \tag{5}$$

Where P denotes as the value of wavelet coefficients obtained before thresholding and T denoted as threshold

$$T = \sigma\sqrt{2\log(N)} \tag{6}$$

N = noise signal length, σ = MAD/0.6745, with MAD representing the exclusive absolute median value estimated over the first scale of the wavelet coefficients.

2.2 Principle Component Analysis (PCA)

The reduction of information of matched variable into an unmatched variable is called principal component analysis (PCA). It consists of the linear combination of variables and the unmatched weight. For the extraction of principal component, the signal p is assumed to be zero mean random processes, i.e., signalized by correlation

$$R_p = E[pp^T] \tag{7}$$

By applying the orthonormal linear transformation on the principle component of signal p i.e. $\psi = [\psi_1\psi_2\psi_3.............\psi_N]$ to p, $w = \psi^T p$. So the PCA vector $w = [w_1w_2w_3........w_N]^T$ happens to be mutually unmatched. The first principle component $w_1 = \psi_1{}^T p$ is a scalar product and the variance of w_1 can be described as

$$E[w_1{}^2] = E[\psi_1{}^T pp^T \psi_1] = \psi_1{}^T R_p \psi_1 \tag{8}$$

The maximum value of constraint is $\psi_1{}^T \psi_1 = 1$ and by choosing the value of ψ_1 as the normalized Eigen vector, the maximum variance is defined. It corresponds to the largest Eigen value of R_p i.e. denoted as λ_1; the deriving variance is

$$E[w_1{}^2] = \psi_1{}^T R_p \psi_1 = \lambda_1 \psi_1{}^T \psi_1 = \lambda_1 \tag{9}$$

The first constraint and second principal component w_1 and w_2 respectively should not be same. By using the value of ψ_2, w_2 is acquired. It reflects the second largest eigenvalue R_p. It is continued till the variance of p totally represented by w. For finding out the N different principle component, R_p needs to be solved.

$$R_p \psi = \psi A \tag{10}$$

Where A is diagonal matrix with Eigen values $\lambda_1......\lambda_N.$. Since R_p is $N \times N$ sample correlation matrix which is defined by

$$\hat{R}_p = \frac{1}{M} P P^T \tag{11}$$

Replaces R_p when the Eigen vector is calculated in (10).

2.3 Independent Component Analysis (ICA)

ICA method is representation of data which is used in different applications, i.e., denoising of ECG signal. ICA of an arbitrary vector r which linear transformation $q=wr$ is independent. By maximizing the value of function $F(q_1, q_2.......q_m)$ the dependency of component is achieved. The observed signal is given by:

$$r_j(k) = \sum_{i=1}^{n} q_i(k)a(i, j) + n_j(k) \tag{12}$$

Where j =1,2,3... m indicates the number of observations whereas i =1, 2, 3.... n indicates no of independent component, n(k) represents noise samples.

$$r(k) = [r_1(k)r_2(k).........r_m(k)]^T, r = [a_1a_2........a_n] \tag{13}$$

The independent components are large and the perceived linear mixtures m should be as large like independent n, i.e., m ≥ n.

2.4 Bionic Wavelet Transform (BWT)

BWT which is explained known by Yao and Zhang, i.e., used in human bio-system giving the necessary results about speech and its processing [11]. If the signal is analyzed as f(t) and the BWT is defined as

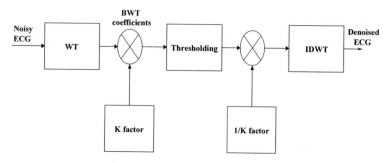

Fig. 1 Block diagram of the BWT denoising technique

$$(BWT_T f)(\tau, a) = < f, \}$$

$$h_T \geq \frac{1}{T\sqrt{a}} \int f(t)\tilde{h}^* \times (\frac{t-\tau}{aT}) \exp(-j\omega_0(\frac{t-\tau}{a}))dt\} \tag{14}$$

where a denotes as scale, τ denotes as the time shift and *, <, >denotes as complex conjugate and inner product, respectively. The mother wavelet $h_T(t)$ is described as

$$h_T(t) = \frac{1}{T\sqrt{a}}\tilde{h}(\frac{t}{\tau})\exp(j\omega_0 t) \tag{15}$$

The envelope function of $\tilde{h}_T(t)$ which relates to the parent wavelet function of the WT h(t) by

$$\tilde{h}(t) = h(t)\exp(-j\omega_0 t) \tag{16}$$

where $\omega_0 = 2\pi f_0$ and $f_0 =$ center frequency of h(t). The T-function is expressed as

$$T(\tau + \Delta\tau) = (1 - \tilde{G}_1 \frac{BWT_s}{BWT_s + |BWT_I(\tau, a)|})^{-1}$$
$$\times (1 + \tilde{G}_2 \left|\frac{\partial BWT_I(\tau, a)}{\partial t}\right|)^{-1} \tag{17}$$

where $\tilde{G}_1$ and $\tilde{G}_2$ are the real active factors, BWT_I is the saturation constant. BWT_1 is the BWT coefficients that can be defined as the average value of $BWT(\tau)$, $BWT(\tau - \Delta\tau)$, $BWT(\tau - 2\Delta\tau)$. $\Delta\tau_S$ is calculation step size.

By using this BWT optimization technique, a model is done, i.e., described in Fig. 1.

BWT coefficients can be calculated by using K factor and wavelet coefficients.

$$BWT_I(\tau, a) = K \times WT_I(\tau, a) \tag{18}$$

$WT_I(\tau, a)$ is wavelet coefficients at time shift τ and scale a. The factor K depends on T and the Morlet function $h(t) = e^{-(\frac{t}{T_0})^2}$ considers as the mother function. K value is equal to

$$\frac{\int_{-\infty}^{\infty} e^{-t^2}}{\sqrt{(\frac{T}{T_0})^2 + 1}} = \frac{1.7725}{\sqrt{(\frac{T}{T_0})^2 + 1}} \qquad (19)$$

For denoising analysis using threshold value, Donoho formula is used which is

$$thr = \sigma \sqrt{2 \log_2 N} \qquad (20)$$

Given thr denotes the value at threshold, N is noise signal length and $\sigma =$ AMFS/0.6745 [9]. The absolute median measured on the first scale (AFMS) of the coefficients is the bionic wavelet.

3 Baseline Reduction

Low-frequency constituent i.e. present in ECG signal occurs due to the offset voltage present in the movement of body, respiration process and electrodes. Due to this, there is having a problem in the enumeration of ECG waveform. Baseline wandering is of two type i.e. dc component present in ECG signal which is not present in zero level and low-frequency interferences. For the estimation level of disintegration of the signal, the center-frequency (f_0)and number of scale r is required. In [11] the value of f_0 is taken as 15165.4Hz. Due to the increment of scale 'r', the f_0 value will be decreased in preceding way:

$$f_r = \frac{f_0}{q^r}, q > 1 \qquad (21)$$

The high f_0 is not required for the ECG signal, so that different value of f_0 is taken. For less distortion on the processed signal, the frequency value should be limited in between 360Hz–500Hz. There is having the credence between r and q. By using the Eq. (22), the value of q is found out and the same procedure is continued for the next scale. If the baseline wandering is the function of sinusoidal, then by using the signal spectrum the 'f' of the sine function is calculated. For the ECG signal which is not the function of sinusoidal, the estimation of f_w is calculated representing the signal in time-frequency domain. The analysing window using r scale is cantered at f_w. After finding out the value of f_w, thresholding rule in Eq. (20) is used for next three consecutive scales. For multi-adaptive purpose 'q' is required. The distance between the two ECG, one which is baseline wandered and other which is baseline corrected is used for calculation of q and that maximum distance should be maximized. The maximum distance can be calculated is given by the following expression:

Fig. 2 Noisy ECG signal 1 [8]

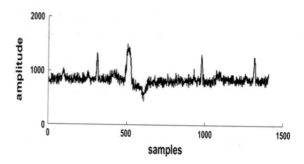

$$q_{opt} = \arg\max\{(|MABWT_x(\tau, a) - MABWT_{\hat{x}}(\tau, a)|\} \tag{22}$$

where x, $\hat{x}$ are the original and processed signals respectively, r denotes the scale number. The ECG signals with two type of baseline are considered for the effectiveness of the proposed technique and using WT, PCA, ICA and MABWT, the baseline wandering reduction has done.

4 Simulation Results

The performance comparison is performed by using the software MATLAB R2016a. The performance of WT, PCA, ICA, and WT techniques are analysed and the improvement of SNR values of signals is also discussed here. All the data were taken from MIT-BIH arrhythmia database [16]. In this paper, signal 101.mat from the database is considered as signal 1 for the analysis purpose. The main focus is to denoise the noisy signal and the dc component reduction. The different parameters for wavelet transform and bionic wavelet transform are given in Appendix.

Figure 2 represents the noisy input ECG signal 1. Different data are derived from the MIT-BIH arrhythmia database [19]. Figures 3 and 4 represent the ECG signal along with noise which is analyzed by using WT (hard) thresholding and WT (soft) thresholding. It can be observed from Fig. 4 that the denoised signal using WT (soft) thresholding gives a better picture than WT (hard) thresholding. Figures 5 and 6 represent ECG signal along with noise which is analyzed by using PCA and ICA technique. It can be observed from Fig. 6 that the denoised signal using ICA gives a better picture than PCA. Figures 7 and 8 represent ECG signal containing noise which is analyzed by using MABWT (hard) thresholding and MABWT (soft) thresholding. It can be observed from Fig. 8 that the noise content in ECG signal using MABWT soft thresholding is less as compared to MABWT hard thresholding, PCA, ICA and WT thresholding techniques.

Fig. 9 represents the noisy ECG signal with dc component and low interference. The data is derived from MIT-BIH arrhythmia database [19]. Figs. 10 and 11 represents the baseline with DC component and low interference reduction is done by

Fig. 3 Denoised ECG signal 1 [8] using WT (hard) thresholding

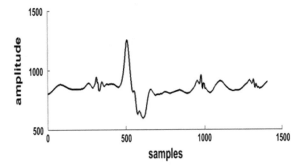

Fig. 4 Denoised ECG signal 1 [8] using WT (soft) thresholding

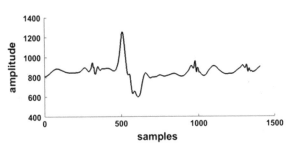

Fig. 5 Denoised ECG signal 1 [8] using PCA

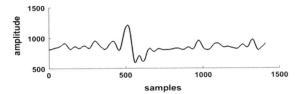

Fig. 6 Denoised ECG signal 1 [8] using ICA

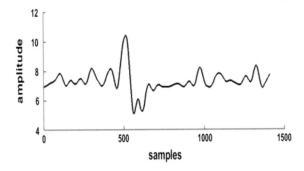

using WT hard thresholding and WT soft thresholding. It is observed from Fig. 11 the DC component reduction using WT soft thresholding gives a better picture than WT hard thresholding. Figs. 12 and 13 represents the baseline with DC component and low interference reduction is done by PCA and ICA tech-nique. The DC component part in ECG signal is less in ICA as compared to PCA. Figs. 14 and 15.

Fig. 7 Denoised ECG signal 1 [8] using MABWT (hard)

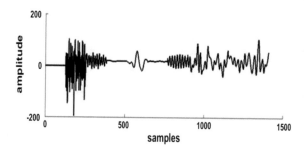

Fig. 8 Denoised ECG signal 1 [8] using MABWT (soft) thresholding

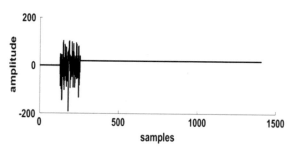

Fig. 9 ECG signal with DC component and low interference

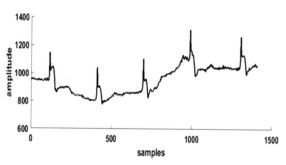

Fig. 10 DC component reduction using WT hard

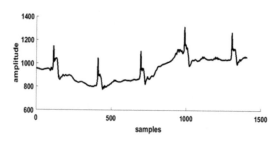

represents the baseline with DC component and low interference reduction is done by MABWT hard thresholding and MABWT soft thresholding. The DC component value of ECG signals is less in MABWT soft thresholding as compared to MABWT hard thresholding, PCA, ICA and WT thresholding.

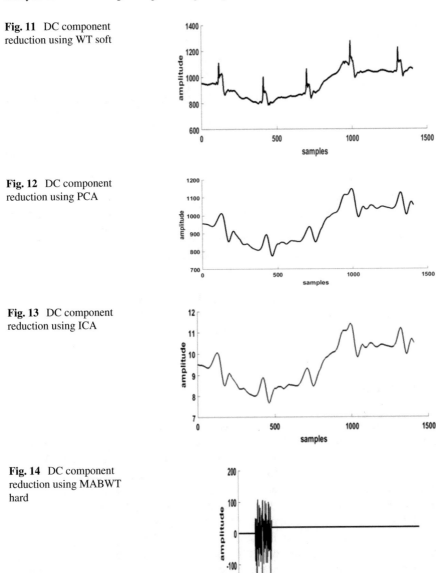

Fig. 11 DC component reduction using WT soft

Fig. 12 DC component reduction using PCA

Fig. 13 DC component reduction using ICA

Fig. 14 DC component reduction using MABWT hard

Table 1 represents the different SNR values of ECG signals in different compu-tational techniques. The data is extracted from MIT-BIH arrhythmia database. From the bold part of Table 1, it is observed that the SNR value of ECG signals in MABWT soft thresholding technique is better than other techniques. If the SNR values are 10-

Fig. 15 DC component reduction using MABWT soft

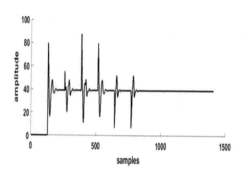

Table 1 SNR improvement of ECG signals

Signal	WT (hard)	WT (soft)	PCA	ICA	MABWT (hard)	MABWT (soft)
Signal 1	2	4	3	17	24	25

15dB then it is allowed to be minimum so that and unreliable connection can be established. If it is 16–24 dB then it is usually considered as Poor. If it is 25–40dB, then it is considered as good and for 41 dB or higher, it is considered as excellent.

5 Conclusion

In this paper, different techniques, i.e., wavelet transform, PCA, and ICA, are discussed for denoising of ECG signals. But a new technique, i.e., bionic wavelet transform (BWT), is used for denoising process. This technique adjusts the center frequency using the method of scale change in an adaptive manner, so that the T-function is also varied and its effects on the thresholding value. The major difference between WT and BWT is that the mother wavelet function is fixed in WT, but in BWT it is not fixed. From the comparative analysis, it is observed that the SNR values of signals in BWT technique are more than the other computational intelligence technique, i.e., signal power is more than the noise power in case of BWT. This multi-adaptive technique is not only used for denoising purpose but also it is used for baseline wandering reduction. Baseline with DC component and low interference using WT, PCA, ICA techniques also. The proposed technique MABWT can correctly denoise the ECG signal and eliminate the baseline wandering. This proposed technique is reliable and simple as compared to other computational intelligence techniques.

Appendix

WT technique: $a = a_0{}^m$, $b = nb_0a_0{}^m$, σ=MAD/0.6745; BWT technique: $\tilde{G}_1 = 0.87$ and $\tilde{G}_2 = 45$, $BWT_s = 0.8$; Simulink model: $e^{t^2} = 1.7725$.

References

1. N.V. Thakor, Y.S. Zhu, Applications of adaptive filtering to ECG analysis: noise cancellation and arrhythmia detection. IEEE Trans. Biomed. Eng. **38**(8), 785–794 (1997)
2. P. Laguna, R. Jane, O. Meste et al., Adaptive filter for event-related bio electric signals using an impulse correlated reference input: comparison with signal averaging techniques. IEEE Trans. Biomed. Eng. **39**(10), 1032–1044 (1992)
3. A.K. Barros, A. Mansour, N. Ohnishi, Removing artifacts from electrocardiographic signals using independent components analysis. J. Neurocomput. **22**(1–3), 173–186 (1998)
4. R. Vigario, J. Sarela, V. Jousmiki, M. Hämäläinen, E. Oja, Independent component approach to the analysis of EEG and MEG recordings. IEEE Trans. Biomed. Eng. **47**(5), 89–593 (2000)
5. M.Z.U. Rahman, R.A. Shaik, D.R.K. Reddy, Efficient and simplified adaptive noise cancelers for ECG sensor based remote health monitoring. IEEE Sens. J. **12**(3), 566–573 (2012)
6. D.L. Donoho, De-noising by soft-thresholding. IEEE Trans. Inf. Theory **41**(3), 613–627 (1995)
7. M. Popescu, P. Cristea, A. Bezerianos, High-resolution ECG filtering using adaptive Bayesian wavelet shrinkage, in *Proceedings of Computers in Cardiology*, Cleveland, Ohio, USA, Sept 1998, pp. 401–404
8. P. M. G. Agante Da Silva, J. P. Marques De Sa, ECG noise filtering using wavelets with soft-thresholding methods, in *Computers in Cardiology*, Hannover, Germany, September 1999, pp. 535–538
9. J. Yao, Y.T. Zhang, Bionic wavelet transform: a new time-frequency method based on an auditory model. IEEE Trans. Biomed. Eng. **48**(8), 56–863 (2001)
10. J. Yao, Y. T. Zhang, From otoacoustic emission modeling to bionic wavelet transform, in *22nd Annual International Conference of the IEEE Engineering in Medicine and Biology Society*, vol. 1, Chicago, Ill, USA, July 2000, pp. 314–316
11. J. Yao, Y.T. Zhang, The application of bionic wavelet transform to speech signal processing in cochlear implants using neural network simulations. IEEE Trans. Biomed. Eng. **49**(11), 1299–1309 (2002)
12. D.L. Donoho, I.M. Johnstone, Adapting to unknown smoothness via wavelet shrinkage. J. Am. Stat. Assoc. **90**(432), 1200–1224 (1995)
13. K. L. Park, K. J. Lee, H. R. Yoon, Application of a wavelet adaptive filter to minimise distortion of the ST-segment, Med. Biol. Eng. Comput. **36**(5), (1998), 581–586
14. K. Daqrouq, ECG baseline wander reduction using discrete wavelet transform. Asian J. Inform. Tech. **4**(11), 989–995 (2005)
15. J. A. Van Alste, T. S. Schilder, Removal of base-line wander and power-line interference from the ECG by an efficient FIR filter with a reduced number of taps. IEEE Trans. Biomed. Eng. **32**(12), 1052–1060 (1985)
16. M. Sifuzzaman, M.R. Islam, M.Z. Ali, Application of wavelet transform and its advantages compared to Fourier transform. J. Phys. Sci. **13**, 121–134 (2009)
17. M.T. Johnsona, X. Yuanb, Y. Rena, Speech signal enhancement through adaptive wavelet thresholding, in *Conference*, Elsevier, 2007, pp. 123–133
18. C.J. James, C.W. Hesse, Independent component analysis for biomedical signals. Physiol. Meas. **26**(1), R15–R39 (2005)
19. The MIT-BIH Arrhythmia Database, http://physionet.ph.biu.ac.il/physiobank/database/mitdB/

20. O. Sayadi, M. B. Shamsollahi, Multi-adaptive bionic wavelet transform: Application to ECG denoising and baseline wandering reduction. EURASIP J. Adv. Signal Process. **1**, 1–11 (2007)

Prediction of Secondary Structure of Proteins Using Sliding Window and Backpropagation Algorithm

Shivani Agarwal, Vijander Singh, Pankaj Agarwal and Asha Rani

Abstract Prediction of protein secondary structure plays a vital role in structural biology. Computational methodology is the initial step in bioinformatics to predict the 3-D secondary structure from a primary sequence and structure homology. This problem lies in the category of NP problem, and thus its time and space complexity is very high. In this paper, in a model for secondary structure prediction of proteins using sliding window and MADALINE, a multilayer feedforward network is proposed. The algorithm starts with encoding of amino acid sequence, which after passing through window is given as input to the neural network. The resultant data is in numeric format and translated back to actual secondary structure. It is observed from the results that the proposed technique provides better prediction with an accuracy more than 75%.

Keywords Primary protein sequence · MADALINE learning · Backpropagation Error minimization · Sliding window

S. Agarwal (✉) · P. Agarwal
Department of Computer Science, IMS Engineering College, Ghaziabad, India
e-mail: kasishivani@gmail.com

P. Agarwal
e-mail: pankaj7877@gmail.com

V. Singh · A. Rani
Division of Instrumentation and Control Engineering, NSIT,
University of Delhi, New Delhi, India
e-mail: vijaydee@gmail.com

A. Rani
e-mail: ashansit@gmail.com

© Springer Nature Singapore Pte Ltd. 2019
H. Malik et al. (eds.), *Applications of Artificial Intelligence Techniques in Engineering*, Advances in Intelligent Systems and Computing 698,
https://doi.org/10.1007/978-981-13-1819-1_50

1 Introduction

The core problem of bioinformatics is protein secondary structure prediction (PSP) using amino acid residues sequence and structure homology [1, 2]. This issue may be well addressed using deterministic computational methods, optimization algorithms, data mining, and machine learning approaches [3, 4]. The knowledge of proteins and their structures is also essential in designing a drug. Earlier, this information was acquired with the help of X-ray crystallographic or nuclear magnetic resonance (NMR) techniques. This method allows 100% accurate prediction of protein structures and helps to understand the three-dimensional structure of a protein along with the valuable subcellular locations. However, these conventional methods are highly time-consuming; therefore, in the present work, a fast alternative technique is proposed. Literature reveals the presence of various algorithms used for prediction of secondary structure of proteins [5]. Borguesan et al. [6] predicted three-dimensional protein structure using angle probability list-based metaheuristics. Genetic algorithms and particle swarm optimization were utilized for the desired prediction. Dubey et al. [7] identified 2D triangular hydrophobic-polar lattice model using a parallel programming approach for PSP. The results obtained to verify the improvement in prediction performance with reduced execution time. Ramyachitra and Ajeeth [8] used diversity controlled self-adaptive cuckoo algorithm for protein structure prediction. The technique is an improvement of self-adaptive differential evolution, for balanced exploration and exploitation. Ding et al. [9] extracted protein structure features using long-range and linear correlation information from position-specific score matrix. The efficiency of developed technique is evaluated using jackknife tests on benchmark database. Lim et al. [10] designed an evolutionary approach-based multimodal memetic algorithm for PSP. Three memetic algorithms are successfully tested on protein data bank and the results obtained are found similar to experimental structures.

It is revealed from the literature that accurate identification of secondary structure of protein is an important problem to diagnose diseases and deficiencies. Therefore, in the present work, a prediction model is proposed using sliding window and multilayer feedforward network, MADALINE. The remaining paper is organized as follows: The subsequent section introduces the methodology and Sect. 3 provides the details of proposed algorithm. Results and discussion are given in Sect. 4, and work is concluded in Sect. 5.

2 Methodology

Artificial neural network is inspired from biological neural network [11, 12]. It is a computational model that can be used for solving PSP problems. It is a group of artificial neurons (Fig. 1) connected in such a way that each neuron gives output to every neuron in the next layer. It is basically a three-layered architecture: input

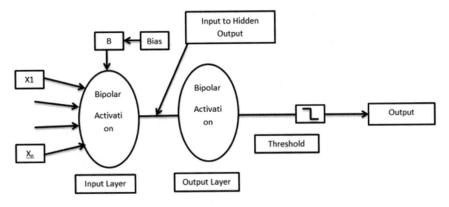

Fig. 1 Artificial neural network [14]

layer, hidden layers, and output layer, where the number of hidden layers may be varied as per the problem under consideration. The neural network is evaluated in two modes, i.e., feedforward and feedback mode. In case of feedforward mode, inputs with relevant weights emanate from input layer through hidden layer to the output layer. The output layer has a threshold to check the emitting power of neuron, whether it is in firing state or not. The process of learning or adjustment of weights is used in inhibitory state. The output for each neuron is calculated using bipolar sigmoid function. The network is trained using backpropagation technique.

2.1 MADALINE

The prediction model is a feedforward MADALINE with sliding window as illustrated in Fig. 2. The selection of inputs in input layer of neural network is made using sliding window as shown in Fig. 3. In this work, the window size is considered to be 5. Each residue of the input layer is encoded by encoding function and information is sent from input to output through hidden layer. The output of neural network indicates the type of secondary structure, i.e., helix, strand, or coil [13]. If the fired value produces a hit, then training is stopped and therefore no updating of weight matrices and bias value takes place. If the fired value produces a miss, training of the network is continued. Otherwise, trained network is used for the prediction of secondary structure.

The objective of the algorithm is to correctly identify the predicted secondary structure of amino acid. The sequence is a known sequence or depends on an error. The error E is expressed as the root mean square error of the total number of incorrect predictions by the output units. The training is stopped if the error is the same as previous error or zero.

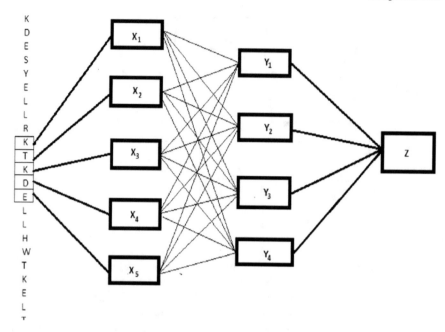

Fig. 2 MADALINE applied to primary protein structure

Fig. 3 Sliding window for selecting the inputs

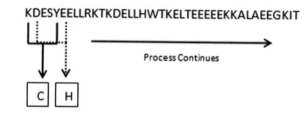

3 Proposed Algorithm

The steps followed for protein structure prediction using the methodology discussed in Sect. 2 are given as follows:

***Step* 0**: Input values (x), weight matrices "W", and bias values are defined. "α" learning rate is also initialized which is a random value lying between −0.1 and 0.1. x_i, the first sample of training input data.

***Step* 1**: Each neuron of the input layer receives input signal $x_i (i = 1$ to $5)$ from the sliding window for a window size 5.

***Step* 2**: $y_{inj}(j = 1$ to $4)$ is evaluated for hidden layer neurons as

$$y_{inj} = w_{oj} + \sum_{i=1}^{n} x_i \cdot w_{ij} \qquad (1)$$

The output from hidden layer neurons is calculated as

$$y_j = f\left(y_{inj}\right) \tag{2}$$

where $f(x) = (1 - e^{\lambda x})/(1 + e^{\lambda x})$

Step 3: The output received from hidden layer neurons ($k = 1$) is passed as an input to the output layer neurons. The output z_k is calculated using the equations

$$Z_{ink} = w_{ok} + \sum_{j=1}^{p} Z_j \cdot V_{jk} \tag{3}$$

$$Z_k = f(Z_{ink}) \tag{4}$$

Step 4: The value from output layer is compared with target value to check whether the neuron is fired or not which depends on the threshold value.

Step 5: If a hit is received, i.e., output value is same as the target value, no further training is required, and weight matrices and bias values are stored in an array; otherwise, both values are updated using backpropagation algorithm.

Step 6: The error in the output value is given as

$$E = (t - z)^2 \tag{5}$$

Step 7: The output unit $z_k (k = 1)$ receives a target value from the pattern corresponding to the input training pattern to evaluate the error correction term.

$$\delta_k = (t_k - z_k)f'(z_{ink}) \tag{6}$$

Here,

$$f'(Z_{ink}) = \left(\frac{\lambda}{2}\right)\left[(1 + f(z_{ink}))/(1 - f(Z_{ink}))\right] \tag{7}$$

Step 8: The output weight matrix and bias value are updated as follows:

$$\Delta V_{jk} = \alpha \cdot \delta_k \cdot y_j \Delta V_{ok} = \alpha \cdot \delta_k \tag{8}$$

The error correction term δ_k is sent to the hidden layer backward.

Step 9: The δ inputs for every neuron in the hidden layers are evaluated as

$$\delta_{inj} = \sum_{k=1}^{m} \delta_k \cdot w_{jk} \tag{9}$$

The error term is calculated as $\delta_j = \delta_{inj} \cdot f'\left(y_{inj}\right)$. The hidden layer weight matrix and bias value are thus updated as follows:

Table 1 Updated weight matrix of hidden layer

0.101320	−0.101131	0.008115	−0.018680	0.101320
−0.100643	0.100551	−0.09908	0.099357	−0.100643
0.048704	−0.048889	0.051851	0.051296	0.048704
−0.049374	0.049929	−0.050951	0.050666	0.049339

Table 2 Updated bias values for hidden layer

0.081150	0.099189	0.118513	0.080490

Table 3 Updated weight for matrix of output layer

0.081585	−0.087402	−0.118342	0.032701

$$\Delta w_{ij} = \alpha \cdot \delta_j \cdot x_i \tag{10}$$

Step 10: The new bias and weight values for output layer neurons are given by

$$V_{jk}(new) = V_{jk}(old) + \Delta V_{jk} \tag{11}$$

$$V_{ok}(new) = V_{ok}(old) + \Delta V_{ok} \tag{12}$$

Step 11: The weights and bias for each hidden unit are updated using the following equations:

$$w_{ij}(new) = w_{ij}(old) + \Delta w_{ij} \tag{13}$$
$$w_{oj}(new) = w_{oj}(old) + \Delta w_{oj} \tag{14}$$

Step 12: Check the stopping condition, that is, if the previous epoch error is same as the current epoch error, then training is stopped. If not, pick the next sample of training input data x_i and repeat from step 1.

In this work, fixed length encoding is considered. Each of the 20 amino acids is represented as a bit pattern of window size 5. As an example, consider the case of a protein tryptophan and its amino acid sequence:

KDESYEELLRKTKDELLHWTKELTEEEKKALAEEGKIT

The encoding scheme is applied to this sequence as input to the neural network. The error correction term is evaluated as 0.003363. Tables 1, 2, and 3 show the updated weight matrix of hidden layer, updated bias values for hidden layer, and updated weight matrix of output layer, respectively.

Table 4 Secondary structure using different tools

No.	Amino acid Sequence	Secondry structure using different tools		
1	IVGGYTCEESLPPYQVSLNSGSHFCGGSL IJSEQWVVSAAHCYKTRIQVRL GEHNIKVLEGNEQFINAVEI	CCCCEEOCCCCCCCE EEEEEECCCEEEEEEEEC CCEEEEEEEECCCCCCCC EEEEEEEEEEECCCCCCC CCCCEEEEEEECCCCCCC CCEEECCCCCCCCCCCEE EEE EECEEECCCCCCCCC CCEEEECCCCCHHHHH HHCCCCCCCCEEEEECC CCCCCCCCCCCCCCEEEC CEEEEEEEECCCCCCCCC CCEEEEEHHHHHHHHHH HHHHHHCC CCHHHHHHHHHHHHHH HHHHHHHHHCCHHHHH HCCCCCCCC	--- --- --- -----EEEEEE- --- EEEEEEE--- EEEEE- --- --- EEEEEEEEEEE--- -- - EEEEEEE--	cccceecccccccccee eecccccccccceeeecc eeeeeecccceeeee cccccchhhhhcchh hhceeec
2	KDESYEELLRKTKDELLHWTELT EEEKKALAEEGKIT	CCHHHHHHHHHHHH HHHHHHHHHHCCHH HHHHCCCCCCCC	--- HHHHHHHHHHHH HHHHHHH---	ccccchhhhhhhhhh hhhhhhhhhhhhhh hhhhcceeec

3.1 Datasets Considered for Experimentation

Plentiful primary protein sequences are generally folded into secondary structures in the form of α-helices and β-sheets in three-dimensional or eight-dimensional configuration. In the present work, three-dimensional configuration, α-helix, is considered which is the most lavish type of secondary structure in proteins having H bond formed in every fourth residue (1.5–11 turns). β-sheets are formed by H bonds at an average of 5–10 consecutive amino acids in one portion of the chain with another 5–10 farther down the chain [14]. The interacting regions may be adjacent, having a short loop, or far apart from other structures. It is more difficult to predict the location of β-sheets than α-helices. If region of secondary structure is not a helix or sheet, then it is recognized as a turn, commonly referred as a coil.

The actual datasets of primary amino sequence of proteins and their secondary structures, respectively, are taken from protein data bank. There are 135694 biological macromolecular structures present in protein data bank. Totally, 20 primary amino acid sequences are considered for the present training and validation purpose. Some of these are shown in Table 4. The datasets contain primary sequences of tryptophan and TITIN, respectively.

4   Results and Discussion

Various techniques are available in literature for prediction [15, 16] however, in this work, artificial neural network technique is used for the purpose. The experimentation is performed by considering the complete range of window sizes. Table 5 shows the variations in error for variable window sizes. It is observed that the error in prediction is reduced up to seven decimal places. The biological methods used earlier like Chou-

Table 5 Variation of error with window size

Window size	Error
3	0.00235
5	0.00141
7	0.00100
9	0.00078
11	0.00064
13	0.00054
15	0.00058
17	0.00066

Fig. 4 Error variations

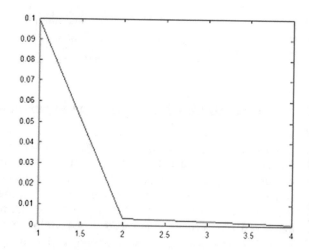

FASMAN and GOR for the purpose provide accuracy up to the level of 63–66%. However, the proposed technique has an error of 0.000234689, and therefore this method provides better prediction that is more than 75% (Fig. 4).

The proposed method provides an efficient way of prediction as compared to other tools used for PSP. It also provides the rectification of error to a maximum extent. The existing SVM-based PSP [17] provides 75.14% which is comparable to the proposed method.

5 Conclusion

The basic problem of bioinformatics is the prediction of secondary structure of protein sequences. The accurate identification of secondary structure of protein is necessary so as to diagnose diseases and deficiencies correctly. It is also a key component to determine the functions of a protein and its interaction with DNA, RNA, and enzymes. The proposed algorithm is used to minimize the error of prediction of secondary structure of proteins by computational model. The preprocessed data are given to the network that comes from the sliding window. The reliability and flexibility of the architecture in the form of sliding window and error minimization allow the introduction of more datasets. The existing prediction model based on SVM provides 75.14% accuracy, whereas the proposed technique also has accuracy of more than 75%. Hence, it is concluded that the proposed technique provides the fast and accurate alternative for PSP. The prediction of protein structure can be extended to a large dataset, and methods based on multiple sequence alignment, voting schemes, etc. may also be developed in future.

References

1. H. Hasic, E. Buza, A hybrid method for prediction of protein secondary structure based on multiple artificial neural networks, IEEE, 22–26 May, 2017. https://doi.org/10.23919/mipro.2 017.7973605
2. B. Al-Lazikani, J. Jung, Z. Xiang, B. Honig, Review: protein structure prediction. Curr. Opin. Chem. Biol. 5(1), 51–56 (2001)
3. M. Dorn, M.B. Silva, L.S. Buriola, L.C. Lamb, Review three-dimensional protein structure prediction: methods and computational strategies. Comput. Biol. Chem. Part B 53, 251–276 (2014)
4. L. Kong, L. Zhang, J. Lv, Accurate prediction of protein structural classes by incorporating predicted secondary structure information into the general form of Chou's pseudo amino acid composition. J. Theor. Biol. 344(7), 12–18 (2014)
5. A.E. Márquez-Chamorro, G. Asencio-Cortés, C.E. Santiesteban-Toca, J.S. Aguilar-Ruiz, Soft computing methods for the prediction of protein tertiary structures: a survey. Appl. Soft Comput. 35, 398–410 (2015)
6. B. Borguesan, M.B. Silva, B. Grisci, M. Inostroza-Ponta, M. Dorn, APL: an angle probability list to improve knowledge-based metaheuristics for the three-dimensional protein structure prediction. Comput. Biol. Chem. Part A 59, 142–157 (2015)
7. S.P. Dubey, N.G. Kini, M.S. Kumar, S. Balaji, Ab initio protein structure prediction using GPU computing. Perspect. Sci. 8, 645–647 (2016)
8. D. Ramyachitra, A. Ajeeth, MODCSA-CA: a multi objective diversity controlled self adaptive cuckoo algorithm for protein structure prediction. Gene Rep. 8, 100–106 (2017)
9. S. Ding, S. Yan, S. Qi, Y. Li, Y. Yao, A protein structural classes prediction method based on PSI-BLAST profile. J. Theor. Biol. 353(21), 19–23 (2014)
10. L. de Lim Corrêa, B. Borguesan, M.J. Krause, M. Dorn, Three-dimensional protein structure prediction based on memetic algorithms, Comput. Oper. Res. 91, 160–177 (2018)
11. S.N. Sivanandam, S.N. Deepa, Principles of Soft Computing (Wiley, 2011)
12. Rashid et al., Protein secondary structure prediction using a small training set (compact model) combined with a complex-valued neural network approach. BMC Bioinform. (2016). https:// doi.org/10.1186/s12859-016-1209-0
13. A.A. Ibrahim, Using neural networks to predict secondary structure for protein folding. Sci. Res. 5(1) (2017)
14. J. Dongardive, Reaching optimized parameter set: protein secondary structure prediction using neural network. 28(8), 1947–1974 (2017). ACM
15. J. Yadav, A. Rani, V. Singh, B.M. Murari, Levenberg-Marquardt based non-invasive blood glucose measurement system. IETE J. Res. https://doi.org/10.1080/03772063.2017.1351313
16. A. Rani, V. Singh, J.R.P. Gupta, Development of soft sensor for neural network based control of distillation column. ISA Trans. 52(3), 438–449 (2013)
17. S. Agarwal et al., Prediction of secondary structure of protein using support vector machine, in ICACEA, IJCA (5), 1–4 (2014)

Multiclass Segmentation of Brain Tumor from MRI Images

P. K. Bhagat and Prakash Choudhary

Abstract A brain tumor is a fatal disease which takes thousands of lives each year. Thus, timely and accurate treatment planning is a critical stage to improve the quality of life. MRI is a very novel method of diagnosis of the brain which shows a fine level of details of the brain tumor. For the treatment of brain tumor, accurate segmentation of the tumor part is highly desirable. The manual tumor segmentation is a challenging and time-consuming process. Thus, automatic tumor segmentation can play a very important role in the process to speed up the treatment process. There exist different types of MRI sequences, each with its own merits and showing varying levels of information. We experimented with T2-weighted (T2), T1 with enhancing contrast (T1c), and FLAIR MRI images of the BRATS 2013 dataset and try to show that one particular MRI sequence is very useful for segmentation of one particular class of tumor than other. We have used thresholding and K-means algorithms along with a set of preprocessing and postprocessing methods for the segmentation of tumor regions. We extracted three classes of tumor: whole tumor, tumor core (whole tumor except edema), and active tumor region which is unique to high-grade (HG) cases from FLAIR, T2, and T1c, respectively. The obtained result according to Dice coefficient matrix is 0.81, 0.54, and 0.61 for the whole tumor, the tumor core, and the active tumor, respectively.

Keywords Brain tumor · Tumor segmentation · MRI · Thresholding
K-means · Multiclass tumor

P. K. Bhagat (✉) · P. Choudhary
Department of Computer Science and Engineering, National Institute
of Technology Manipur, Imphal, India
e-mail: pkbhagat22@gmail.com

P. Choudhary
e-mail: choudharyprakash87@gmail.com

© Springer Nature Singapore Pte Ltd. 2019
H. Malik et al. (eds.), *Applications of Artificial Intelligence Techniques
in Engineering*, Advances in Intelligent Systems and Computing 698,
https://doi.org/10.1007/978-981-13-1819-1_51

1 Introduction

Magnetic Resonance Imaging (MRI) is a novel method of image acquisition which is broadly used and very powerful imaging modality for the diagnosis of the brain. MRI shows an excellent level of details of the human brain and has been proven to be very useful to study the human brain [1]. Its noninvasive nature and no observed side effect on the human body made it even more popular modality of image acquisition. Manual segmentation of MRI image is time-consuming, and it varies from person to person also.

The excellent level of details of the human brain shown by MRI is very helpful for diagnosis. The presence of brain tumor can be ascertained by a radiologist or by an automatic system. The MRI image segmented by different human beings may sometimes result in a different interpretation. Also, manual segmentation is a time-consuming process. The subjectivity and time-consuming nature of human segmentation are not suitable for treatment. For accurate treatment planning, the brain tumor must be identified and segmented accurately in early stage or as early as possible. An automatic segmentation method can be used to minimize human intervention. The automatic system for segmentation can process a large number of images efficiently and can produce faster results which may ultimately result in faster delivery of services and will assist in accurate treatment planning.

There exist various methods for brain tumor segmentation like manual segmentation, semi-automatic segmentation, fully automatic segmentation, and hybrid segmentation [2]. While manual segmentation is error-prone and time-consuming, semi-automatic segmentation requires users to outline region of interest. We have used K-means clustering and thresholding methods for segmentation. K-means clustering [3] is applied on FLAIR image to extract the whole tumor region, and thresholding [4, 5] is employed on the T2 and T1c image to obtain the tumor core and the active tumor region, respectively.

A different set of preprocessing method is used for different segmentation methods. We have also applied some postprocessing after tumor segmentation. These preprocessing and postprocessing help in enhancing the tumor regions. We have experimented with high-grade (HG) image of brain tumors from BRATS 2013 training dataset.

The organization of the paper is as follows. Section 2 explains the segmentation methods used in this work. Section 2 also explains the used preprocessing and postprocessing methods. In Sect. 3, the obtained results and discussions are presented followed by conclusion in Sect. 4.

2 Method

A complete overview of the proposed system is presented in Fig. 1. There are three main stages of the proposed system: preprocessing, segmentation, and postprocessing.

2.1 Preprocessing

We have used different preprocessings for different MRI sequence images. As bias field distortion is the most common type of noise present in the MRI images [6], we have applied Gaussian filtering [7] for FLAIR and T1c images. The Gaussian smoother operator has been used to denoise the images. The Gaussian filtering uses a kernel window which is a weighted window which anchors the filtering. Because of kernel window, a Gaussian operator shifts its focus more toward center pixel, provides soft smoothing, and preserves edges better than any other filter. When the Gaussian filter is used as a preprocessing technique in edge-sensitive images, it may arise the problem of edge displacement, edge vanishing, and phantom edges; we need to select appropriate kernel size based on the value of σ and vice versa. The used Gaussian mask is given in Eq. (1).

$$G(x, y) = \frac{1}{2\pi\sigma^2} \times e^{-\frac{(x^2+y^2)}{2\sigma^2}}. \tag{1}$$

We have applied K-means clustering on FLAIR images and thresholding segmentation on T1c images. The used Gaussian filter has a 3×3 kernel having $\sigma = 0.29$. We have extracted the whole tumor from FLAIR images and the active tumor form T1c images. For T2 images, we have used Gaussian filter in a 3×3 kernel having $\sigma = 0.45$. We have extracted tumor core region from T2 images using thresholding segmentation algorithm. Different types of preprocessing are necessary because various tissues appear uniquely in different types of MRI images sequence. Figure 2 shows images before and after preprocessing of FLAIR, T2, and T1c images.

2.2 Segmentation

We have applied two different segmentation methods. The K-means clustering [8] is applied to extract the whole tumor region from MRI FLAIR images. We applied thresholding segmentation [9] method on T2 and T1c images to obtain the tumor core and active tumor regions, respectively.

(1) *K-means clustering*: when the number of clusters, K, is fixed, the algorithm classifies a given image into K number of cluster. Clustering is the process of

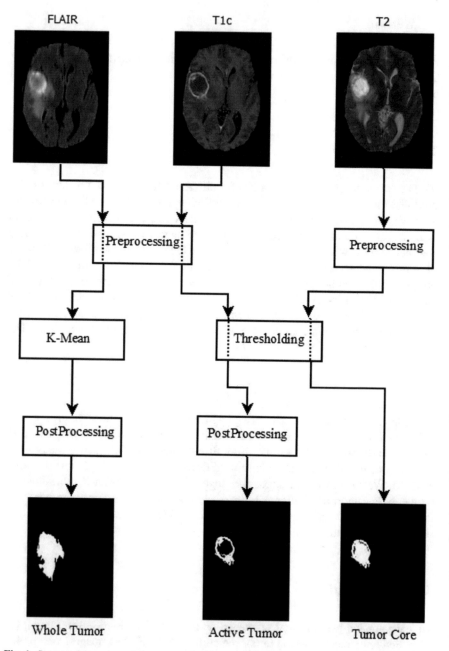

Fig. 1 Structural overview of the proposed system

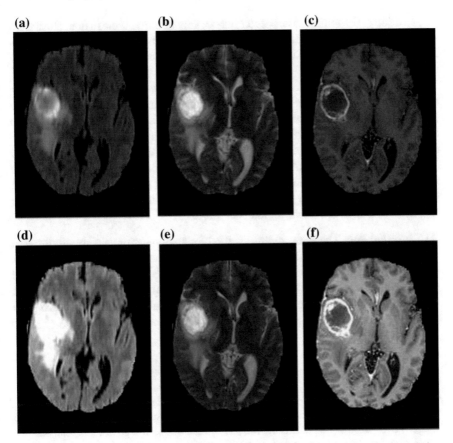

Fig. 2 The first row (subfigure **a**, **b**, and **c**) shows the FLAIR, T2, and T1c image before preprocessing respectively. Subfigure **d**, **e**, and **f** are the results of **a**, **b**, and **c**, respectively, after performing preprocessing

finding groups of objects such that the pixels in one cluster will be almost similar to one another and different from the pixels in other clusters. Let us consider an image which has to be clustered into K number of cluster. The applied algorithm of the K-means clustering is as follows:

1. Initialize K as number of clusters and corresponding cluster center.
2. For each pixel of image I, calculate its distance (city block) from each cluster center.
3. The pixel is designated as part of one of the clusters using calculated distances.
4. After all pixels have been assigned, calculate the mean of each cluster and this mean will be considered as new cluster center.
5. Repeat the process until desired error condition is satisfied.

Selection of the initial cluster center plays a very important role in final clustering. After careful inspection of the input dataset and experimental analysis with various initial centers, the four cluster centers chosen are based on the maximum intensity of the image.

The K-means is a simple unsupervised method for segmentation based on clustering technique. The whole tumor region contains edema, enhancing core, necrotic (or fluid-filled) core, and non-enhancing (solid). Each FLAIR image is clustered into four clusters. The selection of the initial cluster center is automatic based on the intensity of the image. When we apply K-means on MRI FLAIR images, the whole tumor (complete tumor) regions are automatically clustered in one cluster. As K-means algorithm works on the intensity of the image, it also produces some undesirable regions. To get rid of these undesirable regions, we perform some postprocessing after which final whole tumor region is obtained.

(2) *Thresholding*: The thresholding method for segmentation is a straightforward segmentation method. The pixels are partitioned based on their gray values. For segmenting images on space region to separate the regions of background and object, thresholding is considered as a powerful approach [4]. Let us consider an input image I of size $N \times N$, and T is the threshold value. Then, for each pixel $I(i, j)$ of input image, the output pixel $P(i, j)$ is obtained using Eq. (2).

$$P(i, j) = \begin{cases} 1, & I(i, j) \geq T \\ 0, & I(i, j) < T. \end{cases} \tag{2}$$

The suitable value for the threshold can be chosen based on the histogram of the image. After careful inspection of the histogram of the images of our dataset, we found that the desired tumor region has an intensity which is close to the highest intensity of the images. Based on this idea and after few experiments, we found a fixed value of threshold T which produced very good results for all the images of our dataset.

Thresholding is the easiest method of segmentation which includes histogram and gray-level features based on their intensities with one or more thresholding values [10]. To extract the tumor core regions, the algorithm divides the image into two different regions using one threshold value. The tumor core region includes all tumors except edema. In MRI T2-weighted image, the tumor core region has a high intensity as compared to other regions of the image. Hence, the algorithm automatically extracts the tumor core region from the T2-weighted image. We need not perform any postprocessing operation as the obtained region is almost accurate and complete.

To extract the active tumor region, we have applied thresholding segmentation on T1c images. The active tumor region only contains the enhancing core structures that are unique to high-grade (HG) cases. The thresholding segmentation method extracts enhancing core from T1c images along with some undesirable regions. Hence, we performed postprocessing to remove those undesirable regions.

2.3 Postprocessing

As K-means and thresholding segmentation methods directly operate on the gray value of the image, and due to the intensity variation of the image, we have some small undesirable regions in whole tumor region and active tumor region. These undesirable regions are small–small connected components which need to be removed from the final output. We have applied erosion as postprocessing to remove these small connected components. The erosion (Eq. (3)) of an image A by a structuring element B, denoted by $A \ominus B$, is the set of all elements x of E^N for which B translated to x is contained in A [11].

$$A \ominus B = \left\{ x \epsilon E^N | (B)_x \subseteq A \right\} \tag{3}$$

For erosion, we have used Matlab function bwareaopen, having a connectivity of 8, given in Eq. (4). Connectivity may be visualized as a structuring element B of Eq. (3) which slides across the image A. bwareaopen (A, P) removes all bounded components from image A having fewer than Pnumber of connected pixels. This results in another binary image X.

$$X = bwareaopen(A, P). \tag{4}$$

The method that we applied removes all connected components smaller than some predefined P number of pixels. P has been selected experimentally. We have performed various experiments for various values of P, and finally, we were able to select a fixed value of P which satisfies for almost all images. Separate value of P is chosen for K-means and thresholding algorithm. Figure 3 shows before postprocessing and after postprocessing images for K-means and thresholding segmentation methods.

3 Results and Discussions

In this section, we first explain the details of used dataset. After that, we present our experimental results followed by discussion.

3.1 Dataset

The dataset used in the experiment is taken from BRATS 2013 database [12]. The BRATS dataset contains the clinical image data and synthetical image data. The clinical image data includes high-grade (HG) and low-grade (LG) cases. We have used high-grade (HG) cases of clinical image data. Each patient image dataset consists of four MRI contrast images—T1-weighted image (T1), T1-weighted contrast-

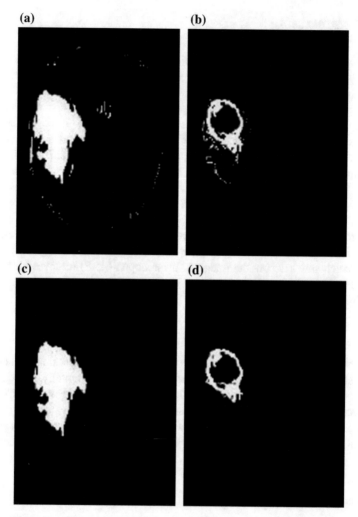

Fig. 3 The first row (subfigure **a** and **b**) shows the whole tumor and the active tumor before postprocessing, respectively. Subfigure **c** and **d** are the results of **a** and **b** after performing postprocessing, respectively

enhanced image (T1c), T2-weighted image (T2), and T2-weighted FLAIR image (FLAIR). In our experiment, we have used T2, T1c, and FLAIR images of each present. A total of 283 images are used in this experiment. Each image of the analysis contains all three types of tumors—the whole tumor, the tumor core, and the active tumor.

3.2   Experimental Results

We have used Dice score to evaluate the performance of the proposed method. We have calculated Dice score [13] for all the three types of tumor. Along with Dice score, we have also calculated sensitivity and specificity for each type of tumor using ground truth (expert consensus truth) given along with training dataset.

Let the prediction $P \in \{0, 1\}$ of our proposed algorithm and the ground truth is given by $T \in \{0, 1\}$, then Dice score is calculated using Eq. (5).

$$DiceScore = \frac{|P_1 \wedge T_1|}{(|P_1| + |T_1|)/2},$$ (5)

where P_1 means all the pixels of the predicted binary image having pixel value 1. T_1 means all the pixels of the ground truth binary image having pixel value 1. The symbol $\wedge$ stands for logical AND operation and $|x|$ is the size of the set x.

The sensitivity and specificity are calculated for each one of the three different types of the tumor using Eq. (6) and Eq. (7), respectively. The sensitivity and specificity are also known as true positive rate and true negative rate.

$$Sensitivity(P, T) = \frac{|P_1 \wedge T_1|}{|T_1|}$$ (6)

$$Sensitivity(P, T) = \frac{|P_0 \wedge T_0|}{|T_0|},$$ (7)

where P_0 and T_0 indicate pixel value 0 in predicted and ground truth binary images respectively.

The average Dice score produced by the proposed method on BRATS 2013 dataset are 0.81, 0.54, and 0.61 for the whole tumor, tumor core, and active tumor regions, respectively. The sensitivity and specificity of the whole tumor, tumor core, and active tumor regions are 0.89 and 0.97, 0.45 and 0.99, and 0.49 and 0.99, respectively. A comparative analysis of the accuracy of the proposed method with other methods that used the same dataset is presented in Table 1 [12].

3.3   Discussion

During the experiment, we observed that it is best to extract whole tumor region from FLAIR image. The whole tumor can also be extracted from T2 images. When we used K-means algorithm on both T2 and FLAIR images to extract the whole tumor region, FLAIR images gave best results in almost all cases. The reason behind this may be due to the intensity similarity between edema and other parts of the brain. To extract the tumor core region, a T2-weighted image is the best suitable. In a T2 image, the tumor core has very high intensity. The higher intensities are not found in other parts of the image except for the tumor part. Hence, this tumor part can be

Table 1 The comparison between on-site test results of BRATS 2013 challenge [12] and our method (last row)

BRATS 2013			
Clinical images Dice score (in %)	Whole tumor	Tumor core	Active tumor
Cordier	84	68	65
Doyle	71	46	52
Festa	72	66	67
Meier	82	73	69
Reza	83	72	72
Tustison	87	78	74
Zhao(II)	84	70	65
Our method	81	54	61

extracted from a T2 image. The active tumor core surrounds necrotic core, and in the T1c necrotic core has a very low intensity and the active tumor has high intensity. This difference plays a very active role while extracting active tumor region from the T1c image.

4 Conclusion

The paper presented an experimental analysis of a simple method for the brain tumor segmentation from BRATS 2013 dataset. Although the whole tumor region can be segmented with greater accuracy, extraction of tumor core and active tumor regions are more challenging. Although we have not applied any sophisticated segmentation methods, our observation is very important for the future experiments. Based on our observation, we applied simple segmentation methods which produced satisfactory results at this point. We hope that, in future, our current work will help us in designing methods for tumor segmentation with user satisfactory results.

Acknowledgements Author would like to thank BRATS 2013 to provide the open access database. This database has been utilized for the study in this paper. Authors would also like to thank the Department of Computer Science and Engineering, National Institute of Technology Manipur, Imphal, India to provide the platform and equipments for the study so that authors able to perform this study.

References

1. S. Bauer, R. Wiest, L.P. Nolte, A survey of mribased medical image analysis for brain tumor studies, in *IEEE Conference Record Nuclear Science Symposium and Medical Imaging Conference*, vol. 58 (2013), pp. 97–129
2. S. Saladi, A. Prabha, A comprehensive review: segmentation of MRI imagesbrain tumor. Int. J. Imaging Syst. Technol. **26**, 295–304 (2016)
3. J. Selvakumar, A. Lakshmi, T. Arivoli, Brain tumor segmentation and its area calculation in brain MR images using k-mean clustering and fuzzy c-mean algorithm, in *IEEE-International Conference on Advances in Engineering, Science and Management (ICAESM-2012)* (2012), pp. 186–190
4. M. Kass, A. Witkin, D. Terzopoulos, Snakes: active contour models. Int. J. Comput. Vis. 321–331 (1988)
5. W.X. Kang, Q.Q. Yang, R.P. Liang, The comparative research on image segmentation algorithms, in *2009 First International Workshop on Education Technology and Computer Science*, vol. 2 (2009), pp. 703–707
6. S. Pereira, A. Pinto, V. Alves, Brain tumor segmentation using convolutional neural networks in MRI images. IEEE Trans. Med. Imaging **35**(5), 1240–1251 (2016)
7. G. Deng, L.W. Cahill, An adaptive gaussian filter for noise reduction and edge detection, in *IEEE Conference Record Nuclear Science Symposium and Medical Imaging Conference* (1993), pp. 1615–1619
8. M. Gupta, M.M. Shringirishi, Implementation of brain tumor segmentation in brain MR images using k-means clustering and fuzzy c-means algorithm. Int. J. Comput. Technol. **5**, 54–59 (2013)
9. N. Sharma, L.M. Aggarwal, Automated medical image segmentation techniques. J. Med. Phys. **35**, 3–14 (2010)
10. Y.-C. Sung, K.-S. Han, C.J. Song, Threshold estimation for region segmentation on MR image of brain having the partial volume artifact, in *WCC 2000 - ICSP 2000. 2000 5th International Conference on Signal Processing Proceedings, 16th World Computer Congress 2000*, vol. 2 (2000), pp. 1000–1009
11. R. Haralick, S. Sternberg, X. Zhuang, Image analysis using mathematical morpholog. IEEE Trans. Pattern Anal. Mach. Intell. PAMI-9, 532–550 (1987)
12. B.H. Menze, A. Jakab, S. Bauer, The multimodal brain tumor image segmentation benchmark (brats). IEEE Trans. Med. Imaging **34**, 1993–2024 (2014)
13. L.R. Dice, Measures of the amount of ecologic association between species. Ecology **26**, 297–302 (1945)

Real-Time Resource Monitoring Approach for Detection of Hotspot for Virtual Machine Migration

Yashveer Yadav and C. Rama Krishna

Abstract Cloud computing is a service delivery model that provides computing resources (RAM, CPU, Memory, etc.), which are highly elastic and can grow or shrink dynamically according to the need of client. Virtual machine migration (VMM) plays very important role to provide the resource elasticity to cloud environment. VMM generates considerable amount of overheads which can degrade overall performance of cloud environment. So, it becomes important to decide when migration should take place. In this paper, we have introduced real-time resource monitoring (RTRM) system for detection of a hotspot host. We have developed three modules: (1) monitoring module, (2) analysis module, and (3) VMM module for detection of hotspot. We have compared the result of proposed RTRM with sandpiper and artificial intelligence technique such as neural network predictor. Experimental result shows that RTRM is able to detect the hotspot in first 20 s of time with the accuracy of 80% as compared to sandpiper with 64% accuracy and neural network predictor with 76% accuracy in the same time frame.

Keywords Virtualization · Hotspot detection · Virtual machines
Virtual machine migration · Real-time resource monitoring

Y. Yadav (✉) · C. Rama Krishna
Applied Science Department (Computer Applications), I. K. Gujral Punjab
Technical University, Punjab, India
e-mail: yadav.yashveer@gmail.com

C. Rama Krishna
e-mail: ramakrishna.challa@gmail.com

C. Rama Krishna
Department of Computer Science and Engineering, NITTTR Chandigarh, Chandigarh, India

© Springer Nature Singapore Pte Ltd. 2019
H. Malik et al. (eds.), *Applications of Artificial Intelligence Techniques in Engineering*, Advances in Intelligent Systems and Computing 698,
https://doi.org/10.1007/978-981-13-1819-1_52

1   Introduction

Cloud computing has redefined the information technology industry and business sector. Cloud computing is not a new technology paradigm as it seems to be originated after the computer network diagrams that represent the Internet like a cloud. Cloud computing relies on sharing of computing resources for achieving economy and coherence of scale. The foundation of cloud computing is the concept mainly converged infrastructure and shared pool of resources and services. Cloud computing provides computing power to clients as per their requirement rather than building and maintaining computing infrastructure.

In [1], authors have described cloud computing as a large pool of virtualized resources such as development platforms, hardware, and services. These resources are scalable at runtime to adjust dynamic load that results in optimal resource utilization. The available resources are used on the basis of pay-per-use model in which quality of service (QoS) is maintained by means of customized service-level agreement (SLA) as provided by infrastructure provider.

The fundamental concept of cloud computing is virtualization which leverages server virtualization technology to form pools of computing resources from the physical infrastructure. Virtualization is an abstraction that creates a virtual version of physical resources. Server virtualization is the virtualization technology that enables single physical machine to host multiple virtual machines [2, 3]. Virtualization provides high availability, kernel development, efficient computation, and peer-to-peer streaming.

The remaining paper is organized as follows: Sect. 2 describes the related work. The proposed RTRM system is presented in Sect. 3. Section 4 presents results and discussion. Section 5 finally concludes the paper.

2   Related Work

In cloud computing environment, each physical host has its own resources. These resources are less physical and more virtual like virtual CPU (vCPU) and virtual RAM (vRAM). VMs are created on physical host and use its resources. The client can increase or decrease the resources of VM at any point of time. Increase in the need of resources leads to physical host out of resources that further leads to need of VMM. This situation is also called as hotspot where physical machine is not able to fulfill the resource requirement for running virtual machines.

Hotspot detection is a continuous process in which we detect a hotspot after regular intervals of time. It is required for smooth running of VMs and also maintains the quality of service (QoS) according to SLA. Cloud service providers monitor the resource utilization of the various cloud resources like CPU, memory, and network, and identify bottleneck based on these utilizations. In [4], authors have proposed a model for hotspot detection and have monitored the resources allotted to VM at the hypervisor level. Mishra and Das [5] used the static threshold value to detect the hotspot. They have monitored resources using an algorithm at hypervisor level. According to this algorithm, if the level of resources go below a static threshold then the machine is marked as hotspot.

Shaw and Singh [6] proposed load prediction algorithm for detection of hotspot. Load prediction algorithm uses time series forecasting technique to predict the future load on CPU. This algorithm kept record of CPU at the physical host and the length of the data was 10. Based on the prediction value, hotspot is detected if the CPU utilization reaches the threshold value. This technique required minimum 10 rounds of data to detect the hotspot. However, in this scheme, when a new VM starts on physical host, it uses the maximum CPU and RAM. Resource usage increases exponentially until VM loads onto the CPU and RAM. Once the VM is loaded into memory, it starts releasing load on CPU and RAM but this technique detects the hotspot, because the resource utilization increases exponentially in recent observation. C. Zou et al. [7] have developed migration triggering module (MTM) to initiate the VMM process. MTM keeps monitoring source host for workload at regular intervals of time. MTM initiates source host to start VMM only when the host load meets three predefined requirements. First, the average value of load in N continuous observation should be greater than or equal to the threshold. Second, in N periods, at least n $(0 < n)$ values of the load monitored need to be greater than the threshold value. Third, the workload trend of this component is incline, which means the value of trend is not going down. If these three conditions are not satisfied simultaneously, host does not initiate a migration process. This method takes too much time, while physical machine remains overloaded and performance of the overall cloud is degraded. The system is not able to detect the hotspot if the resource utilization dips regularly after every two intervals of time.

Goebel and Whlster [8] have developed hotspot removal system using artificial neural network predictor (NNP). Their NNP uses the time series data to predict the future usage of resources. Their resource prediction function is shown in Eq. 1.

$$R_{i,t+1} = f(R_{i,t}, \ldots\ldots R_{i,t-1}, R_{i,t-j}, \ldots\ldots, R_{i,t-d-1}) \qquad (1)$$

where r is the resource and optimal f is a nonlinear function to trained NN, $R_{i,t-j}$ is the jth delay resource usage and d is delay index. They also use standard deviation (SD) function to make this model more accurate. The SD (error) is represented by using σ_i^2 notation, which is the variance of past prediction errors. Their resources usage prediction on physical machine, PM_i, denoted as $P_{i,t+1}$ is defined as

$$P_{i,t+1} = R_{i,t+1} + \sigma_i^2 \qquad (2)$$

The output of the Eq. 2 is converted into 0 or 1 according to Eq. 1. The output of Eq. 2 defines the hotspot where 1 output shows hotspot detected and 0 shows no hotspot situation. Equation 2 generates output in range 0–1. If the value is above the 0.5, it means that hotspot has been detected; otherwise, no hotspot would be considered as shown in Eq. 3.

$$Pi, t+1 = \begin{cases} 1, & > 0.5 \\ 0, & otherwise \end{cases} \tag{3}$$

From the above discussion, we can summarize that the existing approaches [4–7] calculated allotted vCPU and vRAM of all running VMs to spot hotspot. However, do not consider how much resources of physical machine are actually consumed at a time. Most of the techniques use the time series functions to predict the future utilization of resources. These techniques do not consider the reason behind the increase of resource utilization and are not process or interrupt sensitive. To overcome above-said issues, in this paper, we proposed real-time resources monitoring system to monitor the resources in real time.

3   Proposed Real-Time Resource Monitoring (RTRM) System

In this section, we describe the findings from data analysis and then define the proposed RTRM system architecture with algorithm.

We have analyzed data provided by Google [9] and Bitbrains [10], and the findings from the data and literature are described as follows:

On the basis of our observation, we can define hotspot as the situation when resources utilization of the physical machine is continuously above the safe threshold value. We have found that whenever a new task initiates in the physical machine it has three phases: (1) initialization phase, (2) execution phase, and (3) termination phase. In initialization phase, the resource utilization increases exponentially. Once the task acquires all the required resource, it enters in the execution phase. In this phase, it releases access resources which are required in initialization phase and hold only minimum required resources. In termination phase, the resource usage increases for tiny amount of time and then it starts shrinking.

In [6, 7, 11] techniques, time series data is used to predict the next utilization of hotspot. We have found that most of these techniques detect hotspot when a new VM is initialized at the physical host because in initialization phase, resource utilization is increased exponentially. Most of these techniques gather the data after interval of every 10 s [12] for each instance. This 10-s delay is static and remains same even after VM enters the hotspot phase.

The number of occurrence when resource utilization remains exceed the safety threshold is static in mostly techniques. This leads to unnecessary delay in triggering VMM.

On the basis of data analysis through experiments carried out, we have proposed RTRM system. The architecture of RTRM is shown in Fig. 1. RTRM have three modules: (1) monitoring module, (2) analysis module, and (3) VMM module.

a. **Monitoring module**: Monitoring module is responsible for monitoring the resource utilization at OS level. It monitors the data and sent it to analysis module.

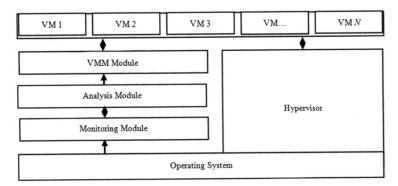

Fig. 1 Proposed RTRM system

Monitoring module monitors the data after every T seconds of time. The default value for T is 5 s. The value of T can be dynamically set by analysis module in hotspot situation. Monitoring module is directly connected to the operating system and monitors the number of processes running on it so this module reports for initialization of new process or task with detail of it to the analysis module.

b. **Analysis module**: Analysis module receives data directly from monitoring module and processes it. It checks the data for hotspot scenario and sets the value of T variable accordingly. Whenever it finds hotspot situation, it sets the value of $T=1$ to monitor the data after the interval of 1 s. It further analyzes the cause of increment in resources by the tasks reported by the monitoring module and sets the value of delay accordingly. Delay is a numeric value that defines time required to complete a particular task. Analysis module maintains a table which defines task and its corresponding delay value. Delay value is the time required to finish a particular task in normal condition. These table entries are static in nature. If the continuous resource utilization is above the threshold value for delay number of times, then only proposed RTRM system triggers the VMM process.

c. **VMM module**: This module is responsible for selecting the victim VM, selecting destination host, and migrating the VM to destination host.

RTRM does not monitor the usage of resources at the hypervisor level. It directly takes input from the underline OS. The monitoring module monitors the current usage of CPU and RAM. It further sends this data to the analysis module. In existing literature techniques, use the static time interval for data monitoring in normal execution conditions and under hotspot execution conditions. In RTRM system, monitoring interval time is set by analysis module dynamically. Monitoring module senses the data and sends it to the analysis module. Monitoring modules send two types of data to analysis module: (1) resource monitoring data and (2) report the new task and interrupts in the system. The monitoring module is the main module of the proposed RTRM system. Monitoring module is responsible for setting the delay, setting the monitoring time, and detecting the hotspot. It maintains a table which defines task and their relative execution time. Whenever a hotspot occurs at first instance, analysis module reacts first. It first sets the interval time (T) to 1. It further sets the delay value that indicates that how long this hotspot occurs by using the data provided by monitoring module and its self-maintained table. If it finds sufficient number of occurrence, then it triggers the VMM process and sends control to the VMM module. VMM module selects the victim VM and sends it to the appropriate physical host. The working of the proposed RTRM system is shown in Algorithm 1.

Algorithm 1: Detection of Hotspot using RTRM system

List of Variables:
Res_req: resource requirement on physical host at a time
Res_thr: Threshold set for resource
RAM: total amount of primary memory of physical host
CPU: total amount of CPU of physical host
RAM_{th}: threshold value for RAM
CPU_{th}: threshold value for CPU
Current_CPU: current CPU utilization of the physical host
Current_RAM: current RAM utilization of the physical host
Obs: number of observations set by analysis module
C[]: array for keeping record of CPU utilization
R[]: array for keeping record of RAM utilization
k : counter variable
T : data monitoring interval

```
1: If Res_req>Res_thr then
2:      Set value of obs using analysis module
3:      Set T = 1 and k = 0
4:      While Res_req>Res_thr and k <= obs
5:              C[k] = Current_CPU
6:              R[k] = Current_RAM
7:              K = K+1
8:              Wait for T time
9:      End While
10:     If Res_req>Res_thr then
11:     set Used_CPU = avg(C) and Used_RAM = avg(R)
12:             if (Used_CPU >= CPUth and Used_RAM>= RAMth) then
13:                     trigger CPU and RAM based migration
14:             else if (Used_CPU>= CPUth) then
15:                     trigger CPU based migration
16:             else if (Used_RAM>= RAMth) then
17:                     trigger RAM based migration
18:             end if
19:     else
20:             set T = 5
21:     end if
22: End if
```

In the proposed RTRM system, RAM and CPU utilization is collected after interval of 5 s. We have performed experiment with the time interval of 5, 10, 15, and 20 s and found that proposed RTRM system performs best with time interval of 5 s. If utilization of CPU and RAM reaches the threshold value, the proposed system starts monitoring the data after interval of 1 s. We keep increasing the counter till the new monitoring data exceed the threshold value. Analysis module maintains table which defines task with corresponding required time for completion. Whenever a hotspot occurs, the monitoring value reports its cause to analysis module. Analysis module checks its table for the task and sets the delay value. RTRM system uses the dynamic delay value for each hotspot to minimize the false trigger percentage.

We have reserved 10% [5, 12] of the total computing resources (RAM, CPU) as the threshold value. Monitoring module monitors resources at physical level which provides the actual usage in real time.

4 Results and Discussion

To validate proposed RTRM system, we have developed an experimental setup. In this, we have in total 28 physical hosts. All the hosting devices are equipped with 4–64 GB of RAM and have dedicated 320 GB to 1 TB of Hard Disk. We have used Cisco 2960x Ethernet switch to establish a connection between devices. Nodes run NFS server to share an image file of VM to other nodes.

We have deployed total 140 virtual machines in which half of the machines are from Windows family and other are from Linux family. VMs have virtual memory from 4 GB to 512 MB depending upon OS type. The critical workload is collected on 140 virtual machines. The detail setup information for workload trace is given in Table 1.

To test the proposed RTRM system, we conducted experiments with empirical data. We set up an experimental environment with predefined workload on running VM. To generate load on running VMs, we have used workload benchmark. We have calculated the normal execution time of all of these benchmarks to set the delay value as shown in Table 2.

We have run totally 516 different test cases with all different combinations of these benchmarks. In these test cases, we try to develop hotspot situation on physical machine and check the performance of proposed RTRM system. We found hotspot 410 times out of 516. The proposed system is able to detect hotspot 328 times out of 410. In our results, we found that proposed RTRM system is able to detect hotspot in first 20 s on time with the accuracy of 80%. RTRM system gives best results on VM creation and ideal VM but performs worst in case of SPECjvm 2006 and KernalCompile benchmark.

To compare and evaluate the performance of our proposed technique, we have set the delay value equal to 10 to compare the results with the sandpiper [13] and artificial intelligence technique such as neural network predictor (NNP) [8] technique.

Sandpiper and NNP techniques give the best results of 76% and 84%, respectively, with the delay of 60 s to 70 s of historical data. In RTRM, we observed best results in the delay of 20 s with the accuracy of 80% as shown in Table 3. We can increase

Table 1 Critical workload trace collected in this work

Number of physical host	Number of virtual machines	Period of data collection	Storage type	Total RAM (GB)	Total cores
28	140	120 h	SAN and NAS	456	212

Table 2 Benchmarks used in our experiments with their workload type and respective delay

Benchmark name	Workload type	Delay
VM creation	CPU and RAM centric workload	20 s
Ideal VM	Constant use of CPU and RAM	20 s
SPECjvm2006	Generates load on CPU and RAM depending upon the passing arguments	40 s on passing value 10
Memtester	RAM intensive workload	40 s
Bzip2	CPU centric workload	20 s
SPECcpu2006	CPU centric workload	20 s
Kernel Compile	CPU and RAM centric workload	35 s

Table 3 Comparison between sandpiper, NNP, and proposed RTRM (1 delay = 10 s)

Technique	Delay							
	2 (%)	3 (%)	4 (%)	5 (%)	6 (%)	7 (%)	8 (%)	9 (%)
Sandpiper [29]	64	68	72	72	76	76	68	68
NNP [26]	76	76	80	80	84	84	80	80
Proposed RTRM	80	82	84	85	87	88	90	93

the accuracy of the proposed RTRM system by increasing delay value but it will also increase the delay of VMM trigger.

5 Conclusion

Cloud computing has taken the utility computing on the next level and opens new doors to the industry. Like all other technologies, it has its own limitations; VMM is one of them. Existing techniques detect the hotspot with the static interval time and static delay. In this paper, we have proposed RTRM technique, which does not use static interval time for data monitoring. The interval time changes from 5 s to 1 s in first occurrence of hotspot. We further set the value of delay according to the type of task. In our experimental setup, we have selected tasks which use intensive CPU, RAM, or both. We also proposed a table method which can be used to define the nominal time for each task running on physical host. To validate our proposed RTRM system, we set the value of delay to compare the results with existing sandpiper [13] and NNP [8] systems. Results show that RTRM system able to detect the hotspot in first 20 s with 80% of accuracy as compared to sandpiper 64% and NNP 76%. In future, we tend to increase the entries in the table and make these entries dynamic.

References

1. L.M. Vaquero, L. Rodero-Merino, J. Caceres, M. Lindner, A break in the clouds: towards a cloud definition, in *ACM SIGCOMM Computer Communication Review*, vol. 39, no. 1, pp. 50–55 (2009), http://doi.acm.org/10.1145/1496091.1496100
2. K. Ramakrishnan, P. Shenoy, J. Van der Merwe, Live data center migration across wans: a robust cooperative context aware approach, in *Proceedings of the 2007 SIGCOMM Workshop on Internet Network Management*, pp. 262–267 (2007), http://dx.doi.org/10.1145/1321753.13 21762
3. S.U. Muthunagai, C.D. Karthic, S. Sujatha, Efficient access of cloud resources through virtualization techniques, in *International Conference on Recent Trends in Information Technology*, pp. 174–178 (2012), http://dx.doi.org/10.1109/ICRTIT.2012.6206761
4. C. Chen, H. Zhang, Z. Yu, Y. Fan, L. Liu, A new live virtual machine migration strategy. Inf. Technol. Med. Educ. **1**, 173–176 (2012). https://doi.org/10.1109/ITiME.2012.6291274
5. M. Mishra, A. Das, Dynamic resource management using virtual machine migrations. IEEE Commun. Mag. **50**(9), 34–40 (2012). https://doi.org/10.1109/MCOM.2012.6295709
6. S.B. Shaw, A.K. Singh, Use of proactive and reactive hotspot detection technique to reduce the number of virtual machine migration and energy consumption in cloud data center. Comput. Electr. Eng. **47**, 241–254 (2015). https://doi.org/10.1016/j.compeleceng.2015.07.020
7. C. Zou, Y. Lu, F. Zhang, S. Sun, Load-based controlling scheme of virtual machine migration. Cloud Comput. Intell. Syst. **1**, 209–213 (2013). https://doi.org/10.1109/CCIS.2012.6664398
8. R. Goebel, W. Wahlster, Effective hotspot removal system using neural network predictor, in *Asian Conference on Intelligent Information and Database Systems*, vol. 9622, pp. 478–488 (2016), https://doi.org/10.1007/978-3-642-36543-0_49
9. www.Google.com, Google Data, https://commondatastorage.googleapis.com/clusterdata-201 1-2/SHA256SUM. Accessed 02 Aug 2017
10. Bitbrains, The Grid Workloads Archive, http://gwa.ewi.tudelft.nl/datasets/gwa-t-12-bitbrains. Accessed 10 Aug 2017
11. C. Chen, K. He, D. Deng, Optimization of the overload detection algorithm for virtual machine consolidation, in *Software Engineering and Service Science*, no. 1, pp. 207–210 (2016), http://dx.doi.org/10.1109/ICSESS.2016.7883050
12. S. Liyanage, S. Khaddaj, J. Francik, Virtual machine migration strategy in cloud computing, in *Distributed Computing and Applications for Business Engineering and Science*, pp. 147–150 (2016), https://doi.org/10.1109/dcabes.2015.44
13. T. Wood, P. Shenoy, A. Venkataramani, M. Yousif, Sandpiper: black-box and gray-box resource management for virtual machines q. Comput. Netw. **53**(17), 2923–2938 (2009). https://doi.org/10.1016/j.comnet.2009.04.014

SWS—Smart Waste Segregator Using IoT Approach

Payal Srivastava, Vikas Deep, Naveen Garg and Purushottam Sharma

Abstract In today's scenario in India, overloaded and unattended dump sites are a major concern, as compared to earlier times, as it is causing calamity and contamination to people living nearby. One way to solve this crisis is to segregate the waste at the dump sites and recycle/reuse it to increase the economy of the country, and reduce the load on the site. This paper describes an approach to create a device which shall be used to segregate the waste according to its possible use, known as Smart Waste Segregator (SWS) using Internet of Things (IoT) approach.

Keywords SWS · IoT · IR sensors · Inductive sensor · Capacitive sensor

1 Introduction

Any material which is not required by the proprietor, maker, or processor is waste. For the most part, the waste is characterized as toward the finish of the item life cycle and is discarded in landfills. Most organizations characterize waste as "anything that does not make esteem." In a typical man's eye, anything that is undesirable or not valuable is rubbish or waste [1].

However, logically taking all things considered on the planet are not wasted. All the parts of strong waste have some potential in the event that it is changed over or treated in a logical way. Thus, we can characterize strong waste as "Natural or inorganic waste materials created out of family or business exercises, that have lost

P. Srivastava (✉) · V. Deep · N. Garg · P. Sharma
Department of Information Technology, Amity University Uttar Pradesh, Noida, India
e-mail: srivastavapayal28@gmail.com

V. Deep
e-mail: vikasdeep8@gmail.com

N. Garg
e-mail: er.gargnaveen@gmail.com

P. Sharma
e-mail: psharma5@amity.edu

© Springer Nature Singapore Pte Ltd. 2019
H. Malik et al. (eds.), *Applications of Artificial Intelligence Techniques in Engineering*, Advances in Intelligent Systems and Computing 698,
https://doi.org/10.1007/978-981-13-1819-1_53

their incentive according to the primary proprietor yet which might be of incredible incentive to another person." There are around 593 locales and roughly 5,000 towns in India. The quantum of waste created in Indian towns and urban areas is expanding step by step because of its expanding populace and expanded GDP. The yearly amount of strong waste produced in Indian urban areas is expanding from six million tons in 1947 to 48 million tons in 1997 with a yearly development rate of 4.25%, and it is required to increment to 300 million tons by 2,047. Waste generation is inescapable in each home howsoever huge or less. Since the beginning of progress, humankind has continuously digressed from nature and today there has been an extreme change in the way of life of human culture. Coordinate impression of this change is found in the nature and amount of refuse that a group produces. We can arrange the waste or reuse the waste and can win cash through legitimate administration. Indian urban areas which are quick contending with worldwide economies in their drive for quick financial improvement have so far neglected to successfully deal with the immense amount of waste created. [1]

Also, it is known globally that India is the second most populous country in the world after China, but the seventh largest country in the world. It must be noted that in the current scenario, the scarcity of land has increased as well. It has been a norm in India and most developing countries that whatever wastes have been collected from various areas are to be dumped at the nearby designated sites. These sites have been unattended for a long time. Due to the presence of nonbiodegradable elements, the dumped waste is not able to decompose as well. As a result of these sites which are causing various epidemics to population surrounding it, also, overloading of these sites causes hazardous reactions which can result in landslides. Recently, the Government of India has allocated around 1,700 billion to segregate the waste collected, under the "Swachh Bharat Abhiyan" scheme. If we only use human beings to segregate the waste, the time and the cost for this will increase. Hence, it is necessary to optimize the following solution (Fig. 1).

Fig. 1 Overloaded dumpsite [2]

Several solutions have been proposed for the collection, segregation, and location of the garbage bins, but, ultimately, the waste is thrown at the dumpsite only. The aim of this paper is to describe an approach for making a device, SWS, through which wastes can be segregated in such a manner that it can be recycled, reused, or reduced, instead of discarding, thus utilizing the dumping site for better purpose. SWS shall be used to ease the workload of waste sorting and hence, saving the time as well.

Section 2 explains about some work and contribution done in the related field of waste segregation.

Section 3 describes the objective of the paper written.

Section 4 describes the proposed approach to resolve the mentioned problem statement and fulfill the given objective.

Section 5 mentions the advantage of the proposed automated system over the current manual solution.

Section 6 explains about the conclusion of the paper and the future scope of the proposed system, that is, how the system can be more efficient using another technique.

2   Literature Survey

Waste segregation is critical issue from last few decades. So, various techniques for waste management and segregation have been proposed at different levels in different parts of the world.

Yann Glouche and Paul Couderc (2013) mentioned the inefficient waste management in Europe. They proposed an approach in which smart bins are created to identify the type of waste materials and the required bin will open automatically. The waste must be brought in front of the bins, and it will be segregated based on its material type [3]. But the limitations felt were that the mechanism used for identifying the waste is the RFID tag, and RFID tags cannot be used, as wastes in India do not have those tags associated with any product.

Ruveena Singh, Dr. Balwinder Singh (2015) have implemented a smart dustbin in which waste is segregated into biodegradable and nonbiodegradable type with the help of sensors [4]. This bin can be used for classrooms, office, etc. Again, the authors did not mention how to recycle or utilize the waste.

Kumar, N. S., Vuayalakshmi, B., Prarthana, R. J., and Shankar A. (2016) defined a mechanism of waste management by alerting the municipal sector about the level of garbage present in the locality [5]. But, here again, RFID tags are used. (Could not be implemented in India), no waste classification and the waste collected cannot be utilized further and hence, it will be dumped at some site.

Bharadwaj, A. S., Rego, R., and Chowdhury, A. (2016) have proposed a waste management system using IoT approach. The dustbins associated with GPS are kept at some location and waste is collected. IR sensors are used to determine if the dust bin is full and need to be emptied or not, and the status of the dustbin is determined using the application administered by the municipality [6].

Though it mentions that it has different dustbins for different types of waste, it does not guarantee that correct waste will necessarily go to its designated bin. It is not mentioned in the paper that the segregated waste will go to its recycling and hence, the waste will ultimately become a garbage dump. Also, the following proposal is not for indoors.

Bharadwaj, B., Kumudha, M., and Chaithra, G. (2017) have provided an approach to collect dry waste and wet waste separately and store it underground. It is mainly used for apartment-type buildings and/or industries where the garbage is collected underground and the bins are emptied periodically using conveyer belt [7].

But the reason behind collecting dry waste and wet waste is not specified clearly. Even though the bins are emptied in a periodic manner automatically, it is not mentioned that how the underground waste will be recycled or utilized.

3   Objective

The objective is to automate the waste segregation technique using the IoT approach in such a way that 95% of it can be recycled and reused in different processes, thus reducing the amount of waste dumped at the site and utilizing that land for other purposes.

4   Proposed System

The device SWS shall have the following components: The architecture of the SWS device is shown in Fig. 2.

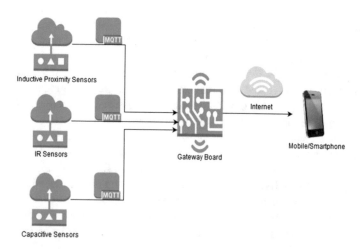

Fig. 2 Basic architecture of SWS

Sensors: There will be three sensors used. The object shall be categorized based on its physical properties.

- **Inductive proximity sensors**—It detects only metallic objects.
- **Capacitive sensor**—It detects metallic as well as nonmetallic objects.
- **IR sensors**—It detects only opaque objects.

Gateway (Intel Edison Board/Arduino Board): The data sent by the sensors cannot be sent to the Internet, and it travels in low networks such as LOWPAN, COAP, etc. The purpose of the gateway is to send the input given by the sensors from low network to the Internet. The gateway enables connection between two different networks.

Front-end software: Software shall be used to process the data collected from the sensors via Internet.

Mobile application: After processing of the data, it shall be presented to the user with the help of a mobile application.

Below is the flowchart of how the SWS device shall perform the operation (Fig. 3).

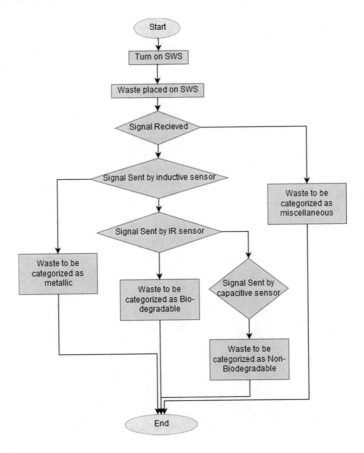

Fig. 3 Flowchart

5 Advantages Over Manual Solution

The list of advantages of using the device over manual segregation are as follows (Table 1 and Fig. 4).

6 Conclusion and Future Scope

Due to rise in urbanization, the waste is increasing very fast. Therefore, waste management is the vital need to protect the environment. The technological growth and innovation can contribute to this vital aspect to achieve environmental sustainability [4]. Various techniques have been proposed for segregating waste at the level where

Table 1 Manual versus SWS

Manual (using manpower)	SWS
Not easy to identify the category of the waste	The category of the waste will be shown as soon as it is kept on the SWS device
More humans will be required to segregate the waste	Less human will be required to segregate the waste
More time will be taken for waste segregation	Less time will be taken for waste segregation
Risk of contamination to the people segregating the waste will increase	Since people are spending less time on segregating the waste, the contamination risk will also decrease
Due to increased number of people working, time taken, and health hazards, the cost of implementing the process will also increase in the long term	Though there will be some cost involved in providing the SWS device as well as the people operating it, in terms of long term, the cost of implementation will be less comparatively

Fig. 4 Attempt to segregate waste manually [8]

it is collected. But the waste, which has been ignored for a long time at the landfills, cannot be ignored today as it is causing harm to the entire society in the guise of epidemics. Most of the landfills are continuously being overloaded which has resulted in harmful chemical reactions within it and thus, landslides occur affecting the daily life of people residing nearby. If these issues are now ignored, then it will cause more harm to our surroundings.

The device SWS, which is IoT based, will identify the waste objects based on its physical properties. It is a well-known fact that IoT is one of the recent technologies to be introduced in the market. But it is not necessary that SWS will be a foolproof device. There are no known sensors available in the market as of now, which will exclusively identify the waste type. For example, a waste object which was supposedly nonbiodegradable was shown miscellaneous by the SWS, decreasing the possibility of its utilization. Hence, based on the frequency of the waste object found at the dumpsite or landfills, the concept of image processing can be introduced. Using this approach, a camera shall be fitted on the SWS, which will capture the waste image pattern and process it accordingly. Based on the result of the pattern, category of the waste shall be computed and shown to the user on the mobile application. In this way, the amount of miscellaneous waste can be reduced significantly.

References

1. H.O. Egware, O.T. Ebu-nkamaodo, G.S. Linus, Experimental determination of the combustion characteristics of combustible dry solid wastes. Res. J. Eng. Environ. Sci. 1(1), 154–161 (2016)
2. http://indiatoday.intoday.in/story/dumping-garbage-overflow-delhi-dda-mcd-south-northcorpo ration/1/346378.html
3. Y. Glouche, P. Couderc, in *A Smart Waste Management with Self-Describing Objects*, ed. by W. Leister, H. Jeung, P. Koskelainen. The Second International Conference on Smart Systems, Devices and Technologies (SMART'13), June 2013, Rome, Italy. 2013
4. R. Singh, D.B. Singh, Design and Development of Smart Waste Sorting System (2015)
5. N.S. Kumar, B. Vuayalakshmi, R.J. Prarthana, A. Shankar, in *IOT Based Smart Garbage Alert System Using Arduino UNO*. Region 10 Conference (TENCON), 2016 IEEE. IEEE (2016, November), pp. 1028–1034
6. A.S. Bharadwaj, R. Rego, A. Chowdhury, in *IoT Based Solid Waste Management System: A Conceptual Approach with an Architectural Solution as a Smart City Application*. India Conference (INDICON), 2016 IEEE Annual. IEEE, (2016, December), pp. 1–6
7. B. Bharadwaj, M. Kumudha, G. Chaithra, in *Automation of Smart Waste Management using IoT to support "Swachh Bharat Abhiyan"-a practical Approach*. 2017 2nd International Conference on Computing and Communications Technologies (ICCCT). IEEE (2017, February), pp. 318–320
8. http://www.hindustantimes.com/delhi-news/methane-trapped-beneath-makes-ghazipurlandfill-a-ticking-time-bomb/story-dByD7XfaiL9FPhh42oR26J.html

Securing Onion Routing Against Correlation Attacks

Saba Khanum, Sudesh Pahal, Aayush Makkad, Akansha Panwar
and Anshita Panwar

Abstract In this paper, we present a network model based on traditional onion routing model but engineered to make correlation attacks much more difficult. The design of the network makes traditional packet counting and volume attacks less feasible. We assume a passive attacker working with partial knowledge of the network traffic trying to use various vulnerabilities and attacks to find communication partners in an onion routing network circuit. Our approach uses a change in the design of the network model which results in enhanced security and anonymization against an attacker using correlation to find communication partners.

Keywords Onion routing · Tor network · Anonymity · Correlation attack
Timing attack · Packet counting

1 Introduction

1.1 Anonymity Over the Internet

Anonymous as defined by Cambridge dictionary means "made or done by someone whose name is not known or made public." Over the entire Internet, users are known by the computer they use or more specifically by their physical and logical address of the computer they use. So, in order for a user to be anonymous over the Internet his physical and logical addresses must not be known to someone he needs to contact with.

It sounds simple but is really complex when it comes to implementing a protocol to anonymize a user over the Internet; there are certain protocols/tools available to do so; one of them is onion routing.

S. Khanum (✉) · S. Pahal · A. Makkad · A. Panwar · A. Panwar
MSIT, New Delhi, Delhi, India
e-mail: sabakhan.msit@gmail.com

© Springer Nature Singapore Pte Ltd. 2019
H. Malik et al. (eds.), *Applications of Artificial Intelligence Techniques
in Engineering*, Advances in Intelligent Systems and Computing 698,
https://doi.org/10.1007/978-981-13-1819-1_54

1.2 Onion Routing

Onion routing was devised by Goldschlag, Reed, Syverson in [1, 2] as a solution for anonymous connections. Onion routing is a technique for anonymous communication over a computer network. In an onion network, a message transmits through a number of hops described by directory node but the way it differs from any other routing mechanism is that the message is encrypted into n layers of encryption where n is the number of hops (onion router). Each router on receiving the message is able to peel off a single topmost layer of encryption through public–private key cryptography, which is enough to tell it just about the succeeding and the preceding hop in circuit. At the last hop in the circuit, last router is able to understand which server to contact to as all layers of encryption were stripped off [3, 4].

The elegance of onion routing is that each cell in the circuits looks exactly the same, so no node knows if the preceding node was the initial node or an intermediary.

Usually, a symmetric key exchange algorithm like is used, as symmetric key exchange is faster than asymmetric key exchange.

Each hop exchanges a public key with entry node, which wraps up the entire message in layers of encryption which are further decrypted using private key at each respective hop. Consider there are n nodes so there are n public keys {pk1, pk2, pk3 ... pkn}, the message "m" is encrypted in the fashion {pkn, p[k(n − 1) ... pk2, pk1} (Fig. 1) [5].

Fig. 1 Onion routing network model

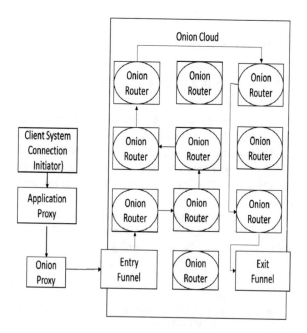

1.3 Correlation Attack

If an attacker somehow gets hold of our knowledge of who the entry and exit nodes of an onion circuit are, he can observe traffic to conclude who is talking to which server which completely de-anonymize a user.

Correlation attacks are done by analyzing network traffic and finding patterns in frequency, timing, volume, etc., to find communication partners.

There are a number of attacks/ways through which an attacker can correlate.

The attacks are as follows:

Message coding attacks
By decrypting the message, an attacker can get to know about the contents of the message and de-anonymize the user but in onion routing (OR) message is encrypted in layers of encryption which makes it extremely difficult to encrypt all layers of encryption to find out who is talking to whom.

Timing attacks
By observing the time taken to evaluate a cryptographic function, an attacker can cripple the entire network. Fortunately, there are already mechanisms available at disposal. One of them is given in the paper by Joan Feigenbaum, Aaron Johnson, and Paul Syverson [6].

Packet volume attack
If the anonymity system allows packets of different sizes, the attacker can downsize the set of possible receivers of a packet to these nodes which received a packet with the same size [7].

Packet counting attack
While counting packets sent and received by the users of an anonymity system, the attacker might link communication partners [7].

Slotted packet counting
Try to correlate by measuring alteration of sent and received packets over time [7].

2 Literature Survey

By going through a number of research papers, we were able to find methodologies to prevent some of the attacks mentioned:

Timing attacks:
Using Joan Feigenbaum, Aaron Johnson and Paul Syverson methodology timing attacks can become less feasible [6].

Attacks made less feasible by our model:

1. Packet volume attacks,
2. Packet counting attacks, and
3. Slotted packet counting attacks (total packets).

3 Our Contribution—Secure Model

To understand the secure model, first traditional onion routing must be understood in which

- Client sends a GET/PUT request to server which passes through three hops on the network.
  A key exchange takes place between client and each hop which provides client with three keys K1, K2, and K3. Client encrypts the messages K3, then K2, and then K1 which are further decrypted at their respective hops, so each intermediary hop just knows about the node which provided him the messages and the node next to it and nothing else.
- The same procedure is repeated for reply from the server (Fig. 2).

  The way our secure model works is as follows:
  The whole procedure is same till the very last hop but at the last hop the incoming bytes get divided into two different buffers based on if they are even or odd indexes, these buffers are sent as messages as two parallel hops, and IP address of third hop is divided into two parts and sent in encrypted format (Fig. 3).

3.1 Packet Volume Attacks

If the anonymity system allows packets of different sizes, the attacker can downsize the set of possible receivers of a packet to these nodes which received a packet with the same size. If the anonymity system allows packets of different sizes, the attacker can downsize the set of possible receivers of a packet to these nodes which received a packet with the same size [7].

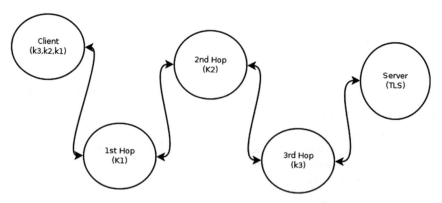

Fig. 2 Traditional onion routing network model

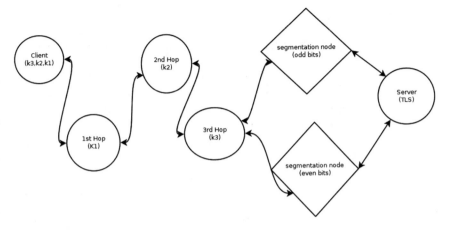

Fig. 3 Secure model

Table 1 Packet volumes (secure model)

2nd hop	3rd hop	4th hop (parallel)	5th hop (parallel)
5	5	3	3

Table 2 Packet volumes (traditional onion routing)

1st hop	2nd hop	3rd hop
5	5	5

Message "hettt" was sent by client, and following conclusions were noted in their length/volume (Table 1):

And in a traditional OR model, volume of packets was as follows (Table 2):

So, it is quite clear that secure model is relatively way more secure than traditional OR model from packet volume attacks.

3.2 Packet Counting Attacks

Traditional packet counting attacks are not anymore effective when a passive attacker is assumed [7].

Common packet counting attacks work on the principle of finding some similarity between the packets sent and received in a network to find communication partners. These approaches are not used as described in the paper [7].

But if we include time as a factor and divide it into time slots and then work on correlating, then that is called as slotted packet counting as described in paper [7].

Since we are taking just a single circuit under consideration, each and every packet incoming or outgoing will be observed and bifurcated as per time slots.

Table 3 Packet count (secure model)

X	N1	Server	N4	N5
T1	1	7	3	3
T2	2	13	6	8
T3	14	0	11	7
T4	3	0	0	2
T5	2	0	0	0
T6	1	0	0	0
T7	3	0	0	0
T8	5	0	0	0
T9	2	0	0	0

Table 4 Correlation coefficients (secure model)

Correlation	N1	Server	N4	N5
N1	1			
Server	−0.26647	1		
N4	0.75524	0.37560	1	
N5	0.47868	0.67709	0.89945	1

Observations for secure model were as follows: each cell entry depicts total number of packets by that node in that specific time slot (Table 3):

Their correlation coefficients are used to evaluate strength and direction of the linear relation of two variables and are evaluated as per the formula [7]:

$$\partial = \frac{cov(X, Y)}{\sigma x \sigma y} = \frac{E(X - \mu x)(Y - \mu y)}{\sigma x \sigma y}$$

where it is in the range [−1, 1], 0 being no correlation at all and 1, −1 being high correlation. In the formula, μi stands for expected value (Table 4).

Same attack for traditional OR model resulted in the following observations (Table 5):

Table 5 Packet count (traditional OR)

X	N1	Server	N3
T1	1	1	1
T2	2	2	2
T3	1	2	4
T4	2	2	0
T5	1	2	0
T6	0	0	0
T7	0	0	0
T8	0	0	1
T9	2	0	1

Table 6 Correlation coefficients (traditional OR)

Correlation	N1	Server	N3
N1	1		
Server	0.57735	1	
N3	0.21821	0.37796	1

Graph 1 Comparison of correlation coefficient

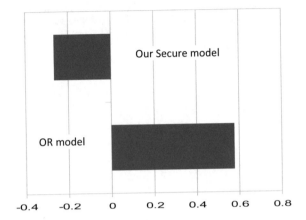

Correlation coefficients for traditional OR model are as follows (Table 6 and Graph 1):

As it is clearly visible that correlation coefficient for first node and server was −0.26647 in secure model and 0.57735 in traditional OR, we can safely say that our secure model is almost 53% more secure here.

4 Conclusion and Future Work

The results safely conclude that our secure model is relatively more secure than traditional OR model against different methodologies for correlation. The model is still weak against slotted packet counting attacks in which sent and received packets outside of anonymity system are considered separately. One way to improve is to shift packets in a time slot to another time slot by randomization.

References

1. D.M. Goldschlag, M.G. Reed, P.F. Syverson, in *Hiding Routing Information*, ed. by R.J. Anderson. Proceedings of the 1st International Workshop on Information Hiding (Cambridge, 1996), pp. 137–150. LNCS 1174
2. P.F. Syverson, D.M. Goldschlag, M.G. Reed, in *Anonymous Connections and Onion Routing*. IEEE Symposium on Security and Privacy (Oakland, California, 1997) pp. 44–54

3. R. Dingledine, N. Mathewson, P. Syverson, Tor: The Second-Generation Onion Router (2004)
4. S. Mauw, J. Verschuren, E.P. de Vink, A formalization of anonymity and onion routing. Proc. ESORICS **2004**, 109–124 (2004)
5. H. Mathakar, A.S. Engineer, A survey on Darknet and various techniques of onion routing, vol. 6, no. 1, pp. 14–18 (2017)
6. F. Joan, J. Aaron, S. Paul, Preventing active timing attacks in low-latency anonymous communication. https://doi.org/10.1007/978-3-642-14527-8_10
7. V. Fusenig, E. Staab, U. Sorger, T. Engel, Slotted packet counting attacks on anonymity protocols. Conf. Res. Pract. Inf. Technol. Ser. **98**(January), 53–59 (2009)

Preference-Oriented Group Recommender System

Nirmal Choudhary and K. K. Bharadwaj

Abstract Group recommender systems (GRSs) are most notable web personalization application for a group of users to satisfy preferences of all users optimally with a sequence of items. To generate the effective recommendation for a group of users, the system must fulfill preferences of all members equally. Our proposed preference-oriented group recommender system (POGRS) analyzes each member's preference in the group for the goodness of group recommendation. The idea behind our approach is that each member in a group would try to fulfill his/her own preferences in decision-making process. We incorporated argumentation process to realize the functioning of recommender system in order to make common consensus point to achieve more realistic way of recommendation. Experiments are performed on MovieLens dataset, and the results of comparative analysis of proposed POGRS with baseline GRSs techniques clearly demonstrate the supremacy of our proposed model.

Keywords Recommendation mechanism · Group consensus · Group decision-making · Preference-oriented group recommender systems

1 Introduction

There are an enormous number of applications of web personalization in which recommender system (RS) [18, 19] is the most remarkable application, which will make straightforward the content detection and information access in most valuable way that would be appropriate to users.

N. Choudhary (✉) · K. K. Bharadwaj
School of Computer and Systems Sciences, Jawaharlal Nehru University,
New Delhi 110067, India
e-mail: nirmal32_scs@jnu.ac.in

K. K. Bharadwaj
e-mail: kbharadwaj@gmail.com

© Springer Nature Singapore Pte Ltd. 2019
H. Malik et al. (eds.), *Applications of Artificial Intelligence Techniques in Engineering*, Advances in Intelligent Systems and Computing 698,
https://doi.org/10.1007/978-981-13-1819-1_55

Generally, the aim of a conventional RS [1] is to study the activities of users and according to that predict the rating of items and then propose the items having maximum rating that users might like [3]. But there are various activities that involve a group of users, i.e., watching a movie [16], going to a restaurant [14], purchasing a book [12], etc. In such circumstances, RSs should consider the taste and preferences of each member of the group [3, 19]. This type of system is called a group recommender system (GRS). This is an active research area in the field of RSs. Recently, large number of GRSs has been developed in order to deal with the challenges of making recommendations for a group of members.

Some acknowledged GRSs are *Polylens* [16], *MusicFX* [15], *Intrigue* [2], and *Travel Decision Forum* [14].

The major objective of group recommender system is to identify each user's preferences and then find out a concession point on which all the members of group agree equally [5]. This concession point can be achieved either by integrating the recommendations or by generating the group profile for each individual in the group. Therefore, the success of a GRS is determined based on how individual preferences are aggregated into group preferences while considering the behavior of each member. To come up with the same preference, we have developed a preference-based model. The idea behind our proposed work is that each member in a group would try to fulfill his/her own preferences while reaching a mutually acceptable agreement. According to influential theory, each member gives highest weight to his/her top preferences compared to the other members' preferences in the group. In this paper, we focus on the excellence of recommendation as well as the precision of the predicted ratings. The experimental results of computational experiments are presented that establish the superiority of our proposed POGRS model over baseline GRS techniques for MovieLens dataset.

The next section includes existing group recommendation methods. Proposed POGRS model to enhance the existing group recommendation is detailed in Sect. 3. The experimental results are given in Sect. 4. At last, Sect. 5 summarizes our work and specifies some ongoing future research directions.

2 Related Work

The major research on GRS is to identify each user's preferences and then come up with a concession point that equally satisfies all the members of group. The purpose of group recommendation is to generate and aggregate the preferences of all individual users [7]. As explained in [4, 9, 10], the major approaches to generating the preference aggregation are (a) merging the individual recommendations, i.e., creating a joint user profile for all the members and make recommendation for whole group as group profile (b) aggregation of individual rankings, i.e., aggregate the recommendations made for each individual in the group.

GRSs are exploited to work in various domains: movies, restaurant, TV program [20], music, etc. *Polylens* [16], an extension of *MovieLens*, is used for recommending movies for a group of users by trying to satisfy the least satisfied member of the group, i.e., by using least misery strategy. Another notable content-based recommender system is *Pocket Restaurant Finder* [14], which is used for recommending restaurants to a group of users based on their area and culinary behavior. Some other models are *FlyTrap* [6] that recommends music to be played in public place, or *LET'S BROWSE* [13] which suggests web pages to a group of individuals who are browsing the web together. All the aforementioned models overlook the priorities of individual users in the group, i.e., they consider all user's preferences in the same way. In our proposed model, we consider the impact of most influential user in the group decision-making process. However, majority of the GRSs aggregate the individual preferences without considering the preference order of group members. Another work in group consensus is to provide recommendation to arrange tours, and visits for a group of travelers. *Travel Decision Forum* [14] and *Intrigue* [2] facilitate a group of tourists to plan their vacations jointly. The system takes into account the interaction among individuals in the group to build recommendation. It allows the users to communicate in order to reach a mutually acceptable agreement. For the goodness of recommendation, we use the status of users by exploiting the preferences of every user in the group. An interesting work proposed by Baltrunas et al. [3] to examine the impact of group size and group homogeneity or the consequences of different weighting models to build group consensus.

In our proposed work, we use random and homogeneous groups with different sizes. According to influential theory, each member gives highest weight to his/her top preferences compared to the other members' preferences in the group. Hence, we have considered preferences of users to enhance the quality of group recommendation. The status of user specifies the influence of that user in the group. The user with higher status is the most influential user in the group. Our proposed work relies on group size and homogeneity with the inclusion of status of each individual in the group. To the best of our knowledge, we have not found any work that integrates these factors in group decision-making process. Therefore, we believe that our proposed model incorporates argumentation process to realize the functioning of recommender system in order to make common consensus point which achieves more realistic way of recommendation. This is the main objective of our work.

3 Proposed Preference-Oriented Group Recommender System (POGRS)

This section presents the details of proposed algorithm to find recommended preferred item list for group of users. Our proposed model involves twofold. Before presenting the details, we have described some preliminaries for better understanding the problem description.

Preliminaries

Let G be a group of n users $U = \{u_1, u_2 \ldots u_n\}$ and m items $I = \{i_1, i_2, \ldots, i_m\}$. Let X be the user–item rating matrix where the users express their ratings for items. A preference function $P_{u_j}(I)$ specifies the preference of jth user for the items in I [19]. Specifically, if jth user prefers i_a over i_b, then it is denoted as $P_{u_j}(i_a) > P_{u_j}(i_b)$. If user j has same preferences for two items i_a and i_b, then it is denoted as $P_{u_j}(i_a) = P_{u_j}(i_b)$. Therefore, item having highest rating represents the top preference of the item in that users' preference list.

The detailed description of our proposed POGRS model is as follows:

Step 1: First, we have calculated the preference list of each individual in the group based on the rating provided by the user to the items and after prioritizing preferences of each user, user preference matrix Y is constructed.

Step 2: Next, we have assigned a weight factor to top 3 items from user preference matrix [17].

Step 3: An influential user is chosen based on the status of user in the group and after selecting the influential user, the preference list of that user is recommended to the group.

3.1 Calculation of Individual Users' Preference

The rating given by a user to an item specifies the interest of the user in that item. Individual preferences are drawn from the user–item rating matrix X. Let $L_{u_j}^i$ denote the ith preferred list of user u_j. Each user u_k has a ranked list of items L_{u_j} to express his/her preferences. The user preference matrix X specifies the degree of interest of each user in U for each item in I, **where** rows represent users and the columns represent the order of preferred items of each user. The ratings are given in a scale of $[1, 5]$, where

$$A = \begin{cases} 5 & StronglyAgree \\ 4 & Agree \\ 3 & Neutral \\ 2 & Disagree \\ 1 & StronglyDisagree \end{cases} \tag{1}$$

3.2 Assignment of Weight Factor

According to influential theory, each member gives highest weight to his/her top preferences compared to the other members' preferences in the group. We classify three top preferred items in the group to be most preferred item (*MP_Item*), preferred item (*P_Item*), and least preferred item (*LP_Item*) as follows:

$$MP\_Item = \left\{ u_k : w(X_j^1) = 1 \right\} \tag{2}$$

$$P\_Item = \left\{ u_k : w(X_j^2) = 0.5 \right\} \tag{3}$$

$$LP\_Item = \left\{ u_k : w(X_{u_j}^3) = 0.25 \right\} \tag{4}$$

As aforementioned, the most preferred item will contribute with weight factor of 1, whereas the least preferred item will contribute with weight factor of 0.25.

3.3 Computation of Status

Here, higher status of a user implies that the user has more similar preferences with other users in the group. The similar users boost the status of each other in the group. The status of every user in the group is calculated in threefold:

The status of most preferred item is computed by the following equation:

$$S_{u_j}^{MP} = \frac{\sum_{i=1}^k score\ of\ user\ j\ in\ the\ 1st\ preferred\ list\ of\ all\ users}{total\ number\ of\ users} * 1 \tag{5}$$

where status of user *j* denotes the number of occurrences of user u_j and $S(u_j)$ symbolizes the status of *j*th user in the group. The status of preferred item is computed as

$$S_{u_j}^{P} = \frac{\sum_{i=1}^k score\ of\ user\ j\ in\ the\ 2nd\ preferred\ list\ of\ all\ users}{total\ number\ of\ users} * 0.5 \tag{6}$$

The status of least preferred item is computed as

$$S_{u_j}^{LP} = \frac{\sum_{i=1}^k score\ of\ user\ j\ in\ the\ 3rd\ preferred\ list\ of\ all\ users}{total\ number\ of\ users} * 0.25 \tag{7}$$

The overall status of a user is given as the sum of status of most preferred item, preferred item, and least preferred item.

$$S(u_j) = S_{u_j}^{MP} + S_{u_j}^{P} + S_{u_j}^{LP} \tag{8}$$

The user having highest status in the group is considered as the most influential user and ranked list according to that user is recommended to the group as group profile.

4 Experiments

4.1 Dataset Description

We have conducted experiments on MovieLens[1] dataset which consist of 100,000 ratings on 1682 movies provided by 943 users, to measure the performance of our proposed model. The ratings are in the range of $[1, 5]$ where 1 and 5 indicate "strongly disagree" and "strongly agree" rating, respectively. Every user has rated minimum 20 movies in the dataset. We have applied our proposed model for each group to generate the group recommendation. We have considered only three highest preferences of every user while making group profile [17]. In our experimental setup, we have randomly selected 30 users from dataset and divided them into three groups of sizes 2, 4, 6, 8, and 10.

4.2 Evaluation Metrics

We have employed two evaluation metrics to examine the performance of our proposed group recommendation model, that is, *Mean Absolute Error (MAE)* and *Normal Discounted Cumulative Gain (nDCG)*.

Mean Absolute Error: The MAE measures the deviation of generated group profile from the actual preferences of users [11]. The MAE for a user is computed as

$$MAE(u_j) = \frac{1}{C_i} \sum_{l=1}^{|C_i|} \left| pr_{j,l} - r_{j,l} \right| \qquad (9)$$

where C_i is the cardinality of rating set and j is the item. $pr_{j,l}$ is the predicted preference for an item, and $r_{j,l}$ is the true preference for an item in the preference matrix. The total MAE for the group of n users is computed as follows:

$$MAE = \frac{1}{n} \sum_{i=1}^{n} (u_i) \qquad (10)$$

Normal Discounted Cumulative Gain: Group recommendation techniques generate a ranked list of items, i.e., $r_1, r_2, \ldots, r_k$. The true rating of user u for item i_j is given by r_{ui_j}. *IDCG* specify to highest achievable gain value for user u. Discounted cumulative gain (*DCG*) and *nDCG* at kth rank are calculated respectively as follows [3]:

$$DCG_k^u = r_{ui_1} + \sum_{i=2}^{k} \frac{r_{ui_j}}{log_2(i)} \qquad (11)$$

---

[1]http://movielens.org.

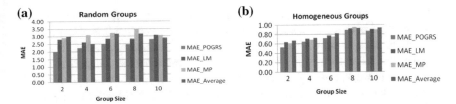

Fig. 1 MAE of preference-oriented group recommendation. **a** Random groups. **b** Homogeneous groups

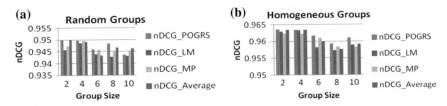

Fig. 2 nDCG of preference-oriented group recommendation. **a** Random groups. **b** Homogeneous groups

$$nDCG = \frac{DCG_k^u}{IDCG_k^u} \qquad (12)$$

We performed an experiment to compare the effectiveness of our proposed model for groups with varying sizes (random groups) and for groups with high inner group similarity (homogeneous groups) against least misery (LM) strategy, most pleasure (MP) strategy, and average strategy [3]. The results shown in Fig. 1 evidently demonstrate that the MAE of proposed model is much lesser than that of other GRS techniques. Also, as the size of the group increases, the deviation between actual and predicted recommendation also increases. This deviation is bigger in random groups as compared to groups with high inner group similarity.

Figure 2 shows the nDCG for random groups and groups with high inner group similarity. The gain of smaller size groups is higher than bigger size groups because it is harder to find out a concession point for bigger size groups. In addition, high similarity groups have better gain value as compared to random groups. Moreover, the gain value of proposed model is higher than other GRS techniques.

5 Conclusion and Future Directions

In this paper, we have proposed a preference-oriented group recommender system (POGRS) for making group recommendation while considering each user's preference for the items. Our proposed POGRS approach progresses from "individual welfare maximization" to "social welfare maximization". The results presented on

MovieLens dataset provide experimental evidence that exhibits the efficiency of proposed method in terms of enhanced accuracy. We observed that it is easier to make recommendation for homogeneous groups than for random groups. Also, it is simple to make good recommendations for smaller size groups rather than bigger size groups. As for further work, it would be interesting to analyze the performance of our model with other datasets. In this work, we employed LM, MP, and average strategy to generate group recommendations. In future, we would like to explore other aggregation strategies to investigate the effectiveness of our proposed approach. Additionally, we plan to incorporate trust–distrust strategies [4, 8] during the group decision-making process to further improve the proposed method.

References

1. G. Adomavicius, A. Tuzhilin, Toward the next generation of recommender systems: a survey of the state-of-the-art and possible extensions. IEEE Trans. Knowl. Data Eng. **17**(6), 734–749 (2005). https://doi.org/10.1109/TKDE.2005.99
2. L. Ardissono, A. Goy, G. Petrone, M. Segnan, P. Torasso, Intrigue: personalized recommendation of tourist attractions for desktop and hand held devices. Appl. Artif. Intell. **17**(8–9), 687–714 (2003). https://doi.org/10.1080/713827254
3. L. Baltrunas, T. Makcinskas, F. Ricci, Group recommendations with rank aggregation and collaborative filtering, in *Proceedings of the Fourth ACM Conference on Recommender Systems*, 2010, pp. 119–126. https://doi.org/10.1145/1864708.1864733
4. K.K. Bharadwaj, M.Y.H. Al-Shamri, Fuzzy computational models for trust and reputation systems. Electron. Commer. Res. Appl. **8**(1), 37–47 (2009). https://doi.org/10.1016/j.elerap.2008.08.001
5. I. Christensen, S. Schiaffino, M. Armentano, Social group recommendation in the tourism domain. J. Intell. Inf. Syst. 1–23 (2016). https://doi.org/10.1007/s10844-016-0400-0
6. A. Crossen, J. Budzik, K.J. Hammond, Flytrap: intelligent group music recommendation, in *Proceedings of the 7th International Conference on Intelligent user Interfaces* (ACM, Jan 2002), pp. 184–185. https://doi.org/10.1145/502716.502748
7. I. Garcia, S. Pajares, L. Sebastia, E. Onaindia, Preference elicitation techniques for group recommender systems. Inf. Sci. **189**, 155–175 (2012). https://doi.org/10.1016/j.ins.2011.11.037
8. J. Golbeck, Generating predictive movie recommendations from trust in social networks. Trust Manag. 93–104 (2006). https://doi.org/10.1007/11755593_8
9. A. Jameson, B. Smyth, Recommendation to groups. *The Adaptive Web* (Springer, Berlin, 2007), pp 596–627
10. A. Jameson, More than the sum of its members: challenges for group recommender systems, in *Proceedings of The Working Conference on Advanced Visual Interfaces* (ACM, 2004), pp. 48–54. https://doi.org/10.1145/989863.989869
11. V. Kant, K.K. Bharadwaj, Fuzzy computational models of trust and distrust for enhanced recommendations. Int. J. Intell. Syst. **28**(4), 332–365 (2013). https://doi.org/10.1002/int.21579
12. J.K. Kim, H.K. Kim, H.Y. Oh, Y.U. Ryu, A group recommendation system for online communities. Int. J. Inf. Manage. **30**(3), 212–219 (2010). https://doi.org/10.1016/j.ijinfomgt.2009.09.006
13. H. Lieberman, N. Van Dyke, A. Vivacqua, Let's browse: a collaborative browsing agent. Knowl. Syst. **12**(8), 427–431 (1999). https://doi.org/10.1145/291080.291092

14. J.F. McCarthy, Pocket restaurant finder: a situated recommender system for groups, in *Workshop on Mobile Ad-Hoc Communication at ACM Conference on Human Factors in Computer Systems*, 2002.
15. J.F. McCarthy, T.D. Anagnost, MusicFX: an arbiter of group preferences for computer supported collaborative workouts, in *Proceedings of ACM Conference on Computer supported Cooperative Work*, 1998, pp. 363–372. https://doi.org/10.1145/289444.289511
16. M. O'connor, D. Cosley, J.A. Konstan, J. Riedl, PolyLens: a recommender system for groups of users, in *ECSCW*, 2001, pp. 199–218. https://doi.org/10.1007/0-306-48019-0_11
17. L. Quijano-Sanchez, J.A. Recio-Garcia, B. Diaz-Agudo, G. Jimenez-Diaz, Social factors in group recommender systems. ACM Trans. Intell. Syst. Technol. (TIST) **4**(1), 8 (2013). https://doi.org/10.1145/2414425.2414433
18. P. Resnick, H.R. Varian, Recommender systems. Commun. ACM **40**(3), 56–58 (1997). https://doi.org/10.1145/245108.245121
19. A. Salehi-Abari, C. Boutilier, Preference-oriented social networks: group recommendation and inference, in *Proceedings of the 9th ACM Conference on Recommender Systems* (ACM, Jan 2015), pp. 35–42. https://doi.org/10.1145/2792838.2800190
20. Z. Yu, X. Zhou, Y. Hao, J. Gu, TV program recommendation for multiple viewers based on user profile merging. User Model. User-Adap. Interact. **16**(1), 63–82 (2006). https://doi.org/10.1007/s11257-006-9005-6

Optimal Fuzzy C-Means Algorithm for Brain Image Segmentation

Heena Hooda, Om Prakash Verma and Sonam Arora

Abstract Segmentation plays a vital role in medical image processing. Manual image segmentation is very tedious and time-consuming. Also, results of manual segmentation are subjected to errors due to huge and varying data. Therefore, automated segmentation systems are gaining enormous importance nowadays. This study presents an automated system for segmentation of brain tissues from brain Magnetic Resonance Imaging (MRI) images. Segmentation of three main brain tissues is carried out, namely, white matter, gray matter, and cerebrospinal fluid. In this work, we performed the initialization step for Fuzzy C-means (FCM) clustering algorithm using Ant Colony Optimization (ACO). Also, Mahalanobis distance metric is used instead of Euclidean distance metric in clustering process to avoid any relative dependency upon the geometrical shapes of different clustering classes. The results of the system are evaluated and validated against the ground truth images for both real and simulated databases.

Keywords Magnetic Resonance Imaging · Image segmentation · Ant Colony Optimization · Mahalanobis distance

1 Introduction

Image segmentation plays a vital role in different image processing techniques. Segmentation is the process of splitting an image into its constituent regions depending upon certain property or characteristics [1, 2]. Each pixel in a region is similar to another in certain aspect such as intensity, texture, color, etc. The basic goal of

H. Hooda (✉) · O. P. Verma · S. Arora
Department of IT, Delhi Technological University, New Delhi 110042, Delhi, India
e-mail: heenahooda@gmail.com

O. P. Verma
e-mail: opverma.dce@gmail.com

S. Arora
e-mail: sonamarora.92.01@gmail.com

© Springer Nature Singapore Pte Ltd. 2019
H. Malik et al. (eds.), *Applications of Artificial Intelligence Techniques in Engineering*, Advances in Intelligent Systems and Computing 698,
https://doi.org/10.1007/978-981-13-1819-1_56

591

segmentation is to extract the meaningful information and make things easier to analyze and visualize. The process of segmentation has spread its roots into different domains such as medical imaging, face recognition, machine vision, expert system, automatic traffic control system, and also to deal with satellite images. Medical image segmentation is of great focus nowadays due to its various practical applications. In its application to brain MRI, difference purposes can be dealt such as extraction of small specific structures such as brain tumor, or some main tissues (White matter, gray matter, and cerebrospinal fluid) [3–5], or partitioning the brain into anatomical structures, etc. [6]. It is a very challenging task as medical images exhibit higher complexity due to the presence of various artifacts such as the presence of noise, intensity variation known as bias field, and also partial volume effect may further add on to its complex nature. Medical Resonance Imaging (MRI) is an advanced technique being used in the field of medical imaging as it provides abundant information regarding the anatomy of human soft tissues. MRI helps in diagnosis and evaluation of any abnormal change such as tumors, or any other focal lesions, in bodily parts and helps in early detection of diseases. MRI helps to achieve varying image contrast by making use of different pulse sequences and changing imaging parameters. Manual segmentation of MRI images [7] is very cumbersome as well as time-consuming. It also involves variability depending upon the individual examining the results [8]. It may vary from one observer to another and also within same individual/observer. Though manual segmentation by an expertise has proven to be of superior quality, automated methods can be very advantageous to deal with such variations and to handle large data. So there is a need to develop appropriate automated or semi-automated system to perform segmentation of medical images as per the requirement. In this paper, we first discuss the recent work in brain image segmentation in Sect. 2. The proposed approach is described in Sect. 3. The experimental results of the proposed algorithms are shown in Sect. 4. The paper is concluded in Sect. 5.

2 Related Work

Different image segmentation techniques, both supervised and unsupervised, have been proposed and applied to numerous applications in real world [9]. Unsupervised segmentation techniques require less human intervention in obtaining clinically useful results. We are basically concerned with unsupervised techniques based on soft clustering in which an object can belong to more than one cluster/class with varying degree of membership. The most commonly studied soft clustering algorithm is FCM which was introduced by Bezdek in 1981 [10]. This approach is simple to implement and perform clustering in an efficient way. But in the presence of noise FCM does not cope up with the expected results. Since it takes into account random values in its initialization step and also no spatial information is considered in complete algorithm, several works have been done to include spatial data/information [11, 12]. Mahmood et al. proposed a framework [13] by integrating Bayesian-based adaptive mean shift, a priori spatial tissue probability maps, and FCM. Ahmed et al.

[14] introduced FCM-S in which the objective function was changed to include information of gray levels but this approach was computationally expensive. Kalaiselvi et al. also modified the FCM algorithm by incorporating spatial parameter for minimizing the objective function of conventional FCM and new weighting parameter for centroid initialization [15]. Chen and Zhang [12] proposed another variant of FCM in which the neighboring term for each data point is computed well in advance to avoid computational delays. Also, the Euclidean measure is used in FCM to calculate the distance between data points and cluster centers. Euclidean distance always takes into account the data points in spherical shape from the point being examined. It does not consider the correlation among the data points. The possibility of data point belonging to same cluster is not only dependent upon the distance but also on the direction. Mahalanobis distance introduced by P. C. Mahalanobis in 1936 considered both these issues. Kim and Krishnapuram [16] tried to show that Mahalanobis distance cannot be directly applied to any of the clustering processes. Kessel and Gustafson (GK) [17] used Mahalanobis distance with fuzzy terminology, i.e., introduction of fuzzy covariance matrix was done in this work. Also, meta-heuristic approaches are introduced to deal with NP-hard problems and the problems for which the data is uncertain and not readily available. These optimization techniques include particle swarm optimization, Genetic Algorithms (GA), and ACO [18]. Thomas A. Runkler showed the extension of simplified ant colony system to be compatible with FCM. Yucheng Kao and Kevin Cheng [19] introduced ACOA in which array-based graph is constructed and ants are moved randomly to from the solution set.

3 Proposed Approach

In this work, an automatic framework for segmentation of brain tissue classes, namely, white matter, gray matter, and cerebrospinal fluid, has been proposed. The brain MRI images are used for the purpose of segmenting these tissues. The results of segmentation are often dependent upon the initialization step. MRI images are often subjected to some random noise. Segmentation algorithm such as FCM is very sensitive to noise. To avoid any stuck in local optimal results, ACO technique is used. ACO is used to determine the value of initial cluster centers. The centers thus obtained are fed into the system to perform segmentation. In modified FCM, Mahalanobis distance is used instead of Euclidean distance as Euclidean distance takes into account only the super-spherical shapes about the center of mass for clustering the data points, whereas data points belonging to same cluster may not be located in that area only. Also, the local neighborhood information is also considered as neighboring pixels are more likely to belong to same cluster.

3.1 Initialization of FCM Using ACO

ACO is the probabilistic-based optimization technique inspired by the behavior of ants in this real world. Various optimization problems can be reduced to finding the shortest path from source/nest to destination/food source depending upon the pheromone concentration on that path. In case of noisy problems, [20] FCM clustering algorithm can easily get stuck into local optimal solution as FCM is very sensitive to noise. One can run the algorithm many times with different initial values each time to get the best solution by comparing the results each time with the best solution [21]. But this process is very cumbersome. Therefore, meta-heuristic approach can be used to solve this problem [22]. Main purpose of using ACO technique is to get the global optimal solutions without getting trapped in local optimal results. Movement of ants from one node to another is dependent upon the probability of choosing that path. The higher the probability, the more likely is the path to be selected. The movement of an ant from node i to node j is judged by the probabilistic equation. Find the probability of belongingness of each pixel to a particular cluster using the following equation:

$$p_{i,j} = (\tau_{i,j}^{\alpha})(\eta_{i,j}^{\beta})/\left(\sum_{\Omega \in allowed} (\tau_{i,j})(\eta_{i,j})\right) \tag{1}$$

where

$\tau_{i,j}$ denotes the amount of pheromone/trail laid by an ant on edge (i, j)
$\eta_{i,j}$ denotes the amount of heuristic matrix whose value is typically 1/d(i, j)
α regulates the impact of $\tau_{i,j}$
β regulates the impact of $\eta_{i,j}$

Updation in pheromone is done according to the equation:

$$\tau_{i,j} = (1 - \rho) * \tau_{i,j} + \Delta\tau_{i,j} \tag{2}$$

where ρ is the rate of evaporation of pheromone

$$\Delta\tau_{i,j}^{k} = \begin{cases} 1/L_k, & \text{if } k\text{th ant travels/covers edge}(i,j) \\ 0, & \text{otherwise} \end{cases}$$

where L_k denotes the length of the kth ant travel/tour. Basically, ACO is used to obtain the initial values of clusters required for segmenting the brain MRI images as explained in algorithm 1.

Algorithm 1: Requirements: Fix the number of clusters=3; initialize all the parameters as $\alpha=\beta=1.5$, max_iteration=150, total_NumberOfAnts=10, Rho (ρ)=0.1

Step 1: Input data matrix/image say X = {$x_1, x_2, ..., x_n$} where n is the total number of pixels in an image.

Step 2: Calculate another matrix X' of same dimension as X, such that value of each pixel in X' is the mean value of 4 * 4 neighborhood of corresponding pixel in X.

Step 3: For iter = 1 to max_iteration do.

Step 4: For ants = 1 to total_NumberOfAnts do.

Step 5: Randomly initialize cluster center values for all clusters "c".

Step 6: Calculate the heuristic matrix η, where $\eta = 1/D$ (D is the distance of pixels from all cluster centers).

Step 7: Find the probability of belongingness of each pixel to a particular cluster using the Eq. (1).

Step 8: Based on the indexing of each pixel into different clusters calculated using the above probability, cluster centers are then updated using the formula:

$$v_j = \left(\sum_{i \subseteq S} (u_{i,j})^m x_i + \alpha * x_i' \right) / \left(\sum_{i \subseteq S} (1 + \alpha) * (u_{i,j})^m \right), \forall j = 1, 2 \ldots nc$$

$$(3)$$

where S is the set of pixels having similar index, i.e., pixels having same cluster numbers, and m is the fuzzifier constant having value 2.

Step 9: Calculate the Euclidean distance using the updated cluster centers.

Step 10: Calculate the objective function using the above-calculated distance:

$$J = \sum_{i=1}^{n} \sum_{j=1}^{c} (u_{i,j})^m d^2(x_i, c_j) + \alpha \sum_{i=1}^{n} \sum_{j=1}^{c} (u_{i,j})^m d^2(x_i', c_j) \qquad (4)$$

where $d^2(x_i, c_j)$ is the Euclidean distance between data point x_i and cluster center c_j.

Step 11: Compare the value of objective function with the best fitness. Our goal is to minimize the fitness function. Update best fitness, best centers, and corresponding index for each pixel of an image.

Step 12: END for (Loop at Step4).

Step 13: Calculate the distance using the best centers after each iteration.

Step 14: Update the membership matrix using the updated distance as

$$u_{i,j} = (d^2(x_i, c_j) + \alpha d^2(x_i', c_j))^{(1/m-1)} / \sum_{k=1}^{nc} (d^2(x_i, c_k) + \alpha d^2(x_i', c_k))^{(1/m-1)}$$

$$(5)$$

Step 15: Calculate the best index of each pixel using the best solution obtained after each iteration.

Step 16: Update the pheromone matrix using the best indexes obtained in the previous step:

$$\tau_{i,j} = (1 - \rho) * \tau_{i,j} + \Delta\tau_{i,j} \tag{6}$$

$\Delta\tau_{i,j}$ equals to 1/distance corresponding to same location where updation of pheromone is being done.

Step 17: END for (Loop for max_iteration Step3).

Step 18: Return the best centers obtained and membership matrix corresponding to those best centers.

3.2 Clustering Algorithm

3.2.1 Mahalanobis Distance

P. C. Mahalanobis introduced this distance in 1936 [23]. It gives the measurement of distance of a data point say M and a distribution say D. It is also defined as the distance of the test point p from the center of mass q divided by the width of ellipsoid measured in the direction of test point.

$$D(p, q) = \sqrt{(p - q)^T C^{-1}(p - q)} \tag{7}$$

where C is the covariance matrix of vector P.

The Mahalanobis distance between data point x_j and cluster center c_i involving fuzzy logic is described by G&K [15] and is given as

$$d^2(x_j, c_i) = (x_j - c_i)^T C_i(x_j - c_i) \tag{8}$$

$$C_i = \left|\sum_i\right|^{1/p} \left|\sum_i\right| \tag{9}$$

$$\sum_i = \sum_{j=1}^{N}\sum_{i=1}^{nc} u_{i,j}^m (x_j - c_i)(x_j - c_i)^T / \sum_{j=1}^{N}\sum_{i=1}^{nc} u_{i,j}^m \tag{10}$$

3.2.2 Spatial Clustering

As we know that MRI images are subjected to random noise during its acquisition, it sometimes becomes difficult to view its anatomy or analyze the brain main tissues in the presence of noise in an image. FCM converges to local solution in case of noise. So, performing clustering of data points using FCM will not be effective. As FCM does not take into account the spatial information of the data points to be clustered, the pixels surrounding the particular pixel are more likely to be segmented into same class. The steps for clustering using Mahalanobis distance and spatial information is described in Algorithm 2.

Algorithm 2: *Requirements: Cluster centers and membership matrix returned by Algorithm1, total number of iterations(max_iter= 100), error (10^{-6})*

Step 1:   For ij = 1:max_iter.

Step 2:   Calculate the Mahalanobis distance using the equations in Sect. 3.2.1.

Step 3:   Update the membership degree/matrix using the equation given below:

$$u_{i,j} = (d^2(x_i, c_j) + \alpha d^2(x_i', c_j))^{(1/m-1)} / \sum_{k=1}^{nc} (d^2(x_i, c_k) + \alpha d^2(x_i', c_k))^{(1/m-1)}$$

$$(11)$$

Step 4:   Update the cluster centers as

$$v_j = \left(\sum_{i=1}^{N} (u_{i,j})^m x_i + \alpha * x_i' \right) / \left(\sum_{i=1}^{N} (1 + \alpha) * (u_{i,j})^m \right), \forall j = 1, 2 \dots nc$$

$$(12)$$

Step 5:   If $\| U^{(ij+1)} - U^{(ij)} \| <$ error, then go to Step 8.

Step 6:   END If

Step 7:   END for (loop for maximum iteration).

Step 8:   Return U, perform clustering using degree of belongingness of each pixel in clusters.

Unlike Euclidean distance, Mahalanobis distance takes into consideration the co-relation among data points or datasets. We also include the information of 4 * 4 neighboring pixels surrounding the pixel being examined for clustering in the algorithm. This spatial information is the average of the neighboring pixels in the defined window and is included in the process of clustering or assigning labels to pixels [14].

4   Experimental Results

The real-time database of brain MRI images has been taken from Insight Journal. One of the Insight Journals is MIDAS. Midas community includes National Alliance for medical image computing (NAMIC) which presents the data for two autistic and two normal children (male and female) [24]. Three types of MRI scanning are presented, i.e., T1-weighted, T2-weighted, and PD-weighted images. Coronal slices are obtained with slice thickness of 1.5 mms. The quantitative analysis of MRI brain images shown in Fig. 1 is done in comparison with the ground truth images and is presented in Tables 1 and 2.

Dice coefficient is the volume overlap metric that evaluates the segmentation results quantitatively given the segmentation volumes pairs. The value of dice coefficient lies between 0 and 1. 0 signifies no match/overlap and 1 signifies complete

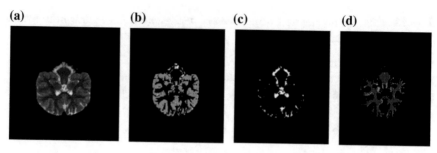

Fig. 1 **a** Autistic female T2-weighted MRI (slice 140), **b** Gray matter, **c** CSF, **d** White matter

Table 1 This table presents the value of dice coefficient for three brain tissues, namely, gray matter, white matter, and CSF using our approach (ACO-FCM) and standard FCM

Patient	Dice coefficient							
	ACO-FCM				FCM			
	GM	WM	CSF	Avg.	GM	WM	CSF	Avg.
P11	0.928	0.889	0.755	0.857	0.917	0.848	0.57	0.778
P12	0.86	0.927	0.785	0.857	0.838	0.848	0.68	0.788
P21	0.902	0.933	0.878	0.904	0.849	0.83	0.784	0.821
P22	0.899	0.806	0.759	0.821	0.872	0.751	0.617	0.746
P31	0.887	0.802	0.796	0.828	0.841	0.793	0.652	0.762
P32	0.903	0.888	0.798	0.863	0.852	0.817	0.716	0.795
P41	0.924	0.94	0.859	0.907	0.887	0.846	0.838	0.857
P42	0.954	0.814	0.853	0.873	0.918	0.744	0.697	0.786
Average	0.907	0.874	0.81	0.863	0.871	0.809	0.694	0.791

Table 2 This table presents the value of Jaccard's overlap ratio for three brain tissues, namely, gray matter, white matter, and CSF using our approach (ACO-FCM) and standard FCM

Patient	Jaccard's overlap ratio							
	ACO-FCM				FCM			
	GM	WM	CSF	Avg.	GM	WM	CSF	Avg.
P11	0.865	0.8	0.606	0.757	0.846	0.736	0.398	0.66
P12	0.754	0.863	0.646	0.754	0.721	0.736	0.515	0.657
P21	0.821	0.874	0.782	0.825	0.737	0.709	0.644	0.696
P22	0.816	0.675	0.611	0.7	0.773	0.601	0.446	0.606
P31	0.796	0.669	0.611	0.692	0.725	0.657	0.483	0.621
P32	0.823	0.798	0.663	0.761	0.742	0.69	0.557	0.663
P41	0.858	0.886	0.752	0.832	0.796	0.733	0.721	0.75
P42	0.912	0.686	0.743	0.78	0.848	0.592	0.534	0.658
Average	0.83	0.781	0.676	0.762	0.773	0.681	0.537	0.663

match/overlap. It is observed from Tables 1 and 2 that the significant improvement can be seen in the result obtained from our approach (ACO-FCM) as compared to standard FCM. The classification of brain tissues is more promising in case of ACO-FCM as there are more number of correctly classified pixels. Our approach uses spatial information which helps in better classification of such tissues. Similar to dice coefficient is the Jaccard's similarity measure. Both these parameters are not sensitive to volume overestimations and underestimations. However, dice coefficient is more famous than Jaccard's ratio as it may sometimes result in a mismatch when there is a strong volumetric overlap. In Table 1, patient numbers refer to the slices of dataset. It can be noticed that classification/segmentation accuracy is highly dependent upon the classification of cerebrospinal fluid. The accuracy of both gray matter and white matter tissues is also improved but significant change can be seen in the case of segmentation of CSF tissue. CSF is a very complex tissue and it is sometimes difficult to segment such a flowing matter from brain MRI. Therefore, there is a requirement to get better classification of data points belonging to CSF class. This improvement can be seen in our approach as compared to standard FCM.

The ACO-FCM approach for segmenting brain MRI images is also applied and validated for simulated 3D brain MRI images with varying level of noise from brain web database. The simulated dataset from brain web is provided by McGill University and can be obtained with different file extensions [25]. It contains normal anatomical brain structures with size of each image of 181 * 217. The results are tested and validated for T1-weighted images with 3 and 5% noise levels and slice thickness of 1 mm as shown in Figs. 2 and 3. It can be noticed from Table 3 that ACO-FCM performs better for both levels of noise as compared to standard FCM. The efficiency of FCM is reduced with higher level of noise. This can be seen from the slice 100 corrupted with both 3 and 5% noises. Accuracy of FCM reduces significantly, whereas ACO-FCM performs efficiently in higher level of noise as well.

(a)   **(b)**   **(c)**   **(d)**   **(e)**

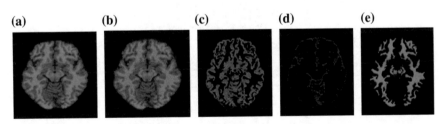

Fig. 2 **a** Normal brain (T1-weighted slice 60), **b** Normal brain (noise 3%), **c** Gray matter, **d** CSF, **e** White matter

(a)   **(b)**   **(c)**   **(d)**   **(e)**

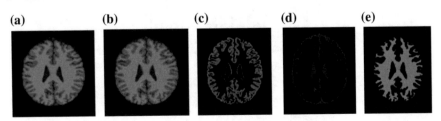

Fig. 3 **a** Normal brain (T1-weighted slice 99), **b** Normal brain (noise 5%), **c** Gray matter, **d** CSF, **e** White matter

Table 3 This table presents the value of dice coefficient for three brain tissues, namely, gray matter, white matter, and CSF using our approach (ACO-FCM) and standard FCM for brain web database

Slice#	Noise (%)	GM	WM	CSF	Avg.	Slice#	Noise (%)	GM	WM	CSF	Avg.
60	3	0.962	0.911	0.812	0.895	60	3	0.947	0.845	0.675	0.822
88	3	0.929	0.963	0.890	0.927	88	3	0.879	0.933	0.822	0.878
100	3	0.900	0.970	0.837	0.902	100	3	0.880	0.948	0.793	0.873
Average		0.930	0.948	0.846	0.908	Average		0.902	0.908	0.763	0.857
99	5	0.915	0.967	0.841	0.907	99	5	0.900	0.955	0.818	0.891
126	5	0.934	0.902	0.874	0.903	126	5	0.896	0.841	0.849	0.862
100	5	0.902	0.962	0.835	0.899	100	5	0.876	0.948	0.764	0.862
Average		0.917	0.943	0.850	0.903	Average		0.890	0.914	0.810	0.871

5   Conclusion and Future Work

In the work presented, image segmentation is done using "Fuzzy Logic" technique. Step-by-step methodology for automated brain MRI image segmentation is presented. Classification of three main tissues of brain is performed. These tissues include gray matter, white matter, and cerebrospinal fluid. The ant colony optimization technique is used to obtain the results of segmentation as ACO is a meta-heuristic approach that tends to give close to optimal solution. Also, inclusion of spatial information of pixels ensures better clustering as neighboring pixels are more likely to belong to the same class/cluster. Consideration of geometrical shape of clustering classes is also taken into account. Therefore, Mahalanobis distance is included to solve the purpose. Brain MRI acquisition is also subjected to random noise. Our work ensures better classification in the presence of such noise as well. The results of our work have been evaluated and validated against ground truth. Both real-time database and simulated database with varying level of noise are used to test the accuracy. Significant improvement in correct classification of brain tissues can be seen as compared to standard FCM. In future, the work of this paper can be extended to further improve the accuracy and efficiency of the algorithm. Also, this work dealt with noise and partial volume effect, so it can be extended to work for images with intensity inhomegeneity as well.

References

1. M. Rastgarpour, J. Shanbehzadeh, Application of AI techniques in medical image segmentation and novel categorization of available methods and tools, in *Proceedings of the International MultiConference of Engineers and Computer Scientists 2011, IMECS 2011*, vol. I, Hong Kong, 16–18 Mar 2011
2. W.X. Kang, Q.Q. Yang, R.R. Liang, The comparative research on image segmentation algorithms, in *IEEE Conference on ETCS* (2009), pp. 703–707
3. S. Bricq, C. Collet, J.P. Armspach, Unifying framework for multimodal brain MRI segmentation based on Hidden Markov Chains. Med. Image Anal. **12**(6), 639–652 (2008)
4. A. Mayer, H. Greenspan, An adaptive mean-shift framework for MRI brain segmentation. IEEE Trans. Med. Imaging **28**(8), 1238–1250 (2009)
5. Benoit Caldairou, Nicolas Passat, Piotr A. Habas, A non-local fuzzy segmentation method: application to brain MRI. Biomed. Image Comput. Group Pattern Recogn. **44**, 1916–1927 (2011)
6. Z. Ji, Y. Xia, Fuzzy local gaussian mixture model for brain MR image segmentation. IEEE Trans. Inf. Technol. Biomed. **16**(3) (2012)
7. C.R. Noback, N.L. Strominger, R.J. Demarest, D.A. Ruggiero, *The Human Nervous System: Structure and Function*, 6th edn. (Humana Press, 2005)
8. W. Marian, An Automated Modified Region Growing Technique for Prostate Segmentation in Trans-Rectal Ultrasound Images, Master's Thesis, Department of Electrical and Computer Engineering, University of Waterloo, Waterloo, Ontario, Canada (2008)
9. A. Nakib, H. Oulhadj, P. Siarry, A thresholding method based on two-dimensional fractional differentiation. Image Vis. Comput. **27**, 1343–1357 (2009)
10. J.C. Bezdek, *Pattern Recognition with Fuzzy Objective Function Algorithms* (Kluwer Academic Publishers, Norwell, MA, USA, 1981)

11. D. Pham, Fuzzy clustering with spatial constraints, in *Proceedings of the International Conference on Image Processing*, vol. 2, NewYork, USA (2002), pp. II-65–II-68
12. S. Chen, D. Zhang, Robust image segmentation using FCM with spatial constraints based on new kernel-induced distance measure. IEEE Trans. Syst. Man Cybern. Part B, Cybern. **34**, 1907–1916 (2004)
13. Q. Mahmood, A. Chodorowski, M. Persson, Automated MRI brain tissue segmentation based on mean shift and fuzzy c-means using a priori tissue probability maps. IRBM **36**(3), 185–196 (2015). ISSN 1959-0318. https://doi.org/10.1016/j.irbm.2015.01.007
14. M. Ahmed, S. Yamany, N. Mohamed, A. Farag, T. Moriarty, A modified fuzzy c-means algorithm for bias field estimation and segmentation of MRI data. IEEE Trans. Med. Imaging **21**, 193–199 (2002)
15. T. Kalaiselvi, P. Nagaraja, V. Ganapathy Karthick, Improved fuzzy c-means for brain tissue segmentation using T1- weighted MRI head scans. Int. J. Innov. Sci. Eng. Technol. **3**(7) (2016)
16. R. Krishnapuram, J. Kim, A note on the Gustafson Kessel and adaptive fuzzy clustering algorithms. IEEE Trans. Fuzzy Syst. **7**, 453–461 (1999)
17. D.E. Gustafson, W.C. Kessel, Fuzzy clustering with a fuzzy covariance matrix, in *IEEE Conference on Decision and Control including the 17th Symposium on Adaptive Processes*, vol. 17, SanDiego, CA, USA (1978), pp. 761–766
18. T.A. Runkler, Ant colony optimization of clustering models. Int. J. Intell. Syst. **20**, 1233–1251 (2005)
19. Y. Kao, K. Chang, An ACO based clustering algorithm, in *ANTS 2006*. LNCS 4150 (2006), pp. 340–347
20. B. Biswal, P.K. Dash, S. Mishra, A hybrid ant colony optimization technique for power signal pattern classification. Expert Syst. Appl. **38**, 6368–6375 (2011)
21. A.N. Benaichouche, H. Oulhadj, P. Siarry, Improved spatial fuzzy c-means clustering for image segmentation using PSO initialization, Mahalanobis distance and post-segmentation correction. Digit. Signal Proc. **23**, 1390–1400 (2013)
22. M. Karnan, T. Logheshwari, Improved Implementation of Brain MRI image Segmentation using Ant Colony System, 978-1-4244-5967-4 (2010 IEEE)
23. P.C. Mahalanobis, On the generalized distances in statistics: Mahalanobis distance. J. Soc. Bengal **XXVI**, 541–588 (1936)
24. http://www.insight-journal.org/midas/community/
25. http://www.bic.mni.mcgill.ca/brainweb

A New Image Segmentation Method Using Clustering and Region Merging Techniques

Nameirakpam Dhanachandra and Yambem Jina Chanu

Abstract Clustering technique is one of the commonly used image segmentation methods. However, all of the clustering existing techniques require user-specific parameters as input. The result of segmentation depends on these parameters; therefore, the selection of an optimal value of these parameters is very crucial. In this paper, we have introduced image segmentation method based on new clustering algorithm where there is no need for initialization of the user-specific parameter. The proposed clustering technique is based on the density estimation of the surrounding pixel values. The recursive approach is used for the density estimation. After the segmentation using clustering, small segments can be present. The presence of the small segments may result in lower performance of the segmentation output. Therefore, each of the segments is merged with another neighbor segment by choosing the best matching segment. For every iteration, one of the segments will be merged with another segment and the iteration with the best performance will be the final segmentation output. The proposed method is compared with the other clustering-based segmentation methods by calculating the performance evaluation indices, and it validates the effectiveness of the proposed algorithm.

Keywords Image segmentation · Clustering · Region merging · Variance

1 Introduction

There are different methods of image segmentation [1–4]. Clustering is one of the widely used image segmentation approaches. There has been a large number of clustering algorithms like k-means, fuzzy c-means, spectral clustering, expectation, maximization etc. [5–9]. Different clustering algorithms are proposed from differ-

N. Dhanachandra (✉) · Y. J. Chanu
NIT Manipur, Langol, Imphal, Manipur, India
e-mail: dhana.namei@gmail.com

Y. J. Chanu
e-mail: jina@nitmanipur.ac.in

ent perspectives and are designed for different purposes. But existing clustering algorithm does require user-specified parameters as input. And the clustering performances depend highly on these user-specific parameters. One example of such parameter is the number of clusters, which is required by many clustering algorithms, e.g., k-means, fuzzy c-means, etc. But there are some other algorithms which can be obtained the number of clusters by themselves, e.g., subtractive clustering, DBSCAN, etc. However, they also required other parameters to define as input. In subtractive clustering, cluster radius and hypersphere radius need to define as input. In DBSCAN, it needs neighborhood radius and the minimum number of data.

Many methods have been proposed to determine these user-specific parameters. Many authors proposed some methods to determine the number of clusters, and some other authors proposed methods to determine the cluster centroid and so on [10–15]. Therefore, many researchers are trying to propose different methods to solve the different problems of the clustering algorithm. Recently, many works on the automatic image segmentation have been proposed. Lai and Chang [16] proposed a hierarchical evolutionary algorithm. It automatically segments the image and it can be viewed as a variant of the traditional genetic algorithm. Haiyang et al. [17] proposed a new hybrid method of dynamic particle swarm optimization and k-means. Shi et al. proposed [18] a novel clustering-based image segmentation known as ICDP algorithm which is based on the density peak algorithm. They have used the variants of integral channel feature along with the clustering method to find the density. Chang et al. [19] introduced a dynamic niching clustering algorithm which is based on the individual connectedness. It automatically calculates the optimal number of clusters and cluster centers using the adaptive selection of the compact k-distance neighborhood algorithm. Hou et al. [20] proposed a parameter-independent clustering algorithm. They have used the dominant set to determine the input parameters of the DBSCAN clustering algorithm. Although there have been so many improved methods, these methods are either complicated or only some of the problems are overcome. Thus, further studies are necessary to solve these problems of user-specific as input, especially the depending on the number of clusters and the initial centroids. Therefore, the clustering algorithm with the automatic determination of parameters is still an open problem.

2 Materials

In this paper, we have introduced a new image segmentation method and it consists of two phase. In the first phase, the image is segmented using the clustering technique, and in the second phase the small segments are merged to the nearest or the best matching segment.

Steps for proposed algorithm: (1) Input image, (2) image segmentation using clustering technique, (3) merging of regions, and (4) output image.

2.1 Clustering Technique

The proposed method is based on the density of the surrounding pixel values of the given image. The proposed method does not require a priori knowledge of the number of clusters. It automatically generates the cluster centroid. In the first step, the density of every pixel values is calculated and the pixel value with the maximum density is declared as the first cluster centroid. The density of surrounding pixels of a pixel is calculated using recursive density estimation method [21]. Let us consider an image having a size of $M \times N$ and each pixel is denoted by $x_{i,j}$, where $i = 1, 2, 3,$ ..., M and $j = 1, 2, 3, ..., N$.

$$D_{ij} = \frac{1}{1 + \left\| x_{ij} - \mu \right\|^2 + X - \left\| \mu \right\|^2} \tag{1}$$

x_{ij} is the pixel value of the image, μ and X are defined as the mean of the pixel values and scalar product of the pixel values.

$$\mu = \frac{1}{M \times N} \sum_{j=1}^{N} \sum_{i=1}^{M} x_{ij} \tag{2}$$

$$X = \frac{1}{M \times N} \sum_{j=1}^{N} \sum_{i=1}^{M} x_{ij}^2 \tag{3}$$

After calculating the density of each pixel value, the pixel value with the maximum density has been considered as the centroid of the cluster. A cluster is defined by a centroid and cluster radius. In other words, it can be defined as a group of pixels in a close proximity to the cluster centroid and the closeness is defined by the cluster radius. Therefore, the cluster radius can be defined as the scaling function of the cluster. If the value of the cluster radius is large, then the cluster will have a large number of pixel values. Again, if the value of cluster radius is small, a little number of pixel values will be there in the cluster. Therefore, the optimal value of the cluster radius will result in more effective approach of clustering. The cluster radius is calculated as square root of the average value of the distance between the current cluster centroid and all the other pixel values. The pixels within the radius from the cluster centroid will form a cluster. The cluster thus formed is defined as an intra-cluster, the pixels as intra-cluster pixels, and the remaining pixels as uncluster pixels.

N. Dhanachandra and Y. J. Chanu

$$r_k = \left[\frac{1}{M \times N} \sum_{i=1}^{N} \sum_{i=1}^{M} \|x_{ij} - C\|^2 \right]^{1/2} \tag{4}$$

where m is the mean of the pixel values of the image, C is the centroid of the cluster, and x_{ij} represent the pixel value of the image. However, the radius is calculated based on the average distance between the centroid and all the pixels of the given image and cluster radius. Therefore, dependence on all the pixels of the given image might affect the similarity property between the intra-cluster pixels. Because of this, the pixels which are not similar might be present in the intra-cluster. Therefore, the cluster radius is updated based on the intra-cluster pixels. We checked if there is any pixel in the intra-cluster which is not within the updated radius from the centroid. If it is there we removed it from the intra-cluster and put it in the uncluster pixel group. Thus, the cluster is updated and the updated intra-cluster is defined as the first cluster. Similarly, in the next iteration, the uncluster pixels are used to form the next cluster by calculating the density, the means, and the scalar product using Eqs. 1–4 and updating again using the intra-cluster pixels. This process continues until there remains a single pixel in the uncluster pixel. Thus, at the end, this will generate k number of the segments.

Algorithm for Image Segmentation

1. Load the image
2. Calculate the mean, the scalar product of the pixels of the given image, and the density of each pixel using the Eqs. 2, 3, and 1, respectively.
3. Find the maximum density and then the corresponding pixel is declared as cluster centroid.
4. Calculate the cluster radius using Eq. 4.
5. Find the pixel values within cluster radius from the centroid and form intra-cluster.
6. Update the cluster centroid and the cluster radius using the pixels of the intra-cluster using Eqs. 1–4.
7. Update the intra-cluster using the updated cluster radius and the cluster centroid.
8. The updated intra-cluster will be declared as the first final cluster, and the remaining clusters are defined as uncluster pixels.
9. The uncluster pixels are used to form next cluster.
10. Repeat steps 2–8 until there remains only one pixel in the uncluster pixel.

2.2 Region Merging

Sometimes, the output of the segmented output can contain small segments which lead to poor segmentation quality. So we need to merge the small segment to another segment which is closely matched, to get a better segmentation result. There are many approaches for region merging [22–24]. But we are using the region merging approach using the variance of the region because of its simplicity and effectiveness. A segment will be merged with another segment which is best matched to the segment. In order to find the best matching segment, the distance between the candidate segments and the other segments is calculated using the variance of segment. The variance of the segments is calculated using Eq. 5.

$$\sigma_k^2 = \frac{1}{n} \sum_{i=1}^{n} (x_i - m_k)^2 \tag{5}$$

where m_k is the mean of the kth segment, n is the number of pixels in the kth segment, and σ represents the variance of the kth segment. The difference between two segments is calculated using the Euclidean distance.

$$d_{k_i} = \sqrt{\left| \sigma_k^2 - \sigma_i^2 \right|^2} \tag{6}$$

where σ_k^2 represent the variance of the small segment and σ_i^2 represent the variance of neighbor segments. From Eq. 6, the minimum distance is calculated and the region with the minimum distance is declared as the best matching segment.

Algorithm for Merging Small Segments

1. Initialize the threshold t.
2. Calculate the variance of every segment using Eq. (5).
3. Sort the number of pixels in the segments.
4. Select the segment with minimum number of pixels.
5. The best matching segment of the selected segment is found using Eqs. (6) and (7).
6. Change the label of the best matching segment with the label of the segment.
7. Repeat steps 2–5 for all the other remaining segments.

$$d_{best} = \min(d_{k_i}) \tag{7}$$

For k segment where $k = 1, 2, \ldots, n$, the best matching bth segment is found using Eqs. 6 and 7. Upon finding the best segment, the label of the bth segment is changed with the label of the kth segment. Thus, this process will continue for other remaining

Table 1 Number of pixel in each segment

Region	Number of lengths
1	68
2	417
3	2996
4	4704
5	12505
6	12804
7	86506

segments also. After merging all the segments, there remains only one segment at last.

3 Results and Discussion

In this section, the result of the proposed method is analyzed and discussed. In the segmentation process, k number of segments are generated. But the number of pixels in each segment can be very low and it can affect the quality of the segmentation. The performance of the segmentation can be varied based on the level of merging of the regions. Therefore, each of these small segments is merged with each corresponding best matching cluster. The merging of small region may improve the segmentation performance. For experimental purpose, the proposed image segmentation algorithm has been applied to Berkeley image database [25]. After the segmentation process using the clustering technique, the number of pixels in each segment is shown in Table 1. The segment with the smallest number of pixels is taken as the first segment to be merged and it has been merged with the best matching segment with it.

Moreover, after merging the segment, the segments have been updated so that there will be no wrong detection of the segment which had been already merged. Thus, this process continues for the rest of the other segments. As we can observe in Table 1, there are seven segments and for each iteration a segment is merged with a neighbor segment which is the best matched. The output image after merging each segment is shown in Fig. 1. After merging each segment, the output image at last iteration would have only one-pixel value as shown in Fig. 1. Different output images are obtained for each iteration, and therefore six different segmentation outputs are generated. The performance of the six different images of each iteration is calculated using the performance indices, root mean square error (RMSE), and peak signal-to-noise ratio (PSNR) as shown in Table 2. As we can observe in Table 2, the value of RMSE is lowest and PSNR is highest at the 4th iteration. In other words, we can say that the performance of the segmentation output at 4th iteration is best as compared to other iterations. Therefore, the segmentation output at the 4th iteration is the final output of the segmentation. The graph plot for RMSE and PSNR values are shown in Fig. 2.

Fig. 1 **a** Original image, **b** segmentation result after clustering technique, after merging **c** first segment, **d** second segment, **e** third segment, **f** fourth segment, **g** fifth segment, **f** sixth segment

Table 2 Performances after each iteration

Iteration	RMSE	PSNR
After clustering	0.0083	68.30
1st iteration	0.0083	68.30
2nd iteration	0.0081	68.31
3rd iteration	0.0079	68.20
4th iteration	0.0073	68.84
5th iteration	0.0085	68.24
6th iteration	–	–

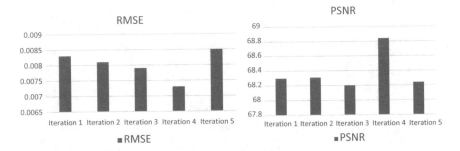

Fig. 2 Graph plot for RMSE and PSNR

Moreover, the segmentation using the conventional methods like k-means, fuzzy c-means, expectation–maximization, subtractive, and kernelized subtractive have been considered for the state-of-the-art comparisons with the proposed method. Segmentation results of the proposed method and the other methods are shown in Fig. 3.

1. K-mean: The number of clusters has been taken same as the number of segments generated in the proposed method. The centroids are initialized randomly.

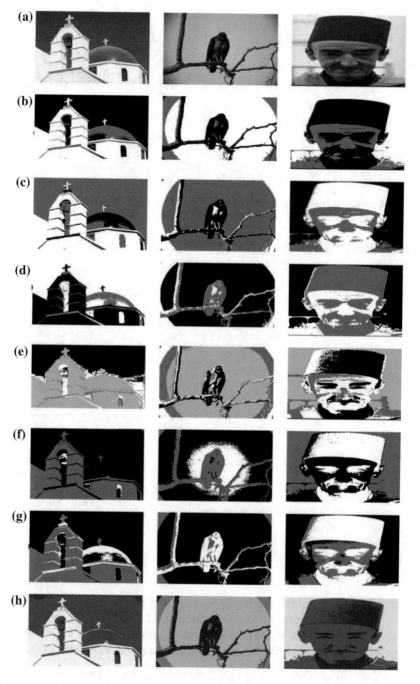

Fig. 3 **a** Original image, **b** k-means, **c** FCM, **d** IFCM, **e** EM, **f** Subtractive, **g** KS, **h** Proposed method

Table 3 Comparison of the proposed method with other methods for church

Methods	Number of clusters	ρ	J
K-means	3	0.97	0.48
FCM	3	0.97	0.97
IFCM	3	−0.74	0.63
EM	3	0.64	0.60
Subtractive	3	0.77	0.39
KS	3	0.68	0.45
Proposed	3	0.98	1

2. Fuzzy c-means (FCM): The number of clusters has been taken same as the number of segments generated in the proposed method. The membership functions are initialized randomly. The value of modification factor is taken as 2.
3. Intuitionistic FCM (IFCM): Initializations are done same as FCM, except that there are some modifications in the objective function of the conventional FCM.
4. Expectation–maximization (EM): The number of clusters has been taken same as the number of clusters generated in the proposed method. The cluster centroid is initialized randomly, and the Gaussian distribution function is used.
5. Subtractive: The number of clusters and the centroid are generated by itself. However, the hypersphere radius is taken as 1.2 and the penalty hypersphere radius as 0.5.
6. Kernelized subtractive (KS): The initial values are taken same as the values which are taken in the subtractive methods, except that the Kernel function is used in the calculation of the potential and revised potential function of the subtractive clustering.

For the quantitative analysis, there are many performance evaluation indices. But we have used evaluation indices, namely, correlation coefficient and Jaccard index (J). Table 3, 4 and 5 shows the comparison between the proposed method and other clustering methods over different evaluation indices. From Table 3, 4 and 5, we can observe that the performance of the proposed method outperformed the other methods when correlation coefficient and Jaccard index are compared for all the images. As we know, the execution time is an important factor in the application like the medical imaging where it is used for the diagnosis of many patients, face recognition, etc. Therefore, many algorithms are trying to reduce the execution time at the cost of the performance of the segmentation. As we can observe in Table 6, the execution time of the proposed methods has very less time as compared to the other methods. Thus, the proposed method has the performance values which are better as compared to the other methods, and execution time is quite low as compared to the other methods and at the same time, it is an automatic image segmentation without any user prior initialization.

Table 4 Comparison of the proposed method with other methods for man

Methods	Number of clusters	ρ	J
K-means	3	0.96	0.79
FCM	3	0.93	0.92
IFCM	3	−0.96	0.66
EM	3	0.49	0.70
Subtractive	3	−0.41	0.70
KS	3	−0.94	0.60
Proposed	3	0.96	1

Table 5 Comparison of the proposed method with other methods for bird

Methods	Number of clusters	ρ	J
K-means	3	0.96	0.84
FCM	3	0.96	0.94
IFCM	3	−0.55	0.25
EM	3	0.61	0.85
Subtractive	3	−0.09	0.37
KS	3	−0.96	0.22
Proposed	3	0.97	1

Table 6 Execution time in second

Image	K-mean	FCM	IFCM	EM	Subtractive	KS	Proposed
Church	8.76	57.11	85.42	17.96	228.9	5440	5.89
Man	6.63	95.6	120.9	14.02	233.3	6959	6.53
Bird	8.46	84.3	79.78	19.33	221.2	5128	7.45

4 Conclusion

In this paper, a new clustering-based image segmentation is proposed, which is based on the density of the surrounding pixel values. This proposed method is an automatic image segmentation method. It automatically defines the number of clusters and its corresponding center of the cluster which are the common problem encountered in others clustering-based image segmentation methods. Consequently, if the small segments are present in the segmented output, then it has been merged with another segment which is the best match to it. Moreover, the proposed method is compared with other clustering methods using the evaluation indices and it validated the effectiveness of the proposed method. At the same time, the execution time of the proposed method has been proved far better than the other methods. Although the proposed method has yielded good performance indices and execution time, there is still some aspect that needs to be improved. In the future work, we plan to optimize the cluster radius using the optimization techniques. Besides, we will try different techniques for merging small regions to yield a more effective region merging.

References

1. D.L. Pham, X. Chenyang, J.L. Prince, Current methods in medical image segmentation. Annu. Rev. Biomed. Eng. **2**, 315–337 (2000)
2. C. Pantofaru, M. Hebert, *A comparison of Image Segmentation Algorithm* (The Robotics Institute, Carnegie Mellon University, Pittsburgh, Pennsylvania, 2005)
3. H. Zhang, J.E. Fritts, S.A. Goldman, Image segmentation evaluation: a survey of unsupervised methods. Comput. Vis. Image Underst. **110**(2), 260–280 (2008)
4. Y.-H. Wang, Tutorial: Image Segmentation, Graduate Institute of Communication Engineering. National Taiwan University, Taipei, Taiwan, ROC
5. R. Xu, D. Wunsch, Survey of clustering algorithm. IEEE Trans. Neural Network. **16**(3), 645–678 (2005)
6. K. Hammouda, F. Karray, A Comparative Study of Data Clustering. University of Waterloo, Ontario, Canada (2000)
7. K.M. Bataineh, M. Naji, M. Saqer, A comparison study between various Fuzzy Clustering Algorithms. Jordan J. Mech. Ind. Eng. **5**(4), 335–343 (2011)
8. A.S. Abdul Nasir, M.Y. Mashor, Z. Mohamed, Color image segmentation approach for detection of malaria parasites uing various colour models and k-means clustering. WSEAS Trans. Biol. Biomed. **10**(1), 41–53 (2013)
9. M.Y. Chong, W.Y. Kow, Y.K. Chin, L. Angelin, K.T. Kin Teo, Image segmentation via normalized cuts and clustering algorithm, in *IEEE International Conference on Control System, Computing and Engineering*
10. M.E. Celebi, H.A. Kingravi, P.A. Vela, A comparative study of efficient initialization methods for the clustering algorithm. Expert Syst. Appl. **40**, 200–210 (2013)
11. E.A. Zanaty, Determining the number of cluster for kernalized fuzzy c-means algorithm for automatic medical image segmentation. Egypt. Inf. J. **13**, 39–58 (2012)
12. A.N. Benaichouche, H. Oulhadj, P. Siarry, Improved spatial fuzzy c-means clustering for image segmentation using PSO initialization, Mahalanobis distance and post-segmentation correction. Digit. Signal Process. **23**(5), 1390–1400 (2013)
13. G. Evanno, S. Regnaut, J. Goudet, Detecting the number of clusters of individual using software structure: a simulation study. Mol. Ecol **14**(8), 2611–2620 (2005)
14. S.S. Khan, A. Ahmad, Cluster center initialization algorithm for k-means cluster. Pattern Recognit. Lett. **25**(11) 1293–1302 (2004)
15. HesamIzakian and Ajith Abraham, Fuzzy C-means and fuzzy swarm clustering problems. Experts Syst. Appl. **38**, 1835–1838 (2011)
16. C.-C. Lai, C.-Y. Chang, A hierarchical evolutionary algorithm for medical image segmentation. Expert Syst. Appl. **36**(1), 248–259 (2009)
17. H. Li, H. He, Y. Wen, Dynamic particle swarm optimization and k- means clustering algorithm for image segmentation. Optik **126**, 4817–4822 (2015)
18. Y. Shi, Z. Chen, Zhiquan, F. Meng, L. Cui, A novel clustering based image segmentation via density peak algorithm with mid-level feature. Neural Comput. App. (2016
19. D. Chang, Y. Zhao, L. Liu, C. Zheng, A dynamic niching clustering algorithm based on individual-connectedness and its application to color image segmentation. Pattern Recogn. **60**, 334–347 (2016)
20. J. Hou, H. Gao, Xu. Li, DSets-DBSCAN: a parameters free clustering algorithm. IEEE Trans. Image Process. **25**-7, 3182–3193 (2016)
21. P. Angelov, Fundamentals of probability theory, in *Autonomous Learning System* (Wiley, 2012), pp. 17–36

22. R. Nock, F. Nielsen, Statistical region merging. IEEE Trans. Pattern Anal. Mach. Intell. **26**, 11 (2004)
23. F. Calderero, F. Marque, Region merging techniques using information theory statistical measures. IEEE Trans. Image Process. **19**, 1567–1586 (2010)
24. Z.F. Muhsin, A. Rehman, A. Altameem, T. Saba, M. Uddin, Improved quadtree image segmentation approach to region information. Imaging Sci. J. **62** (2014)
25. D. Martin, C. Fowlkes, The Berkeley segmentation database and benchmark. Computer Science Department, Berkeley University 2011

Forecasting of Nitrogen Dioxide at One Day Ahead Using Nonlinear Autoregressive Neural Network for Environmental Applications

Vibha Yadav, Satyendra Nath and Hasmat Malik

Abstract In this paper, short-term forecasting of nitrogen dioxide (NO_2) at one day ahead is performed using nonlinear autoregressive neural network. For this, 491 measured time series data are utilized. The presented results with root mean square error of 0.0456 validate accuracy and effectiveness of the proposed nonlinear autoregressive neural network forecasting model for NO_2.

Keywords Forecasting · NO_2 · Nonlinear autoregressive neural network

1 Introduction

In addition to water and land, air is also the most important resource for survival of life. In current time, the problem of air pollution due to nitrogen dioxide is a well-known issue. Nitrogen dioxide is one of the most hazardous pollutants associated with health impacts. Atmospheric nitrogen combines with oxygen at high temperatures, as generated during fuel combustion to form nitric oxide. The nitric oxide further combines with atmospheric nitrogen to form nitrogen oxide. It is reddish brown with highly reactive gas. NO provides an important role in tropospheric ozone formation. The effect of NO_2 irritates the lungs and causes pneumonia and bronchitis . Eventually, a small concentration of any pollutants present in atmosphere

V. Yadav (✉) · S. Nath
Department of Environmental Sciences and NRM, College of Forestry, Sam Higginbottom University of Agriculture, Technology and Sciences, Allahabad 211007, Uttar Pradesh, India
e-mail: yvibha3@gmail.com

S. Nath
e-mail: satyendranath2@gmail.com

H. Malik
Electrical Engineering Department, IIT Delhi, New Delhi, India
e-mail: hmalik.iitd@gmail.com

© Springer Nature Singapore Pte Ltd. 2019
H. Malik et al. (eds.), *Applications of Artificial Intelligence Techniques in Engineering*, Advances in Intelligent Systems and Computing 698, https://doi.org/10.1007/978-981-13-1819-1_58

is harmful to human. The adverse impacts are not only restricted to human but also to animals and plants. Therefore, the clean and pure air is essential for survival. Hence, the forecasting of significant pollutant like NO_2 can play a crucial role to check the critical concentration of particular pollutant to prevent air pollution. Several authors have used different ANN models for NO_2 prediction [1–9]. In this study, nonlinear autoregressive neural network is developed to forecast NO_2 at one day ahead.

The paper is organized as follows. Section 2 presents a methodology. Results and discussion are given in Sect. 3 followed by conclusions in Sect. 4.

2 Methodology

2.1 Data Measurement

The daily value of measured time series data of NO_2 from January 1, 2016 to October 13, 2017 for Talkatora (latitude 26.83°N, longitude 80.89°E) in Lucknow, India are used in model. Data are taken from Central Pollution Control Board (CPCB) monitoring station New Delhi India [10]. The daily variation of NO_2 and its normalized values are shown in Figs. 1 and 2. Out of 491 normalized values of NO_2, 392 values are used for training and rest of the values are used for testing the model.

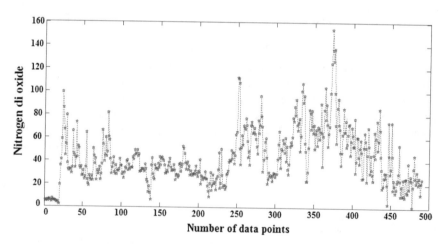

Fig. 1 Time series value of NO_2

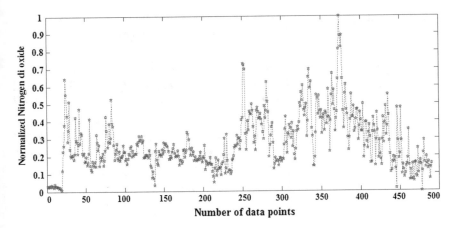

Fig. 2 Normalized value of NO$_2$

2.2 *Nonlinear Autoregressive Neural Network (NANN)*

Forecasting is a kind of dynamic filtering (DF) in which historical values of one or more time series are utilized to forecast future values. DF incorporates tapped delay used for forecasting and nonlinear filtering. The developed NANN in this study is shown in Fig. 3.

The implementation of NANN is shown in Fig. 4. The NO$_2$ forecasting accuracy is determined by mean absolute error (MAPE) that is calculated by Eq. (1):

$$RMSE = \left(\frac{NO_{2_{i(NANN)}} - NO_{2_{i(measured)}}}{n} \right)^2 \tag{1}$$

where n is used for training and testing NANN model, $NO_{2_{i(measured)}}$ is measured NO$_2$ for i day, and $NO_{2_{i(NANN)}}$ is forecasted NO$_2$ for i day.

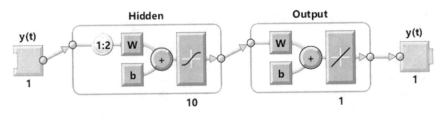

Fig. 3 Nonlinear autoregressive neural network

Fig. 4 Proposed algorithm

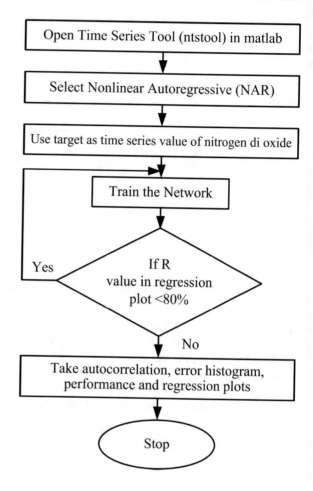

3 Results and Discussion

The performance plot of NANN model presents that mean square error (MSE) decreases with increase in number of epochs (Fig. 5). The epoch denotes complete cycle of training, testing, and validation. The validation and test set errors have similar characteristics, and no major overfitting happens near epoch 4. The correlation coefficient (R-value) shows the correlation between outputs and target. The R-value is more than 0.92 as shown in Figs. 10, 11, 12, and 13 proving that NANN model developed with nstool predicts NO_2 close to measured values (Fig. 9). The error histogram, autocorrelation, and fit plots are shown in Figs. 6, 7 and 8 showing authentication of NANN for forecasting of NO_2. Root mean square error (RMSE) values for training and testing are found to be 0.0971 and 0.0456, respectively.

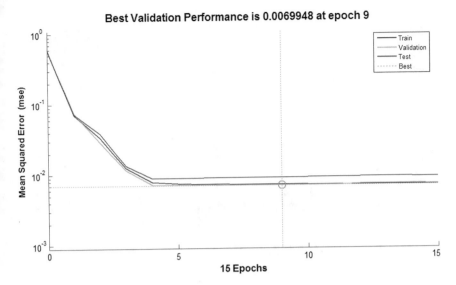

Fig. 5 Performance plot

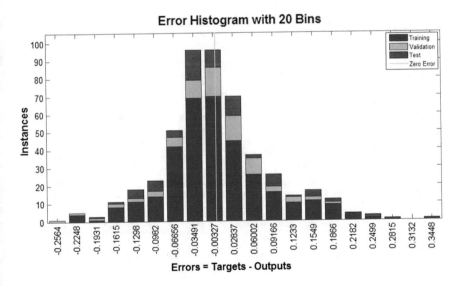

Fig. 6 Error histogram plot

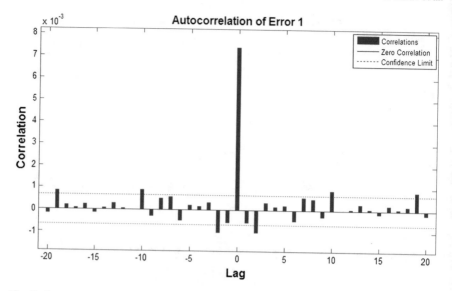

Fig. 7 Autocorrelation plot

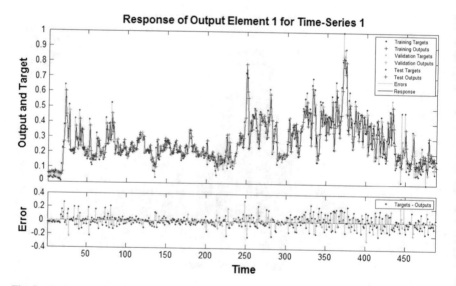

Fig. 8 Fit plot

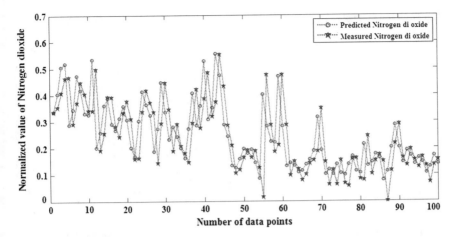

Fig. 9 Comparison between forecasted and measured NO₂

Fig. 10 Regression plot for training

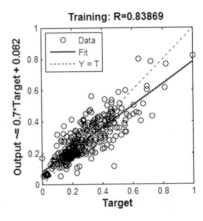

Fig. 11 Regression plot for validation

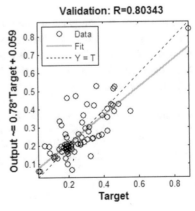

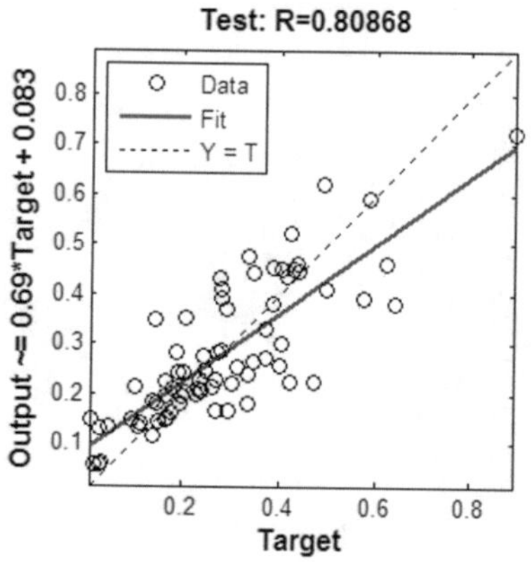

Fig. 12 Regression plot for testing

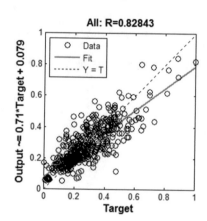

Fig. 13 Overall regression plot

4 Conclusion

The results of this study show accuracy and effectiveness of the proposed new technique for NO_2 forecasting. The proposed model, i.e., nonlinear autoregressive neural network forecast NO_2 for one day ahead with root mean square error of 0.0456. Moreover, the results revealed that NANN models produce higher coefficients of correlation and less error.

Acknowledgements The authors would like to acknowledge Department of Science and Technology, New Delhi-110016, India for providing inspire fellowship with Ref. No. DST/INSPIRE Fellowship/2016/IF160676. We would also like to thank Central Pollution Control Board, New Delhi India for providing online time series data for this study.

References

1. Z.W. Lu, W.J. Wang, Y.H. Fan, T.Y.A. Leung, B.Z. Xu, K.C.J. Wong, Prediction of pollutant levels in causeway bay area of hong kong using an improved neural network model. J. Environ. Eng. **128**(22), 1147–1157 (2001)
2. K.H. Elminir, A.H. Galil, Estimation of air pollutant concentration from meteorological parameters using Artificial Neural Network. J. Electr. Eng. **57**, 105–110 (2006)
3. L. Barbes, C. Neagu, L. Melnic, C. Iliec, M. Velicum, The use of artificial neural network (ANN) for prediction of some airborne pollutants concentration in urban areas. J. Rev. Chem. **60**(3): 301–307 (2009)
4. A. Soni, S. Shukla, Application of neuro-fuzzy in prediction of air pollution in urban areas: IOSR. Int. J. Eng. **2**(5), 1182–1187 (2012)
5. A. Russo, F. Raichel, G.P. Lind, Air quality prediction using optimal neural networks with stochastic variables. J. Atmos. Environ. **79**, 822–830 (2013)
6. S.M. Baawain, A.S.A. Serihi, Systematic approach for the prediction of ground-level air pollution (around an industrial port) using an artificial neural network. Aerosol Air Qual. Res. **14**, 124–134 (2014)
7. M.A. Elangasinghe, N. Singhal, K.N. Dirks, A. Jennifer, J.A. Salmond, Development of an ANN–based air pollution forecasting system with explicit knowledge through sensitivity analysis. Atmos. Pollut. Res. **5**, 696–708 (2014)
8. D.A. Syafei, A. Fujwara, J. Zhang, Prediction model of air pollutant levels using linear model with component analysis. Int. J. Environ. Sci. Dev. **6**(7), 1–7 (2015)
9. D. Mishra, P. Goyal, Development of artificial intelligence based NO2 forecasting models at Taj Mahal. Agra. Atmos. Pollut. Res. **6**, 99–106 (2015)
10. http://www.cpcb.gov.in/CAAQM/frmUserAvgReportCriteria.aspx

Wearable Arm Robot Steering (WARS) Device for Robotic Arm Control Used in Unmanned Robotic Coconut Tree Climber and Harvester

Rajesh Kannan Megalingam, Shiva Bandyopadhyay, Vamsy Vivek Gedala, K. G. S. Apuroop, Gone Sriteja, N. Ashwin Kashyap and Md. Juned Rahi

Abstract As there is a severe shortage in human coconut tree climbers, there is a dire need to help the coconut tree farmers and the consumers who are suffering a lot due to inability to harvest coconuts in time, which has led to severe increase in price of coconuts. In this research work, we propose to develop a wearable device to be worn over the human arm to control a robotic arm which is fitted with the unmanned robotic coconut tree climber to harvest the coconuts. We present the design, development, and application of this wearable device in a detailed way in this paper. The user can control the robotic arm via the wearable device with the help of the robotic operating system based controller from the ground with wireless interface. The algorithm we developed can estimate the position of the arm of the user wearing the device in three-dimensional space and also map it to the workspace of the robotic arm. The experimentation and the results in our lab setup show that this wearable device can be user-friendly and simple to use apart from playing significant role as the controller of the arm.

R. K. Megalingam (✉) · S. Bandyopadhyay · V. V. Gedala · K. G. S. Apuroop · G. Sriteja
N. Ashwin Kashyap · Md. Juned Rahi
Department of Electronics and Communication Engineering,
Amrita Vishwa Vidyapeetham, Kollam 690525, Kerala, India
e-mail: rajeshkannan@ieee.org

S. Bandyopadhyay
e-mail: shiva1bandyopadhyay2@gmail.com

V. V. Gedala
e-mail: vamsyvivek.394@gmail.com

K. G. S. Apuroop
e-mail: apuroopkgs@gmail.com

G. Sriteja
e-mail: sriteja.gone97@gmail.com

N. Ashwin Kashyap
e-mail: n.ashwin.k@ieee.org

Md. Juned Rahi
e-mail: junedrahi@gmail.com

© Springer Nature Singapore Pte Ltd. 2019
H. Malik et al. (eds.), *Applications of Artificial Intelligence Techniques in Engineering*, Advances in Intelligent Systems and Computing 698,
https://doi.org/10.1007/978-981-13-1819-1_59

Keywords Agriculture · Robotic operating system · Coconut harvesting Robotic climber · Wearable device

1 Introduction

The agricultural industry is shrinking day by day due to the increase in the number of people moving into other job sectors. There are many factors like literacy, job security, and higher pay/profit which are playing a crucial role in this. The coconut harvesting job is usually taken up by the community in the lower level of the economic spectrum. Due to increase in literacy rate and awareness, the current generation of men from this community is moving toward white-collar jobs because of high pay compared to the lower pay and risk involved in coconut tree climbing. A fall from the tree could prove to be fatal and, in some cases, this could be the only breadwinner in the family. In such cases, the entire family is affected as there is no proper insurance coverage for these coconut tree climbers. Some mechanical coconut tree climbers are developed in due course of time to address this issue in which a human can sit or stand and climb to the top of the tree to harvest coconuts. An alternative solution could be the unmanned robotic coconut tree climber and harvester. The harvester is usually a robotic arm and has to be controlled from the ground for harvesting the coconuts. The controller of this robotic arm should be simple and user-friendly. In this context, we propose our Wearable Arm Robot Steering (WARS) device which the user can wear and control the robotic arm from the ground.

2 Related Works

The authors have discussed designing an autonomous mobile robot using ROS which does not require any specific control method for operating the robot [1]. Paper [2] talks about the origin of ROS. ROS is not an operating system in the traditional sense of process management and scheduling; rather, it provides a structured communication layer above the host operating systems of a heterogeneous compute cluster. Over the years, researchers have come up with several types of interfaces for control of the robotic arm for various applications. Electrooculogram-based control [3], fuzzy logic-based control [4, 5], brain–computer interface [6–8], electromyogram control [9, 10], tongue-based control [11], admittance-based control [12], and torque-based control [13] are some of the control methods that researchers have come up with. Though the applications of these control methods are for a robot or a robotic arm in general, the areas of applications include biomedical, industrial, cooperative robots, prosthesis, skill learning, etc. When it comes to agricultural applications, we do not find much research work in this direction. There are robots that are used for weed detection [8], sweet pepper harvesting [14], kiwi fruit picking [15], etc. Being this the situation, there is not much research work to develop new control methods for control robots or robotic arms for agricultural applications. Several of the wearable devices presented in the literature are for biomedical applications. In the research work presented in [16], the authors discuss wearable intelligence to help demented elders. A real-time monitoring of the blood pressure and electrocardiogram using a wearable device is proposed in [17]. [18, 19] also discusses wearable device for health monitoring. We could find that there is not much prior work in terms of control methods used for agricultural applications and also most of the research work done on wearable devices are for health monitoring or biomedical applications. Papers [20] and [21] talk about various ways to implement and control a robotic arm. Paper [22] is also about a robotic arm but with minimal cost. Inverse kinematics of a dual linear actuator pitch/roll heliostat is discussed in paper [23]. In this work, we propose a closed loop control of the 6-DoF robotic arm using inverse kinematics for agricultural applications.

3   System Architecture

Figure 1 shows the overall system architecture including the WARS device. The ROS controller/host processor is chosen to be a Raspberry Pi 3 model B. The proposed WARS device has wireless connectivity with the ROS controller. Similarly, the robotic arm is also controlled via ROS controller using wireless interface. The robotic arm shown in Fig. 1 is part of the robotic coconut tree climber and harvester.

3.1   Wearable Arm Robotic Steering (WARS) Device

The objective of the WARS device is to calculate and publish the position vector and orientation of the user's hand in 3D space as seen from a particular reference. WARS has two parts IMU_tracker and IMU_ref. The IMU_tracker can find the position vector and the orientation of the user's hand, and the IMU_ref acts as the reference device to help in calculating the position vector and the orientation. The WARS device architecture is shown in Fig. 2.

3.2   IMU_ref and IMU_tracker

The IMU_ref and the IMU_tracker are the only sensor modules present in the WARS device. The difference between these obtained positions and orientations gives the position vector and orientation of the user's hand in a deterministic 3D space defined by the position of the user, thus minimizing any error which may arise due to the movement of the user.

Fig. 1 Overall system architecture

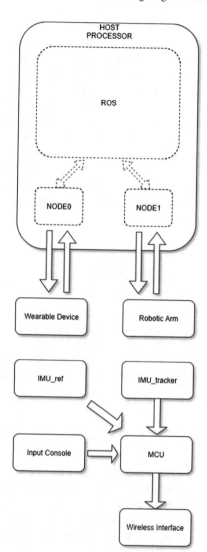

Fig. 2 WARS device architecture

3.3 MCU and Input Console

The MCU module is responsible for all the manipulations performed on the data obtained from the IMU sensors, handling inputs from the input console, and all the transactions between the WARS device and the robotic arm. Apart from the IMU sensors, the MCU is interfaced with an input console also. The input console allows the user to access important functions local to the WARS device like calibration command, calibration status, workspace calibration, power status, and play/pause button. For safety reasons, the input console has a master kill switch too, which overrides the request–reply communication protocol followed between the WARS device and the host processor.

3.4   Orientation Estimation

The MCU obtains raw accelerometer and gyroscope data from both the IMU sensors. The angular velocity from the gyroscope is obtained by dividing the raw data by the gyroscope sensitivity. The orientation of an IMU sensor around any axis can be thus determined by integration of the gyroscope data over time.

$$\theta(t) = \int_0^t \dot{\theta}(t)dt \approx \sum_0^{nT_s} \dot{\theta}(t)T_s$$

where $\dot{\theta}$ is the scaled gyroscope data given by

$$\frac{GYRO(axis, t)}{GYROSCOPE\_SENSITIVITY}$$

and T_s is the reciprocal of the sampling rate. However, a drift error is introduced in this step due to the digital approximation of the orientation. To minimize the drift error, the accelerometer can be used. The accelerometer is capable of providing accurate orientation estimates when the sensor is in slow or uniform motion, or at rest. Thus, the gravity vector estimate can be given by

$$acc = atan\left(\frac{y}{x}\right) \times \frac{180°}{\pi}$$

where y and x are the acceleration along respective axes. For minimizing the drift error, the orientation estimates obtained can be passed through a complementary filter where the orientation estimate θ is updated as

$$\theta = (1 - \alpha) \times (\theta) + \alpha \times acc$$

where α is a small quantity (<1) determined experimentally for good accuracy. Orientations of both the IMU sensors are determined, and the relative orientation $O_{tracker}$ of IMU_tracker as seen by IMU_ref is calculated as

$$\dot{O}_{tracker} = O_{tracker} - O_{ref}$$

where O represents the orientation matrix of the particular sensor.

3.5 Position Estimation

Data from the accelerometer of a sensor and its orientation estimate can be used for calculating its position estimate. The component of earth's gravity, g, is calculated along each axis as

$$g_{axis} = g/cosec(\varnothing)$$

$$a_{axis} = 9.81g$$

where $\varnothing$ is the tilt of the particular axis, obtained from the orientation estimate. The component of gravity along each axis is removed. Thus, the displacement d along each axis can be obtained by double integrating the acceleration.

$$d_{axis} = \iint_0^t a_{axis}$$

The resultant position vector $\dot{D}$ of the IMU_tracker is given by the difference between the position vectors of the IMUs.

$$\dot{D}_{tracker} = D_{tracker} - D_{ref}$$

3.6 Workspace Transformation

Due to the difference in size, orientation, and joint capabilities between a human arm and the robotic arm, the effective workspace of the robotic arm and the user is different. The position estimate obtained from the wearable device is scaled by a factor μ defined as the ratio between the reach of robotic arm and the reach of the user's arm.

3.7 Robotic Arm

A 6-DoF articulated robotic arm is custom built for testing purposes as shown in Fig. 3. An MCU along with a servo driver forms the control unit of the robotic arm. The control unit subscribes to the joint angles published by ROS. Whenever a new set of joint angles is received, the control unit maps each joint angle to a range appropriate for the servomotors. The mapped servo data is written into the servo driver, which drives the servomotors at each joint. The control unit can also publish robot end-effector position and orientation upon request from ROS. The end-effector

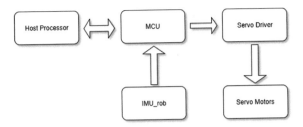

Fig. 3 6-DoF robotic arm control

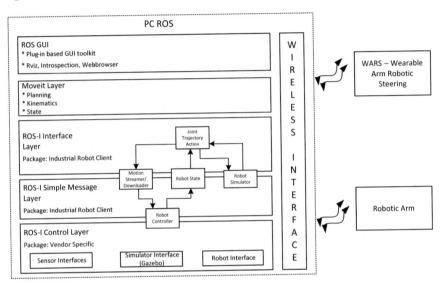

Fig. 4 Architecture of ROS

position and orientation is calculated using an IMU sensor placed at the end effector. The architecture diagram of ROS implementation is shown in Fig. 4. The ROS plays a major role in interfacing the robotic arm and the WARS device. To maintain the least lag time, the ROS takes up the responsibility to synchronize the entire flow of operation using necessary acknowledgements and handshake signals. All the services and actions are enabled with this rosbridge protocol by default and can be activated with suitable shell commands. The ROS being a versatile platform also provides its users a feature to model any robot model under any environment with real-time objects and physics. RViz is the simulator that we are using to realize our slave robot model in the virtual environment. To model the robot model, a Universal Robot Description Format (URDF) file of the robot is needed and this file can be generated from a standard SolidWorks file. All the parameters in this tool can be accessed using the GUI with the help of the rosbridge protocol.

3.8 System Architecture of ROS

See Fig. 4.

4 Implementation

4.1 WARS Device and Robotic Arm

The microcontroller module used in the wearable device is an Arduino Pro Mini. It takes inputs from IMU sensors, which are MPU6050 modules, via I2C interface, and the input console which is custom built. The IMU sensors are located at addresses 0X68 and 0X69. The connection with the host processor, Raspberry Pi 3 Model B, is established via wireless interface. The wireless module used here is the ESP8266 WiFi module. The control unit of the robot arm consists of an Arduino UNO, an Adafruit PCA9685 16-channel 12-bit PWM servo controller, and an MPU6050 IMU sensor. The Arduino UNO communicates with the servo driver and the IMU sensor via I2C bus. The IMU sensor is located at address 0X68, and the servo driver is located at the address 0X69.

4.2 WARS Device and Robotic Arm

The microcontroller module used in the wearable device is an Arduino Pro Mini. It takes inputs from IMU sensors, which are MPU6050 modules, via I2C interface, and the input console which is custom built. The connection with the host processor, Raspberry Pi 3 Model B, is established via wireless interface. The control unit of the robot arm consists of an Arduino UNO, an Adafruit PCA9685, and an MPU6050 IMU sensor. Upon power-up, the wearable device prompts the user for a series of motion calibration tasks.

4.3 ROS Implementation

The master wearable arm is to be worn by the user. Once the initial setup is achieved, the user is expected to take his/her hand to the point where the slave arm is desired to go. The operating flow can be described as follows. After the initial startup sequence, ROS requests for position estimate from the IMU sensors based wearable device mounted and secured on the user's arm. The wearable device replies with the position estimate of the user's arm. All the ROS communications happen among nodes over topics. The MCU publishes the data over a user-defined topic and then the ROS

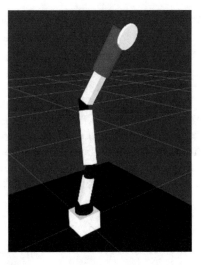

Fig. 5 6-DOF designed arm

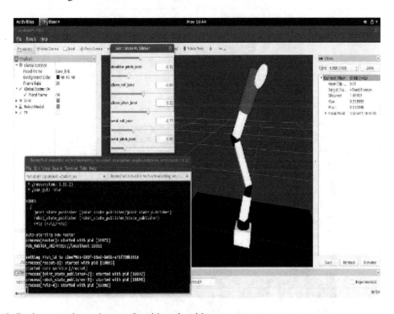

Fig. 6 Package works on inverse Jacobian algorithm

subscribes the data from the node over a different topic. RViz provides an Inverse Kinematic (IK) solver to solve the robot pose from a given end coordinate. Thus, the robotic arm subscribes to the node and receives the data and thus moves accordingly. The display in the Rviz GUI is shown in Fig. 5. The motion planning of the designed arm is shown in Fig. 6.

5 Experiments and Results

A serial six degrees of freedom robotic manipulator was made with seven rigid links. The details, which is used in the designing process is included in Table 1. Figure 7 shows the 6-DoF arm as implemented by us as the test setup. Figure 8 shows one of the team members of this project wearing the WARS device during a test. This WARS device in Fig. 8 is designed and tested by the team member in the Humanitarian Technology (HuT) Labs of Electronics and Communication Dept., Amrita School of Engineering, Amrita Vishwa Vidyapeetham, Kerala under the supervision of the Head of HuT Labs, Dr. Rajesh Kannan Megalingam. The Denavit-Hartenberg (DH) parameters of a robotic arm parameter variation are shown in Table 2 (Table 3).

Table 1 Arm design parameters

Link length (cm)	Joint	Motor	Torque (Stall) @ 6v kg cm	Gear ratio (Gr)	Output torque (Gr × Torque)
3.5	1	HS-785HB	13.2	7:1	92.4
4.8	2	HS-785HB	13.2	7:1	92.4
6.8	3	TS-500HD	27.3	Nil	27.3
2.3	4	RKI-1211	16	Nil	16
8	5	RKI-1211	16	Nil	16
2.2	5	RKI-1211	16	Nil	16

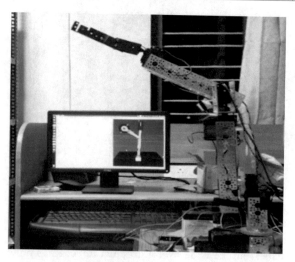

Fig. 7 6-DoF arm control using ROS platform

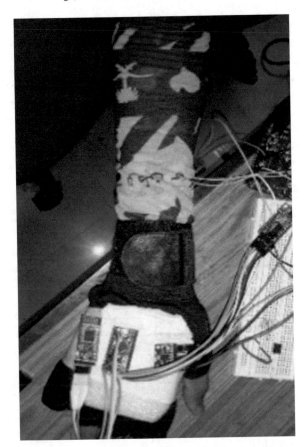

Fig. 8 A user wearing WARS device during testing

Table 2 DH parameters of 6-DoF robotic arm

Joint No.	Joint Offset (b) (cm)	Joint Angle (θ) (°)	Link Length (a) (cm)	Twist Angle (α) (°)
1	12.5	$\Theta 1$	0	−90
2	−4	$\Theta 2$	0	90
3	22	$\Theta 3$	0	−90
4	3	$\Theta 4$	0	90
5	25	$\Theta 5$	0	−90
6	0	$\Theta 6$	10	0

5.1 Reaction Time Calculations

Four different users were chosen to wear the WARS device and control the 6-DoF robotic arm. For each of the gestures, the video of the arm reaction is captured and

Table 3 Planning time and execution time

Planning time (s)	Execution time in Software simulation (s)	Execution time in Hardware (s)
0.017	6.8	7.52
0.020	5.22	5.39
0.023	6.0	7.13
0.016	5.2	5.83
0.017	6.89	7.50
0.020	4.8	6.15

Table 4 Reaction time—Individual users

S. No	Users	Gestures	Reaction time (ms)
1	User 1	G1	943
2		G2	883
3		G3	969
4		G4	920
5	User 2	G1	918
6		G2	901
7		G3	923
8		G4	957
9	User 3	G1	882
10		G2	898
11		G3	882
12		G4	908
13	User 4	G1	971
14		G2	892
15		G3	898
16		G4	908

analyzed for the reaction time. The variation in delay comes only how fast the sensors respond to the hand gesture and give the signal to the MCU at the user HCI and how fast this MCU converts these signals into commands. Table 4 lists this reaction time which is measured from the video when the success/failure tests are carried out. The reaction time for all the four gestures for four different users is within the range of 800–980 ms with a timing difference of 180 ms. The average reaction timings listed in Table 4 points to the fact that for the chosen gestures the reaction time is well within 1 s.

6 Discussion

Innovative methods using latest technologies are crucial for the agricultural industry. In this research work, we have proposed a wearable control method for the robotic arm control to be used for cutting coconuts. In this context, we have come up with Wearable Arm Robot Steering (WARS) device which can be worn by the user in the hand for the control of the robotic arm.

A mathematical modeling is developed based on accelerometer and gyroscope data from both the IMU sensors to minimize the drift error and increase the accuracy. ROS architecture was defined to implement the mathematical model and the 6-DOF robotic arm simulation is carried out. A hardware model of the 6-DOF arm was developed and tested with ROS to work in synchronous fashion with that of the simulated arm. Finally, the wearable device, the ROS simulation of the arm, and the hardware model of the arm were integrated and tested for the verification of the WARS device. The planning time and the execution time for both hardware and software were tabulated and found that the software simulation time is 5 s, whereas time taken by the hardware to reach the destination is 5.2 s on an average for the arm to move from position A to position B. The WARS device was tested with four different users to find the reaction time, and the results are found to be satisfactory. We could see that our proposed WARS device is simple and user-friendly device which would benefit the farmers struggling with the extreme shortage of people to climb the coconut trees and harvest the coconuts.

7 Conclusion

In this paper, we have presented the design, implementation, and evaluation of a wearable device WARS over an ROS operating system, which can be used to control the robotic arm used in coconut harvesting. WARS is simple and user-friendly wearable device which anyone can get used to, in short time. It has wireless interface and hence comfortable for wearing and using. A test setup at our lab facility for verifying the correctness of the algorithm implemented shows the possibility of the WARS device as an efficient, simple, and user-friendly control method.

Acknowledgements Author would like to thank Dept. of Electronics and Communication Engineering, Amrita Vishwa Vidyapeetham, Kollam, Kerala-690525, India to provide the necessary equipment and platform for setup the experiment in the Humanitarian Technology (HuT) Labs under the supervision of the Head of HuT Labs, Dr. Rajesh Kannan Megalingam.

638

R. K. Megalingam et al.

References

1. M. Köseoğlu, O.M. Çelik, Ö. Pektaş, Design of an autonomous mobile robot based on ROS, in *2017 International Artificial Intelligence and Data Processing Symposium (IDAP)*
2. M. Quigley, K. Conley, B. Gerkey et al., ROS: an open-source robot operating system, in *ICRA Workshop on Open Source Software* (2009)
3. A. Úbeda, E. Iáñez; J.M. Azorín, An integrated electrooculography and desktop input bimodal interface to support robotic arm control. IEEE Trans. Hum.-Mach. Syst. **43**(3), 338–342 (2013)
4. A. Chatterjee, R. Chatterjee, F. Matsuno, T. Endo, Augmented stable fuzzy control for flexible robotic arm using LMI approach and neuro-fuzzy state space modeling. IEEE Trans. Ind. Electron. **55**(3), 1256–1270 (2008)
5. S.-L. Wu, Y.-T. Liu, T.-Y. Hsieh, Y.-Y. Lin, C.-Y. Chen, C.-H. Chuang, C.-T. Lin, Fuzzy integral with particle swarm optimization for a motor-imagery-based brain–computer interface. IEEE Trans. Fuzzy Syst. **25**(1), 21–28 (2017)
6. A.F.P. Vidal, M.A.O. Salazar, G.S. Lopez, Development of a brain-computer interface based on visual stimuli for the movement of a robot joints. IEEE Lat. Am. Trans. **14**(2), 477–484 (2016)
7. H.G. Moorman, S. Gowda, J.M. Carmena, Control of redundant kinematic degrees of freedom in a closed-loop brain-machine interface. IEEE Trans. Neural Syst. Rehabil. Eng. **25**(6), 750–760 (2017)
8. S. Bozinovski, A. Bozinovski, Mental states, EEG manifestations, and mentally emulated digital circuits for brain-robot interaction. IEEE Trans. Auton. Ment. Dev. **7**(1), 39–51 (2015)
9. P. Shenoy, K.J. Miller, B. Crawford, R.P.N. Rao, Online electromyographic control of a robotic prosthesis. IEEE Trans. Biomed. Eng. **55**(3), 1128–1135 (2008) (Cited by: Papers (102))
10. M. Ison, I. Vujaklija, B. Whitsell, D. Farina, P. Artemiadis, High-density electromyography and motor skill learning for robust long-term control of a 7-DoF robot arm. IEEE Trans. Neural Syst. Rehabil. Eng. **24**(4), 424–433 (2016)
11. D. Johansen, C. Cipriani, D.B. Popović, L.N.S.A. Struijk, Control of a robotic hand using a tongue control system—a prosthesis application. IEEE Trans. Biomed. Eng. **63**(7), 1368–1376 (2016)
12. J. Meuleman, E. van Asseldonk, G. van Oort, H. Rietman, H. van der Kooij, LOPES II—design and evaluation of an admittance controlled gait training robot with shadow-leg approach. IEEE Trans. Neural Syst. Rehabil. Eng. **24**(3), 352–363 (2016)
13. F. Petit, A. Dietrich, A. Albu-Schäffer, Generalizing torque control concepts: using well-established torque control methods on variable stiffness robots. IEEE Robot. Autom. Mag. **22**(4), 37–51 (2015)
14. H. Nejati, Z. Azimifar, M. Zamani, Using fast fourier transform for weed detection in corn fields, in *2008 IEEE International Conference on Systems, Man and Cybernetics* (2008), pp. 1215–1219
15. A.J. Scarfe, R.C. Flemmer, H.H. Bakker, C.L. Flemmer, Development of an autonomous kiwifruit picking robot, in *2009 4th International Conference on Autonomous Robots and Agents* (2009), pp. 380–384
16. Q. Xu, S.C. Chia, J.-H. Lim, Y. Li, B. Mandal, L. Li, MedHelp: enhancing medication compliance for demented elderly people with wearable visual intelligence. Sci. Phone Apps Mob. Devices **2**(3) (2016)
17. F.A. Ferreira Marques, D.M.D. Ribeiro, M.F.M. Colunas, J.P. Silva Cunha, A real time, wearable ECG and blood pressure monitoring system, in *2011 6th Iberian Conference on Information Systems and Technologies (CISTI)*. Print ISBN: 978-1-4577-1487-0
18. Z. Zhou, W. Dai, J. Eggert, J.T. Giger, A real-time system for in-home activity monitoring of elders, in *31st Annual International Conference of the IEEE EMBS Minneapolis, Minnesota, USA, 2–6 Sept 2009*. https://doi.org/10.1109/iembs.2009.5334915
19. Y. Lee, YonseiUniv, Implementation of accelerometer sensor module and fall detection monitoring system based on wireless sensor network, in *29th Annual International Conference of*

the *IEEE Engineering in Medicine and Biology Society, 2007. EMBS 2007*, 22–26 Aug 2007, pp. 2315–2318. https://doi.org/10.1109/iembs.2007.4352789

20. R.K. Megalingam, G.V. Vivek, S. Bandyopadhyay, M.J. Rahi, Robotic arm design, development and control for agriculture applications, in *2017 4th International Conference on Advanced Computing and Communication Systems (ICACCS)* (2017), pp. 1–7

21. R.K. Megalingam, S. Bandyopadhyay, G.V. Vivek, M.J. Rahi, Hand gesture based wireless robotic arm control for agricultural applications. IOP Conf. Ser.: Mater. Sci. Eng. **225** (2017)

22. C. Nutakki, A. Vijayan, H. Sasidharakurup, B. Nair, K. Achuthan, S. Diwkar, Low-cost robotic articulator as an online education tool: design, deployment and usage, in *International Conference on Robotics and Automation for Humanitarian Applications, RAHA 2016—Conference Proceedings* (2017)

23. J. Freeman, B. Shankar, G. Sundaram, Inverse kinematics of a dual linear actuator pitch/roll heliostat, in *AIP Conference Proceedings*, vol. 1850, Article number 030018, 27 June 2017

24. E. Vitzrabin, Y. Edan, Changing task objectives for improved sweet pepper detection for robotic harvesting. IEEE Robot. Autom. Lett. **1**(1), 578–584 (2016) (Cited by: Papers (1))

Author Index

© Springer Nature Singapore Pte Ltd. 2019
H. Malik et al. (eds.), *Applications of Artificial Intelligence Techniques in Engineering*, Advances in Intelligent Systems and Computing 698,
https://doi.org/10.1007/978-981-13-1819-1

Printed in the United States
By Bookmasters